Williamson on Modality

Timothy Williamson is one of the most influential living philosophers working in the areas of logic and metaphysics. His work in these areas has been particularly influential in shaping debates about metaphysical modality, which is the topic of his recent provocative and closely argued book *Modal Logic as Metaphysics* (2013). The present book comprises 10 essays by metaphysicians and logicians responding to Williamson's work on metaphysical modality. The authors include some of the most distinguished philosophers of modality in the world, as well as several rising stars. Each essay is followed by a reply by Williamson. In addition, the book contains an original essay by Williamson, 'Modal science', concerning the role of modal claims in natural science.

This book was originally published as a special issue of the *Canadian Journal of Philosophy*.

Juhani Yli-Vakkuri is a Humboldt Research Fellow at the Department of Philosophy, University of Bielefeld, Germany.

Mark McCullagh is an associate professor in the Department of Philosophy, University of Guelph, Canada.

Previous titles published from the *Canadian Journal of Philosophy*:

New Essays on Thomas Reid
Edited by Patrick Rysiew

New Essays on the Nature of Propositions
Edited by David Hunter and Gurpreet Rattan

Belief, Action and Rationality over Time
Edited by Chrisoula Andreou and Sergio Tenenbaum

Williamson on Modality
Edited by Juhani Yli-Vakkuri and Mark McCullagh

Williamson on Modality

Edited by
Juhani Yli-Vakkuri and Mark McCullagh

Routledge
Taylor & Francis Group
LONDON AND NEW YORK

First published 2017 by Routledge

2 Park Square, Milton Park, Abingdon, Oxfordshire OX14 4RN
52 Vanderbilt Avenue, New York, NY 10017

Routledge is an imprint of the Taylor & Francis Group, an informa business

First issued in paperback 2019

British Library Cataloguing in Publication Data
A catalogue record for this book is available from the British Library

ISBN 13: 978-1-138-73949-9 (hbk)
ISBN 13: 978-0-367-22988-7 (pbk)

Typeset in MyriadPro
by diacriTech, Chennai

Publisher's Note
The publisher accepts responsibility for any inconsistencies that may have arisen during the conversion of this book from journal articles to book chapters, namely the possible inclusion of journal terminology.

Contents

Citation Information vii
Notes on Contributors xi

1 Modal science 1
Timothy Williamson

2 Higher-order free logic and the Prior-Kaplan paradox 41
Andrew Bacon, John Hawthorne and Gabriel Uzquiano

Reply to Bacon, Hawthorne and Uzquiano 90
Timothy Williamson

3 Williamson on Fine on Prior on the reduction of possibilist discourse 96
Kit Fine

Reply to Fine 119
Timothy Williamson

4 First-order modal logic in the necessary framework of objects 132
Peter Fritz

Reply to Fritz 158
Timothy Williamson

5 Williamson on necessitism 161
Jeremy Goodman

Reply to Goodman 188
Timothy Williamson

6 Plurals and modals 202
Øystein Linnebo

Reply to Linnebo 225
Timothy Williamson

7 On Williamson and simplicity in modal logic 231
Theodore Sider

Reply to Sider 247
Timothy Williamson

8 Models and reality 257
Robert Stalnaker

Reply to Stalnaker 275
Timothy Williamson

9 An A-theory without tense operators 283
Meghan Sullivan

Reply to Sullivan 307
Timothy Williamson

10 Williamsonian modal epistemology, possibility-based 314
Barbara Vetter

Reply to Vetter 344
Timothy Williamson

11 Epistemicism and modality 351
Juhani Yli-Vakkuri

Appendix to Juhani Yli-Vakkuri's 'Epistemicism and modality' 384
Peter Fritz

Reply to Yli-Vakkuri 387
Timothy Williamson

Index 401

Citation Information

The chapters in this book were originally published in the *Canadian Journal of Philosophy*, volume 46, issues 4–5 (2016). When citing this material, please use the original page numbering for each article, as follows:

Chapter 1

Modal science
Timothy Williamson
Canadian Journal of Philosophy, volume 46, issues 4–5 (2016) pp. 453–492

Chapter 2

Higher-order free logic and the Prior-Kaplan paradox
Andrew Bacon, John Hawthorne and Gabriel Uzquiano
Canadian Journal of Philosophy, volume 46, issues 4–5 (2016) pp. 493–541

Reply to Bacon, Hawthorne and Uzquiano
Timothy Williamson
Canadian Journal of Philosophy, volume 46, issues 4–5 (2016) pp. 542–547

Chapter 3

Williamson on Fine on Prior on the reduction of possibilist discourse
Kit Fine
Canadian Journal of Philosophy, volume 46, issues 4–5 (2016) pp. 548–570

Reply to Fine
Timothy Williamson
Canadian Journal of Philosophy, volume 46, issues 4–5 (2016) pp. 571–583

Chapter 4

First-order modal logic in the necessary framework of objects
Peter Fritz
Canadian Journal of Philosophy, volume 46, issues 4–5 (2016) pp. 584–609

Reply to Fritz
Timothy Williamson
Canadian Journal of Philosophy, volume 46, issues 4–5 (2016) pp. 610–612

Chapter 5

Williamson on necessitism
Jeremy Goodman
Canadian Journal of Philosophy, volume 46, issues 4–5 (2016) pp. 613–639

Reply to Goodman
Timothy Williamson
Canadian Journal of Philosophy, volume 46, issues 4–5 (2016) pp. 640–653

Chapter 6

Plurals and modals
Øystein Linnebo
Canadian Journal of Philosophy, volume 46, issues 4–5 (2016) pp. 654–676

Reply to Linnebo
Timothy Williamson
Canadian Journal of Philosophy, volume 46, issues 4–5 (2016) pp. 677–682

Chapter 7

On Williamson and simplicity in modal logic
Theodore Sider
Canadian Journal of Philosophy, volume 46, issues 4–5 (2016) pp. 683–698

Reply to Sider
Timothy Williamson
Canadian Journal of Philosophy, volume 46, issues 4–5 (2016) pp. 699–708

Chapter 8

Models and reality
Robert Stalnaker
Canadian Journal of Philosophy, volume 46, issues 4–5 (2016) pp. 709–726

Reply to Stalnaker
Timothy Williamson
Canadian Journal of Philosophy, volume 46, issues 4–5 (2016) pp. 727–734

Chapter 9

An A-theory without tense operators
Meghan Sullivan
Canadian Journal of Philosophy, volume 46, issues 4–5 (2016) pp. 735–758

Reply to Sullivan
Timothy Williamson
Canadian Journal of Philosophy, volume 46, issues 4–5 (2016) pp. 759–765

Chapter 10

Williamsonian modal epistemology, possibility-based
Barbara Vetter
Canadian Journal of Philosophy, volume 46, issues 4–5 (2016) pp. 766–795

Reply to Vetter
Timothy Williamson
Canadian Journal of Philosophy, volume 46, issues 4–5 (2016) pp. 796–802

Chapter 11

Epistemicism and modality
Juhani Yli-Vakkuri
Canadian Journal of Philosophy, volume 46, issues 4–5 (2016) pp. 803–835

Appendix to Juhani Yli-Vakkuri's 'Epistemicism and modality'
Peter Fritz
Canadian Journal of Philosophy, volume 46, issues 4–5 (2016) pp. 836–838
Reply to Yli-Vakkuri
Timothy Williamson
Canadian Journal of Philosophy, volume 46, issues 4–5 (2016) pp. 839–851

Notes on Contributors

Andrew Bacon is an assistant professor in Philosophy at the University of South California, Los Angeles, CA, USA. His recent work is concerned with the probabilities of conditionals, vagueness, and the semantic paradoxes.

Kit Fine is University Professor and Silver Professor of Philosophy and Mathematics at the Department of Philosophy, New York University, New York, NY, USA. His areas of research include logic, metaphysics, and the philosophy of language.

Peter Fritz is a postdoctoral fellow at the Department of Philosophy, Classics, History of Arts and Ideas, University of Oslo, Norway. His primary research interest is in modality.

Jeremy Goodman is an assistant professor in the Department of Philosophy at the University of Southern California, Los Angeles, CA, USA. He works mainly on metaphysics, epistemology, the philosophy of mind, and philosophical logic.

John Hawthorne is Professor of Philosophy at the University of South California, Los Angeles, CA, USA. He is a leading academic figure in metaphysics, epistemology, and the philosophy of language.

Øystein Linnebo is Professor at the Department of Philosophy, Classics, History of Arts and Ideas, University of Oslo, Norway. His specialities include the philosophy of logic, language, and mathematics, as well as metaphysics.

Theodore Sider is Andrew W. Mellon Chair at the Department of Philosophy, Rutgers University, New Brunswick, NJ, USA. His research interests include metaphysics and the philosophy of language.

Robert Stalnaker is Professor of Philosophy at the Department of Linguistics and Philosophy, Massachusetts Institute of Technology, Cambridge, MA, USA. His research interests are philosophical logic, the philosophy of mind, and the philosophy of language.

Meghan Sullivan is an associate professor at the Department of Philosophy, University of Notre Dame, Notre Dame, IN, USA. Her areas of interest are metaphysics, philosophical logic, and the philosophy of religion.

Gabriel Uzquiano is Professor of Philosophy at the University of South California, Los Angeles, CA, USA. His research specialities include philosophical logic, metaphysics, and the philosophy of language.

Barbara Vetter is Professor at the Institut für Philosophie, Friedrich-Alexander-Universität Erlangen, Germany. Her primary research interest is metaphysics.

Timothy Williamson is Wykeham Professor of Logic at the Faculty of Philosophy, University of Oxford, UK. His main research interests are in philosophical logic, epistemology, metaphysics, and philosophy of language.

Juhani Yli-Vakkuri is a Humboldt Research Fellow at the University of Bielefeld, Germany. He works in the philosophy of language, the philosophy of mind, philosophical logic, epistemology, and metaphysics.

Modal science

Timothy Williamson

Wykeham Professor of Logic, University of Oxford, Oxford, UK

ABSTRACT
This paper explains and defends the idea that metaphysical necessity is the strongest kind of objective necessity. Plausible closure conditions on the family of objective modalities are shown to entail that the logic of metaphysical necessity is S5. Evidence is provided that some objective modalities are studied in the natural sciences. In particular, the modal assumptions implicit in physical applications of dynamical systems theory are made explicit by using such systems to define models of a modal temporal logic. Those assumptions arguably include some necessitist principles.

Too often, philosophers have discussed 'metaphysical' modality — possibility, contingency, necessity — in isolation. Yet metaphysical modality is just a special case of a broad range of modalities, which we may call 'objective' by contrast with epistemic and doxastic modalities, and indeed deontic and teleological ones (compare the distinction between objective probabilities and epistemic or subjective probabilities). Thus metaphysical possibility, physical possibility and immediate practical possibility are all types of objective possibility. We should study the metaphysics and epistemology of metaphysical modality as part of a broader study of the metaphysics and epistemology of the objective modalities, on pain of radical misunderstanding. Since objective modalities are in general open to, and receive, natural scientific investigation, we should not treat the metaphysics and epistemology of metaphysical modality in isolation from the metaphysics and epistemology of the natural sciences.

In what follows, Section 1 gives a preliminary sketch of metaphysical modality and its place in the general category of objective modality. Section 2 reviews some familiar forms of scepticism about metaphysical modality in that light. Later sections explore a few of the many ways in which natural science deals with questions of objective modality, including questions of quantified modal logic.

1. The category of objective modality

Let 'n' name the actual number of inhabited planets. There are exactly n inhabited planets, as our stipulation guarantees. Since our planet is inhabited, we know that $n \geq 1$. However, even though we know for sure that there are no fewer than n inhabited planets, there could have been fewer than n, because there could have been *no* inhabited planets. Such a sense in which things could have been otherwise is *objective* rather than *epistemic*. It is not a matter of what any actual or hypothetical agent knows, or believes, or has some other psychological attitude to; nor is it a matter of what any actual or hypothetical agent *ought* to be or do, either morally or in order to achieve a given purpose. Conversely, some epistemic possibilities are not objective possibilities of any kind. For instance, since we do not know whether other planets are inhabited, it is in some sense both epistemically possible for us that $n \geq 2$ and epistemically possible for us that $n < 2$. But the ordering of the natural numbers is non-contingent: either $n \geq 2$ and there is no objective possibility of any kind that $n < 2$, or else $n < 2$ and there is no objective possibility of any kind that $n \geq 2$. Objective modalities also differ in conception from *deontic* and *teleological* modalities, which concern how things ought or may permissibly be, either absolutely or for a given purpose; but since there is much less danger of confusion in that respect, no more need be said about deontic and teleological modalities.

Objective modalities are non-epistemic, non-psychological, non-intentional. Thus they are not sensitive to the guises under which the objects, properties, relations and states of affairs at issue are presented. For instance, suppose that there are in fact exactly 29 inhabited planets; we may read 'n' and '29' as simply two directly referential names of the same object. Then, for each kind of objective necessity, since it is necessary that $29 = 29$, it is necessary that $n = 29$; that is just the same objective necessity in other words. By contrast, epistemic modalities presumably *are* sensitive to guise. For instance, it is epistemically necessary that $29 = 29$, but not epistemically necessary that $n = 29$. The matter is admittedly delicate, because on coarse-grained views of the individuation of propositions the proposition that $29 = 29$ just *is* the proposition that $n = 29$, so the latter proposition is epistemically necessary if the former is. But even such coarse-grained views have to make the relevant epistemic distinctions *somehow*. In some sense, the identity is epistemically necessary under the guise of the sentence '$29 = 29$' but not under the guise of the sentence '$n = 29$'. Then the pertinent contrast is that objective modality admits no such non-trivial relativization to guises. The identity is simply objectively necessary irrespective of the guise under which it is presented.

In linguistics, the distinction between epistemic and non-epistemic modality is widely taken to be fundamental to the taxonomy of modal constructions in natural languages, such as the almost omnipresent auxiliary verbs like 'can' and 'must'.[1] The present category of objective modality corresponds roughly to Angelika Kratzer's 'root' or 'circumstantial' modals and to Paul Portner's 'dynamic'

$I(\alpha) = 1$ for all faithful interpretations I (see also Williamson 2013a, 106). One can show that S_i is automatically a normal modal logic in the usual sense, by the assumption that L_i commutes with conjunction. That is, all truth-functional tautologies are theorems, as are all formulas of the form $\Box(\alpha \to \beta) \to (\Box\alpha \to \Box\beta)$ (axiom schema K), and the theorems are closed under the rules of modus ponens, uniform substitution and necessitation (whenever α is a theorem, so is $\Box\alpha$).

We now postulate four closure principles about O:

(1) **Identity**: There is an $L_1 \in O$ for which $L_1p = p$ for all propositions p.

L_1 is in effect a propositional truth operator, or double negation; we can consider it as an objective necessity operator by recalling that it is the limiting case of the successive application of *n* objective necessity operators for $n = 0$ (these are all the same function on propositions). Trivially, L_1 commutes with conjunction. On a corresponding account in terms of Kripke models, the accessibility relation for L_1 would simply be the identity relation between worlds.

(2) **Composition**: For all $L_i, L_j \in O$ there is an $L_{ij} \in O$ for which $L_{ij}p = L_iL_jp$ for all propositions p.

Composing a pair of objective necessity operators yields an objective necessity operator. $L_{ij}p$ commutes with conjunction because L_i and L_j do. In terms of Kripke models, the accessibility relation for L_{ij} would be the composite of the accessibility relations for L_i and L_j. The joint effect of (1) and (2) is that composing a sequence of length *n* of objective necessity operators yields an objective necessity operator for any natural number *n*.

(3) **Conjunction**: For all $O' \subseteq O$ there is $L_{O'} \in O$ for which $L_{O'}p = \wedge_{i \in O'} L_ip$ for all propositions p.

The conjunction of any objective necessity operators is itself an objective necessity operator. It commutes with conjunction because each of the conjoined operators does so. In terms of Kripke models, the accessibility relation for $L_{O'}$ would be the union (disjunction) of the accessibility relations for each operator $L_i \in O'$. One might also conjecture that the intersection (conjunction) of the accessibility relations for a set of objective necessity operators would itself be the accessibility relation for an objective necessity operator, but that conjecture is not needed for what follows.

(4) **Reversal**: For all $L_i \in O$ there is an $L_{/i} \in O$ for which p entails $L_iM_{/i}p$ for all propositions p.

Here p entails q just in case $p \wedge q = p$. Of the four postulates, (4) has the least claim to obviousness. The claim is that if p holds, then it must$_i$ be possible$_j$ in some objective sense *j* for p to hold. A helpful analogy is with the relations between past and future operators: if p holds then it will always hold that p once held, and it always held that p would one day hold. Thus we might add to the consequent of (4) that p entails $L_{/i}M_{i}p$, although that conjecture is not needed for what follows. In terms of Kripke models, the accessibility relation for $L_{/i}$ would be the converse of the accessibility relation for L_{i}, and *vice versa*.

By postulate (3), there is a *strongest* objective necessity operator L_{O}, the conjunction of all objective necessity operators (let $O' = O$). Hence $L_{O}p$ entails $L_{i}p$ for every proposition p and objective necessity operator L_{i}. We identify L_{O} with *metaphysical necessity*. Thus metaphysical necessity implies every objective kind of necessity, and dually every objective kind of possibility entails metaphysical possibility. Then, given postulates (1), (2) and (4), the logic S_{O} of metaphysical necessity is at least, and presumably exactly, S5.[2] For we can show that the characteristic axiom schemas of S5 belong to S_{O} thus:

Axiom T $\quad \Box\alpha \rightarrow \alpha$ (what is necessarily so is so)

For since L_{O} is maximal in O, and $L_{1} \in O$ by (1), $L_{O}p$ entails $L_{1}p$, which is p. In terms of Kripke models, the accessibility relation for L_{O} is reflexive because it includes identity. We could have imposed satisfaction of axiom T as a constraint on *all* objective necessity operators, and dispensed with postulate (1), although that strategy would have required a restriction on postulate (3) to nonempty sets of operators, since $L_{\{\}}p = 1$ for all propositions p, including false ones. However, there is no clear need to impose T on all objective necessity operators, so for the sake of generality it is better not to. Thus an objective necessity operator might range only over some counterfactual possibilities.

Axiom 4 $\quad \Box\alpha \rightarrow \Box\Box\alpha$ (what is necessarily so is necessarily necessarily so)

For since L_{O} is maximal in O, and $L_{OO} \in O$ by (2), $L_{O}p$ entails $L_{OO}p$. In terms of Kripke models, the accessibility relation for L_{O} is transitive because it includes its composition with itself. It would not have been plausible to impose satisfaction of axiom 4 as a constraint on all objective necessity operators, for local forms of objective necessity concern only close worlds, and closeness is not transitive.

Axiom B $\quad \alpha \rightarrow \Box\Diamond\alpha$ (what is so is necessarily possibly so)

For since L_{O} is maximal in O, and $L_{/O} \in O$ by (4), $L_{O}p$ entails $L_{/O}p$, so $M_{/O}p$ entails $M_{O}p$, so $L_{O}M_{/O}p$ entails $L_{O}M_{O}p$ by normality, but p entails $L_{O}M_{/O}p$ by (4), so p entails $L_{O}M_{O}p$. In terms of Kripke models, roughly, if the accessibility relation for L_{O} contains its own converse then it is symmetric. Consequently, such accessibility is an equivalence relation. It would not have been plausible to impose satisfaction of axiom B as a constraint on all objective necessity operators: for instance, even if I shall in fact cross the bridge tomorrow, in some objective sense there may still be a practical possibility of its being destroyed later today, making it practically impossible for me to cross it tomorrow.

Thus, even though some objective necessity operators satisfy only much weaker logics, metaphysical necessity itself satisfies the strong modal logic S5, thanks to its role as the strongest objective necessity operator combined with plausible closure principles on the family of objective necessity operators. The result is a tidy picture of metaphysical modality: the modal status of a proposition as metaphysically necessary, metaphysically contingent, or metaphysically impossible is never itself metaphysically contingent.

An attractive conjecture consistent with the foregoing account, including the coarse-grained view of propositions, is that the metaphysical necessity operator is simply defined by these equations: $L_O p = 1$ if $p = 1$; $L_O p = 0$ otherwise. We already know that the first half of that holds because it does so for all objective necessity operators, as we saw; the crux is that $L_O p = 0$ whenever $p \neq 1$. If we assume that commuting with conjunction is sufficient as well as necessary for being an objective necessity operator, the conjecture follows, for the operator defined by those equations can easily be shown to commute with conjunction, and to be the strongest operator to do so.

The schematic characterization of metaphysical modality as the maximal objective modality leaves many questions about it unanswered. In particular, how far does it satisfy Kripke's seminal account (Kripke 1980)? Since the guise under which an object is presented has no bearing on an objective modality, we presumably have the necessity of identity: $x = y \rightarrow \Box\, x = y$.[3] Given that metaphysical modality obeys the principles of S5, we can thence derive the necessity of distinctness: $x \neq y \rightarrow \Box\, x \neq y$ (Prior 1956). These principles already suffice for some distinctive examples of the necessary a posteriori, as we should expect of an objective modality. For instance, it is metaphysically necessary, but not knowable a priori, that Socrates is distinct from Plato. But such general structural principles are typically neutral with respect to specific essentialist claims.

Nathan Salmon has argued in detail on essentialist grounds that the 4 axiom schema fails for metaphysical modality (Salmon 1982, 1989, 1993). If the essence of an artefact permits small but not large variations in its original constitution, then we should expect the accessibility relation for metaphysical modality to be non-transitive, because many small differences can add up to a large one. Under the reading of 'metaphysical modality' as meaning the maximal objective modality, Salmon's argument against the 4 axiom must fail. He rejects that reading, and indeed his argument may be sound under some alternative readings of 'metaphysical modality' as meaning various non-maximal objective modalities with non-transitive accessibility relations. One might even use Kripkean claims as to how different a given object could or could not have been as paradigms with which to explain an alternative sense for 'metaphysical modality'. For present purposes, however, the question is where Salmon's argument goes wrong when read, contrary to his intentions, with respect to the maximal objective modality. One option is to go for a much stricter form of essentialism, on which ordinary claims that an artefact could have had a slightly different original constitution are dismissed as loose talk (Chisholm 1973). Another option is to go for a much less strict form of essentialism, on which ordinary claims that an artefact could not have had a very different original constitution are interpreted as concerning only quite restricted types of possibility (Mackie 2006). There are also intermediate options (Williamson 2013b, 126–143). We need not decide between these options here.

More generally, the conception of metaphysical modality as the maximal objective modality leaves open a wide range of theoretical options. At one extreme, metaphysical modality might involve only a bare minimum of structural constraints, such as the principles of S5 and the necessity of identity and distinctness. At the other extreme, it might involve rich essentialist constraints. It has not even been excluded that metaphysical modality coincides with nomic modality. These questions should be decided by detailed theoretical investigation, not by stipulation. Our present interest is in the whole range of objective modalities, not just in their maximum.

2. Scepticism about objective modalities

Objective modalities are envisaged as out there in the world, independently of us. There is a long philosophical tradition of scepticism about such modalities. Its patriarch is of course David Hume. He is usually, and most interestingly, interpreted as calling into question the very idea of objective necessity. On this reading, he denies not merely that we can know that billiard balls *must* rebound from the cushion as they do, but even that we can use the word 'must' to express any idea of objective necessity — as opposed to something psychological in ourselves, such as an expectation that they will so behave. Of course, since correlative types of objective necessity and possibility are interdefinable duals, he is calling into question the idea of objective possibility just as much as the idea of objective necessity. Hume's scepticism targets not only metaphysical modality; his arguments are just as relevant to more restricted objective modalities, such as nomic modality, which may be more appropriate to the motion of billiard balls.

Humean arguments remain surprisingly influential in the philosophy of modality, despite (or even because of) their seeming reliance on a priori crudely empiricist assumptions. In particular, it is often still taken for granted that the contents of perception are non-modal. Yet, as I write, I can see that, from where I sit, I *can* reach the computer screen but not the window. Of course, such modal contents might conceivably be the conclusions of inferences from the contents of one's perception in some stricter sense to be explained, combined with background beliefs about one's body, but such an interpretation is not obviously correct or even very plausible. A responsible empiricist should at least consider scientific alternatives such as Gibson's theory of perceptual affordances (1979) and its recent successors, on which sense perception has inherently modal contents.[4] After all, from an evolutionary perspective, it would be highly adaptive for perception to present such information about possibilities for action directly to us, rather than leaving us to get there through such time-consuming and troublesome inferences as occur to us. Fast reactions to new perceptual information are often crucial to the success of action, and to survival.

Recent metaphysics has witnessed a less extreme critique of objective modality, which concedes its intelligibility but denies its fundamentality. Where Quine (1953) dismissed quantified modal logic as incoherent, his student David Lewis found a way of interpreting it more charitably within a similar broadly Humean framework, using counterpart theory on the assumption that there are many concrete worlds (maximal connected spatiotemporal systems) other than our own (1968). Lewis formulates counterpart theory in a first-order non-modal language to which Quineans can hardly object, even though they may of course deny that there is more than one world in Lewis's sense. But although his counterpart-theoretic translation presents the quantified modal language as *meaningful*, it also presents it as far from *perspicuous*. The messy complexities of Lewis's translation scheme make the surface forms of quantified modal sentences a misleading guide to their underlying logical relations. From this perspective, it is better to do one's theorizing in the language of counterpart theory itself, free of modal operators, since the latter tend to obscure the deep structure of the metaphysical issues. In his later work, Lewis explicitly did just that, bypassing modal formulations to work directly in the language of counterpart theory (1986). So-called modal realism may just as well be regarded as a form of anti-realism about the modal. Of course, Lewis-style modal realism has never been a majority position. Nevertheless, it has encouraged the tendency not to take modal distinctions at face value, but instead to suspect them of distracting attention from deeper issues.

A more recent motive for downgrading modality is less Humean than Aristotelian in spirit. Disappointingly, distinctions drawn in modal terms have often turned out to be too coarse-grained to do the metaphysical work initially hoped of them. A prominent case was the attempt to use the notion of supervenience to explain the relation between the mental and physical: no mental difference without a physical difference (Davidson 1970). Despite Davidson's Quinean qualms about modality, supervenience is naturally defined in modal terms (no *possible* mental difference without a physical difference), although making it precise reveals that many subtly inequivalent modal definitions are available. However, on all the most attractive modal definitions, supervenience is not an asymmetric relation, and even one-way supervenience may hold between families of properties that seem to be 'on a level' with each other. Moreover, where supervenience does hold, one wants to know *why* it holds; the suspicion is that the real metaphysical action will be in answering the latter question. Thus just saying in a precise modal sense that the mental supervenes on the physical, even if true, clarifies the dependence of the mental on the physical much less than had been hoped (see Kim 1993 for discussion). Similarly, following Kripke (1980), it was widely accepted that what it is for a property to be essential to an object can be explained in modal terms: necessarily, if the object exists then it has the property. But Kit Fine (1994) argued persuasively that any such modal definition of essence is too coarse-grained to capture the difference between essential and accidental properties. Such disappointments have contributed to a view of modal distinctions as shallow and inadequate substitutes for metaphysically deeper distinctions concerning essence, grounding, fundamentality, naturalness, constitution, real definition, ontological explanation, or whatever it may be, even if the modal distinctions have some practical utility as a stopgap convenience (Kment 2014 develops a view of this sort).

The focus of such criticism is often specifically on *metaphysical* modality, for instance on whether it is metaphysically fundamental and unified or just a miscellaneous ragbag of disparate elements (Sider 2011). However, metaphysical modality is only one member of the extended family of objective modalities. Arguments for scepticism about metaphysical modality tend to generalize to other objective modalities, irrespective of the theorist's intentions. For instance, the epistemological challenge 'If something is non-actual, how do you know whether it is possible?' arises for *any* non-trivial objective modality, not just for the metaphysical sort — which is not to say that the challenge cannot be met. Likewise, Quine's logical qualms about quantifying into the scope of modal operators do not depend on whether those modal operators are interpreted as metaphysical, nomic, or practical. If Lewis's counterpart theory is used to interpret metaphysical modality, it should also be used to interpret the other non-trivial objective modalities, which are restrictions of metaphysical modality. Similarly, substituting another objective modality for metaphysical modality just exacerbates the problem of coarse-graining, for the trouble is that metaphysical necessity comes too cheap, and other forms of objective necessity come even cheaper.

Suppose that *some* non-trivial objective modality, A-modality, has the virtues critics deny to metaphysical modality. It is intelligible, but not to be explained in counterpart-theoretic terms, our knowledge of it is reasonable though far from complete, it cuts at a joint, and so on. Perhaps we should not *identify* metaphysical modality with A-modality, because some more general objective modality has all those virtues too. Still, given the virtues of A-modality, standard critiques of metaphysical modality are clearly missing something of crucial importance. Those who seek to disarm metaphysical modality had better disarm the whole family of objective modalities, lest other family members exact their revenge.

One response is that *nomic* modality is a non-trivial objective modality that does not stand or fall with metaphysical modality, because it can be independently explained in terms of natural science: to be nomically possible is to be consistent with the laws of nature, and natural science is our best source of knowledge about those laws.

What does 'consistent' mean there? It might mean *logically consistent*. But 'Hesperus ≠ Phosphorus' is logically consistent with the laws of nature, for their formulation involves nothing as parochial as the names 'Hesperus' and 'Phosphorus', and 'Hesperus = Phosphorus' by itself is no truth of logic. For the same reason, 'Hesperus is a quark' is also logically consistent with the laws of nature. On the proposed account, therefore, it is nomically possible that Hesperus is distinct from Phosphorus and nomically possible that Hesperus is a quark. That is not an attractive view of nomic possibility. Indeed, the overall view is not even consistent. For it entails that nomic modalities are objective modalities, and one mark of objective modality is that it does not block the substitution of co-referring names. Thus if it is nomically possible that Hesperus is distinct from Phosphorus, it is also nomically possible that Hesperus is distinct from Hesperus, in which case the view requires 'Hesperus ≠ Hesperus' to be logically consistent with the laws of nature, which it is not, because it is not even logically consistent with itself. The difference between the names 'Hesperus' and 'Phosphorus' is a difference in our representations that corresponds to no difference in the states of affairs represented or their objective modal status. The same problems arise if one appeals to 'conceptual consistency' (whatever that is) instead of logical consistency. To avoid the problems, in defining nomic possibility one would have to conjoin the laws of nature with all true claims of identity and distinctness, such as 'Hesperus = Phosphorus', and true claims of kind membership and non-membership, such as 'Hesperus is a planet' and 'Hesperus is not a quark'. But those are just the sorts of move philosophers make in trying to reductively define metaphysical modality itself. It is an illusion that one can define a nomic objective modality without running into the issues that beset metaphysical modality. One might as well admit that nomic possibility is metaphysical compossibility with the laws of nature, as suggested earlier. If metaphysical modality is in trouble, so is nomic modality. Like other objective modalities, it depends on metaphysical modality. Indeed, for many purposes, though presumably not all, we may even be able to work with the hypothesis that metaphysical modality coincides with nomic modality. At least, nomic modality is a good approximation to metaphysical modality.

The sceptic may respond: if the other objective modalities depend on metaphysical modality, so much the worse for them, and in particular for nomic modality. Even though the appeal to natural science is necessary for nomic modality, it is not sufficient. According to such a sceptic, natural science can in principle be done without reliance on objective modalities; the science does not vindicate their specifically modal aspect. In a scientistic climate, such an assumed lack of connection to natural science makes the objective modalities look suspiciously ill-grounded.

It is notable how minor a role natural science plays in current discussion of the epistemology of modality. The main emphasis is on folk methods of knowing whether something is possible, perhaps by imaginative means, described in one way or another (Williamson 2007a). One would expect such folk methods to be primarily geared to quite restricted forms of practical modality, though philosophers usually want to discuss knowledge of metaphysical modality. One might get the impression that philosophers have taken some practically convenient everyday ways of thinking ('can'), drastically generalized them ('metaphysically possible'), perhaps far beyond their domains of reliability, and on those tenuous foundations erected a shaky castle of philosophical theory. Implicit in this picture is that science itself has no essential objective modal aspect, so its track record of success offers no support to the enterprise of objective modal theorizing. For instance, Ted Sider claims that 'modality is unneeded for the most fundamental inquiries' (2011, 267).

There is no need to disparage folk methods of gaining knowledge about various types of objective possibility and necessity, perhaps including metaphysical possibility and necessity. Such methods are easy to underestimate. But they are not the focus of this paper. Rather, its concern will be with more scientific methods of learning about objective modalities. Of course, logic is itself a science, in some ways the most rigorous science of all, and the study of quantified modal logic with respect to objective interpretations of the modal operators is a branch of that science. Arguably, it is best pursued by abductive methods of theory choice similar to those used in the natural sciences (Williamson 2013a). But this chapter does not take that view of modal logic for granted. Instead, it asks to what extent there is an implicit (or explicit) objective modal dimension to what are ordinarily counted as natural sciences, and to ordinary mathematics as applied in those sciences.

The mere definition of 'nomic possibility' as compossibility with the laws of nature poses no threat to the picture of natural science itself as non-modal, since it does not imply that the idea of nomic possibility plays any essential role in scientific attempts to identify the laws of nature. For all that the definition shows, modal ideas might be merely epiphenomenal in the scientific process. Likewise with the deduction of nomic possibility from actuality: if natural science discovers that there are black holes, we can of course deduce that it is nomically possible for there to be black holes, but that offers natural scientific help only where it is least needed, since the hard question is how far the possible extends *beyond* the actual.

We could go through numerous articles in journals of natural science and list all the places where modal expressions are used, in plainly objective senses, but we are unlikely to achieve much just by doing so. For such articles are written in quasi-natural language, and one can expect authors often to fall into such everyday ways of expressing themselves, even where they are not strictly needed. How might objective modality play a more essential role in natural science?

An apparent reason for pessimism is the increasing extent to which, as natural scientific theories become more rigorous, their core is expressed in equations or other mathematical formulas. For the language of mathematics is non-modal. It does not contain symbols like $\Box$ and $\Diamond$, at least not to mean necessity and possibility.[5] Of course, the absence of modal expressions *within* the formulas does not preclude us from ascribing nomic or metaphysical necessity to them from the outside. If '$5 + 7 = 12$' and '$E = mc^2$' are purely non-modal statements, we may still affirm that it is metaphysically necessary that $5 + 7 = 12$ and nomically necessary that $E = mc^2$. But the danger is that such modal claims are merely philosophers' exogenous honorific glosses, functionless within the science itself. It is as if a philosopher went round sticking gold labels on his favourite machines, reading 'This machine has been approved by a qualified metaphysician'. The label may look good, but it makes no difference to the working of the machine. More specifically, if the modal glosses are merely external to the science, then they draw no abductive support from the explanatory successes of the science. Can we find cases where instead the modal glosses reflect some endogenous need of the science? The next section starts to come to grips with that question.

3. Laws support counterfactuals

One obvious starting-point is the near-platitude in the philosophy of science that laws support counterfactuals. If it *is* a law that all Fs are Gs, then if there had been an F, there would have been a G. Even if the universal generalization 'All Fs are Gs' itself contains no modal element, in claiming that it is a law we licence its application to (at least some) counterfactual circumstances. Surely we want to use our scientific theories, including our mathematical theories, in reasoning about how things could have been, as well as about how they are. In such reasoning, we engage a specifically objective modal dimension. For those purposes, the core theory itself need not be cast in modal terms. It is enough that sometimes our legitimate applications of it assign it a modal status. What's the problem?

Not all reasoning from a false hypothesis is counterfactual in the sense relevant to objective modality. To use a standard example, the uncontentious truth 'If Oswald didn't shoot Kennedy, someone else did' is an ordinary indicative conditional, even though Oswald did in fact shoot Kennedy. It is uncontentious because, for sure, *someone* shot Kennedy, so if it wasn't Oswald, it was someone else. It is not equivalent to the so-called subjunctive conditional 'If Oswald hadn't shot Kennedy, someone else would have', which suggests another back-up conspirator lying in wait. Unlike the subjunctive conditional, the indicative conditional does not involve an objective modality. The indicative conditional, but not the subjunctive conditional, can be reasonably inferred from the non-modal statement 'Someone shot Kennedy'. Indicative and subjunctive conditionals interact differently with modally rigidifying devices, such as 'actually' and 'in this world' (Williamson 2006). 'If Oswald didn't shoot Kennedy, someone else did' is equivalent to 'If Oswald didn't shoot Kennedy, someone else did in this world'. By contrast, 'If Oswald hadn't shot Kennedy, someone else would have' is not equivalent 'If Oswald hadn't shot Kennedy, someone else would have in this world', since the former is true and the latter false in the scenario where Oswald's shot pre-empts an efficient backup assassin. In that respect, subjunctive conditionals pattern like objective modals while indicative modals do not: although the epistemic modal sentences 'Oswald may have missed' and 'Oswald may have missed in this world' are more or less equivalent, the objective modal sentences

'Oswald could have missed' and 'Oswald could have missed in this world' are not; indeed, the former is true and the latter false (where in both cases 'in this world' is read as within the scope of the modal verb). In considering applications of scientific theories (including mathematical theories) to hypothetical situations, we must be careful about whether they really require subjunctive conditionals rather than indicative ones, even when it is natural for us to articulate them in terms of subjunctive conditionals.

Imagine that we are assessing a plan A for building a bridge. We ask 'What would happen if we were to build the bridge according to plan A?'. We apply our relevant theories, and come to the conclusion 'If we were to build the bridge according to plan A, it would fall down'. Consequently, we decide not to build the bridge according to plan A (so the antecedent of the conditional is false). We reasoned with subjunctive conditionals, and it was quite natural to do so. But there was no real need to do so. We could just as well have reasoned with indicative conditionals, asking 'What will happen if we build the bridge according to plan A?', applying our relevant theories as before, and concluding 'If we build the bridge according to plan A, it will fall down'. If we know that indicative conditional in those circumstances, we have reason enough not to build the bridge according to plan A. It can be natural to articulate an application of a theory in subjunctive terms even when there is no real need to do so. One might wonder for a moment whether an objective modal dimension, by contrast with an epistemic modal dimension, is ever really needed in practical applications.

Consider learning from mistakes. We see a bridge fall down. We ask the subjunctive conditional question 'What would have happened if the bridge had been built according to plan B?'; we apply our relevant theories, and come to the subjunctive conditional conclusion 'If the bridge had been built according to plan B, it would not have fallen down'. Consequently, we do better ourselves next time we have to build a bridge. In this case, indicative conditionals will not do just as well. We do not assert 'If the bridge was built according to plan B, it did not fall down', for we know for sure that it did fall down, whether or not it was built according to plan B. Even if we treat the indicative conditional as truth-functional, true simply because its antecedent is false (we have concluded that the bridge was not built according to plan B), we should regard it as too conversationally misleading to assert. It is the subjunctive conditional that carries the relevant information. For such applications, we need an objective modal dimension.

Causal hypotheses are also far more strongly connected to subjunctive conditionals than to indicative conditionals, even though it may well be over-optimistic to expect strictly necessary and sufficient conditions for causal hypotheses in terms of subjunctive conditionals, or strictly necessary and sufficient conditions for subjunctive conditionals in terms of causal hypotheses.

Here is a toy example. Suppose that we are wondering whether there is a causal relationship between the variable X, whose value we set at time t, and the variable Y, whose value we observe at time $t + 1$. We set $X = 1$ and observe $Y = 0$.

Clearly, just on that basis, we are not in a position to conclude that $Y=0$ (causally) *because* $X=1$. But suppose that a well-confirmed scientific theory, formulated in austerely mathematical, non-modal and non-causal terms, entails that $Y=1-X$. That surely does much to confirm the causal hypothesis. A reasonable story about how it does so is that the theory supports subjunctive conditionals, and so in particular the subjunctive conditional $X \neq 1 > Y \neq 0$ (if X had not been 1, Y would not have been 0): without the putative cause, there would not have been the putative effect. Varying the antecedent, we can derive other subjunctive conditionals similarly. Such patterns of counterfactual dependence are closely connected to causal hypotheses, even if the connection falls short of strict implication (see Woodward 2003 for one discussion; the literature is vast).

The material conditional $X \neq 1 \rightarrow Y \neq 0$ is obviously no substitute for the subjunctive conditional, since we already had the material conditional just from the initial observations. However, it is also inappropriate to assert the indicative conditional 'If $X \neq 1$ then $Y \neq 0$' in the absence of the theory, since even if we are wrong about how we set X, that by itself casts no doubt on our observation of Y. To eliminate this feature of the example, tweak it so that we no longer observe Y directly; instead, we set a reliable alarm to go off if and only if $Y=1-X$. In both cases, the alarm goes off, so we can assert the indicative conditional without need of the theory. It is the subjunctive conditional that makes the difference.

Thus, even if we start with an austerely formulated scientific or mathematical theory, free of modal and causal vocabulary, applying the theory for practical purposes or to reach causal conclusions often depends on its supporting subjunctive conditionals. When those applications are successful, part of what their success abductively supports are those subjunctive conditionals.

Notoriously, the subjunctive conditional $\alpha > \beta$ is in general much weaker than the corresponding objective strict conditional $\Box(\alpha \rightarrow \beta)$, where $\Box$ expresses metaphysical necessity. In some sense of 'nearby', if α is true at some nearby possible worlds, and β is true at all nearby possible worlds at which α is true, but α is true and β false at some more distant possible world, then $\alpha > \beta$ is true and $\Box(\alpha \rightarrow \beta)$ false. One might therefore suspect that a theory can support subjunctive conditionals without supporting attributions of objective necessity. Fortunately, that danger is largely avoided.

Let us be a little more precise. We start with a scientific theory T in a non-modal language L with at least the truth-functors $\neg$ and $\rightarrow$ (the material conditional) obeying the usual classical laws. We are supposing that T supports counterfactuals. Even if T *entails* only formulas of L, T may *support* formulas of a modal extension L^+ of L. L^+ includes the subjunctive conditional >, and dual operators $\Box$ and $\Diamond$ for metaphysical necessity and possibility and $\Box_N$ and $\Diamond_N$ for nomic necessity and possibility. A reasonable assumption is that what T supports is closed under entailment. That is, if a formula α of L supports a formula β of L^+, and β entails another formula γ of L^+, then α also supports γ.[6] A simple interpretation of 'T supports subjunctive conditionals' is simply this:

(S1) Whenever T entails $\alpha \to \beta$, T supports $\alpha > \beta$.

Note that $\alpha \to \beta$ is a formula of both L and L^+, $\alpha > \beta$ a formula only of L^+. But from (S1), we can derive (S2):

(S2) Whenever T entails α, T supports $\Box\alpha$.

Thus T supports claims of metaphysical necessity after all. As a special case, when T entails a material conditional, it supports the corresponding strict conditional as well as the subjunctive conditional.

The argument from (S1) to (S2) is simple. Suppose that (S1) holds and that T entails α. By classical propositional logic, α entails $\neg\alpha \to \alpha$. Hence, by the transitivity of entailment, T entails $\neg\alpha \to \alpha$. Therefore, by (S1), T supports $\neg\alpha > \alpha$. But $\neg\alpha > \alpha$ surely entails $\Box\alpha$, for a formula subjunctively implies its own contradictory only if it does so vacuously; if $\neg\alpha$ were metaphysically possible, it would not be that if $\neg\alpha$ were true, so would be α.[7] Thus, by the principle that what T supports is closed under entailment, T supports $\Box\alpha$.

The conclusion (S2) is reasonable when T is a theory of pure mathematics. In effect, it tells us that we cannot use mathematics freely in subjunctive reasoning unless mathematics is assumed to be metaphysically necessary. But for most theories T in the natural sciences, (S2) looks too strong. Plausibly, the most to be claimed for them is nomic rather than metaphysical necessity, given that the latter modality is stronger than the former.[8] In such cases, the problem is not with the reasoning from (S1) to (S2) but with (S1) itself. On the assumption that α is nomically impossible, we should not expect T to yield correct information about what would happen if α were true; T is only concerned with the realm of the nomically possible. In such cases, we should expect exceptions to (1). Thus, for a theory T of natural science, we should weaken (S1) by a restriction to nomic possibility:

(S1*) Whenever T entails $\alpha \to \beta$, T supports $\Diamond_N \alpha \to (\alpha > \beta)$.

The reasonable assumption underlying (S1*) is that, for nomic possibilities, if they were to obtain, they would obtain in (actually) nomically possible ways, even if it is also metaphysically possible for them to obtain in (actually) nomically impossible ways: nomically possible worlds are 'closer' than nomically impossible ones to the actual world.[9] From (S1*) we can derive (S2*):

(S2*) Whenever T entails α, T supports $\Box_N\alpha$.

The argument is a modification of that from (S1) to (S2). For suppose that (S1*) holds while T entails α. As before, T entails $\neg\alpha \to \alpha$, so T supports $\Diamond_N\neg\alpha \to (\neg\alpha > \alpha)$ by (S1*). But $\neg\alpha > \alpha$ surely entails $\neg\Diamond_N\neg\alpha$; if $\neg\alpha$ were nomically possible, it would not be that if $\neg\alpha$ were true, so would be α. Hence $\Diamond_N\neg\alpha \to (\neg\alpha > \alpha)$ entails $\neg\Diamond_N\neg\alpha$, which is equivalent to $\Box_N\alpha$. Thus, by the principle that what T supports is closed under entailment, T supports $\Box_N\alpha$. Even the qualified way in which scientific theories arguably support subjunctive conditionals requires them to support claims of nomic necessity too.

In brief, if we want to apply a scientific theory freely in the scope of subjunctive reasoning about nomic possibilities, the theory had better be at least nomically necessary, even if the content of the theory itself is purely non-modal. Furthermore, if such modal applications have a track record of success, it provides abductive confirmation for the relevant claims of nomic necessity. It is a serious mistake to picture those objective modal claims as supported by nothing more than folk habits of thought and metaphysical speculation. Nevertheless, that point does not eliminate the suspicion that conceptualizing matters in such supposedly objective modal terms somehow misleads us about the underlying fundamental joints in nature. Thinking in terms of natural language subjunctive conditionals may look prescientific. These concerns are fuelled by the absence of objective modal constructions from the language L, in which by hypothesis the scientific theory T was formulated. Such suspicions can best be answered by considering cases where the content of the scientific theory itself is objectively modal. That is the task for the rest of the paper.

4. Objective probabilities

Talk of probabilities is, of course, widespread in the natural sciences. Whilst casual uses of the word 'probably' may merely express caution, explicit quantification of probabilities — for instance, in the interpretation of statistics — presupposes some form of modality, for any probability distribution is defined over a probability space of mutually exclusive, jointly exhaustive 'possibilities': in any given circumstances, all but one of them is counterfactual. Probabilities are assigned to all members of a field of 'events', that is, subsets of the set of all those possibilities. The possibilities behave like possible worlds, and the events behave like coarse-grained propositions, sets of possible worlds. The probability of a proposition is a measure of its closeness to necessity. If the number of events is finite, all nonempty propositions may be assigned nonzero probability, in which case probability 1 corresponds to necessity, because it is equivalent to 'truth everywhere' in the space, and nonzero probability corresponds to possibility, because it is equivalent to 'truth somewhere' in the space. If the number of events is infinite, there are technical obstacles to assigning all nonempty propositions nonzero probability; probability 1 is equivalent only to 'truth almost everywhere' in the space.[10] But even in the latter case, probability 1 and nonzero probability still behave logically like dual modal operators in a finitary modal language.[11] Moreover, necessity and possibility can still be defined in a natural way directly over the probability space itself as 'truth everywhere' and 'truth somewhere', respectively. A one-way connection still holds between probability and possibility: whatever has nonzero probability is possible in the corresponding sense, even though the converse fails. Moreover, probabilistic distinctions resemble modal distinctions in being coarse-grained: just as truth-functionally equivalent formulas are necessarily equivalent, they also have the same probability, as a consequence of the standard axioms for probability.

For present purposes, however, not any old probabilities will do. Only *objective* probabilities are appropriately related to objective possibilities. Often the probabilities discussed in science are epistemic, dependent on an evidence base, and so not suitable here. Subjective probabilities (credences, degrees of rational belief) help still less. Indeed, not even all objective probabilities are interesting for our purposes, since some of them are in effect distributions only over sets of actual cases. In particular, we are not concerned with probabilities understood as actual frequencies, even though they are objective at least in being agent-independent. But frequentist interpretations of probability are in any case unpromising, because actual frequencies may happen to be utterly wayward, in principle even over a very long run. A fair coin can come up heads any number of times in succession. Frequencies are better understood as good *evidence* for underlying probabilities that explain, and so should not be identified with, the frequencies.

The most familiar genuine objective probabilities are *chances*. Consider some physical system of scientific interest, perhaps the whole universe. We can ask: given that the system is in a maximally specific state s_0 at a time t_0, what is the *chance* that it will be in a maximally specific state s_1 at a later time t_1? If the system is deterministic, the answer will be 1 or 0. But if the system is indeterministic, the answer may, more interestingly, be some intermediate real number. The most celebrated example of indeterminism and intermediate chances in science is of course in quantum mechanics, under some interpretations. It is widely accepted that the probabilities in the formulation of quantum mechanics are not merely epistemic or subjective. However, given the notorious difficulties of interpreting quantum mechanics, we shall leave discussion of its probabilities to the experts.

Significantly, the very distinction between deterministic and indeterministic systems itself involves objective modality. Suppose that you are given the entire history of a system, past, present and future, all described in purely non-modal terms, and that the history contains no recurrences: the system is in each maximally specific state at most once. There is no way of reading off from the history whether the system is deterministic or indeterministic. That depends on whether there are two *possible* total histories of the system, h and h^*, maximally specific states s_0 and s_1, and times t_0 and t_1, such that in h the system is in s_0 at t_0 and in s_1 at t_1, while in h^* the system is in s_0 at t_0 but is not in s_1 at t_1. If so, the system is indeterministic (its state at one time does not determine its state at another time); if not, the system is deterministic. The type of possibility at issue is nomic and objective. Chance is a measure of closeness to a timebound sort of objective necessity, such as nomic necessity conditioned on the circumstances at the time.

Non-trivial objective probabilities may also arise for deterministic systems. For scientists may explain some general features of the system's actual total history by showing them to be *typical* of its possible histories. That is, it is highly probable that the system will have a total history with those features. The relevant probabilities here are not chances given the state of the system at a time but rather something like probabilities over initial conditions. For the explanation to work properly, those probabilities should be objective. If we were merely told that it would be rational for someone in a particular evidential situation to be confident that the system would have a total history with the features at issue, we should be unsatisfied, because such a hypothetical agent is quite extraneous to what was to be explained. A better explanation would strip out the irrelevant material about the agent, and isolate the relevant facts about the system itself that the non-objective 'explanation' was clumsily attempting to communicate. An example of such an explanation of the general behaviour of a deterministic system in terms of objective probabilities over its initial conditions is the derivation of standard thermodynamic principles from classical statistical mechanics (see Loewer 2001; Maudlin 2007 for discussion).

For illustrative purposes, a toy example will suffice instead. Suppose that a coin was tossed 1000 times. It came up heads approximately 500 times (the explanandum); why? A potential explanation is that the coin was fair and the tosses mutually independent (the explanans). Once the relevant calculations are made, the explanans gives a reasonable explanation of the explanandum. It is a piece of proto-science. Probability enters the explanation in at least two ways. First, the explanans itself is implicitly probabilistic: the coin is said to be fair in the sense that the probability of heads on a given toss is ½, and the tosses are said to be mutually independent in the sense that the unconditional probability of an outcome of a given toss equals its probability conditional on given outcomes of other tosses. Second, the connection between the explanans and the explanandum is also probabilistic, since the explanans does not entail the explanandum — the explanans is consistent with the coin's coming up heads everyone of the 1000 times — but instead only makes the explanandum probable (in the same sense of 'probable'). The relevant probabilities are not subjective or epistemic, since the degrees of belief or evidential situation of an actual or ideal agent played no relevant role in the event to be explained. They are quite extraneous to the explanandum and should not figure in the explanans.[12] Nor are the relevant probabilities frequentist. For consider any given toss in the long run over which such frequencies would have to be calculated. If the toss is one of the 1000 in the explanandum, that would make for circularity in the explanation, but if the toss is not one of the 1000, then it played no role in bringing about the explanandum. Either way, it should be excluded. The example is best understood as involving a reasonable proto-scientific explanation in terms of objective probabilities.

The example does not require the physics underlying coin-tossing to be indeterministic. Instead, each cell of the macroscopic probability space may correspond to one equivalence class of a coarse-grained macroscopic partition of possible microscopic deterministic histories that differ from each other on the past and present as well as on the future; microscopically different ways of tossing the two coins lead deterministically to macroscopically different outcomes. Such possibilities are just as objective as indeterministic chances; no 'initial conditions' were nomically necessary. We should not suppose that an explanation in terms of the detailed microscopic histories of the actual tosses would in principle be better. For an explanation of the latter sort involves a drastic loss of generality: its microscopic explanans would obtain in only a tiny fraction of the cases in which the explanandum (as characterized above) would obtain. To capture the generality of the explanandum, we need the generality of the macroscopic objectively probabilistic explanans. A proper microscopic explanation would involve objective probabilities over different microscopic possibilities that realize the initial conditions of the coin-tossing.

5. State spaces

Probability is far from the only form in which objective modalities become the object of natural scientific inquiry. It is standard practice to study a physical system by analysing its state space or phase space, the abstract space of its *possible* states. The system may be as large as the universe or as small as a few interacting fundamental particles. The type of possibility is objective, more or less nomic. The states are maximally specific. This way of thinking is widespread in science. As a recent historian of the idea of phase space puts it, 'Listen to a gathering of scientists in a hallway or a coffee house, and you are certain to hear someone mention phase space' (Nolte 2010, 33).

State spaces have played a philosophically significant role in various connections. For example, in his critique of Hartry Field's nominalizing programme (1980), David Malament objects that if Field's method of nominalization is applied to various theories of mechanics, its effect is to replace quantification over abstract objects by quantification over the '"possible dynamical states" (of particular physical systems)', to which, he argues, a nominalist is not entitled (1982, 533). Aidan Lyon and Mark Colyvan have taken the latter objection further, arguing that attempts to nominalize standard phase-space theories in physics would result in a loss of explanatory power; as they explain, 'phase spaces are spaces of possible, but mostly non-actual, initial conditions' (2008, 227). In none of these cases are the possibilities subjective or epistemic; they are aspects of the physical domain under study, not of any real or ideal physicist's state of knowledge or belief. Rather, they are in some sense objective possibilities. For present purposes, our concern is not with the prospects for nominalization.[13] Rather, it is with the objectively modal dimension of the physics.

To develop the point, we may consider for a case study the theory of dynamical systems (Strogatz 2001). In itself it is a mathematical theory, but it has intended applications in physics, chemistry, biology and engineering, for instance to a pendulum, the solar system, the population growth or decline of predator and prey species, the weather and so on. As the last case suggests, it is a standard framework for the study of chaotic systems.

Mathematically, a dynamical system consists of a set S on which some geometrical or topological structure is defined, a set T (usually either the set of real numbers or the set of integers) with an additive structure, and a family of functions $\{f_t\}_{t\in T}$ indexed by T, obeying the following constraints for all $s \in S$ and $t, t^* \in T$:

(i) $f_0(s) = s$

(ii) $f_t(f_{t^*}(s)) = f_{t+t^*}(s)$

Informally, we understand the formalism thus. S is the set of *instantaneous states* of the target system; they are maximally specific in relevant respects, mutually exclusive and jointly exhaustive. T is the set of *directed lengths of time*; thus +1 and −1 may represent one second into the future and one second into the past respectively, distinct directed lengths of time whose sum is 0 (seconds). The system is assumed to be deterministic in both past and future directions; thus given its state at any one time, the dynamics fixes its state at any directed length of time from then (T includes negative as well as positive lengths of time).[14] Thus it is legitimate to understand $f_t(s)$ as the state of the system a length of time t after an instant when its state was s. For this interpretation to make sense, conditions (i) and (ii) must hold: (i) because zero time after an instant it is still that instant, and (ii) because the instant a length of time t after the instant a length of time t^* after a given instant is just the instant a length of time $t + t^*$ after the given instant.

Dynamical systems may be either *continuous* or *discrete*, depending on the structure of T. For a continuous dynamical system, T is the set of positive and negative real numbers, and the functions f_t are typically continuous with respect to the selected topology on S and implicitly defined by some differential equations. They form a *flow*. For a discrete dynamical system, T is the set of positive and negative integers, and the functions f_t are typically implicitly defined by some difference equations, though they may still be continuous with respect to the designated topology on S.

The mathematical theory of dynamical systems is just a branch of regular, non-modal mathematics. However, most *intended applications* of that mathematical theory are modal, in the sense that S is interpreted as the set of *possible* states of the target dynamical system — not, for instance, just the set of actual past, present and future states of the system. To be more precise, given a dynamical system, let an *orbit* be any set of the form $\{f_t(s): t \in T\}$ for some state $s \in S$, in other words, the set of states which the system goes through at some time or other if at some time it is in s. It is easy to show that if any orbit exhausts

S (so the system sooner or later goes through every state in *S*), then every orbit exhausts *S*. But, typically, no orbit exhausts *S*. Thus some of the possible states in *S* are mutually incompossible given the dynamics, in the strong sense that if the system is *ever* in one of them, then it is *never* in the others. The states in *S* are *possible states*, not all of which are ever actualized. It would be foolish to try to eliminate all the counterfactual states in the system by cutting it down to just its actual orbit, because that would typically destroy the geometrical or topological structure defined over *S*: that structure is crucial to the explanatory power of the theory of dynamical systems. The point is to study the dynamical system of possible states as a whole, exploiting that mathematical structure.

As before, the sort of possibility at issue is not subjective or epistemic. It depends on the nature of the physical system under study, not on the psychological or epistemic states of the theorist who studies it, or of anyone else, real or ideal. It is some sort of objective possibility, usually nomic rather than metaphysical, perhaps even more restricted (it is not nomically necessary for there to be a pendulum). Of course, dynamical systems are mathematical models of complex and often messy natural structures, and as such are likely to involve some degree of simplification, idealization and approximation. But that is just the normal case with natural science. It does not mean that dynamical systems have nothing to tell us about reality.

The possible states in *S* are clearly quite like *possible worlds*. However, since states are instantaneous, they are even more like ordered pairs of a world and a time, such as one evaluates formulas at in some formal theories of semantics for languages with both modal and temporal operators. But not even that comparison is perfectly apt, for nothing in a state specifies *when* the system is in it: indeed, unlike world-time points, states may be *repeatable*: when a dynamical system exhibits cyclic behaviour, it will be infinitely many times in each state that it is ever in (Nietzsche's eternal recurrence). One might try saying that the states in *S* are qualitative in a way that world-time points are not, but even that claim may be misleading, since the mathematical structure on states may require treating the result of spatially rotating, reflecting or translating a given state as a different state, even though they are qualitatively indistinguishable. Still, the states correspond to equivalence classes of some world-time pairs under some relevant equivalence relation. Thus, although states cannot be straightforwardly assimilated into the framework of possible world semantics, they still quite clearly have an objective modal aspect, as well as a qualitative-temporal one.

We can make the modal aspect of dynamical systems explicit by treating them as models over which we evaluate formulas of a propositional modal language. This can be done in a very smooth and natural way, without applying any Procrustean methods. By their structure, dynamical systems ask to be so treated. More specifically, with respect to a given dynamical system, we will evaluate a formula as true or false at a state relative to an assignment of values to variables, just as a formula of an ordinary modal language is evaluated as true or false at a world in Kripke models for modal logic. We write $s, \underline{a} \vDash \alpha$ to mean that the formula α is true at the state s relative to the assignment $\underline{a}$. The variables of the formal language are 'propositional': they take sentence position. There are infinitely many such variables U, V, … Formally, an assignment assigns each of them a set of states, a subset of S. Such sets play the role of propositions in the model, just as sets of worlds (subsets of W) play the role of propositions in Kripke models. The language has the usual truth-functors ¬, v, &, →, ↔, which behave as expected. It has dual modal operators ◇ and □, treated like quantifiers ranging over all states of the system (in this language, they need not express *metaphysical* modalities).[15] It also has the 'tense' operators F ('at some future state'), G ('at every future state'), P ('at some past state'), H ('at every past state') and G_t ('at the state t after the current state'), for each directed length of time $t \in T$. To govern the propositional variables, there are propositional quantifiers ∀ and ∃, ranging over all subsets of S. Further sentence operators are needed to express relevant mathematical aspects of the dynamical system, such as a monadic operator 'Open' to express the openness of a subset of S in the underlying topology of the model.

Here is an explicit definition of truth in a given model:

[atom]	$s, \underline{a} \vDash V$	iff	$s \in \underline{a}(V)$
[¬]	$s, \underline{a} \vDash \neg\alpha$	iff	not $s, \underline{a} \vDash \alpha$
[v]	$s, \underline{a} \vDash \alpha \vee \beta$	iff	$s, \underline{a} \vDash \alpha$ or $s, \underline{a} \vDash \beta$
[&]	$s, \underline{a} \vDash \alpha \,\&\, \beta$	iff	$s, \underline{a} \vDash \alpha$ and $s, \underline{a} \vDash \beta$
[◇]	$s, a \vDash \Diamond\alpha$	iff	for some $s^* \in S: s^*, \underline{a} \vDash \alpha$
[□]	$s, \underline{a} \vDash \Box\alpha$	iff	for all $s^* \in S: s^*, \underline{a} \vDash \alpha$
[F]	$s, \underline{a} \vDash F\alpha$	iff	for some $t \in T, t > 0: f_t(s), \underline{a} \vDash \alpha$
[G]	$s, \underline{a} \vDash G\alpha$	iff	for all $t \in T, t > 0: f_t(s), \underline{a} \vDash \alpha$
[P]	$s, \underline{a} \vDash P\alpha$	iff	for some $t \in T, t < 0: f_t(s), \underline{a} \vDash \alpha$
[H]	$s, \underline{a} \vDash H\alpha$	iff	for all $t \in T, t < 0: f_t(s), \underline{a} \vDash \alpha$
$[G_t]$	$s, \underline{a} \vDash G_t\alpha$	iff	$f_t(s), \underline{a} \vDash \alpha$
[Op]	$s, \underline{a} \vDash$ Open(α)	iff	$\{s^*: s^*, \underline{a} \vDash \alpha\}$ is open
[∃]	$s, \underline{a} \vDash \exists V \alpha$	iff	for some $X \subseteq S: s, \underline{a}[V/X] \vDash \alpha$
[∀]	$s, a \vDash \forall V \alpha$	iff	for all $X \subseteq S: s, \underline{a}[V/X] \vDash \alpha$

Such is the naturalness of the interpretation, these clauses require very little commentary: they are just what one would expect. The dynamics of the system is built into the clauses for the temporal operators, which are therefore not *purely* temporal. We continue to call them 'tense operators' rather than 'dynamical operators' only for the sake of familiarity. The underlying topology is used in the clause for 'Open'. In the quantifier clauses, a[V/*X*] is the assignment like a except for assigning the set of states *X* to the variable V.

A formula is *valid* on a model if and only if it is true at every state on every assignment with respect to the model. It would be mathematically pointless to equip the models with a designated actual state, so we avoid doing so, and therefore could not define validity in terms of truth at the actual state of the model. A formula α is *valid* without qualification if and only if it is valid on every dynamical system model.

We briefly note some valid formulas, most of them standard. The underlying non-modal propositional logic is classical: all truth-functional tautologies are valid, and modus ponens preserves validity. The unrestricted modal operators $\Box$ and $\Diamond$ obey all the principles of the modal system S5.[16] The unrestricted necessity operator $\Box$ entails all the tense operators F, G, P, H and G_t for all $t \in T$. Consequently, if something is ever possible, it is always possible[17]:

$$(P\Diamond V \vee \Diamond V \vee F\Diamond V) \rightarrow (H\Diamond V \,\&\, \Diamond V \,\&\, G\Diamond V)$$

We also have standard principles of tense logic for linear time with no first or last moment, which derive from the additive structure of the real numbers or the integers and conditions (i) and (ii) above.[18] In particular, conditions (i) and (ii) themselves correspond to the respective validity of these two axioms of metric tense logic (compare Prior 1967):

$$G_0V \leftrightarrow V$$

$$G_tG_{t^*}V \leftrightarrow G_{t+t^*}V$$

Other axioms, such as those corresponding to the density or discreteness of the time order, are validated if we restrict validity to continuous or to discrete models. The quantifiers obey the standard principles for propositional quantifiers in a modal setting, including for each state the existence of a proposition true at exactly that state:

$$\exists U\,[U \,\&\, \forall V\,[V \rightarrow \Box[U \rightarrow V]]]$$

Of course, the state may be a recurrent one: in the spirit of dynamical systems theory, the semantic theory does not distinguish between distinct times when the system is in the same state, although we can say that the system will again be in the same state:

$$\forall U\,[U \rightarrow FU]$$

This formula has the right effect because it entails that the atomic proposition true now will be true again.

We can see how the formal language can express characteristic ideas of the theory of dynamical systems by means of an example. The idea of an *attractor* plays an important role in the theory, helping us understand the long-term qualitative behaviour of dynamical systems. Very roughly, an attractor is a region of state space that the system gets pulled towards and stuck close to or in, once it has entered a surrounding region. Not all authors define the term in exactly equivalent ways, but the following definition is fairly standard (see Strogatz 2001, 324). As is typical, it assumes a topological structure on S. An *attractor* is a closed set of states $A \subseteq S$ such that:

(a) for all $s \in A$, $f_t(s) \in A$ whenever $t \geq 0$;

(b) for some open $U \supseteq A$: for all $s \in U$ and open $V \supseteq A$, for some t, $f_{t^*}(s) \in V$ for all $t^* \geq t$;

(c) if $A^* \subseteq A$ and A^* satisfies (a) and (b), then $A^* = A$.

Informally, (a) means that once the system is in A, it stays in A; (b) means that A draws and keeps all trajectories that ever come sufficiently close to it arbitrarily close to it (A attracts such trajectories); (c) means that A is minimal in these respects (it excludes redundant members). The *basin of attraction* of A is the largest set U satisfying the condition in (b); if A is an attractor, there is bound to be such a set. A *strange attractor* is an attractor that exhibits sensitive dependence on initial conditions. Such attractors are often fractal sets. They are important for the theory of chaos.

We can formalize (a)–(c) as a(A)–c(A), respectively (replacing italicized set variables by roman propositional variables):

a(A) $\quad \Box[\mathrm{A} \rightarrow \mathrm{GA}]$

b(A) $\quad \exists \mathrm{U}\, [\mathrm{Open(U)}\ \&\ \Box\, [\mathrm{A} \rightarrow \mathrm{U}]$
$\qquad \&\ \forall \mathrm{V}\, [[\mathrm{Open(V)}\ \&\ \Box\, [\mathrm{A} \rightarrow \mathrm{V}]] \rightarrow \Box[\mathrm{U} \rightarrow \mathrm{FGV}]]]$

c(A) $\quad \forall \mathrm{A}^*[[\Box[\mathrm{A}^* \rightarrow \mathrm{A}]\ \&\ \mathrm{a}(\mathrm{A}^*)\ \&\ \mathrm{b}(\mathrm{A}^*)] \rightarrow \Box[\mathrm{A}^* \leftrightarrow \mathrm{A}]]$

For c(A), note that $\Box[\mathrm{A}^* \leftrightarrow \mathrm{A}]]$ behaves like $\mathrm{A}^* = \mathrm{A}$, because it requires the regions of state space A^* and A to contain exactly the same states. We can now formalize '*A* is an attractor' thus (since a closed set is the complement of an open set):

Attractor(A) $\quad \mathrm{Open}(\neg \mathrm{A})\ \&\ \mathrm{a(A)}\ \&\ \mathrm{b(A)}\ \&\ \mathrm{c(A)}$

To handle 'basin of attraction', we formalize '*A* is an attractor and *B* is its basin of attraction, where b*(U, A) is b(A) minus its initial ∃U:'

$$\text{Basin-of-attraction}(A, B) \quad \text{Attractor}(A) \,\&\, \Box\big[B \leftrightarrow \exists U\, [U \,\&\, b^*(U, A)]\big]$$

Both Attractor(A) and Basin-of-attraction(A, B) are *state-independent* formulas, in the sense that each is true either at every state or at no state (on a given assignment of values to variables): this reflects the lack of reference to the current state in what they formalize. Whether a given region is an attractor and whether one given region is the basin of attraction of another does not depend on where we are in state space. Clearly, any formula of the form $\Box\alpha$ or Open(α) is state-independent, for the variable '*s*' for the state of evaluation does not appear on the right-hand sides of the semantic clauses for $\Box$ and Open. Clearly too, the truth-functors and quantifiers preserve state-independence, because in their semantic clauses '*s*' is used only to express the truth-conditions of their inputs, and so makes no difference when the inputs are state-independent. Since the displayed formulas result from applying truth-functors and quantifiers purely to formulas of the form $\Box\alpha$ and Open(α), they are state-independent. By contrast, the subformulas A and B themselves are typically state-dependent; whether the system is in an attractor, or its basin of attraction, depends on which state it is in. As such examples suggest, the 'tensed' modal language has considerable power to express key ideas in dynamical systems theory.

The language with modal and temporal operators has some advantages over the set-theoretic notation. It renders some ideas more simply and perspicuously: for instance, a(A) is simpler than (a). It also avoids making some distinctions that lack physical significance. For instance, in the set-theoretic framework we must distinguish between members and subsets of *S*; the state *s* is distinct from the singleton region {*s*} of state space. By contrast, no such distinction arises in the modal-temporal language. In that respect, the latter stays closer to what is physically significant.

The definitions of scientifically significant ideas in the modal-temporal language make uninhibited use of its modal resources. For instance, the natural way to express the condition that every state is in a basin of attraction is with this formula:

$$\Box\, \exists A\, \exists B\, \big[B \,\&\, \text{Basin-of-attraction}(A, B)\big]$$

Since b*(U, A) in Basin-of-attraction(A, B) already contains $\Box$, this is an example of how modal operators naturally occur in the scope of further occurrences of those operators in such applications.[19]

It is also notable that the definition of 'Attractor' involves quantification into the scope of a modal operator, in clauses b(A) and c(A). That is the equivalent for quantification into sentence position of *de re* modality, Quine's third and most reprehensible grade of modal involvement (Quine 1966). Since Quine's official methodology involves taking our metaphysics from our best theory of the world, which is supposed to include physics, it is unfortunate for him that our best theory employs something like his bugbear, *de re* modality. Indeed, any dynamical system validates *de re* formulas that attribute incompatible necessary features to different 'propositions', for instance:

$$\exists U \, \Box U$$

$$\exists U \, \Box \neg U$$

Any dynamical system with more than one state also validates this formula:

$$\exists U \, [\Diamond U \, \& \Diamond \neg U]$$

Thus Quine's naturalistic deference to natural science is in tension with his rejection of quantified modal logic. Although the readings of the operator □ associated with dynamical systems theory are weaker than metaphysical necessity, Quine's objections of principle to quantifying into the scope of a modal operator apply here just as much as they do elsewhere; they generalize to any non-trivial objective modal operator. If his arguments fail here, they fail generally.

From the present perspective, the non-modal nature of the mathematics that constitutes dynamical systems theory looks no more metaphysically significant than the non-modal nature of the mathematics that constitutes possible worlds model theory. In both cases, the modal connection is made by the intended applications of the mathematics. It is very convenient to reason in the non-modal language of mathematics, but in many applications we implicitly or explicitly characterize the entities we are reasoning about in modal terms. In metaphysics we may reason about modality by quantifying over possible worlds; in natural science we reason about modality by quantifying over possible states of a physical system. In the former case, the relevant modality is metaphysical; in the latter, it is more like nomic, but in both those cases it is objective. Natural science studies the structure of spaces of objective possibilities just as much as metaphysics does.

6. Necessitism and contingentism in dynamical systems theory

Sections 4 and 5 explained a general connection between the study of objective modalities by modal metaphysics and their study by the natural sciences. The present section will explore a much more specific connection between some contested issues in modal metaphysics and the modal logic underlying intended applications of dynamical systems theory as sketched in Section 5.

In the model theory of first-order modal logic, one key choice-point is between *constant domain semantics*, which interprets the first-order quantifiers as ranging over a fixed domain of individuals, irrespective of the world of evaluation, and *variable domain semantics*, which interprets them as ranging over a domain that depends on the world of evaluation; informally, it is conceived as containing just the individuals that exist in that world. Every model in the constant domain semantics is equivalent to a model in the variable domain semantics, but not *vice versa*. Various formulas are valid (true in all models) on the constant domain semantics but invalid (false in some models) on the variable domain semantics, famously including the controversial first-order Barcan schema and its converse.[20] More simply, this formula is valid on the constant domain semantics, invalid on the variable domain semantics:

NNE $\Box\forall x\Box\exists y\, x = y$

Informally, NNE says that necessarily everything is necessarily something. We read $\Box$ in NNE as expressing some sort of objective necessity. Without relying on the model theory, we can see that NNE raises a significant metaphysical issue. *Necessitists* assert NNE; *contingentists* deny NNE (for metaphysical modality). For necessitists, it is necessary which individuals there are. For contingentists, it is contingent which individuals there are. Normally, a necessitist and a contingentist agree that it is contingent which things are concrete (or are in space-time, or have causes and effects, ...). But the necessitist then adds that, in addition to the concrete things, there are also non-concrete things that merely could have been concrete, so that contingency in what is concrete does not generate contingency in what there is. The contingentist rejects any such way of saving NNE, insisting that a concrete thing is contingent in the strictest sense that there could have been no such thing as it at all. Each side has an internally coherent view; the issue is by no means easy to decide, although I have made a provisional case for necessitism (Williamson 2013a).

The issue between necessitism and contingentism might look like a paradigm of the sort of scholastic metaphysical dispute that utterly fails to engage with anything in natural science. But appearances can mislead. There are in fact quite specific connections.

The 'tensed' modal language for dynamical systems theory in Section 5 lacks individual quantifiers and so does not contain NNE (for whatever objective modality is associated with the interpretation at hand). However, the language does have propositional quantifiers, and so contains the propositional analogue of NNE:

NNE_P $\Box\forall \text{U}\Box\exists \text{V}\Box[\text{U} \leftrightarrow \text{V}]$

Whereas NNE concerns the necessary being of *individuals*, NNE_P concerns the necessary being of *propositions* (represented mathematically by sets of states).[21] Informally, NNE_P says that necessarily every proposition is necessarily some proposition (as already noted, the formula $\Box[\text{U} \leftrightarrow \text{V}]$ is tantamount to $\text{U} = \text{V}$): for propositions, being is state-independent. *Propositional necessitists* assert NNE_P; *propositional contingentists* deny NNE_P (for the given sort of objective modality). Necessitists tend to be propositional necessitists too (Williamson 2013a defends both views). Many contingentists are also propositional contingentists, holding that what propositions there are depends on what individuals there are (Stalnaker 2012). However, the two views do not always go together. Plantinga (1983) seems to defend a combination of contingentism about individuals with necessitism about propositions, on which it is (metaphysically) contingent that there is Socrates, but necessary that there is the proposition that there is not Socrates. Whether such a combination is well motivated is another question (see Williamson 2013a, 267–277). Henceforth we will ignore such hybrid positions, and concentrate on those which are either necessitist about all orders or contingentist about all orders.

Notably, NNE_P is valid on the semantics in Section 5 for the modal language of dynamical systems. Thus the modal logic of dynamical systems embodies a necessitist metaphysics of propositions (for the given sort of objective modality). The logic also validates several other characteristically necessitist principles related to NNE_P, including an unrestricted comprehension schema for propositions, which guarantees that there is (necessarily) a proposition for each formula α of the language:

$COMP_P$ $\quad \exists V \Box[V \leftrightarrow \alpha]$

Here α is any formula in which the variable V does not occur free, though other variables may. Strengthening $COMP_P$ by prefixing it with any sequence of universal quantifiers and $\Box$ operators in any order preserves validity (see Williamson 2013a, 290 for the same schema, interpreted with respect to metaphysical modality). Versions of the Barcan schema and its converse are also valid, with propositional quantifiers in place of individual ones:

BF_P $\quad \Diamond \exists V \alpha \rightarrow \exists V \Diamond \alpha$

CBF_P $\quad \exists V \Diamond \alpha \rightarrow \Diamond \exists V \alpha$

By BF_P, if there could have been a proposition that met a given condition, then there is a proposition that could have met that condition. By CBF_P, if there is a proposition that could have met a given condition, then there could have been a proposition that met that condition. The validity of all these principles is not an artefact of a gerrymandered semantics. Quite the opposite: without extreme gerrymandering, there is no way of making explicit the modal content of intended applications of the mathematics of dynamical systems without validating these necessitist principles. In effect, intended applications of the mathematics of dynamical systems theory take necessitism about propositions for granted (for the relevant sort of objective modality). The modal logic for dynamical systems theory in Section 5 just makes that metaphysical commitment explicit. That is not surprising; logic is not metaphysically neutral in any deep sense (see Williamson 2013a).

What about necessitism and contingentism for individuals? In dynamical systems, individuals are typically not represented as such. To find them, one must look into the internal structure (if any) of the states of the system. Typically, the states are treated as assignments of values to one or more independent variables. The number of variables is the dimension of the state space. For mathematical purposes, each state in an *n*-dimensional state space may be treated as just an *n*-tuple of numbers, if *n* is finite. A state space may also be of infinite dimension. For instance, if one is interested in the dynamics of temperature, it might be convenient to have a temperature variable for each point in a continuous physical space. The variables need not represent individuals: instead, they may represent global features of the target physical system, such as the number of predators and the number of prey. For many purposes, representing individuals one by one would involve needless, perhaps intractable, complexity. However, in some applications of dynamical systems theory, the variables *are* in principle associated with individuals. For a system of *n* particles, we may need to keep track of, say, three independent features (such as position or energy) per particle, so we need 3*n* variables altogether, and so a 3*n*-dimensional state space. In such cases, some distinctions between states correspond to distinctions between individuals: for instance, one state may differ from another only in that the values of the variables associated with a given particle in the former are those of the corresponding variables associated with another particle in the latter, and *vice versa*. Dynamical systems whose variables are associated with distinct individuals are sometimes called *agent-based*.

In many of the systems studied in natural science, individuals are sometimes created or destroyed. Such individuals may be particles, cells, organisms, whatever.[22] It is a temporary matter what individuals there are. It may also be a contingent matter what individuals there *ever* are; a given individual may be created on one trajectory but not on another. This is handled by having variables for each individual that *can* occur in some state or other (it would be horribly messy to handle it any other way). Thus, on the given application, each state *s* in effect encodes the identities of all *possible* individuals, whether or not they are present in *s* or in any other state on the same orbit as *s*. In making generalizations about states and sets of states, as is continually done in applications of dynamical systems theory, one is in effect quantifying over merely *possibly* present individuals, as well as actual ones: possible particles, possible cells, possible organisms, whatever they happen to be. On the face of it, this assumes a necessitist conception of what there is to quantify over rather than a contingentist one.

Of course, in applying the mathematics of state spaces, natural scientists typically have in mind a *type* of physical system, rather than a single token of that type. Thus the dimensions of the state space are not really associated with particular individuals. But the underlying point remains. For although the statements made about the application implicitly or explicitly have that extra level of generality, they are still generalizations about merely possibly present individuals as if they were all there to be generalized about, contrary to the contingentist view. For a necessitist, by contrast, such quantification is unproblematic.

A natural strategy for contingentists is to try to *simulate* in their own terms the necessitist effect of quantification, using crafty combinations of modal operators and quantifiers understood in contingentist terms. That strategy can be taken quite far. More specifically, the contingentist can simulate the necessitist effect of ordinary first-order quantification over possible individuals. Since each possible state is in effect an assignment of values to variables associated with possible individuals, the contingentist may well be able to simulate the necessitist effect of first-order quantification over possible states too. However, the contingentist simulation strategy arguably fails for second-order quantification over properties or sets of possible individuals (Williamson 2013a; 305–375; see also Fritz 2013). The necessitist effect of such second-order quantification cannot always be simulated in contingentist terms. Corresponding problems may therefore arise for a contingentist's attempt to simulate quantification over sets of possible states. Such quantification is ubiquitous in dynamical systems theory. For instance, as seen above, it is used to define basic terminology such as 'attractor' and 'basin of attraction'. The crucial feature of the definitions is that they quantify not only over possible states (with the variable '*s*') but also over sets of possible states (with the variables '*U*' and '*V*'). Without such quantification the definitions make no sense. But the intended effect of such quantification over sets of possible states characterized in terms of possibly concrete individuals is just the sort of necessitist move that contingentists cannot always simulate. They certainly cannot just help themselves to it without explanation.

In some special cases, contingentists can simulate the necessitist effects of higher-order quantification, for instance when all the possibly present individuals are compossible (they can all be present together in the same state). If the variables associated with distinct individuals are wholly independent of each other, then some state will assign them all nonzero values, which presumably means that all those individuals are present at that state. However, that argument does not work if some combinations of assignments are excluded (as they are by the Pauli exclusion principle, for instance). A contingentist simulation also works when the total number of possibly present individuals for the system is finite. But if there are infinitely many possible individuals for the system, while only finitely many of them can be present together, then neither of those special cases applies.

Contingentists could undertake the strategy of trying to show that, in practice, all the cases that arise in real-life natural science admit contingentist simulation, not only in dynamical systems theory but in scientific applications of the state space approach more generally. But that would be to give a very significant hostage to fortune. Pending the successful execution of the strategy, why should we assume without evidence that it will succeed? What is striking is that natural scientists themselves seem to feel no need of such precautions. In applying mathematics to state spaces, they make free use of quantification over possible individuals, possible states, sets of possible states and so on, with no checks on whether they are straying beyond the limits of contingentist simulation. They do not treat the legitimacy of their practice as dependent on the availability of such simulation. In effect, they unreflectively rely on an invisible framework of necessitist modal logic. In that way, necessitism is more hospitable than contingentism to the normal practice of natural science.[23]

The absence of modal expressions from the core language of mathematics does not mean that its applications in natural science are non-modal. Rather, it manifests the necessitist presuppositions on which those scientific applications rely. They are no more non-modal than are the investigations of a metaphysician who reasons freely in a language without modal operators about possible worlds and their inhabitants. Contingentists who want to reconcile their modal metaphysics with scientific practice face a major reconstructive challenge: to vindicate within their own framework the free scientific use of quantification over possible states, sets of possible states and so on. We currently have no good reason to expect that they will be able to meet the challenge.

7. Necessitism and contingentism in probability spaces

Similar metaphysical issues arise for the application of standard probability theory to implicitly modal matters. Suppose that we are reasoning about a counterfactual circumstance *C*, specified just as one in which there would have been exactly two fair coins tossed independently of each other at time *t*; *C* itself is not maximally specific, and in particular does not specify the microscopic details of the coins or the outcomes of the tosses. Our interest is in objective probabilities, such as chances (*t* is in the distant future), not in subjective or epistemic ones. What is the actual probability, conditional on *C*, that the result is one heads, one tails? The standard, correct answer is: ½. For there are four equiprobable possible outcomes: (HH) both coins come up heads; (HT) the first coin comes up heads, the second tails; (TH) the first coin comes up tails, the second heads; (TT) both coins come up tails. Since 'one heads, one tails' results in two of the four equiprobable outcomes, the probability is 2/4 = ½. But familiarity should not make us regard the correctness of that argument as immediate. For an intelligent person without a suitable background in probability could instead have argued thus. There are three equiprobable possible outcomes: (HH) two heads; (H+T) one heads, one tails; (TT) two tails. Since 'one heads, one tails' results in just one of the three equiprobable outcomes, the probability is 1/3. One very intelligent person who made just such a mistake was Leibniz, who claimed that on a throw of two dice, 11 and 12 are equally likely outcomes, because each can be obtained in only one way (a five and a six; two sixes).[24] Kripke uses this very example of the two ways of getting 11 to explain his conception of possible worlds and trans-world identity (Kripke 1980, 16–18).

What did we just mean by 'the first coin' and the 'the second coin'? Obviously, we did nothing to pick out one coin from the other. In effect, we used variables:

'Let *x* be one of the two, and *y* the other'. Reasoning that way is fine when given two objects, no further distinguished from each other. But that is not exactly what we were told to suppose. The supposition was that we are reasoning in an actual circumstance @ about a counterfactual circumstance *C*. The circumstance postulated to contain two coins was *C*, not @. In @, we are not given two coins; we are merely given that in *C* there would be two coins. For necessitists, that difference does not matter. The two coins in *C* are also in @, even if there they are not coins but merely possible coins. They are there in @ to be reasoned about. But for contingentists, the difference is crucial. They cannot assume that in @ there are two possible coins for them to reason about. Consequently, they cannot assume that there are two possible outcomes such as (HT) and (TH) above for them to reason about, because (HT) and (TH) were described in terms of the supposed difference between the two possible coins. Stalnaker (2012), one of the most thoughtful defenders of a contingentist position, is quite clear that in a case like @ there is only a single qualitative possibility, which can be characterized in quantificational terms. That corresponds to (H+T) above. Thus contingentism undermines the standard probability calculation for those cases, because it provides only three possibilities, not four.

The point is easy to miss. For if we imagine reasoning in *C* itself, there are two coins, and the standard calculation is unproblematic. What that shows is that if *C* had obtained, the probability of 'one heads, one tails' *would have been* ½. But that was not the question. What was in question was the *actual* probability of 'one heads, one tails' conditional on *C*. For a contingentist such as Stalnaker, probabilities in @ are distributed over the actual space of possibilities, which may differ from the space of possibilities over which probabilities would have been distributed in *C*, because it is contingent what possibilities there are.

None of this is yet to say that contingentists are *forced* to assign probability 1/3 to 'one heads, one tails'. They might try to argue that some actual way of differentiating between the coins will always be available, for instance in terms of spatial location, though it is doubtful that such tactics will succeed with sufficient generality. Alternatively, in the style of Plantinga rather than Stalnaker, they might insist that there are actually two possibilities such as (HT) and (TH) even though there are not actually two possible coins to distinguish them, and somehow explain why they take such different attitudes to possible states and possible coins. If all else fails, they might say without further explanation that although there are actually only three possibilities, (HH), (H+T) and (TT), they are not equiprobable; (H+T) must have twice the probability that either (HH) or (TT) has. But what contingentists cannot do is simply endorse the standard calculation. It is not available to them just as it stands. At best, they will have to work hard to recover the standard calculation, and it is not clear that they will succeed.

The bearing of the necessitism-contingentism issue on calculations of probability is not confined to toy examples like those above. The choice between treating (HH), (HT), (TH) and (TT) as the four equiprobable possibilities and treating (HH), (H+T) and (TT) as the three equiprobable possibilities corresponds to the choice between Fermi–Dirac statistics and Bose–Einstein statistics in particle physics. For some applications, Bose–Einstein statistics do better; for others, Fermi–Dirac statistics do better. The danger for contingentism is that it may pressure us towards the Bose–Einstein statistics on general metaphysical grounds when the physics requires the Fermi–Dirac statistics. In general, different views in modal metaphysics mandate different ways of individuating objective possibilities, which in turn motivate different assignments of objective probability, and thereby have knock-on effects in natural science. The two enterprises are nothing like as disconnected as many philosophers and many scientists assume.[25]

8. Metaphysical vs. other objective modalities

Appeals to objective modal aspects of natural science seem to have this limitation: they concern at best some form of nomic modality, but not metaphysical modality. How much light do the arguments of Sections 3–7 cast on metaphysical modality?

The gap between nomic and metaphysical may be narrower than is usually thought. Following Kripke (1980), Bird (2007) has argued in detail that laws of nature may be metaphysically necessary. If what it is to be an F involves being a G, then it is metaphysically, not just nomically, necessary that all Fs are Gs. It is a good question how far such arguments can be taken: could not motion have obeyed different laws? In any case, the total assimilation of nomic modality to metaphysical modality is not only rather implausible: it is not even relevant to all the cases discussed above. For the possibilities in a probability space or state space may not even exhaust all nomic possibilities, let alone all metaphysical ones. They may cover just the possible states of a highly contingent system, such as the tossing of coins or the weather on earth. Still, we may assume that in a typical case they *are* nomic possibilities, even if they are not all of the nomic possibilities.

A simple point is that if nomic modality is an objective modality, nomic possibility entails metaphysical possibility, the most general type of objective probability. Thus if science shows something to be nomically possible, it thereby shows it to be metaphysically possible too. In particular, if science shows something *nonactual* to be nomically possible, it thereby shows it to be metaphysically possible. The nomic possibility of various states is built into applications of probability spaces and phase spaces. Curtailing the phase space typically disrupts its mathematical structure and thereby reduces the explanatory power of the theory. Of course, someone might challenge the entailment from nomic possibility to metaphysical possibility. We saw in Section 2 that if nomic possibility is just logical consistency with the laws of nature, it does not entail metaphysical possibility. However, that account of nomic possibility fared very badly. If instead nomic possibility is metaphysical compossibility with the laws of nature, then it trivially entails metaphysical possibility, for a metaphysical impossibility is not metaphysically compossible with anything. Another argument is that if α is nomically possible, then the subjunctive conditional $\alpha > \neg\alpha$ is false, but if α were metaphysically impossible that conditional would arguably be vacuously true (Williamson 2007a). There is no good reason to deny the entailment from nomic possibility to metaphysical possibility.

Although no attempt will be made here to argue in general from nomic necessity to metaphysical necessity, often the main challenge to a claim that something is metaphysical necessary is also by implication a challenge even to the claim that it is nomically necessary. Consider, for instance, the necessitist thesis NNE. Perhaps the strongest contingentist objection to it is of this sort: if my parents had never met, there would have been no such thing as me; therefore, I am a counterexample to the claim that (necessarily) everything is necessarily something (and so is everyone else). Presumably, it is nomically as well as metaphysically possible for my parents never to have met. Thus the putative counterexample tells just as much against NNE on a nomic reading of the modal operators as against NNE on the intended metaphysical reading of them. But scientific evidence *for* the nomic version of NNE (as suggested in Sections 6 and 7) is also evidence that such putative counterexamples do not work against the nomic version, and if they fail against the nomic version they fail against the metaphysical version too. It is evidence for nomically possible people who are not actually people, and for nomically possible particles which are not actually particles. Once such things are accepted, there is little to be gained by holding out against the metaphysical version of NNE; its theoretical virtues triumph in the absence of a compelling counterexample. In cases like these, the main dialectical action is within the realm of nomic possibility. The distinction between nomic and metaphysical necessity, though granted, sometimes makes less difference than might have been expected to the modal upshot of natural science.

Those considerations combine naturally with the more general observations in Sections 1 and 2 of how the objective modalities tend to stand or fall together. It is implausible to treat the objective modal dimension of natural science as merely an artefact of folk cognitive architecture or overheated metaphysical speculation. We have no good evidence that it is a proxy for something else.

We can also draw a more general moral in epistemology. The epistemology of modality cannot be treated in isolation; it is not an autonomous branch of epistemology. Our natural scientific knowledge of objective modality is too tightly integrated with the rest of our scientific knowledge to permit such a division. Not very surprisingly, the abductive methodology of natural science plays a major role in the epistemology of modality.[26]

Notes

1. See Kratzer (1977, 2012, 49–62), Portner (2009, 144–184), and Vetter (2016), for instance. It is not denied that the same word can express an objective modality in one context and an epistemic modality in another: compare 'She could run a marathon in three hours' (objective) with 'Goldbach's Conjecture could be true' (epistemic).
2. There is a plausible argument that, in propositional modal logic, if metaphysical modality obeys *at least* the principles of S5, then it obeys *at most* the principles of S5 (Williamson 2013a, 111).
3. Contingentists may wish to insert a qualification 'if *x* exists' within the scope of the necessity operator to handle the possible non-existence of the objects. What matters is that $x = y$ licenses the inter-substitution of the free variables '*x*' and '*y*' in objective modal contexts; the same object is at issue under different guises.
4. See Strohminger (2015) for a detailed development of the case for perceptual knowledge of nonactual possibilities. This strikes at the Humean assumption that impressions are non-modal in content. Roca-Royes (Forthcoming) makes a more empiricist argument for inductive knowledge of nonactual possibilities via their similarity to perceived actualized possibilities: if you have seen cups break, and thereby know that they *can* break, you may infer that a similar unbroken cup can break (though that does less to confront empiricist worries about how we come to understand 'can' in the first place).
5. It is worth noting that some philosophers of mathematics interpret the language of mathematics itself as implicitly modal: mathematics becomes a science of possible structures (Putnam 1967; Hellman 1989). Despite taking such views seriously, for present purposes I prefer not to rely on philosophical interpretations so distant from the way mathematicians explicitly talk and think. More recent modal interpretations of the language of set theory, such as Linnebo (2013) and Studd (2013) (see also Parsons 1983, 298–341; Fine 2006), have been motivated by a (laudable) desire to avoid Russell's paradox for sets without ad hoc restrictions; that too is quite far from the concerns of most working mathematicians. Moreover, the latter motivation requires a (so far somewhat obscure) non-objective reading of the modal operators, because even the hierarchy of pure sets must involve such a modal aspect (since Russell's paradox arises even for pure sets, and it is generally agreed that the existence of pure sets is *metaphysically* non-contingent). Non-objective modal interpretations of the language of mathematics are not strictly relevant to the concerns of this paper.
6. The principle in the text is a form of *single-premise closure*. We do *not* assume the multi-premise closure principle that if a formula α of L supports the formulas $\beta_1, \ldots, \beta_n$ of L^+, and $\beta_1, \ldots, \beta_n$ jointly entail γ, then α also supports γ. Unlike multi-premise closure, single-premise closure is consistent with an interpretation of 'support β' as 'confer a probability above the threshold c on β', where $0 < c < 1$.
7. See Williamson (2007a, 293–299), for relevant background on the logical relations between counterfactuals and metaphysical modality.
8. As already hinted, it is controversial how much weaker than metaphysical necessity nomic necessity really is (see also Section 8). It is also controversial how much of natural science really aims at nomic necessity. The present remarks about nomic modality should be taken in the spirit of a first approximation.
9. Some Lewisians may object to the assumption because it clashes with the 'small miracle' conception of subjunctive conditionals, but the assumption is nonetheless very plausible.
10. Allowing infinitesimal probabilities does not solve the problem (Williamson 2007b).

11. Let Ω be the set of all possibilities. If we define □E = Ω if E has probability 1 and □E = {} otherwise, and ◇E = Ω if E has nonzero probability and ◇E = {} otherwise, then in the finite case in the text □ and ◇ satisfy the principles of the propositional modal system S5 (=KT45); in the infinite case they satisfy only the principles of the weaker modal system KD45, where the T principle (□E ⊆ E) is weakened to the D principle (□E ⊆ ◇E).
12. If we used evidential or subjective probabilities, the putative explanation would at best show that the explanandum 'was to be expected'. But to show that an outcome was to be expected is not to explain why it occurred, in the relevant broadly causal sense. To revert to the example at the beginning of Section 1, it was trivially to be expected that *n* would number the inhabited planets, since '*n*' was defined to name their number. For instance, given that *n* = 29, to explain causally why there are exactly *n* inhabited planets is to explain causally why there are exactly 29 inhabited planets, but the trivial 'was to be expected' explanation does not advance the latter project. The difference between the two non-obviously co-referential names '*n*' and '29' is epistemically relevant but causally irrelevant.
13. One may question the assumption that merely possible states of a physical system are abstract objects (Malament 1982, 533; Lyon and Colyvan 2008, 233). On the approach of Williamson (2013a, 7), their non-concreteness does not make them abstract. But that point is irrelevant to the present argument.
14. For some purposes we might require directions to be non-negative, which would require only forwards determinism.
15. Of course, this is not the only way of interpreting modal operators over a dynamical system. If a topology is defined over the states, one can interpret □ as the topological interior operator, which is a much more 'local' form of necessity (though it does not involve an accessibility relation between worlds). It yields an S4 modal logic, whereas the present 'global' interpretation of □ yields an S5 modal logic. The two interpretations are not rivals; they simply pick out different aspects of the system for study. The global interpretation is more general, because it does not depend on what kind of mathematical structure is defined over the states. For more on topological interpretations of modal logics on dynamical systems see Artemov, Davoren, and Nerode (1997) and Davoren and Goré (2002).
16. They are unrestricted in the sense of ranging over all states of the system; as already noted, they are typically not equivalent to metaphysical necessity and metaphysical possibility.
17. This 'diamonds are forever' principle is reminiscent of, but not equivalent to, the principle defended by Dorr and Goodman (Forthcoming); the latter concerns metaphysical possibility and a more standard reading of the tense operators.
18. For an introduction to tense logic that explains the relevant background see, for instance, Müller (2011).
19. In unquantified S5, every formula is equivalent to one without such embedded occurrences of modal operators. That is not in general so for quantified S5, even with constant domains; see Fine (1978, 146–151), for the case of first-order S5.
20. See Williamson (2013a) for a discussion in more depth. I am assuming that the rest of the model theory is more or less standard.
21. For reasons explained in Williamson (2013a, 254–261), reading higher-order quantifiers such as those in NNE_P as first-order quantifiers restricted to objects of a special sort (propositions) is ultimately inappropriate: semantically, the difference between name position and sentence position runs deeper than a difference between objects of one kind and objects of another, or between objects in general and objects of a special kind (where every object can in principle be

named). Nevertheless, for present purposes the talk of propositions as objects is a harmless over-simplification.

22. For discussion of the biological case see Gunawardena (2009).
23. The complaint on pp. 286–288 of Williamson (2013a) that various contingentist comprehension principles for second-order modal logic are too weak to serve the purposes of 'modal mathematics' relates to just this point. Consider the free application of non-modal mathematics to an implicitly modal subject matter, as in reasoning about dynamical systems or Kripke models intended for some objective modality. When the intended modal content of the application is made explicit model-theoretically in the manner of Section 5 (which can be extended to other forms of quantification), unrestricted modal comprehension principles such as $Comp_P$ can be proved valid in the model theory by non-modal mathematics (for instance, set theory). Since the free application of non-modal mathematics to the modal subject matter is committed to all such valid modal formulas, in particular it is committed to those unrestricted modal comprehension principles. That is fine for necessitists but not for full-blooded contingentists such as Stalnaker (2012).
24. See Cussens (2014) for a recent discussion of Leibniz's failure to contribute to the mathematics of probability.
25. For a related application of probability to an issue in modal metaphysics see Kment (2012), although I doubt that Kment would endorse the conclusions of this paper.
26. The material in this paper has evolved over several years. Various parts of it were presented as the Ruth Manor Lecture at Tel Aviv University, the Saul Kripke Lecture at City University New York and the Wade Lecture at St Louis University. Earlier versions of the material were presented as talks at conferences on the epistemology of modality at Belgrade University, Aarhus University (where Daniel Dohrn provided a detailed response) and Stirling University, a conference on logic and metaphysics at the University of Southern California, a workshop on modal metaphysics in Montreal, and the Universities of Athens, Connecticut (Storrs), Michigan (Ann Arbor) and Oxford. Embryonic predecessors were presented to workshops at the Centre for the Study of Mind in Nature in Oslo and the Institute of Philosophy in London. I am grateful to all the participants at those events who helped me develop the material with their questions and comments, and for discussion or correspondence on the issues to Kit Fine, Peter Fritz, Peter Godfrey-Smith, Jeremy Goodman, Lloyd Humberstone, Matthias Jenny, Øystein Linnebo, Maurico Suárez and Trevor Teitel. I believe that Saul Kripke envisaged an analogy with states in phase space early in his thinking about possible worlds.

References

Artemov, Sergei, Jennifer Davore, and Anil Nerode. 1997. *Modal Logics and Topological Semantics for Hybrid Systems*. Technical Report MSI 97-05. Cornell University.

Bird, Alexander. 2007. *Nature's Metaphysics: Laws and Properties*. Oxford: Clarendon Press.

Chisholm, Roderick. 1973. "Parts as Essential to Their Wholes." *Review of Metaphysics* 26: 581–603.

Cussens, James. 2014. "Probability and Statistics." In *The Oxford Handbook of Leibniz*, edited by M. R. Antognazza. Oxford: Oxford University Press. Advance online publication.

Davidson, Donald. 1970. "Mental Events." In *Experience and Theory*, edited by L. Foster and J. W. Swanson, 79–101. London: Duckworth.

Davoren, Jen, and Rajeev Goré. 2002. "Bimodal Logics for Reasoning about Continuous Dynamics." In *Advances in Modal Logic*. Vol. 3, edited by F. Wolter, H. Wansing, M. de Rijke, and M. Zakharyaschev, 91–111. London: World Scientific.

Dorr, Cian, and Jeremy Goodman. Forthcoming. "Diamonds Are Forever." *Noûs*.

Field, Hartry. 1980. *Science Without Numbers: A Defence of Nominalism*. Oxford: Blackwell.

Fine, Kit. 1978. "Model Theory for Modal Logic: Part I – The *De Re/De Dicto* Distinction." *Journal of Philosophical Logic* 7: 125–156.

Fine, Kit. 1994. "Essence and Modality: The Second Philosophical Perspectives Lecture." *Philosophical Perspectives* 8: 1–16.

Fine, Kit. 2006. "Relatively Unrestricted Quantification." In *Absolute Generality*, edited by A. Rayo and G. Uzquiano, 20–44. Oxford: Clarendon Press.

Fritz, Peter. 2013. "Modal Ontology and Generalized Quantifiers." *Journal of Philosophical Logic* 42: 643–678.

Gibson, James. 1979. *The Ecological Approach to Visual Perception*. Boston, MA: Houghton Mifflin.

Gunawardena, Jeremy. 2009. "Models in Systems Biology: The Parameter Problem and the Meanings of Robustness." In *Elements of Computational Systems Biology*, edited by H. Lodhi and S. Muggleton, 21–48. Hoboken, NJ: Wiley.

Hellman, Geoffrey. 1989. *Mathematics Without Numbers: Towards a Modal-Structural Interpretation*. Oxford: Clarendon Press.

Kim, Jaegwon. 1993. *Supervenience and Mind: Selected Philosophical Essays*. Cambridge: Cambridge University Press.

Kment, Boris. 2012. "Haecceitism, Chance, and Counterfactuals." *Philosophical Review* 121: 573–609.

Kment, Boris. 2014. *Modality and Explanatory Reasoning*. Oxford: Oxford University Press.

Kratzer, Angelika. 1977. "What 'Must' and 'Can' Must and Can Mean." *Linguistics and Philosophy* 1: 337–355.

Kratzer, Angelika. 2012. *Modals and ConditionalsNew and Revised Perspectives*. Oxford: Oxford University Press.

Kripke, Saul. 1980. *Naming and Necessity*. Oxford: Blackwell.

Lewis, David. 1968. "Counterpart Theory and Quantified Modal Logic." *The Journal of Philosophy* 65: 113–126.

Lewis, David. 1986. *On the Plurality of Worlds*. Oxford: Blackwell.

Linnebo, Øystein. 2013. "The Potential Hierarchy of Sets." *The Review of Symbolic Logic* 6: 205–228.

Loewer, Barry. 2001. "Determinism and Chance." *Studies in History and Philosophy of Science Part B: Studies in History and Philosophy of Modern Physics* 32: 609–620.

Lyon, Aidan, and Mark Colyvan. 2008. "The Explanatory Power of Phase Spaces." *Philosophia Mathematica* 16: 227–243.

Mackie, Penelope. 2006. *How Things Might Have Been: Individuals, Kinds, and Essential Properties*. Oxford: Clarendon Press.

Malament, David. 1982. "Review of *Science Without Numbers: A Defense of Nominalism* by Hartry Field." *Journal of Philosophy* 79: 523–534.

Maudlin, Tim. 2007. "What Could Be Objective about Probabilities?" *Studies in History and Philosophy of Science Part B: Studies in History and Philosophy of Modern Physics* 38: 275–291.

Müller, Thomas. 2011. "Tense or Temporal Logic." In *The Continuum Companion to Philosophical Logic*, edited by L. Horsten and R. Pettigrew, 324–350. London: Continuum.

Nolte, David. 2010. "The Tangled Tale of Phase Space." *Physics Today* 63 (4): 33–38.

Parsons, Charles. 1983. *Mathematics in Philosophy: Selected Essays*. Ithaca, NY: Cornell University Press.

Plantinga, Alvin. 1983. "On Existentialism." *Philosophical Studies* 44: 1–20.

Portner, Paul. 2009. *Modality*. Oxford: Oxford University Press.

Prior, Arthur. 1956. "Modality and Quantification in S5." *The Journal of Symbolic Logic* 21: 60–62.

Prior, Arthur. 1967. *Past, Present and Future*. Oxford: Clarendon Press.

Putnam, Hilary. 1967. "Mathematics Without Foundations." *The Journal of Philosophy* 64: 5–22.

Quine, Willard van Orman. 1953. "Reference and Modality." In *From a Logical Point of View*, edited by Willard Van Orman Quine, 139–159. Cambridge, MA: Harvard University Press.

Quine, Willard van Orman. 1966. "Three Grades of Modal Involvement." In *The Ways of Paradox and Other Essays*, edited by Willard Van Orman Quine, 156–174. New York: Random House.

Roca-Royes, Sonia. Forthcoming. "Similarity and Possibility: An Epistemology of *De Re* Possibility for *Concreta*." In *Modal Epistemology after Rationalism*, edited by B. Fischer and F. Leon.

Salmon, Nathan. 1982. *Reference and Essence*. Oxford: Blackwell.

Salmon, Nathan. 1989. "The Logic of What Might Have Been." *The Philosophical Review* 98: 3–34.

Salmon, Nathan. 1993. "This Side of Paradox." *Philosophical Topics* 21: 187–197.

Sider, Ted. 2011. *Writing the Book of the World*. Oxford: Clarendon Press.

Stalnaker, Robert. 2012. *Mere Possibilities: Metaphysical Foundations of Modal Semantics*. Princeton, NJ: Princeton University Press.

Strogatz, Steven. 2001. *Nonlinear Dynamics and Chaos: With Applications to Physics, Biology, Chemistry, and Engineering*. Boulder, CO: Westview Press.

Strohminger, Margot. 2015. "Perceptual Knowledge of Nonactual Possibilities." *Philosophical Perspectives* 29: 363–375.

Studd, James. 2013. "The Iterative Conception of Set." *Journal of Philosophical Logic* 42: 697–725.

Vetter, Barbara. 2016. "Williamsonian Modal Epistemology, Possibility-Based." *Canadian Journal of Philosophy* 46.

Williamson, Timothy. 2006. "Indicative Versus Subjunctive Conditionals, Congruential Versus Non-Hyperintensional Contexts." In *Philosophical Issues, Volume 16: Philosophy of Language*, edited by E. Sosa and E. Villanueva, 310–333. Oxford: Blackwell.

Williamson, Timothy. 2007a. *The Philosophy of Philosophy*. Oxford: Wiley Blackwell.

Williamson, Timothy. 2007b. "How Probable is an Infinite Sequence of Heads?" *Analysis* 67: 173–180.

Williamson, Timothy. 2013a. *Modal Logic as Metaphysics*. Oxford: Oxford University Press.

Williamson, Timothy. 2013b. *Identity and Discrimination*. 2nd ed. Oxford: Blackwell.

Woodward, James. 2003. *Making Things Happen: A Theory of Causal Explanation*. New York: Oxford University Press.

Higher-order free logic and the Prior-Kaplan paradox

Andrew Bacon, John Hawthorne and Gabriel Uzquiano

Philosophy, University of Southern California, Los Angeles, CA, USA

ABSTRACT

The principle of universal instantiation plays a pivotal role both in the derivation of intensional paradoxes such as Prior's paradox and Kaplan's paradox and the debate between necessitism and contingentism. We outline a distinctively free logical approach to the intensional paradoxes and note how the free logical outlook allows one to distinguish two different, though allied themes in higher-order necessitism. We examine the costs of this solution and compare it with the more familiar ramificationist approaches to higher-order logic. Our assessment of both approaches is largely pessimistic, and we remain reluctantly inclined to take Prior's and Kaplan's derivations at face value.

1. Introduction

A central theme from *Modal Logic as Metaphysics* is the idea that higher-order logic is a fruitful framework for formulating and assessing some of the somewhat elusive debates in the metaphysics of properties and modality. It is clear from reading the book that the choice to theorise in higher-order logic is not incidental to Williamson's project, but rather a substantial theoretical commitment. Within this framework, certain compelling principles can be formulated, which, if taken to be about first-order entities, would straightforwardly collapse into paradox; this is true, for example, of the comprehension principle that plays a central role in motivating higher-order necessitism.

However, Williamson's version of higher-order logic is not obviously a safe haven when it comes to the paradoxes. Indeed, this was something early proponents of higher-order logic were well aware of, and this was a central theme in the philosophy of higher-order logic in the early twentieth century. There were roughly two main schools of thought concerning the proper way to regiment a higher-order language. On a broadly Fregean picture, there's a tight connection between grammatical categories and the type-theoretic hierarchy. There is a type e to which individual constants and variables belong, and whenever $\tau_1, \ldots, \tau_n$ are types, there is a type of n-ary relations, $(\tau_1, \ldots, \tau_n)$, whose arguments are of type $\tau_1, \ldots, \tau_n$, respectively. Note that propositions can be thought of as 0-ary relations, which in this notation would be written $()$. Thus, for example, monadic predicates like 'is happy' have type (e), unary connectives like 'it's not the case that' have type $(())$, binary connectives like 'and' have type $((), ())$, first-order quantifiers like 'most people' have type $((e))$ and quantifiers into sentence position have type $((()))$.

As the last two examples suggest, each logical type in this sense can have its own kind of universal quantifier, which binds variables of that type. Whereas first-order quantifiers simply range over eligible referents for singular terms in the intended interpretation, it's important not to think of higher-order quantifiers as ranging over another domain of entities. For example, it wouldn't do to think of quantification into predicate position as quantifying over eligible 'referents' for predicates. Predicates do not refer, they predicate, and they cannot occupy the position of a singular term. It would be as much of a mistake to put a second-order quantifier in the place of a singular term, as to put a predicate in the place of a singular term. Broadly speaking, this is the framework that Williamson is employing, in line with a venerable tradition tracing back to the likes of Frege, Church, Montague and Prior.

By contrast, there's another school of thought, often attributed to Russell, in which each relational type is further stratified into levels indexed by the natural numbers.[1] That is, instead of a single type of relations $(\tau_1, \ldots, \tau_n)$, for any ramified types, $\tau_1, \ldots, \tau_n$, there is a whole hierarchy of relations of different levels, written $(\tau_1, \ldots, \tau_n)/k$ for $k \in \mathbb{N}$.[2] So instead of a single type of proposition, we have infinitely many ramified propositional types, $()/0$, $()/1$, $()/2$ and so on. One way to think of these types, borrowing a metaphor from Kaplan (1995), is to think of the zero-level propositional type as containing only propositions concerning the distribution of earth, wind, fire and water. The first-level propositional type then concerns matters of earth, air, fire and water, *and* matters concerning propositions about earth, air, fire and water (i.e. matters concerning zero-level propositions). This generalises in the obvious way: at any stage, a new level can be generated in an analogous way. For each of these ramified propositional types, there is a universal quantifier. Note, however, that unlike the Fregean, the Russellian typically treats certain parts of the logical vocabulary as syncategorematic; negation, conjunction and even the quantifiers are not assigned a ramified type, but $\forall v\phi$, $\neg\phi$ and $\phi \vee \psi$ are all well-formed formulas, provided ϕ and ψ are of some propositional type and v is a variable of any type.[3] When we combine the universal quantifier with a propositional variable of ramified type $()/0$ can achieve the effect of quantifying over 0 level propositions: in general, the expression $\forall p_n$, where p_n is of ramified type $()/n$, quantifies over propositions of level n.

This picture does not sit particularly well with the principle of universal instantiation applied at any level (hence the restriction in Church [1976], 751). For example, at sentence level, this principle allows us to move from a universal generalisation like $\forall pA$ to any instance $A(B/p)$, where B could be any sentence:

$$\forall pA \rightarrow A(B/p)^4 \qquad \text{(UI)}$$

Notice that for the stratifier, this principle is doubly schematic in that $\forall p$ is a place holder for a quantifier of a particular level as well as having A and B as place holders for any formula. For example, it could be that every 0-level proposition satisfies A (maybe A is $\exists_0 q\ p = q$, namely, being a 0-level proposition) but not every 1-level proposition does. So, if B expresses a 1-level proposition, then it will not satisfy A. Logics that relax the principle of universal instantiation fall under the umbrella term 'free logic'. Some versions of free logic accept restricted instances of universal instantiation – for example, negative free logics characteristically claim that universal generalisations of atomic formulas imply their instances – but all relinquish unqualified UI.

There is in fact a more principled reason why the Russellian cannot keep the principle of universal instantiation: given any two quantifiers into sentence position satisfying the principle of universal instantiation, and some fairly uncontroversial background logic, one can prove that those quantifiers are logically equivalent. If $\forall$ and Π are two quantifiers taking sentence position satisfying UI, then one can prove $\forall pA \leftrightarrow \Pi pA$.[5] This argument is a generalisation of a theorem due to Harris (1982). While Harris was primarily concerned with first-order quantifiers, his result is general and applies to quantifiers into other grammatical categories, assuming analogous principles of UI.[6] Given UI, any ramification in any grammatical category would be pointless as all the different quantifiers in that category would collapse into one.

It's worth noting that if universal instantiation is relinquished, then the following principle of existential generalisation will have to be given up as well:

$$A(B/p) \rightarrow \exists pA \qquad \text{(EG)}$$

This principle is equivalent to UI – this fact can be easily be seen by taking the contrapositive of the $\neg A$ instance of UI, $\neg\neg A(B/p) \rightarrow \neg\forall p\neg A$, eliminating double negations and applying quantifier duality. Without this rule, one cannot infer from the fact that I'm thinking that snow is white that there is something I'm thinking.

The Russellian picture is supposed to help with a family of paradoxes, though we will in this paper be focusing on a particular species of paradox. As we will become clear, the rejection of UI is the key moving part to the Russellian-style response to at least that species of paradox. Indeed, the principal virtues of this response can be articulated within a theory containing only one quantifier abiding by a restricted version of UI.[7]

Williamson, on the other hand, is primarily concerned not with the intensional paradoxes but rather with the metaphysics of modality as revealed by the interaction of various higher-order quantifiers with the modal operators within a broadly Fregean framework. While Russell-style theorists are forced to abandon UI, the Fregean is free to adopt or reject it. One central theme in Williamson's book is to connect this choice point with the higher-order necessitist–contingentist dispute. Let us focus on the principle of universal instantiation for quantification into sentence position. From this principle, one can prove a necessitist principle for propositional quantification, which we can formalise as $\Box\forall p\Box\exists q\, p \equiv q$. Here, $\equiv$ could stand for propositional identity, or simply necessary equivalence: however, $\equiv$ is interpreted, all one needs to prove this result is UI, universal generalisation, the principle of necessitation plus the logical truth of the principle $p \equiv p$.[8]

There hasn't been much work done to connect the debates about contingentism and necessitism with the literature on the intensional paradoxes. However, the principle of universal instantiation plays a pivotal role in both disputes, giving us reason to think that there are interesting connections to be drawn. It bears emphasis that both the Russellian stratifier and the contingentist reject UI, and *ipso facto*, go in for some kind of free logic. In this paper, we're going to explore whether a free logic motivated by certain intensional paradoxes can undermine the case for necessitism. While necessitism can be formally reconciled with a free logic of this sort, many of Williamson's key abductive arguments for necessitism will not survive in this setting. This is because those arguments rely on logical explanations of certain data which depend on UI and Williamson's favoured comprehension principles, which aren't available within such a free logic.

Section 2 introduces our focal kind of paradox and explains why a free logic can offer a distinctive kind of solution. Section 3 makes vivid the relevance of universal instantiation to the higher-order necessitist/contingentist debate. Section 4 offers a number of free logical models for a family of claims that threatens paradox within the classical setting. Section 5 distinguishes two ideas that get packaged together in Williamson's world view: one idea is that existence at any type does not come and go from world to world (this idea is naturally articulated using a suitably typed version of the Barcan formula). The second idea is that existence comes cheap so that, for example, whenever we have a meaningful sentence, there is an associated proposition. This is naturally articulated using a suitably typed comprehension principle. The two ideas are natural allies for Williamson since both the Barcan formulae and comprehension can be derived from universal instantiation with minimal additional assumption; however, the free logician who rejects UI is free to take a very different view of these two strands of Williamson's world view. In Section 6, we raise a variety of problems for free logical packages. In Section 7, we articulate and critically assess different styles of ramificationist approaches – including one that merely ramifies quantifiers, one that merely ramifies operators and another that ramifies both. In this connection, we introduce a second kind of puzzle due Saul Kripke. Having that puzzle, as well as our focal puzzle, in view allows us to better assess the merits of the various styles of ramification.

2. Priorean paradoxes and free quantificational logic

As we saw above, non-redundant stratification involves dropping the unrestricted principle of universal instantiation. Indeed, as we shall see, it is the rejection of UI that is the key to blocking certain of the intensional paradoxes, whether or not that is motivated or accompanied by a stratificational picture. This is because UI is all by itself sufficient for deriving certain paradoxes that can't be derived without UI. Here, we have in a mind a family of paradoxes associated with Prior and Kaplan.

Let's begin by discussing a puzzling result of Prior (1961). Here, Prior is working in a small fragment of the Fregean hierarchy, consisting of nothing but the propositional calculus augmented with a device for quantifying into the position of sentences, and a unary sentential operator, *Q*, whose interpretation can be left completely open for the purposes of this result.[9] Armed with a pair of assumptions, namely UI for the quantifiers into sentence position, and classical propositional logic, one can prove the following theorem.[10]

PRIOR'S THEOREM: $Q\forall p(Qp \rightarrow \neg p) \rightarrow \exists p(Qp \wedge p) \wedge \exists p(Qp \wedge \neg p)$.

To get a feel for why this theorem is so puzzling, let us look at a particular interpretation for *Q* by reading *Qp* as 'Kaplan believes that *p*'. So interpreted, Prior's theorem says roughly[11] that if Kaplan believes that everything that he believes is false, then he believes something true and also believes something false.

As it happens, the material conditional in question is almost certainly true, for mundane reasons – for example, we doubt very much that Kaplan believes that everything he believes is false, or that he believes only true or only false things. Nevertheless, it is extremely surprising to be informed that this fact has the status of a logical truth. This doesn't seem like the sort of thing that could be figured out from logic alone.

One could learn to live with the necessary truth of this instance of Prior's theorem, maintaining perhaps that belief is closed under logical consequence, so that one can't believe anything false uniquely. It's not clear that these speculations about belief would render the relevant instance of Prior's theorem a logical truth. But more importantly, it doesn't speak to the full generality of the theorem. Other interpretations of *Q* are harder to live with. Consider, for example:

(1) If Kaplan says at midnight that everything Kaplan says at midnight is false, then Kaplan has said a true and a false thing at midnight.
(2) If at midnight, Kaplan writes on a whiteboard that everything Kaplan writes on a whiteboard at midnight is false, then Kaplan writes on a whiteboard a true and a false thing at midnight.
(3) If Kaplan seems to have said at midnight that everything Kaplan seems to have said at midnight is false, then there is a true and a false thing Kaplan seems to have said at midnight.
(4) If at midnight, it seems that Kaplan writes on a whiteboard that everything at midnight it seems that Kaplan writes on a whiteboard is false, then there's something true that at midnight it seems that Kaplan is writing on a whiteboard, and there's something false that at midnight it seems that Kaplan is writing on a whiteboard.

Prior attempted to explain the truth of 1 and 2 in the face of apparent counterexamples by suggesting that even though certain speeches or sentences seem to say something, they in fact fail to do so. Suppose, for example, that Kaplan had at midnight written nothing but the sentence 'everything Kaplan writes on a whiteboard at midnight is false' on the whiteboard. Although it seems as though Kaplan has written that everything he wrote on the whiteboard at midnight is false, Prior suggests that what seems to be so isn't so, and indeed that the inscription on the whiteboard in this case has failed to express anything.[12]

As Prior was aware, his response was far from pain free. However, the strategy just outlined is particularly unsatisfying as an explanation of the truth of 3 and 4. Applying that strategy to these cases involves denying that it even *seems* that Kaplan writes that everything that Kaplan seems to write at midnight is false.[13]

While we may in the end have to grudgingly accept these results, we think that their strangeness provides excellent motivation for thoroughly exploring the alternatives. In this instance, the alternatives are very clear cut: short of relinquishing propositional logic, the only other principle appealed to in the proof of Prior's theorem was the principle of universal instantiation.[14]

Let us now show how we can give a distinctive diagnosis of 1–4 in a setting where universal instantiation is relaxed. Recall that Prior's account of two is that while Kaplan wrote down the sentence 'everything Kaplan writes on a whiteboard at midnight is false', he (i) didn't write at midnight that everything Kaplan writes on a whiteboard at midnight is false, and indeed (ii) he didn't write anything whatsoever at midnight. The free logical approach we're exploring can accept (ii) without thereby accepting (i). For once, universal instantiation is relinquished; the inference from Kaplan didn't write anything at midnight to Kaplan didn't write that everything Kaplan writes on a whiteboard at midnight is false is invalid. Similarly, in the case of four, the inference from 'there's nothing that Kaplan seemed to write at midnight' to it's not the case at midnight that Kaplan seemed to write that everything Kaplan seems to writes on a whiteboard at 5 midnight is false' is invalid as well.

The approach here shares much with a similar free logical approach to the problem of empty names in the first-order context. In that setting, the free logician can maintain that Alice said that Pegasus flies without thereby committing us to inferring that there's something that Alice said flies.[15]

Note that even if Kaplan didn't in fact uniquely write 'everything Kaplan writes on a whiteboard is false' on a whiteboard, it's clear that he could have done so. The free logical approach described above will accommodate the idea that Kaplan could have written that everything Kaplan writes on a whiteboard is false without there being both a true and a false thing written. Let us abbreviate $\forall p(Qp \to \neg p)$ as λ. Then, this amounts to:

$$\Diamond(Q\lambda \wedge \forall q(Qq \to (q \leftrightarrow \lambda))) \tag{K1}$$

But this, notice, conflicts with the necessity of Prior's theorem: $\Box(Q\forall p(Qp \to \neg p) \to \exists p(Qp \wedge p) \wedge \exists p(Qp \wedge \neg p))$. The necessity of Prior's theorem is derivable if we assume that UI is necessary, and that the laws of propositional logic preserve necessity.[16] The free logical approach accommodates K1 by relaxing universal instantiation. Note that this principle is not even *prima facie* plausible unless Q is replaced by certain sorts of attitudes – it is not plausible if we replace Q for 'it is true that', or even 'it is necessary that' or 'it is known that'.

The motivation for K1 seems to generalise to any sentence of English. Roughly speaking, it seems that for any meaningful sentence 'ϕ' of English, Kaplan could have written 'ϕ' on the whiteboard at midnight and nothing else. By a completely analogous thought, it seems that we ought to be able to infer that Kaplan wrote that ϕ.[17] Thus, if we are sympathetic to K1 for these kinds of reasons, one should be strongly tempted to accept a schema that generalises K1:

$$\Diamond(Q\phi \wedge \forall q(Qq \to (q \leftrightarrow \phi))), \tag{K2}$$

where ϕ is any closed sentence.

K1 is just the instance of K2 you get when you substitute ϕ for $\forall p(Qp \to \neg p)$.

The principle K2 is closely related to, but subtly different from a thesis that is like K2 but which universally generalises into sentence position.

$$\forall p\Diamond(Qp \wedge \forall q(Qq \to (q \leftrightarrow p))) \tag{K3}$$

K3 initially looks as though it is straightforwardly stronger than K2 since it talks about all propositions, not just those expressible in a particular language. However, to get K2 from K3, one must employ an instance of universal instantiation (instantiating p with ϕ) which is not available in the free logical setting that we're currently exploring. Indeed, for the same reason, K1 doesn't follow from K3, for that would require instantiating p with λ (i.e. the nonexistent proposition that everything Kaplan says is false).

K3 is itself a straightforward consequence of a thesis sympathetically discussed by Kaplan which replaces the final material equivalence with an identity connective.[18]

$$\forall p \Diamond (Qp \wedge \forall q(Qq \rightarrow (q = p))) \qquad (\mathrm{K3}^{=})$$

We may note in passing that there are principles that strengthen K1 and K2 in a way analogous to the way that K3$^{=}$ strengthens K3. The difference will not matter much in what follows – the models that we consider later will also validate these strengthened principles.

Discussions of K3$^{=}$, beginning with Kaplan's own discussion, often focus on putative explanations of the falsity of that principle that turn on cardinality considerations (e.g. Davies 1981; Divers 2006; Lewis 1986).[19] The key line of thought is that there are more collections of possible worlds than there are possible worlds – it can't be that for every collection of worlds, there is a world at which the proposition corresponding to that collection is uniquely thought.[20] (Kaplan's own discussion focuses on cardinality arguments for the logical falsity of K3$^{=}$. Kaplan was happy with the thought that K3$^{=}$ was false, but unhappy with the thought that it was false as a matter of logic; as he saw things, standard possible worlds' semantics delivers the latter unwanted conclusion and concluded that this semantics is thereby unsatisfactory. We will return to these issues in Section 7.1.) However, cardinality considerations have no direct application to the puzzles we obtained from K1–K3. For unlike K3$^{=}$, for example, K3 doesn't require different worlds to serve as witnesses for propositions with different modal profiles: one and the same possible world could be a witness for the possibility that a certain proposition is thought, and simultaneously a witness for the possibility of thinking a different proposition so long as that proposition has the same truth value of the former at that world (regardless of whether it has the same truth value at other worlds). Moreover, in the case of K1 and K2, there is an even simpler reason why cardinality considerations are beside the point: K2 is a schema with countably many instances, and there are certainly enough possible worlds to witness each instance. Moreover, K1 concerns only *one* proposition, λ! Note, thus, that even K1$^{=}$ and K2$^{=}$ are similarly immune to cardinality worries.

3. Higher-order necessitism and higher-order free logic

Our principle of universal instantiation is also at the centre of the higher-order necessitist/contingentist debate; in this case, as it concerns quantification into sentence position. Given necessitation and the principle UI, one can quickly derive the central theses of higher-order necessitism. For quantification into sentence position, this amounts to:

$$\Box\forall p\Box\exists q\, p \equiv q \tag{NNE}$$

Here, the symbol $\equiv$ is a connective expressing the analogue of identity at the type of sentences. However, we shall leave it open whether this notion can simply be defined in terms of necessary equivalence, where $p \equiv q$ is defined as $\Box(p \leftrightarrow q)$, or perhaps in terms of higher-order quantification into operator position by analogy with Leibniz's law, so that $p \equiv q$ is defined as $\forall Q(Qp \rightarrow Qq)$.[21]

Recall that universal instantiation is equivalent to the principle of existential introduction. One proof of NNE – that relies on the principle of necessitation – starts with the theorem $p \equiv p$,[22] and infers $\exists q\, p \equiv q$ by existential introduction. By necessitation, which allows one to prefix any theorem with a box, one infers $\Box\exists qp \equiv q$. By universal generalisation, we get $\forall p\Box\exists q\, p \equiv q$, and finally by necessitation, again, we get $\Box\forall p\Box\exists q\, p \equiv q$. Instead of relying on a general principle of necessitation, one could start instead with a suitably necessitated version of UI as a premise. This is closer to Williamson's own argumentative path as he is reluctant to rely on the principle of necessitation.[23]

The first-order analogue of NNE – where $\equiv$ is interpreted as first-order identity – is often discussed in connection with the Barcan formula (a first-order analogue of BF below) and its converse (an analogue of CBF).

$$\forall p\, \Box A \rightarrow \Box\forall pA \tag{BF}$$

$$\Box\forall pA \rightarrow \forall p\Box A \tag{CBF}$$

Just as in the first-order setting, NNE is a consequence of CBF which can be seen by taking A to be $\exists q\, p \equiv q$.

Like NNE, CBF is also a consequence of UI, given some weak standard background assumptions about the logic of necessity. BF is not forced on us by the principles we have considered so far, but natural assumptions about the logic of necessity guarantee it. Here, we have in mind the Brouwer axiom: $p \rightarrow \Box\Diamond p$.[24]

Another principle that plays a pivotal role in Williamson's book is a principle of comprehension for each type. For quantification into sentence position, it amounts to[25]:

$$\Box\exists p\Box(p \leftrightarrow \phi) \tag{Comp}$$

One might think that Comp can be derived from NNE alone. Indeed, if we understand $\equiv$ as necessary equivalence, then Comp simply becomes $\Box\exists pp \equiv \phi$ which looks like it just follows from an instance of NNE. However, this latter inference – from $\Box\forall q\Box\exists pp \equiv q$ to $\Box\exists pp \equiv \phi$ – relies on UI. The relationship between NNE and Comp is much like that between K3 and K2: without the help of UI, one can't prove the instance from the universally generalised claim (we will see models which demonstrate this presently).

Since UI entails all of the central theses associated with higher-order necessitism, it's clear that if one wants to be a higher-order contingentist in any reasonable sense, then one has to reject UI.

It's worth noting that although the central theses of necessitism listed above do not themselves require UI, the rule fits well with a wider package of views that Williamson holds.[26] In fact, this wider package plays an important role in Williamson's abductive argument for necessitism: as Williamson notes in Williamson (1998), the result of combining standard quantificational principles (including UI) and standard modal principles has necessitism as a consequence.[27] Thus, although, as we shall presently see, one can be a necessitist without accepting UI, this would undercut one of the primary motivations for being a necessitist. Thus, insofar as one responds to Prior's paradox by adopting a free logic that relaxes UI, one has less reason to be a necessitist.

4. How to model K1, K2 and K3

In what follows, we shall be looking at illustrative models of the language fragment of interest: quantified propositional logic (with propositional constants $\top, \bot$), supplemented with a modal operator, $\Box$ and a sentential operator, Q, and a propositional identity connective $\equiv$. We will be primarily concerned with Kripke-style models of this language: these consist of (i) a set of points, W, (ii) a domain for the propositional quantifiers at each point $w \in W$, consisting of a set of sets of points $D(w) \subseteq \mathcal{P}(W)$, (iii) an accessibility relation $R \subseteq W \times W$ to interpret $\Box$ and (iv) an extension for Q at each point, which can be represented by a function, written $|Q|$, which maps each point to a set of sets of points. In what follows, we shall be concerned exclusively with models that validate S5, in which R is an equivalence relation.

In what follows, we shall adopt a sharp distinction between arbitrary sets of points, which we shall call an *intension*, and a set of points that that belongs to the domain, $D(w)$, of some world w – the latter kind of set corresponds to a *proposition* at that world: an intension that is in the range of the propositional quantifiers there.

$|Q|$ maps each point to the set of intensions that are Q at that point, where an intension is given by any set of points. As noted, $D(w)$ is a set of intensions in this sense representing the propositions that exist at w. Note, however, that the extension of Q at w might include intensions that don't exist at w. Note also that we have not insisted that the intensions that exist or that are Q at w consist only of sets of points accessible to w. In certain places, we will consider hyperintensional models in which distinct intensions are necessarily equivalent, where such a constraint would be inappropriate (in these models, an arbitrary set of points might be more appropriately called a 'hyperintension', although we'll stick to our terminology for consistency). In these kinds of models, two intensions may agree at every accessible point, but nevertheless be represented by different sets of points; the difference may show up when these propositions are embedded under Q and $\equiv$, even when they don't show up when embedded under $\Box$.[28]

To complete the specification of the model, we need the notion of a variable assignment: a function, v, from propositional variables to sets of points. A set of points can now be assigned recursively to every sentence of our language (relative to an assignment) in the following way[29]:

> $[\![P]\!]_v = v(P)$ if P is a propositional variable and $[\![P]\!]$ if an atomic sentence letter.
> $[\![\phi \equiv \psi]\!]_v = W$ if $[\![\phi]\!]_v = [\![\psi]\!]_v$ and $= \emptyset$ otherwise.
> $[\![\phi \vee \psi]\!]_v = [\![\phi]\!]_v \cup [\![\psi]\!]_v$
> $[\![\neg\phi]\!]_v = W \setminus [\![\phi]\!]_v$
> $[\![\Box\phi]\!]_v = \{w \in W \mid R(w) \subseteq [\![\phi]\!]_v\}$
> $[\![Q\phi]\!]_v = \{w \in W \mid [\![\phi]\!]_v \in |Q|(w)\}$
> $[\![\forall p\phi]\!]_v = \{w \in W \mid w \in [\![\phi]\!]_u, \forall u \text{ such that } u[p]v \text{ and } u(p) \in D(w)\}$

Here, we have used $R(w)$ to denote the set of worlds accessible to w ($\{x \mid Rwx\}$) and $u[p]v$ to mean that v and u agree except, possibly, on which set of points they have assigned to p.

As in the case of first-order free logic, the fact that the domain of the quantifiers is non-empty does not come for free, although one can ensure it by requiring models to satisfy all instances of the schema $\forall x\phi \rightarrow \exists x\phi$. Generalising to the propositional case, we want the principle $\forall p\phi \rightarrow \exists p\phi$ to be satisfied in our models, which is guaranteed by requiring that $D(w) \neq \emptyset$ for each $w \in W$. In what follows, we shall refer to models of this type as 'regular models'. Regular models are models of propositionally quantified S5, with a free logic for the quantifiers and the principle $\forall p\phi \rightarrow \exists p\phi$; thus, the existence of a regular model of some principle guarantees the consistency of that principle with this fragment of propositionally quantified modal logic. If the domain of propositions is empty at some world, we shall refer to the model as 'non-regular'.

It should be noted that according to one prominent tradition, the intended model of a higher-order language ought itself be specified in a higher-order metalanguage (Rayo and Uzquiano 1999; Rayo and Williamson 2003; Tarski 1936 §4; Williamson 2003).[30] Similarly, one might also insist that the intended model of a modal language itself be described in a modal metalanguage. Thus, for example, Prior and Fine (see the appendix of Prior [1977]) replace first-order quantification over times and worlds with quantification into sentence position, and simulate worlds and times by (the higher-order equivalent of) talk of propositions that are only ever true at most once, or propositions that are possibly both true and necessitate every truth.

Our model theory above is presented in the first-order language of set theory, and so does not have this form. However, our purpose is not to specify the intended model of $\mathcal{L}$ but to figure out which collections of principles are consistent with one another. Since consistency is a mathematical consequence of the existence of any kind of model, a Kripke-style model theory will suffice for our purposes.

4.1. Finite models

Note that models of K1–K3 are easy to come by on their own. The most flat-footed way to construct a model of these principles is to consider models of a higher-order form of nominalism: suppose that universal quantification into sentence position always yields vacuous truth because loosely speaking, there are no such things as propositions.[31] Then, K3 is vacuously true because it begins with a universal quantifier. As for K1 and K2, one can ensure their truth by having a possible world where $Q\phi$ is true for any interpretation of ϕ – this could be achieved simply by letting the extension of Q at this world include every set of worlds. So, for example, $W = \{w\}, R = \{\langle w, w\rangle\}, D(w) = \emptyset, |Q|(w) = \{\emptyset, \{w\}\}$. (Recall that $D(w)$ is not the domain of accessible worlds; it is the domain of existing propositions – in this case, no propositions exist at w.)

Thus, we get the following proposition:

Proposition 4.1: *K1, K2 and K3 have a finite constant domain non-regular model.*

Note that because this is a non-regular model, we do not get the principle $\forall p\phi \rightarrow \exists p\phi$ which requires the domain to be non-empty. Indeed, one might want to go further and require not only that the propositional domains be non-empty but that they contain both $\top$ and $\bot$. Indeed, if the domain contains only these two propositions, we also get the desirable thesis that the existing propositions are closed under arbitrary Boolean operations.

With this mind, let us consider a slightly more interesting model of K3. Here, we let there be two worlds $W = \{x, y\}$ with a universal accessibility relation, such that at both worlds, the necessary and impossible propositions are the only propositions that exist: $D(x) = D(y) = \{\{x, y\}, \emptyset\}$. Since there are only two existing propositions at either world, it is easy to find an interpretation of Q in which each of those propositions is the only proposition that is Q at some world: $|Q|(x) = \{\{x, y\}\}$ and $|Q|(y) = \{\emptyset\}$. Thus, K3 is validated at both worlds in this model. It should be noted that the propositions that exist at each world satisfy a number of closure conditions: propositions corresponding to $\top$ and $\bot$ exist, and an arbitrary Boolean combination of existing propositions exists. However, existence is not closed under the Q operator. For example, $Q\top$ is true at the set of worlds $\{x\}$, but $\{x\}$ does not exist even though the set of worlds $\top$ corresponds to does.

What about K1 and K2? The set of worlds at which Prior's sentence, λ, is true is $\{y\}$: at x a truth is Q, whereas at y, only the inconsistent proposition is Q. Since $\{y\}$ is not Q at any world, $Q\lambda$ is not true at any world, and so $\Diamond Q\lambda$ is also false at both worlds. This precludes K1 being true, and thus also the schema K2 of which K1 is an instance.

We could validate K1 by changing our model slightly. Let $W = \{x, y, z\}$, and as before, we stipulate that R is universal and moreover that only W and $\emptyset$ exist at each world, and that $|Q|(x) = \{W\}$ and $|Q|(y) = \{\emptyset\}$. Finally, we stipulate that $|Q|(z) = \{\{y, z\}\}$. Notice that K3 is valid in this model for the same reason as before. In this model, λ is true at y and z (it's true at y because only $\bot$ is Q at y, and vacuously true at z because nothing that exists is Q at z). Thus, λ expresses $\{y, z\}$, and thus $Q\lambda$ is true at z. Moreover, $\forall p(Qp \rightarrow \lambda \equiv p)$ is vacuously satisfied at z, so the principle K1 – $\Diamond(Q\lambda \wedge \forall p(Qp \rightarrow \lambda \equiv p)$ – is true at every world. Notice, however, that we can still find instances of K2 that are not validated. For example, $\neg\lambda$ has the singleton $\{x\}$ as its intension, which is not in the extension of Q at any world, so the $\neg\lambda$ instance of K2 is false at every world. Note as before that existence is closed under arbitrary Boolean operations, but not under the Q operator.

Let us see if we can push this strategy further and validate all three K principles. So far, we have only considered models in which one set of worlds appears in extension of Q at each world. But if we drop that constraint, we can generate a three-world model that validates K1–K3. Let $W = \{x, y, z\}$, and let the extension of Q at x and y, the accessibility relation and the existing propositions be as before. Now let the extension of Q at z be $\mathcal{P}W \setminus \{W, \emptyset\}$. K3 is validated as above. K2 is validated because for any sentence that ϕ is true at some but not all worlds is such that (i) its intension is Q at z and (ii) it is vacuously true that every existing proposition at z that is Q is equivalent to it. Meanwhile, if ϕ is true at every world or none, then its extension will be Q without any existing proposition being Q at x and y, respectively.

Proposition 4.2: *K1, K2 and K3 have a finite constant domain regular model containing $\top$ and $\bot$ and closed under arbitrary Boolean operations.*

This model is peculiar: it secures the validity of K2 by having the extension of Q be radically non-unique at z, even though the object language statement of uniqueness is satisfied at z. For example, at z, λ (i.e. the set of points $\{y, z\}$) is Q without any existing proposition being Q, and, likewise, $\neg\lambda$ (i.e. the set of points $\{x\}$) is Q without any existing proposition being Q.

The sense in which λ and its negation are uniquely Q is by no existing proposition being Q alongside them. What we want is to be able to impose a stronger uniqueness requirement; however, this turns out to be hard to express in the free logical setting. By analogy, suppose we wanted to say that someone is imagining Pegasus but not Atlantis or Zeus or .... The '...' cannot be captured quantificationally because a quantificational claim won't entail the instances involving fictional names.[32] However, we could introduce an operator $Q^*\phi$ for this purpose. The interpretation of 'Q^* snow is white' is that it means the same as 'snow is white is Q but grass is green is not Q, nor ... ' where the ellipsis is understood in this open-ended way. Two schemata partially characterising the behaviour of Q^* can be outlined:

$$Q^*\phi \rightarrow Q\phi$$
$$Q^*\phi \wedge Q^*\psi \rightarrow \phi \equiv \psi$$

These schemata must be understood open endedly.[33] Model theoretically, we can interpret Q^* in our models as follows:

$|Q^*|(x) = \emptyset$ if $|Q|(x)$ is not a singleton, and $= |Q|(x)$ otherwise.

Armed with Q^*, there's a sort of uniqueness demand that strengthens K1–K3 that can be articulated as follows:

$$\Diamond Q^*\lambda \qquad (K1^+)$$
$$\Diamond Q^*\phi \qquad (K2^+)$$
$$\forall p \Diamond Q^* p \qquad (K3^+)$$

However, $K1^+$ fails in our present model; for example, $\Diamond Q^*\lambda$ is not true in the present three-world model. In fact, no three-world model will do.[34] This idea generalises to show that no finite world model will do: there can be no finite model in which the set of worlds expressed by each sentence is the one and only member of $|Q|(w)$ for some world w (and thus no finite model of $K2^+$). This holds even when restricting attention to the original language without Q^*.

Proposition 4.3: *If $\langle W, D, R, |Q|\rangle$ is a model such that for each closed sentence ϕ there is a point $w \in W$ such that $|Q|(w) = \{[\![\phi]\!]\}$, then W must be an infinite set of points.*

Proof: We shall show that in any such model, for any $n < m$, $[\![Q^n\top]\!] \neq [\![Q^m\top]\!]$ (this precludes the satisfaction of our condition in a finite model because there are infinitely many sets of worlds expressed by sentences in the language). Suppose that this hypothesis holds for every $n < m$. Suppose for contradiction that for some $n < m + 1$, $[\![Q^n\top]\!] = [\![Q^{m+1}\top]\!]$. n cannot be 0 since this entails that $[\![Q^{m+1}\top]\!] = [\![\top]\!] = W$ so that $[\![Q^m\top]\!]$ is Q at every world. Since $[\![Q^m\top]\!]$ can't be identical to both W and $\emptyset$, either there's no world where W is the only member of the extension of Q or there's no world where $\emptyset$ is the only member of the extension of Q. So $n > 0$, and $[\![Q^n\top]\!] = [\![Q^{m+1}\top]\!]$, which means that for every world w such that $[\![Q^{n-1}\top]\!] \in |Q|(w)$, $[\![Q^m\top]\!] \in |Q|(w)$. By inductive hypothesis, $[\![Q^{n-1}\top]\!] \neq [\![Q^m\top]\!]$, so there's no world where $[\![Q^{n-1}\top]\!]$ is the unique member of the extension of Q.

4.2. Infinite models

It's natural to wonder if there are any infinite models satisfying the constraint. We can construct one in the spirit of the earlier models – involving a fairly frugal domain of propositions – as follows. Let $W = \mathbb{N}$, let the accessibility relation be universal and as before let $D(n) = \{W, \emptyset\}$. Now pick some bijection, σ, between $\mathbb{N}$ and the set of finite and cofinite subsets of $\mathbb{N}$.[35] Given σ, the extension of Q can specified as: $|Q|(n) = \{\sigma(n)\}$.[36] In this model, every finite and cofinite set is the unique member of the extension of Q at exactly one world. Moreover, it can be shown by a simple induction that in this particular construction, every closed sentence of our language expresses a finite or cofinite set.[37] Thus, it follows that in this model, for every set of worlds expressed by some sentence, there is a world where that set is the unique member of the extension of Q. So we have a model of K1–K3, and also a model of $K2^+$ – and $K1^+$ and $K3^+$. Just like in the earlier models, the propositions that exist at each world are closed under all Boolean operations (including infinitary conjunction and disjunction), but are not closed under the Q operator. For example, the set corresponding to $\top$ exists, but $Q\top$ is true at exactly one world, and no singleton set is in the domain of quantification at any world.

Proposition 4.4: *$K1^+$, $K2^+$ and $K3^+$ have an infinite constant domain regular model whose domain contains $\top$ and $\bot$ and is closed under arbitrary Boolean operations.*

The models considered so far might be deemed uninteresting because they only permit quantification over the necessary and impossible proposition: there is both a paucity of propositions (there are only two) and a lack of variety (there are no contingent propositions). It's easy to modify the model so that it allows a few more propositions: for example, we could take a finite number of finite sets, and close them under Boolean operations, and the resulting model would still satisfy $K1^+$–$K3^+$. Of course, this model will only deliver finitely many existing propositions in total since closing finitely many finite sets under Boolean operations will only yield finitely many sets. As a corollary, existing propositions will not be closed under Q in these models: if $\top$ expresses an existing proposition, and existence was closed under Q, then $Q^n\top$ would express an existing proposition for each n. By the result above, any model of this sort that satisfies $K2^+$ or $K3^+$ will be such that $Q^n\top$ expresses a different proposition for each n, so there must be infinitely many existing propositions.[38] Since these models only have finitely many propositions, they therefore cannot simultaneously satisfy closure of existence under Q and either $K2^+$ or $K3^+$.

It's natural to investigate models with an infinite variety of existing propositions, and in particular which satisfy the principle that existence is closed under Q. The preceding model can be adapted to this setting by expanding the domain to include all finite and cofinite sets. As in the earlier models, Qp always expresses a singleton of a world, if p expresses a finite or cofinite set. Thus, if *p* expresses an existing set of worlds, *Qp* expresses a finite and thus existing set of worlds. Notice also that existence is closed under the finitary Boolean operations, but not the infinitary Boolean operations: a finite combination of unions and complements of some combination of finite and cofinite sets will yield either a finite or a cofinite set, although infinite unions of finite sets might be neither finite nor cofinite.

Notice also that this is a model of K3 and K3$^+$: every finite or cofinite set is the unique member of the extension of *Q* at some world.

Proposition 4.5: *K3$^+$ has an infinite constant domain regular model in which the propositional domain is closed under finitary Boolean operations and Q.*

However, the proof that every closed formula expresses a finite or cofinite set breaks down in this setting.[39] It is no longer clear that this model validates K1 or K2. Indeed, we can show that the Prior sentence expresses an intension that is neither finite nor cofinite for any model of this sort.

Proposition 4.6: *Any regular model of K3$^+$ in which the propositional quantifiers range over the finite and cofinite sets of worlds at each world, and only one intension is in the extension of Q at each world, will be one in which the Prior sentence expresses an infinite and coinfinite set of worlds.*

Notice, of course, that by Proposition 4.3, any regular model of K3$^+$ will be infinite.

For the Prior sentence to correspond to a finite set, the propositions corresponding to each singleton set of worlds must be in nearly every case (i.e. in all but finitely many cases) be *Q* at the world contained by the singleton. (After all, in these models, only one intension is *Q* at each world, and so if there are infinitely many cases where a singleton is *Q* at a world that isn't contained by the singleton, then there are infinitely many cases where every proposition that is *Q* at a world is false.) But that leaves only finitely many worlds to take care of the possibility of uniquely thinking each of the remaining finite and cofinite sets, which is impossible (since there are infinitely many that aren't singletons). What about the hypothesis that the Prior sentence corresponds to a cofinite set? To rule this out, let us consider now not the singleton set of each world but its complement. For the Prior sentence to be true at cofinitely many worlds, it must be that in nearly all cases, the complement of a singleton be *Q* at the world contained in the singleton. (After all, if the complement of a singleton is *Q* at a world contained in that complement, then the proposition that is *Q* is true at the world in which it is thought and so λ will be false at such a world.) But if nearly all worlds are such that the complement of their singleton is *Q* at those worlds, then that leaves only finitely many worlds to take care of the possibility that each remaining finite and cofinite set is *Q* without anything else being *Q*, which is impossible.[40] Recall now that it is a feature of our model that the only sets falling in the extension of *Q* are finite or cofinite sets since λ doesn't express either a finite or a cofinite set, the extension of *Q* won't include the set expressed by λ. Therefore, this model won't satisfy K1, and *ipso facto* will not satisfy K2.[41]

4.3. Rich models and ramification

In the above model, we've made space for a plenitude of existing contingent propositions, but offered no means for securing K2. Notice also that in the models of K3 given above, we have shown that the intension corresponding to any given sentence is the unique member of the extension of *Q* at some worlds. This doesn't by itself guarantee that there is a possible world at which the proposition that snow is white and the proposition that grass is green are both *Q* without anything else being *Q*. A further constraint on the models is required if we want to ensure this and similar possibilities. Say that a model is *rich* if and only if, for any variable assignment, *v*, assigning variables to existing propositions, and set of formulae in the language, *X*, there is some world in the model at which the extension of *Q* is given by the set of intensions expressed by each member of *X* relative to *v*.

Here is a recipe for constructing a rich model that also ensures a plenitude of existing contingent propositions, while securing K1–K3 and $K1^+$–$K3^+$. In these kinds of models, we can think of the existing propositions as being arbitrary Boolean combinations of 'basic propositions' concerning only non-intensional matters. Following Kaplan, let's pretend that the basic propositions state facts about the distribution of earth, wind, fire and water (the details don't matter much). We can model these basic propositions as sets of entities we shall call 'proto-worlds'. We shall build up a richer space of possibilities in stages, and hence a space of intensions for formulae, with the proto-worlds corresponding to the possibilities at stage 1. Another layer of possibilities, roughly corresponding to which basic propositions are *Q*, can be generated by considering ordered pairs of proto-worlds with sets of sets of proto-worlds, where the latter component corresponds to the stage 1 intensions that are *Q* at that possibility. Call these stage 2 possibilities. Now there are more sets of stage 2 possibilities that can be *Q* than merely sets of proto-worlds, so we must repeat the process. We can generate a third layer – the stage 3 possibilities – by considering ordered triples consisting of a proto-world, a set of sets of stage 1 possibilities (proto-worlds) and a set of sets of stage 2 possibilities. This iterates in the obvious way, with our complete possibilities (or 'worlds') being infinite (ω-length) sequences where the $n+1$th component of the sequence is sets of sets of stage *n* possibilities, giving us the stage *n* intensions that are *Q* at that stage.

For a technical reason, that will become apparent later, we also need to stipulate that at each stage, we only add sets of intensions that are 'new' – intensions that couldn't have been generated at an earlier stage. A set of *n*-tuples *X* is generated by a set of *m*-tuples *Y* for $m < n$ iff *X* is the set of all n-tuples that extend m-tuples in Y; thus, we will require that the $n+1$th component of our infinite sequences consist only of sets of n-tuples that are not generated by a set of m-tuples for smaller m.

Our complete intensions are just sets of these infinitary sequences. Notice that our possibilities, as we have constructed them, do not directly tell us which sets of complete intensions are Q at each possibility: they give us an infinite sequence of incomplete (i.e. finite stage) intensions, none of which belong to our space of complete intensions. However, each stage n intension can be extended to a complete extension in a unique way: if I is a stage n intension, we can consider the set of all infinite sequences in our space, I^+, that extend the elements of I. (Intuitively, we can think of I^+ as an intension that settles some questions about which stage n propositions are Q, but leaving it completely open what higher stage intensions are Q.) Let us call the extension of a stage n intension to the space of complete intensions a *completed* stage n intension. Finally, call an intension *bounded* if and only if it is a completed stage n intension for some finite n.

We are now in a position to construct a model. Given any set of proto-worlds, we can construct a set of worlds W consisting of infinitary sequences as described above. As usual, we will let our accessibility relation be the universal relation. The domain of the propositional quantifiers, $D(w)$, will be the same at every world: the set of all completions of sets of proto-worlds, i.e. the set of all completed stage 1 intensions. The extension of Q at a world w, will contain, for each stage n intension belonging to the nth component of w, the completion of that stage n intension: $|Q|(w) = \{I^+ \mid \exists n : I \in w(n)\}$. So the extension of Q is a set of sets of infinitary sequences, each such set being a completed stage intension that encodes what is Q at a certain particular stage.

Two things are worth noting about this construction. Firstly, the completed stage n intensions are closed under arbitrary Boolean operations. This means in particular that the propositions that exist at each world – completed stage 1 intensions – are closed under arbitrary Boolean operations. Secondly, if ϕ expresses a completed stage n intension, and is not a completed stage m intension for any $m < n$, then $Q\phi$ will express a completed stage $n+1$ intension.[42] Thus, in particular, the completed stage 1 intensions are not closed under the Q operator.

This model validates K1–K3 and K1$^+$–K3$^+$. We'll show this by demonstrating two things: (i) for every bounded intension, there is some world where that intension is uniquely in the extension of Q and (ii) every closed formula expresses a bounded intension. For (i), we can in fact show something stronger: for any set of bounded intensions, there is some world where they are the only intensions in the extension of Q. This fact will guarantee that our model is 'rich' in the sense defined earlier. Given a set X of bounded intensions, we can construct a world in the following way: pick any proto-world for the first component, and then for the nth component of the sequence simply take the set of n stage intensions whose completions are in X. The proof of (ii) is a straightforward induction on the complexity of formulae, which we leave to a footnote.[43]

Since the propositional quantifiers range over completed first stage intensions, and thus bounded intensions, (i) guarantees that K3 and K3$^+$ are validated. Moreover, given (ii), every closed sentence of the language expresses a bounded intension, so K2 and K2$^+$ (and thus K1 and K1$^+$) are also validated.

Proposition 4.7: *K1$^+$, K2$^+$ and K3$^+$ have a rich constant domain regular model whose domain contains $\top$ and $\bot$ and is closed under arbitrary Boolean operations.*

As noted above, the domain of the propositional quantifiers is closed under arbitrary Boolean operations at each world, but not under the Q operation. It is possible to tweak the above model so that propositional existence is closed under the Q operation, is closed under finitary Boolean operations but is not in general closed under infinitary Boolean operations. This is achieved by iterating the construction to ω_1 instead of ω, and letting the propositional quantifiers range over (completions of) ω-bounded intensions, i.e. intensions that appear at some finite stage. In this model, one can show that (i) every ω_1-bounded intension (i.e. any intension appearing at a countable stage) is uniquely in the extension of Q at some world and (ii) every closed formula expresses an ω_1-bounded intension, guaranteeing the validity of K1–K3 and K1$^+$–K3$^+$. Since the details of this construction are more involved, we omit it.

Proposition 4.8: *K1$^+$, K2$^+$ and K3$^+$ have a rich constant domain regular model containing $\top$ and $\bot$ and closed under finitary Boolean operations and closed under Q.*

The first model we considered above is interesting for another reason: it naturally accommodates the addition of a hierarchy of propositional quantifiers roughly in line with a broadly Russellian approach to higher-order logic. Indeed, for each n, we can introduce quantifiers into the language that range over the completed n-stage intensions. Such quantifiers will satisfy a restricted version of universal instantiation, guaranteeing that the higher quantifiers range over larger domains of propositions:

$$\forall_n p\phi \rightarrow \phi[\psi/p] \qquad (\mathrm{UI}_n)$$

where Q embeds within itself at most n times in ψ.

In these models, we also obtain a generalisation of K3 and K3$^+$ for each propositional quantifier, for example:

$$\forall_n p \Diamond Q^* p \qquad (\mathrm{K3}_n^+)$$

The resulting models bear a family resemblance to Kaplan's description of a construction designed to accommodate K3$^=$. We'll discuss these connections later in Section 7.1.

Augmenting our models in this way thus gives us the following result:

Proposition 4.9: *$K1^+$, $K2^+$ and $K3_n^+$ have a rich constant domain-ramified model containing $\top$ and $\bot$ and closed under finitary Boolean operations and closed under Q.*

5. Higher-order necessitism revisited

It's worth drawing attention to the fact that all of the models we have discussed in this section illustrate a theme introduced in Section 3, namely that higher-order necessitism (as encoded by NNE, BF and CBF) does not entail UI. In all of our models, these three principles are valid because in each model, we chose a constant domain for the propositional quantifiers to range over. However, since in no model was the range of those quantifiers the whole powerset of the domain, we have no reason to think that UI is validated; indeed, it can't be validated, given that UI is inconsistent with principles, such as K1, that are validated in these models.

Notice that in this weaker setting, there is a schematic formulation of necessitism that can't be derived by UI from the principle NNE:

$$\Box\exists p(p \equiv \phi) \qquad (\text{NNE}^S)$$

Informally, it's necessary that the proposition that ϕ exists. This is closely related to a principle of comprehension that is dear to Williamson, and that plays an important role in the book:

$$\Box\exists p\Box(p \leftrightarrow \phi) \qquad (\text{Comp})$$

Indeed, they amount to the same thing when we take $\equiv$ to mean necessary equivalence. This principle is not validated in any of the above models. This is hardly surprising. As noted earlier, UI can be proved from Comp if we assume that identity is necessary equivalence, which is true in all of the above models. Indeed, assuming necessarily equivalent propositions can be distinct, Comp may be true even if certain propositions are contingently existing, provided that a proposition necessarily equivalent to it persists through modal space. So, NNE^S may be a better way of regimenting the thought that is motivating Williamson to accept Comp.

In the absence of UI, one might therefore wonder whether the combination of NNE, BF and CBF is the proper way of regimenting higher-order necessitism. However, we think there's no point quibbling about the proper use of the term 'necessitism', especially given that this isn't a term with a long philosophical history. But it is worth noting that there are two separable ideas implicit in Williamson's metaphysical picture. The first is that existence doesn't come and go across metaphysical possibilities; modelled theoretically, this means that the domain of quantification remains constant between worlds. And this is correctly captured by the combination of NNE, BF and CBF even in the absence of UI. The second, loosely speaking, is that existence (or its higher-order correlates) is extraordinarily easy, and this is correctly captured by NNE^S. If you are thinking that pigs fly, then there's something that you are thinking, and there's some *F* such that you think that pigs *F*. More generally, the kind of meaning an expression such as, say, a singular term has is such that if it is meaningful, then existential generalisation of the appropriate order into that position is permitted. So, in particular, the kind of meaning a sentence has is such that if a sentence is so meaningful, then existential quantification into the position of that sentence is always permitted. This second thought is closely related to the Fregean picture according to which for each grammatical category, there is a kind of quantifier that generalises into it, so that a meaningful occupant of that category will be a fit subject for such generalisation.[44]

Regarding the second thought, one might think that even if the quantifiers introduced above are not 'easy' in this sense, it is always possible to formally introduce quantifiers that are, by stipulating that they satisfy universal instantiation.[45] Similar moves have been made in the context of first-order disputes concerning, for example, the existence of Sherlock Holmes. Were the cogency of such a quantifier automatically conceded, then the main residual dispute would be as to whether the perfectly general quantifier would be interesting, useful or 'cut at the joints'. But we shouldn't be too quick to assume that such a quantifier is possible to introduce. Granted, for any language, $\mathcal{L}$, it is possible to introduce a quantifier which satisfies universal instantiation of any sentence of $\mathcal{L}$. For example, if we introduce infinitary disjunctions into the language, then one can introduce $\Sigma p\phi$ as a short hand for the infinite disjunction $\bigvee_{\psi\in\mathcal{L}}\phi[\psi/p]$. Σ will allow one to existentially generalise sentences of the original language $\mathcal{L}$, provided we have disjunction introduction. However, this is not enough to make good on the idea that existence is easy since we want to be able to generalise into the position that any sentence takes. As defined, there will be sentences in the extended language, containing infinitary disjunctions, that we cannot existentially generalise into using the newly defined quantifier (one could repeat this process, but it generates new gaps at each iteration).[46]

Are there general reasons to think that an 'easy' quantifier cannot be stipulatively introduced? We know from the case of 'tonk' that not any old set of stipulations will succeed in introducing a meaningful operator, and in the current context, there might be legitimate concerns as to whether these stipulations are satisfiable. While not strictly tonk-like, in that no supplementary principles are needed to derive a contradiction beyond those stipulated to govern tonk, a contradiction will follow from an open-ended construal of $K2^+$, in which the new quantifiers are allowed to appear as instances.

The models we considered above make good on the first necessitist idea, that existence doesn't 'come and go' across possibilities, but not the second idea according to which existence is easy. One might wonder to what extent this picture is consistent with the spirit of Williamson's view. We won't attempt to judge the issue, but let us mention two considerations that seem pertinent in this regard. Firstly, one of Williamson's arguments for higher-order necessitism stems from the derivability of NNE, BF and so on from the principles such as Comp, NNE^S and universal instantiation.[47] In other words, Williamson motivates the idea that existence doesn't come and go across possibilities from the thought that existence is easy. Once we've relinquished the latter thought, as we have in these above models, we undercut this way of motivating the former thought, even if we can in fact maintain it in the face of paradox.

The other consideration comes from reflection upon the contingent liar. Consider an ordinary speech involving the word 'says', e.g. 'Everything the Pope says is false' as said by Kaplan at a certain time *t*. There's nothing unusual or self-referential about this speech, and since the Pope does not say anything remotely liar-like, one might think that even if false, in making this speech, Kaplan succeeds in saying something. On the other hand, consider a counterfactual scenario in which Kaplan makes this speech at *t*, and in which Kaplan is the Pope. The usual kind of paradox ensues, and the higher-order free logician, for the kinds of reasons we have been examining at length, will maintain that Kaplan doesn't succeed in saying anything. In particular, since Kaplan says that everything the Pope says is false, the proposition that everything the Pope says is false doesn't exist in this counterfactual scenario. Given that for the necessitist, propositions don't 'come and go' into existence, it follows that there is *no* world at which the proposition that everything the Pope says is false exists. This contradicts the original judgement that the ordinary speech 'everything the Pope says is false' would result in the speaker saying something.

6. Problems

6.1. Paradoxes from closure assumptions

In the models above, we noted that the domain of the propositional quantifiers was often closed under some operations but not others. It is natural to wonder whether this is an accident, or if there is an obstacle to satisfying all of the closure principles that we gestured at.

To sharpen matters, let's look at what these closure conditions amount to in the object language. Consider the following closure principles:

$$\exists p(p \equiv \top) \tag{CL$\top$}$$

$$\exists p(p \equiv \bot) \tag{CL$\bot$}$$

$$\forall p \exists q(q \equiv \neg p) \tag{CL$\neg$}$$

$$\forall p \exists q(q \equiv \Box p) \tag{CL$\Box$}$$

$$\forall p \exists q(q \equiv Qp) \tag{CLQ}$$

$$\forall p \forall q \exists r(r \equiv (p \wedge q)) \tag{CL$\wedge$}$$

$$\forall p \forall q \exists r(r \equiv (p \equiv q)) \tag{CL$\equiv$}$$

$$\forall p \forall q \exists r(q \equiv \phi \rightarrow (r \equiv \forall p \phi)) \tag{CL$\forall$}$$

where ϕ does not contain q or r free.

In all but the last case, the CL$*$ principle corresponds to closure of the domain under the model theoretic operation corresponding to the connective $*$. In the case of CL$\forall$, closure under arbitrary conjunctions suffices for its truth, although there is no condition statable in the object language under consideration that precisely corresponds to the closure of the domain under arbitrary conjunctions.

Each set of connectives, X, the above principles corresponding to X along with the principle $\forall p\phi \rightarrow \exists p\phi$ (a principle ensuring that there's at least one proposition) and the normal axioms governing proposition identity give rise to a corresponding restricted version of universal instantiation:

$$\forall p\phi \rightarrow \phi[\psi/p] \tag{UI(X)}$$

where ψ only contains connectives in X.

None of our models validated all of these principles. There is a fairly systematic reason for this.

Proposition 6.1: *The principles CL$\top$, CL$\neg$, CLQ, CL$\wedge$ and CL$\forall$ are jointly inconsistent with K1 and K3.*

Informally, if one has all of the closure principles, one can prove the existence of the Prior proposition from which one can derive a contradiction, given K1 or K3 and other background principles. Given CL$\neg$, CLQ and CL$\wedge$, it is straightforward to show $\forall p \exists r r \equiv (Qp \rightarrow \neg p)$ (noting that $\rightarrow$ is definable from $\wedge$ and $\neg$). An instance of CL$\forall$ states that $\forall p \forall q \exists r(q \equiv (Qp \rightarrow \neg p) \rightarrow (r \equiv \forall p(Qp \rightarrow \neg p)))$. Then, given the existence of at least one proposition (as guaranteed by CL$\top$), we can infer from these two things that $\exists r(r \equiv \forall p(Qp \rightarrow \neg p))$. Given that the Prior proposition exists, and the usual axioms governing propositional identity, one can prove a contradiction from K1. Another way to see this is that in the proof of Prior's theorem, we only appealed to an instance of UI($\{Q, \neg, \wedge, \forall\}$), and given CL$\top$, we can prove $\forall p\phi \rightarrow \exists p\phi$.

Note here that the fact that we are using identity rather than necessary equivalence in the statements of our closure principles plays an important role in this derivation. It could be that for any Boolean combination of existing propositions, there is an existing proposition necessarily equivalent to it. In particular, there could be a proposition necessarily equivalent to the Prior proposition. But unless this existing proposition is identical to the Prior proposition, we cannot infer that I'm thinking something from the fact that I'm thinking the Prior proposition.

Closure principles are special instances of general principles, telling one how to combine elements from various types to form elements of other types (another example is the principle that for every x and every F, there's a p identical to Fa: $\forall x \forall F \exists p(p \equiv Fa)$.) It's a feature of the free logical approach that it is forced to deny a variety of principles of this sort at the type of propositions that are *prima facie* elegant. Of course K1–K3 are *prima facie* compelling too, but one wonders whether the gain outweighs the cost.

We should note in passing that there are a number of different kinds of pictures that may be used to flesh out the free logical approach in rather different ways. One kind of constraint is suggested by Russell's vicious circle principle, applied to the propositional type: the constraint that a proposition cannot quantify over a domain that includes itself. But a rather different constraint is suggested by the idea that operators corresponding to intentional attitudes can often take one from the existent to the nonexistent. The first constraint is naturally aligned with the rejection of CL$\forall$. Suppose that for every p, the proposition that $\Box p$ exists, then CL$\forall$ entails that $\exists r(r \equiv \forall p \Box p)$ which violates the vicious circle principle: $\forall p \Box p$ is in the range of $\exists r$, and thus also in the range of $\forall p$, meaning that $\forall p \Box p$ quantifies over a domain that includes itself.[48] The second constraint naturally leads one to reject CLQ: when Q corresponds to an intentional attitude, the result of combining Q with an existing proposition may take one to the nonexistent. Of course within this approach, there are all sorts of further decision points. One might think that any sentence containing 'Q' fails to express an existing proposition, or one might instead think that only the paradoxical ones fail to do so. (If one takes the latter approach, one might also very well deny CL$\neg$ since there are sentences for which no paradox attends the assumption that it expresses an existent proposition, but from whose negation a paradox does follow from the correlative existence assumption.) Moreover, the latter approach naturally suggests higher-order contingentism. It will, for example, be a contingent matter whether assuming the existence of the proposition that the guy sitting in room 101 at time t is thinking only falsehoods leads to paradox. A more necessitist friendly version of the second idea will say that any sentence that could have led to trouble in fact doesn't express an existing proposition.

6.2. *Comprehension and supervenience*

Whatever one's motivations are for restricting the realm of the existing propositions, it is natural to assume that there remains a realm of fundamental propositions whose existence is unthreatened by intensional paradoxes, or by vicious circle constraints. These may be facts about the field values at particular space–time points, but in the current context, no harm will be done to once again assume the imagery of a crude version of pre-Socratic physics, in which the fundamental consists facts about earth, air, fire and water. Further, it appears harmless to assume that arbitrary conjunctions and negations of such propositions similarly exist, for they only concern fundamental matters, and moreover harmless to assume that these propositions exist necessarily if they exist at all.

It is further a metaphysically compelling thought that all truths supervene on these fundamental propositions: that no two possibilities can agree with regard to the truth of all earth, air, fire and water facts but disgree about some intensional matter – e.g. whether I'm thinking something, or about whether everything I'm thinking is false. Once supervenience is assumed, then given our Boolean liberalism regarding the fundamental propositions, it is a short step to saying that every sentence is necessarily equivalent to some fundamental proposition. This in turn entails, given the existence of the fundamental propositions, that every sentence is necessarily equivalent to an existing proposition:

$$\exists p\Box(p \leftrightarrow \phi)$$

This is, of course, just the comprehension principle Comp we introduced is Section 3.

If necessary equivalence guarantees propositional identity, then Comp is tantamount to saying that every sentence expresses an existing proposition, which in turn prevents all the free logical manoeuvres for saving K1–K3. But might we move to a hyperintensional setting in which Comp is admitted for supervenience theoretic reasons, but in which K1–K3 are saved by a combination of hyperintensionality and free logic? For example, even given the modal equivalence of some existing proposition, call it 'Fire', with the sentence 'everything I'm thinking is false', our original paradox can be blocked. Given hyperintensionalism and free logic, it's coherent to suppose (i) that I'm thinking that everything I'm thinking is false, (ii) that there's some proposition (i.e. Fire) necessarily equivalent to its being the case that everything I'm thinking is false and yet (iii) there's no proposition that I'm thinking. Now this is indeed a way of saving K1 while accepting Comp. For example, consider a two-world model, $W = \{x, y\}, R = \{\langle x, x\rangle, \langle y, y\rangle\}, |Q|_x = |Q|_y = \{W\}$ $Dom(x) = \{\{\}, \{x\}\}, Dom(y) = \{\{\}, \{y\}\}$. Note that since the accessibility relation is the identity relation, $\Box$ just amounts to truth, and Comp just amounts to the existence of a true and a false proposition, which is satisfied at both worlds (similarly, propositions exist necessarily for the same reason, even though we have a variable domain model). At both worlds, no existing proposition is Q (W does not exist at either world), even though there are existing intensions necessarily equivalent to the nonexistent intension at each world ($\{x\}$ exists and is necessarily equivalent to W at x, similarly for y and $\{y\}$). Since nothing is Q at either world, then $\forall p(Qp \rightarrow \neg p)$ is vacuously true at both worlds, thus has the intension W. Since this is Q at both worlds, K1 (and K1$^+$) is true at both worlds. Indeed, one can construct (infinite) hyperintensional models that validate K2 and K2$^+$, at least, for the language fragment under consideration.[49] We can summarise our results with the following proposition:

Proposition 6.2: *K1$^+$ and K2$^+$ are consistent with Comp and higher-order free logic.*

But it is instructive to see that even in this hyperintensional setting, K3 cannot be salvaged by free logical resources without giving up the assumptions made at the outset. Assume as before that Fire, being a fundamental proposition, exists necessarily, and that it is necessarily equivalent to its being the case that everything I'm thinking is false. K3 requires that it be possible that Fire is thought and that anything else thought has the same truth value as Fire. But whether or not Fire is identical to the proposition that everything that's thought is false, this requirement of K3 is paradoxical. For if Fire is true at the world it is thought, then since Fire is sufficient for it being the case that everything thought is false, then Fire is also false at the world it is thought. And if Fire is false at the world it is thought, and everything thought at that world shares the same truth value as Fire, then everything thought at that world is false, in which case, Fire cannot after all be modally equivalent to it being the case that everything thought is false. So in the current setting, we have to concede that Fire is a counterexample to K3 and cannot be thought without thinking a proposition with a different truth value. More formally, we've sketched an argument for the following proposition:

Proposition 6.3: *K3 is inconsistent with $\Box\exists p(\Box\exists q(q \equiv p) \wedge \Box(p \leftrightarrow \lambda))$ and higher-order free logic.*

Note that our assumption strengthens Comp by assuming that the necessarily equivalent proposition to λ exists necessarily, which is in line with the thought that fundamental propositions like Fire exist of necessity.

With this concession made, the benefits of free logic seem fairly scant. There are a few ways to resist this kind of argument against K3, but all seem a little desperate. One is to deny that arbitrary Boolean combinations of fundamental propositions preserve existence: that is to say, a conjunction, disjunction or negation of earth, fire and water facts might not exist. It could be that every truth is fixed by the distribution of truth values over existing fundamental propositions, but if conjunctions and disjunctions of these propositions needn't exist, it would be unreasonable to suppose that each truth is modally equivalent to any particular existing proposition. Another move would be to deny that the fundamental propositions exist necessarily. For example, supposing Fire exists and is necessarily equivalent to it being the case that everything thought is false. Perhaps at worlds where I think that everything I think is false and nothing else, Fire doesn't exist.

6.3. Paradox without universal instantiation

Free logic provides a way of keeping K1–K3, but K1–K3 are just instances of a wider pattern of intuitive judgements, many of which cannot be saved by the free logical manoeuvres we've been exploring. For example, while K1 guarantees that it's possible to think λ and only thing materially equivalent to λ, one might think it is similarly possible to think *all and only* the propositions that are materially equivalent to λ. We could formalise this idea as follows:

$$\Diamond \forall p(Qp \leftrightarrow (p \leftrightarrow \lambda)) \tag{K4}$$

Of course, if many propositions exist, this requires that one think a lot of propositions since K4 requires that at the witnessing world, every proposition or its negation is thought. But it's tempting to think that K4 represents a metaphysical possibility (and this would be particularly plausible if one adopts a lax interpretation of thinking in which it is closed under logical consequences.) However, K4 is inconsistent with the assumption that a truth and a falsehood exist – which would be guaranteed if we accepted the minimal principles CL$\top$ and CL$\bot$.[50] Notice moreover that we can make similar trouble for the weaker hypothesis that one thinks some proposition materially equivalent to λ and no proposition that is not materially equivalent to λ:

$$\Diamond(\exists p(Qp \wedge (p \leftrightarrow \lambda)) \wedge \forall p(Qp \rightarrow (p \leftrightarrow \lambda)) \tag{K5}$$

If K5 is true, it is because it is possible to think an (existing) truth and only truths or an (existing) falsehood and only falsehoods. To think an existing truth would mean that it's false that everything thought is false, i.e. λ is false, in which case, one wouldn't have succeeded in thinking only existing propositions materially equivalent to λ. Meanwhile, if one thought only existing falsehoods, λ would be true, and we would not have succeeded in thinking an existing proposition materially equivalent to λ (i.e. an existing truth). This informal argument can be turned into a rigorous proof in propositionally quantified free logic. A closely parallel argument can be run against K4 and the additional assumption of the existence of a truth and a falsehood (as guaranteed, if we accept CL$\top$ and CL$\bot$).

A related observation is that assuming the existence of a truth and a falsehood, there will be interpretations of *Q* for which a version of Prior's theorem is still provable. Let's interpret *Q* to mean 'think*', where one thinks* a proposition so long as one thinks a proposition materially equivalent to it. Then, one can prove that if one thinks* that everything one thinks* is false, then one thinks* something true and thinks* something false.

It is worth taking a closer look at the Russellian treatment of these paradoxes. Supposing we read the quantifier $\forall p$ as the Russellian's zero-level quantifier $\forall_0 p$, so that the quantifiers appearing in λ and the wider scope quantifiers appearing in K4 and K5 are both of zero level. As it stands, neither principle violates the vicious circle principle: even though λ may express a first-level proposition, it's still true that there are zero-level propositions materially equivalent to it, and thus that one can divide the zero-level propositions into those materially equivalent to it and those not materially equivalent to it.[51] Given these observations, K4 and K5 – interpreted in terms of zero-level quantification – should look just as appealing to the ramifier as anyone else; yet, they are plainly inconsistent.

Notice that a proponent of ramification might attempt to retreat to a weaker form of K4 and K5 in which the outer quantifiers range over the same level as the level of the proposition λ expresses (which will have to be higher than zero because λ contains zero-level propositional quantifiers). Thus, for example, we might weaken K4 to the following principle:

$$\Diamond\forall_1 p(Qp \leftrightarrow (p \leftrightarrow \forall_0 p(Qp \rightarrow \neg p))) \qquad (\text{K4}^-)$$

Unfortunately, this principle is also problematic. Assume, as customary, that every zero-level proposition is a first-level proposition (that is, the levels are 'cumulative'), and assume also that there is at least one zero-level truth, and at least one zero-level falsehood. Consider the world that witnesses the truth of K4$^-$. Either I'm thinking a zero-level truth or I'm not at this world. If I am, then $\forall_0 p(Qp \rightarrow \neg p)$ is not the case; thus, if I'm only thinking first-level propositions materially equivalent to $\forall_0 p(Qp \rightarrow \neg p)$, I'm only thinking first-level falsehoods. But by assumption, I'm thinking a zero-level truth, and thus a first-level truth since every zero-level proposition is a first-level proposition. On the other hand, suppose that I'm not thinking any zero-level truths. This gives us the truth of $\forall_0 p(Qp \rightarrow \neg p)$, which means, since I'm thinking all and only the first-level propositions materially equivalent to $\forall_0 p(Qp \rightarrow \neg p)$, that I'm thinking every first-level truth, including the zero-level truth, contradicting our assumption.

Given all of these observations, one might wonder just how much we have really achieved by moving to higher-order free logic. The proponent must find a principled background picture that motivates K1 (say) without also motivating the outright inconsistent principles K4 and K5. Here is one way you might attempt to flesh out that picture: one could argue that thinking λ (and nothing else) is an easy matter. All one needs to do to achieve that is to mentally token something corresponding to the sentence 'everything I'm thinking is false', without tokening anything else at the same time. By contrast, it is not so easy to ensure the existence of a proposition by performing these mental tokenings – the existence of propositions is beyond your mental powers. This picture naturally motivates K1 (and indeed K2 if we generalise to other sentences), but note that it does not motivate K5 since K5 requires it to be possible for you to think something existing materially equivalent to λ – tokening λ does not guarantee the existence of anything. Similarly, it's unclear what you could mentally token in order to ensure the truth of K4: since it is beyond your control which propositions exist, merely tokening a number of sentences does nothing to guarantee that you think all and only the existing propositions materially equivalent to λ.

Although one can in principle outline a picture in which K1 and K2 are motivated without also motivating K4 and K5, we feel that once one has stopped taking these kinds of compelling judgements at face value, one wonders whether we shouldn't just accept the original consequences of UI for K1 and K2.

6.4. *Expressive limitations*

Given UI, one can in a higher-order framework make extremely general claims by recourse to higher-order quantification. For example, if we think that knowledge is factive, one can capture this idea quite straightforwardly with the following sentence:

$$\forall p(Kp \rightarrow p)$$

Similarly, one can state that all conjunctions have true conjuncts in a way that is not limited to the conjunctions expressable in any particular language:

$$\forall p \forall q(p \wedge q \rightarrow p)$$

Now even when we relinquish UI, we may write down these things, and remain confident in them. But in a free logical setting, they may feel insufficiently general. For example, supposing there is no such proposition as λ, we may feel the urge to write down generalisations for which the following are special cases:

$$K\lambda \rightarrow \lambda$$

$$\lambda \wedge \forall pp \rightarrow \lambda$$

However, the quantified sentences specified above are unfit for this purpose because they do not entail these instances.

A natural thing to try to go for is a schema rather than a quantified claim. For example, the factivity of knowledge could be captured by the following schema:

$$KA \rightarrow A$$

An initial concern with this strategy is that it won't satisfy our desire to write down a general truth since schemas are neither true nor false (even though they can have true or false instances).[52] Note that this cannot be readily circumvented by instead writing down a metalinguistic generalisation to the effect that all uniform interpretations of the schematic letters deliver truths since this reintroduces propositional quantification and hence may not have the requisite generality.

In order to avoid quantification over interpretations of sentence letters, one might opt for a metalinguistic generalisation that is substitutional in nature, along the following lines: every uniform substitution of a schematic sentence letter by a meaningful sentence in the language yields a truth. But this would lose generality for a different reason since, for example, the generalisation about factivity is supposed to cover cases of knowledge that aren't expressible in our language. A similar concern would attach to any effort to express the required generalisations using infinitary conjunctions or disjunctions of sentences in our finitary language. Note that arguably no infinite conjunction can contain all sentences of the infinitary language as conjuncts, for then it would contain itself as a conjunct and thus be non-well founded.

Thirdly, quite aside from the desire to express bona fide truths, the schematic approach is particularly problematic in certain embedded environments. Take the generalisation that in the classical setting, we would happily express with the following formula:

$$\Diamond \forall p Q p$$

That is, it is possible that every proposition is Q. Now suppose we felt this inadequate because we wished to convey the thought that it's possible to think, alongside all the existing propositions, λ, and $\neg\lambda$ and so on, whether or not the latter corresponds to existing propositions. In this connection, the following schema will not be of any use:

$$\Diamond QA$$

After all, that schema does not require that λ and $\neg\lambda$ be thought together in any single world. This point generalises to occurrences of universal quantifiers in embedded environments where, in the classical setting, the embedded universal quantifier wouldn't commute with the embedding.

One might hope to bypass a number of these concerns by appeal to a device alluded to earlier. One might try to express some of these generalisations using ellipsis in the following kind of way:

> If you know that snow is white then snow is white and if you know that grass is green then grass is green and ...

Note that for this to do the requisite work the '...' cannot merely be a stand in for an infinitary conjunction of sentences in the finitary language, for this lacks the generality we were seeking. The ellipsis must, for example, be understood at least to cover all possible extensions of the language. So let us assume that the object language has been extended to include exactly this kind of ellipsis.

There are a number of pressing concerns here. One is that the more careful one is about the syntax of this device, the more it begins to resemble a quantifier. For example, one will have to introduce a variety of '...'s corresponding to the need for different propositional variables in the original setting. But we'll focus on a particular dilemma. Are '...' thoughts to be understood so as to be under the scope of '...' generalisations? For example, is the claim that if I know that grass in green, then grass is green and ... to include as a special case that if I know that [if I know that grass is green then grass is green and ...], then [if I know that grass is green then grass is green and ...]? If the ellipsis is indeed a legitimate extension of the language, and '...' claims are supposed to cover all legitimate extensions of the language, then '...' claims had better be special instances of '...' claims. An initial worry is that this raises the spectre of non-well foundedness. But more importantly, the '...' claim seems now to behave enough like a classical quantifier as to allow one to reinstate the Priorian paradoxes. Consider K1$^{\cdots}$:

K1$^{\cdots}$ $\Diamond(Q((Q\top \rightarrow \neg\top) \wedge \ldots) \wedge (Q\top \rightarrow (\top \leftrightarrow (Q\top \rightarrow \neg\top) \wedge \ldots)) \wedge \ldots)$

Once one has gotten the hang of ellipsis, K1$^{\cdots}$ should look just as compelling as the original K1. Note that from $\phi(\top) \wedge \phi(\bot) \wedge \ldots$ one can conclude $\phi(\psi)$ for any ψ in our language (whether ψ includes ellipsis or not). This principle is exactly analogous to the principle of universal instantiation that caused problems for K1 in the first place.

6.5. First-order free logic

We've have in effect been developing pessimistic pictures of the prospects for a higher-order free logic that's motivated by the prospect of keeping principles like K1–K3. It's less clear whether the materials of this section can be developed into a case against higher-order free logical approaches that are motivated by other concerns (for example, by the desire to be a contingentist about the existence of properties). Moreover, it's even less clear whether these considerations carry over to the first-order setting. For example, concerns about closure under Boolean operations have no analogue in the domain of individuals nor do concerns about supervenience. Of course expressivity worries, and in particular the challenge of finding suitable ways of formulating perfectly general principles, are likely to be a common theme among those who oppose free logical approaches in general.

We are not going to systematically explore the resources of various free logicians for enhancing expressive power. But we will mention one theme that makes things particularly difficult for one who promotes free logic as a means for saving K1–K3. Suppose, for example, one is a first-order free logician. One might still be attracted to the schema $\Diamond\exists x(x \equiv t)$, for that allows one to use a modalised quantifier to formulate perfectly general claims, for which singular terms are instances. Similarly, one might pursue such a strategy for a higher-order free logic. However, such a strategy is dangerous, to say the least, if ones motivation for higher-order free logic is the preservation of K1–K3: $\Diamond\exists p(p \equiv \phi)$. This is because modalised quantifiers, once accepted, may well come to form the basis for revenge-like versions of Prior's paradox.

7. Kripke, Kaplan and ramification

Let us now compare the free logical approach to the Prior–Kaplan paradox to another prominent family of approaches, based on the Russellian, as opposed to the Fregean, tradition in higher-order logic. As noted at the outset, both approaches raise complications for the principle of universal instantiation. The Russellian tradition deploys the machinery of ramification. In Russell's work, this ramification is relatively thorough going: every relational type over arguments of type $\tau_1, \ldots, \tau_n$ is stratified into an infinite hierarchy of levels indexed by natural numbers, with exceptions for certain logical vocabulary.[53] Complex grammatical expressions are built up in such a way that no non-logical grammatical expression applies to an expression of an equal or higher level – in other words, the hierarchy of levels is *cumulative*: so, for example, an operator with ramified type $(()/4)/5$ can apply to propositions of types $()/i$ for $i \leq 4$. It should be noted that Kaplan himself is working squarely within this tradition.

Of course one might explore more relaxed ramified approaches. We have already noted that some expressions, namely the logical expressions, are not subject to stratification. In what follows, we'll explore a range of less stringent stratificational approaches to the Prior–Kaplan paradoxes, by way of attempting to isolate the benefits of particular stratificational manoeuvres.

There are some important choice points facing the ramifier. One concerns the effects of level mismatch. Suppose, for example, a predicate taking arguments of level two – e.g. a predicate of ramified type $(e/2)/3$ – is combined with a name of level five – i.e. $e/5$. Is the resulting string simply ungrammatical, with no truth value, or is it a grammatical sentence that is simply false (and whose negation is true)? Russell (and Kaplan) takes the former route, but one might explore the falsity route as well. For example, many proponents of the idea of indefinite extensibility argue that there is a hierarchy of domains for the first-order quantifiers, which we might label $e/0, e/1$ and so on. If this kind of theorist is not also stratifying propositions – something we will consider shortly – then relational types need not be stratified (so that if $\tau_1, \ldots, \tau_n$ are types $(\tau_1, \ldots, \tau_n)$ is a type). In one context, we can predicate a property F, of type $(e/0)$, of every member of the initial domain, $e/0$. But from another context, one can quantify more expansively – over $e/1$ individuals, say – and assert that some $e/1$ individuals don't have F. This would require applying F of type $(e/0)$ to a variable ranging over $e/1$ entities and thus the grammaticality (albeit falsity) of a level mismatch. There are also proponents of indefinite extensibility concerning quantification into sentence position, introduced so as to deal with the liar paradox (see Glanzberg 2001; Parsons 1974). If you are stratifying both individual and propositional types, then both the individual and relational types must have levels, and we are in a version of the Russellian framework modulo this interpretation of level mismatch.[54]

It's perhaps worth noting that the ungrammaticalist might try to retain a facsimile of universal instantiation, for they claim that all of its grammatical instances are valid, although this defensive strategy is less promising than it might initially seem (we'll return to this point later).[55]

A second choice point concerns the extent to which we ramify. As noted by Russell, and Kaplan after him, we could ramify all relational types (with the exception of a small list of logical operators). Particularly relevant in the language fragment we outlined in Section 2 is that both the operators Q and $\Box$ will be ramified into an infinite family of operators with different ramified types. The quantifier phrase $\forall p$ will similarly come in a family of different levels depending on the ramified type of the variable p.[56] But less thorough going ramification could also be entertained. One might ramify quantifiers for reasons having to do with the vicious circle principle, without thereby feeling any particular pressure to ramify attitudes or modal operators. The vicious circle principle, as Russell articulates it, requires that 'whatever contains an apparent variable must not be a possible value of that variable. Thus whatever contains an apparent variable must be of a different type from the possible values of that variable; we will say that it is of a *higher* type'. Whatever Russell's own view, one might think that attitudinal operators do not contain 'apparent variables', and therefore will not require ramification.[57]

To pursue this sort of approach, we would simply treat Q and $\Box$ as syncategorematic, in the same way that we treat $\neg$. One important upshot of this relaxed stratificational approach is that we can iterate these operators so that, for example, the formula $\Box\Box(p \rightarrow p)$ is well-formed (and, indeed, true). Since there are now no expressions in our language for which level mismatch can arise, we effectively have a classical language that just contains an infinite number of distinct quantifier phrases $\forall p_0, \forall p_1, \forall p_2, \ldots$.

To illuminate each approach, we'll also see how each applies to some puzzles related to one discussed in Kripke (2011) (and for which Kripke also entertains a stratificational solution). Kripke imagines that at a certain time t, someone undertakes to think uniquely about the set of times, t', at which there is some set of times being thought about at t' that doesn't contain t'. On the one hand, it seems as though that person is uniquely thinking about this set at t. But, on the other there, seems to be a logical difficulty attaching to such a supposition, for there is now no coherent answer to the question whether the set thought about contains t.

To begin, it's worth noting that there are some uninteresting puzzles in the vicinity for which an appeal to ramification seems excessive. Suppose at noon on each day of 16 days I stand on a different square of a large square chessboard. On each day, I'm given the opportunity to touch some or all of the squares on a chess board. Now there seems to be a logical guarantee that on none of the 16 days will I touch all and only those squares which have the property of being not among the squares I touch when I'm standing on it. Nothing is particularly puzzling here – it turns on such mundane facts such as the fact that I can't touch on Tuesday the squares I am not touching on Tuesday – and certainly nothing that calls for anything as high powered as ramification.

There are some observations about thinking that are equally mundane. Take, for example, the wide scope construal of the sentence 'he's thinking about the smallest number that he's never thought about' viz 'there's a unique number, *n*, such that *n* is the smallest number that he's never thought about, and he's thinking about *n*'. It is unsurprising that there's a logical guarantee that he is not thinking this number, and it has nothing to do with thinking – try kicking the smallest policeman you'll never kick!

But let's turn to a more puzzling version of the chessboard puzzle. Let's introduce a relation, touching*, where you touch* a square by touching a predicate that expresses a property possessed by the square. In the new protocol, you have a book of predicates, and you touch one or more predicates from the book whenever you stand on a square. Now suppose on a given day you touch just one predicate which expresses the property of being a square that you don't touch* on that day (suppose it's Tuesday and the book contains the predicate 'being a square that you don't touch* on Tuesday'). For each square, we can derive a contradiction from the hypothesis that you don't touch* it, for then you would satisfy a sufficient condition for touching* it (namely touching a predicate that expresses a property instantiated by the square). And we can also derive a contradiction from the assumption that you do touch* it, and the assumption that you've touched a predicate that doesn't express any other properties, for then you would fail a necessary condition for touching* it. It would be very unsatisfying to conclude from these logical reflections that in scenarios such as this, there would be a force field surrounding the relevant predicate token in the book, so that if you were to try to touch it, you'd be repelled, or suffer other similar physical mishap. Similarly, it would be unsatisfying to conclude, with respect to the puzzles about thinking, that there are force fields that block the appearance of certain mentalese sentences in one's belief box, or in some other cognitive module.

Let's contrast two very different approaches roughly corresponding to differing ways of ramifying gestured at above. On one approach, vicious circle considerations prohibit such properties as the property of being a square that does not instantiate any property expressed by a touched predicate, with the quantifier naïvely construed as ranging unrestrictedly over all properties. For such a property would then quantify over a domain that includes itself. One might thus only tolerate properties that quantify over a limited domain of properties that do not contain themselves. For example, there will be properties like being a square that does not instantiate any colour property expressed by a touched predicate, where that property is not itself a colour property. Supposing one touches a predicate expressing that property and nothing else, there is no puzzle about what to say: in touching that predicate, one thereby touches* all the squares.[58] Note that there is no need here to ramify the relation of expressing; all the work is done by applying vicious circle considerations to quantification over properties.

There is a very different approach that may be considered. To make it vivid, let's unpack the predicate 'being a square that isn't touched* on Tuesday' as follows: 'being a square that doesn't instantiate a property expressed by a predicate that is touched on Tuesday'. Supposing one touches that predicate on Tuesday. For a paradox to arise, we need, among other things, to suppose that 'express' expresses the expressing relation. So one strategy for blocking the puzzle is to deny that in that context the predicate 'expressing' as it appears in the book expresses the expressing relation. Of course, we can allow that it expresses something – perhaps there is another relation, expressing* that 'express' in that context expresses. Suppose that in the book, the predicate 'being a square not possessing any property expressed by a property that Lucifer touches' occurs in the book. Supposing Lucifer touches nothing, there is no immediate paradox in supposing that 'expresses' as it occurs in that predicate expresses the expressing relation, but we may be attracted to a uniformity assumption according to which at a context 'expresses' is uniform with regard to which relation it expresses, and is not at the semantic mercy within a context of its linguistic surroundings. This kind of consideration may well encourage one to think that no token of 'that token of "expresses" expresses expressing' is true.[59]

Of course, these anti-disquotational thoughts are one thing, but the claim there is a family of expressing-like relations that form a hierarchy, of the kind envisaged by typical ramifiers, is a further commitment. Justifying this will require *inter alia* blocking loops (so that, for example, we don't allow that 'expresses' expresses expressing* and 'expresses' expresses* expressing), and infinite descending sequences, thereby securing well-foundedness.[60] Notice that even if a well-ordered hierarchy of expressing-like relations were justified that would still not yet secure some of the standard prohibitions voiced by the ramifiers. For example, a typical ramifier for 'expressing' will contend that whenever two tokens of 'expresses' express the same relation and one occurs within the scope of another, then the result is either false or meaningless. Even given the hierarchy, and even given that no token of 'expressing' ever stands in some relation from the hierarchy to itself, this further commitment is far from secure. Suppose, for example, that one had a coarse-grained conception of propositions. Then, since '2 + 2 = 4' expresses the proposition that 2 + 2 = 4, and since the proposition that 2 + 2 = 4 just is the proposition that 'dog' expresses something or it doesn't (i.e. the one and only necessary proposition), then '2 + 2 = 4' expresses the proposition that 'dog' expresses something or it doesn't. This would then give us a counterexample to the aforementioned ramificationist dictum, one that doesn't require us thinking that 'expresses' ever expresses expressing. In particular, it may be that there is some relation other than expressing, namely expressing*, that is expressed by each occurrence of 'expresses' in the previous example. For these reasons, the postulated hierarchy does not yield any straightforward syntactic constraints connecting embeddings or relative scope to the possibility of truth or meaningfulness (indeed, if one takes a coarse-grained approach, there could not be any such constraints). Note that this hierarchical vision does not involve any requirement that property quantification be somehow restricted or stratified.

Kaplan has suggested that once one has embraced ramification, one should be unhappy with the fundamental structure of possible world semantics. While this is not the place to pursue the matter at length, it's worth noting that this sort of stratification just described does not indicate a fundamental difficulty for possible world semantics. As we see things, Kaplan's concern with naïve possible world semantics is that it assumes the domain of propositions can be given 'all at once' – after all, possible world semantics takes propositions to be identical to or determined by sets taken from an antecedently given domain of worlds.[61] And it is certainly true that once one ramifies propositional quantification, this aspect of possible world semantics is called into question. However, if one leaves quantification alone but instead appeals to a hierarchy of expressing relations, or thinking relations, then no difficulty will arise from the idea that one can propositionally quantify over all propositions at once. So from this second perspective, possible worlds semantics may survive completely unscathed.[62]

7.1. Ramification and Prior–Kaplan puzzles

We've distinguished some varieties of ramification, but how do these varieties interact with K1–K3? Let's examine in turn first the strategy of ramifying the attitudinal operators and then the strategy of quantificational ramification. We'll then examine whether there are special advantages to ramifying along both fronts when it comes to the Prior–Kaplan puzzles.

Suppose first one leaves higher-order quantification unramified, but does ramify operators like 'it is thought that'.[63] We'll focus here on the standard kind of ramifier that prohibits the truth of any sentence that has an attitudinal operator appear within either its own scope, or the scope of an operator of lower type (as noted, there is a further divide among those who think that such occurrences are either false or ungrammatical).[64] Obviously, this framework won't validate K1 on any uniform interpretation of '*Q*', for *Q* appears within its own scope in K1 (and so it is either false or ungrammatical). Generally, when one writes down a schematic principle, one intends the schematic variables to be interpreted uniformly. As applied to K1, it seems as though we shouldn't have been writing it down after all. That said, the present kind of ramifier can offer a diagnosis of why claims like K1 seem attractive. It's natural for the proponent of this view to contend that when, for example, in ordinary thought and talk we say things like 'I thought Bob thought it was Tuesday', we don't automatically say something false or incoherent. Rather, without properly keeping track of what we're doing, the two occurrences of 'thought' correspond to two things at the level of logical form. From this perspective, there's a kind of logical form 'blindness' that ordinary people are susceptible to, and which could explain why they would mistake K1 for a family of well-typed principles that could be true. Indeed, it's easy to find models in which one can uniquely think$_2$ the proposition that everything I think$_1$ is false.[65]

This style of explanation can't be easily extended to the puzzles associated with K3, or the variant Kaplan discusses, $K3^=$: $\forall p \Diamond (Qp \wedge \forall q(Qq \rightarrow q = p))$. Both principles are well-typed by the lights of the ramifier. Moreover, Kaplan's original considerations telling against $K3^=$ seem to be just as compelling in this setting. There are still many more propositions than worlds, so not every proposition can be uniquely thought$_1$ – the fact that there will be a whole hierarchy of additional unsatisfiable principles is of little comfort. As for K3, one cannot run Prior's original argument because certain steps in the proof violate the typing constraints. However, Prior's original argument can be adapted so as to respect those typing constraints, e.g. by saying things like 'there's something identical to the proposition that everything I'm thinking$_1$ is false, and I could have thought$_1$ it and only propositions materially equivalent to it'. One could block this trick by taking seriously the idea that the quantifiers are typed, and thus that the variable flanking the right-hand side of the identity connective cannot appear under the scope of 'thinks$_1$' since it must be of the same level as the sentence on the left-hand side ('everything I'm thinking$_1$ is false' which is of level 2). But this would no longer be the selective kind of ramification under consideration.[66]

What about K2? A natural thing for the ramifier to try is to say that K2 is true for any well-typed instance.[67] A modification of the model construction in proposition 4.7 from Section 4.3 serves to show this is consistent.[68] On this way of understanding K2, K1 isn't an instance of it since it violates the typing constraints. In this setting, the supervenience problems discussed in Section 6.2 are still relevant: supposing that thinking facts supervene on earth, fire, wind and water facts, then for all we know, there's an earth, fire, wind and water sentence that expresses a proposition that's necessary and sufficient for thinking that everything I'm thinking is false (of course, supervenience doesn't guarantee that there is such a sentence). The instance of K2 with this sentence would not violate typing constraints, but as we saw in Section 6.2, this instance would be false. That said, one might, taking a leaf from Kaplan book, have the more modest ambition of saving the logical coherence of K2, rather than the truth, and in particular, its compatibility with various plausible metaphysical hypotheses.[69]

Let us turn to the ramifier who selectively ramifies quantification, but does not ramify the attitudinal operators. There are issues of detail concerning whether it is the quantifiers, variables or lambda operator that are to be ramified here; we won't fuss about this, but for simplicity, we will work with a picture in which it is the variables that get ramified and that determiners like 'all' are no more typed than logical operations such as 'not'. (Moreover, while the abstract idea of typing quantifiers does not force any kind of cumulative hierarchy, we'll imagine, with the standard ramifier, that the domains of these quantifiers are linearly ordered by inclusion.)[70]

By contrast with the last strategy discussed, this kind of ramifier would with Kaplan have legitimate concerns about possible world semantics as it is standardly understood. Possible world semantics encourages one to think that from within one's language can quantify 'all at once' over a complete and unextendable domain of propositions. But for this kind of ramifier, this is an illusion: each quantifier ranges over a domain of propositions, each of which may have maximally strong elements. For each level of quantifier, these maximally strong elements become stronger and stronger with no end – while possible worlds may be the limit of these elements, there is no way of simulating quantification over these limits.[71]

What happens to K1–K3 on this interpretation? As noted in Section 1, if the different levels of quantification are not to collapse into one another, then each quantifier must abide by a free logic (this is due to Harris's theorem).[72] What bears emphasis is the tight connection between the ramified quantifier picture and the discussion of free logic in the earlier sections of this paper. For note that on any interpretation of the quantifiers, universal instantiation – and coordinately existential generalisation – will fail in just the kind of way discussed in those sections. Thus, if I'm saying that everything_1 I'm saying is false, there's nothing_1 (although something_2) that I'm saying; and similar kinds of results can be produced for the quantifier 'something_2' and so on. So each quantifier is governed by a positive free logic. Note that in those sections, as here, we are not typing 'thinks' in such a way that prohibits it appearing within its own scope.

It is worth noting that in Section 4.3, we provide a model that explicitly accommodates a hierarchy of quantifiers of the sort envisaged here (proposition 4.9). And in that context, we provided a model which satisfies K2 (on any interpretation of its quantifiers), and even the stronger $K2^+$. Note that these schemata now include instances that involve the new typed quantifiers that we have introduced to the language in this environment. It thus follows that for any interpretation of the quantifiers in K1 that version of K1 will be an instance of our schema K2 (and a similar relation holds between $K1^+$ and $K2^+$). As for K3 and $K3^+$, there will be a different version of the principle for each quantifier, which is also validated in the model described in the proof of proposition 4.9.

Given these results, it is clear enough that as far as Prior–Kaplan puzzles are concerned, there is no immediate benefit to ramifying both the quantifiers and the attitudes. Indeed, the dual ramification strategy would seem to have significant costs in this setting, *viz* K1–K3, after all if one merely ramifies quantifiers, one can easily write down K1–K3 allowing that it is governed by the standard convention that 'thinks' has a uniform interpretation. But once one has ramified the attitudes, an endorsement of K1 appears much more problematic. Of course there may be other philosophical motivations for a more thorough going ramification. However, considerations pertaining to the Prior–Kaplan paradoxes seem very weak in this regard.

8. Concluding thoughts

Williamson is operating within a Fregean higher-order logic that combines universal instantiation with the thesis of higher-order necessitism (articulated using higher-order correlates of the Barcan formulae). Indeed, universal instantiation, and especially one of its consequences, comprehension, are a key argumentative tool in defending necessitism. In that setting, two strands of his thinking come together very naturally. First that existence doesn't modally come and go and second, that existence comes cheap. In this paper, we have been concerned to explore frameworks in which there is no such natural alliance. On the one hand, we have looked at Fregean frameworks which reject universal instantiation. Such frameworks will inevitably reject the idea that existence comes 'cheap'. They may, however, accept necessitism, but not on the argumentative basis that Williamson offers.

One might wonder how these distinctions look in a Russellian framework. In a setting of thorough going ramification, universal instantiation isn't even an option. We mentioned earlier that the Russellian might hope for a good facsimile of universal instantiation, one according to which a universally quantified claim entails all of its grammatical instances. Thus, for example, if it's true that for all$_1$ p, God wonders$_2$ whether p, no counterexample is provided by the sentence 'God doesn't wonder$_2$ whether he wonders$_2$ whether snow is white' because this not a grammatical sentence since it violates the typing constraints.[73] Nevertheless, given the cumulative hierarchy, the proposed salvage of universal instantiation is unsuccessful.[74] After all, an nth-level expression can grammatically apply to anything that is a lower level, including levels more than one below. Suppose a third-level predicate is true of all first-level things, but not true of all second-level things: $\forall x_1 F_3 x_1 \wedge \exists y_2 \neg F_3 y_2$. Suppose that t_2 denotes a second-level object that is not F_3, so that $\neg F_3 t_2$. Then, $F_3 t_2$ is grammatical, but one cannot infer it from $\forall x_1 F_3 x_1$; thus, universal instantiation even construed in the way just described fails. In short, universal instantiation is a bad idea for the Russellian, even if one's ramification is maximally thorough going. And since the idea that existence comes cheap is tightly allied to existential generalisation, a companion principle to universal instantiation, the Russellian necessitist cannot be motivated by that idea, nor by the logical principles that subserve it.

We've been exploring frameworks, Russellian and Fregean, that unlike Williamson disavow universal instantiation, but which like Williamson may still endorse necessitist principles for each quantifier. To give things focus, we've been considering a particular cluster of motivations for rejecting universal instantiation, namely: the desire to maintain the truth, or at least the logical coherence, of principles like K1–K3. Our exploration has however been rather pessimistic – although we hope instructive – and our own inclination is to take Prior's result at face value, and to reject principles like K1–K3 on that basis. There still remains the question of how to implement that discovery. Suppose on the heels of Prior's principle we grudgingly accept that it could never be that Tim Williamson seems to be saying that everything Tim Williamson seems to be seems to be saying is false, without there being something else Tim Williamson seems to be saying. It would nevertheless be unsatisfactory to suppose that if Tim Williamson at the point of death had never spoken, and was on the verge of seeming to be saying that everything he's seemed to say is false, then logic would somehow clamp his larynx, or else logic would somehow prolong his life until he seems to say something else, or render his audience deaf and so on. In this connection, certain of the other puzzles discussed in this paper are suggestive, and may force us to grudgingly accept at least some aspects of the ramified vision. In particular, we may be forced to accept a family of thinking relations, that may take on at least some of the structural features of the Russellian hierarchy, and we may offer a diagnosis of our initial confusion in terms of our blindness to the proliferation of such relations. That said, it's a substantial further step to impose syntactic constraints typical of the standard ramifier. And even more importantly, it remains far from clear whether there is important philosophical work done by ramifying the quantifiers themselves – existence may yet come cheap.

Notes

1. See Kaplan (1995), Kripke (2011), Ramsey (1960), Russell (1908) and Tucker and Thomason (2011).
2. Note that there are many different ways of implementing the ramificationist approach (it's very natural to only allow types $(\tau_1, \ldots, \tau_n)/k$ where k is greater than the levels of $\tau_1, \ldots, \tau_n$). In what follows, we shall be following Church (1976) in considering only cumulative ramified type theories: informally, a relation that accepts arguments of a given level, accepts arguments of any lower level. For the connection between this and the various versions of Russell's original theory, see footnote 2 of Church (1976).
3. We assume that applying the logical connectives does not take us to a higher level.

4. Provided B contains no free occurrences of p, and no free variables that would become bound if substituted for p in A, in what follows, we will take this condition to be tacitly assumed.
5. Without loss of generality, we give the proof of one direction (up to a change of variables):
   (1) $(\forall pA \rightarrow A(q/p))$ (instance of UI for $\forall$.)
   (2) $\Pi q(\forall pA \rightarrow A(q/p))$ (by universal generalisation for Π.)
   (3) $\forall pA \rightarrow \Pi qA(q/p)$ (by the principle $\Pi p(A \rightarrow B) \rightarrow (A \rightarrow \Pi pB)$ when p is not free in A.)
6. See also Sher and Tieszen (2000) for a discussion of the second-order variant.
7. It's arguable that the rejection of UI is also the key part to a Russell-style treatment of many other paradoxes, but we shall not defend the broader claim here.
8. Suppose that $p \equiv p$. Then, by EG, we can infer $\exists q p \equiv q$. We then have $\Box\exists q\, p \equiv q$ by necessitation, $\forall p\Box\exists q\, p \equiv q$ by universal generalisation and finally $\Box\forall p\Box\exists q\, p \equiv q$ by necessitation again.
9. Whereas Prior's original discussion is formulated in Polish notation and used the symbol d for the relevant unary operator, Kaplan used more standard notation and used the symbol Q. For the sake of typographical uniformity, we shall use Q throughout. Had Prior not been using Polish notaton, we suspect his paper would have entered the philosophical canon much earlier.
10. A complete axiomatisation of the logic of quantification into sentence position would include several other axioms governing the propositional quantifiers; however, this argument only requires UI, at least given certain non-trivial decisions about which truth functional connectives to take as primitive.
11. We say 'roughly' because we are using the English 'if then' to interpret the material conditional, we are using singular quantification instead of quantification that takes sentence position and finally, we are employing truth and falsity predicates that make no appearance in Prior's actual theorem.
12. Obviously this isn't the only way to reconcile Prior's theorem with the apparent data. One could also insist that despite appearances, Kaplan wrote two or more things at midnight (see, e.g. Bradwardine 2010; Dorr 2012; Slater 1986). Or one might insist that while Kaplan didn't say what he seemed to say, he succeeded in saying something else (Bacon 2014; Smith 2006). The challenge for these kinds of responses is to offer some guidance as to what the additional or alternative contents are.
13. A similar worry arises for the alternative and multiple proposition views. The alternative proposition view has to deny what seems obvious, namely: the antecedent of four, while the multiple proposition view must deny something else that seems obvious, namely: the negation of the consequent of four.
14. We should mention in passing another kind of paradox where denying universal instantiation might provide a path to a distinctive kind of solution. Here is a principle about propositional identity that is prima facie compelling, and particularly compelling if you think of propositions in a structured way: if Q and Q' are propositional operators with different extensions, then the proposition that something is Q and the proposition that something is Q' are not identical (more formally: $(\exists p Qp \equiv \exists p Q'p) \rightarrow \forall p(Qp \leftrightarrow Q'p)$). But now a higher-order analogue of the Russell–Myhill paradox threatens to show that this is inconsistent. The derivation of this essentially involves the principle of universal instantiation into sentential position (Uzquiano 2015). It is beyond the scope of this paper to explore the promise of this response. However, we should note

that the models in this paper are generally not useful for modelling structured propositions since, while they allow modally equivalent propositions to be distinct, they treat logically equivalent propositions as identical.

15. Both the first-order free logician described above, and the higher-order free logician will therefore have to be *positive* free logicians. That is, they will reject the inference of $\exists x\phi$ from $\phi[t/x]$ even for atomic instances of ϕ.
16. One could also prove this by directly appealing to a general rule of necessitation; however, this rule is not uncontroversial in this context as it is relinquished, for example, in Kaplan's logic of demonstratives in Kaplan (1977), and the status of necessitation in Williamson's book is complicated. The above proof does not rely on necessitation – in general, most appeals to necessitation can be eliminated in favour of slightly stronger (necessitated) premises.
17. We are, of course, taking a few liberties here: for example, if the report is made after the writing, then we would use different tense morphologies.
18. Kaplan's sympathy is not directed at the truth of this sentence, but at its logical coherence.
19. It should be noted that most of these authors, while attributing the paradox to Kaplan, did not have access to the details of Kaplan's text since it had not been published.
20. Of course if you're a hyperintensionalist, there won't be any such thing as *the* proposition corresponding to a set of worlds. But if there are more propositions than collections of possible worlds, that only makes the situation worse for the principle.
21. Given universal instantiation, the latter definition satisfies schematic Leibniz's law, but without it, we cannot derive it (we can derive quantified LL either way); thus, the latter definition only seems appropriate assuming UI.
22. This is a theorem under either definition of $\equiv$, as it amounts to $\Box(p \leftrightarrow p)$ and $\forall Q(Qp \rightarrow Qp)$, respectively.
23. To be more precise, Williamson relies on a suitably necessitated version of a consequence of UI, namely: Comp. See below.
24. The proofs of these are well known (see Prior [1956]). In fact, much weaker axioms suffice for this purpose; for example, the principle $\Box(p \rightarrow \Box p) \rightarrow (\Diamond p \rightarrow p)$(see Bacon 2011).
25. Notice that in Williamson's system, there is a subtle distinction between comprehension principles for quantification into sentence position, and quantification into, say, monadic predicate position. The first is derivable from UI, but derivation of the second from UI requires additional assumptions about what counts as a predicate. In Williamson's system, the only monadic predicates are simple monadic predicates (whereas it's obviously not true that all sentences are simple), or at any rate, he doesn't explicitly license a device for converting any old open sentence into a monadic predicate (such as λ-abstraction). Absent such a device, one cannot, for example, derive $\exists F\Box\forall x(Fx \leftrightarrow (Gx \wedge Hx))$. And this is essentially because in order to use existential quantifier introduction to move from $\Box(Gx \wedge Hx \leftrightarrow Gx \wedge Hx)$ to $\exists F\Box\forall x(Fx \leftrightarrow Gx \wedge Hx)$, one would need to first convert the open sentence $Gx \wedge Hx$ into a type suitable for monadic second-order existential generalisation. Given that the focus of this paper is on propositional necessitism, these subtleties are not very relevant to the main discussion.
26. That NNE, BF and CBF do not entail UI will be demonstrated by the models we consider in Section 4. To see that Comp does not entail UI, we must consider a model in which propositions are not individuated by necessary equivalence.

For example, in the model $W = \{x,y\}$, $R = \{\langle x,x\rangle, \langle y,y\rangle\}$, $D(w) = \{\{w\},\emptyset\}$, $[\![Q]\!]_w = D(w)$. $x \in [\![\forall pQp]\!]$, $x \notin [\![Qp]\!]_{v[p/\{y\}]}$. Note that if propositions are individuated by necessary equivalence, then UI can be proven from Comp.

27. In (2013), Williamson focuses instead on abductive arguments for Comp which in turn guarantee many of the necessitist theses. There Williamson describes a sound (but incomplete) axiomatisation of higher-order modal logic. In that system, Comp is not an axiom, but a theorem derived in the above way from UI. This shift of focus from UI to Comp perhaps seems reasonable, given his simplifying assumption that higher-order entities are individuated by necessary equivalence.
28. It should be noted that it would also be inappropriate to restrict attention to universal accessibility relations – as is often done when, like us, one is dealing with models of S5 – but in the current context, that is unwise since we wish to allow models in which *Q* is hyperintensional.
29. A richer language might also include propositional letters, each of which denotes a set of points under an interpretation, but we will have no need for them here.
30. For example, one might take Frege's 'concept horse' problem to be a result of not properly acknowledging this idea; see Williamson (2003) for this and other reasons for dissatisfaction with first-order models.
31. Note that this is a model of higher-order nominalism and moreover, we don't use higher-order quantification in the articulation of the models. However, the model itself is of course not nominalistically acceptable in a broader sense because the models are constructed out of abstract objects, and so admit first-order objects that nominalists have historically rejected.
32. One could attempt to state this using substitutional quantification, but this may have limitations. Suppose the object language lacked Meinongian quantifiers, but the theorist had them. Then, the theorist could express the shortcomings by pointing out that there may be fictional things in the broad sense for which there were no names, and that imagining one of those alongside Pegasus would a counterexample to the intended meaning of the elliptical claim. Exactly the same point applies if one attempts to use substitutional quantification to capture the claim 'I'm thinking that snow is white, but I'm not thinking that grass is green or ...'.
33. See McGee (2000). Roughly, a schema is understood open endedly if it holds under any extension of the language.
34. To satisfy $K2^+$, then there must be two distinct worlds that make $Q^*\top$ and $Q^*\bot$ true, respectively. $Q^*\top$ and $Q^*\bot$ are therefore both contingent and different. $\top, \bot, Q^*\top$ and $Q^*\bot$ require four different worlds as witness to the relevant instances of $K2^+$, but there are only three worlds to go around.
35. We know that such a bijection exists. Consider, for example, the bijection that maps each finite set *X* to $2.\prod_n p_n^{x_n}$ where p_n is the *n*th prime and x_{n+1} is the *n*th largest member of *X* and x_0 is the size of *X*. And we map $\mathbb{N} \setminus X$ for a finite *X* to $2\prod_n p_n^{x_n} + 1$.
36. Given the bijection defined in footnote 35, $|Q|(0) = \{\emptyset\}$ and $|Q|(1) = \{W\}$.
37. We shall show that $[\![\phi]\!]_v$ expresses a finite or cofinite set whenever *v* assigns only finite or cofinite sets to the propositional variables (thus, every closed sentence expresses a finite or cofinite set relative to *any* assignment). Suppose $[\![\phi]\!]_v$ and $[\![\psi]\!]_v$ are finite or cofinite sets relative to every such assignment *v*. Then, any Boolean combination of them will be finite or cofinite. Given that *R* is universal, $[\![\Box\phi]\!]_v$ will be either *W* or $\emptyset$ and thus cofinite or finite, respectively. Similarly,

for $[\![\phi \equiv \psi]\!]_v$. If $[\![\phi]\!]_v$ is finite or cofinite for every v assigning finite or cofinite sets, then $[\![Q\phi]\!]_v$ is a singleton, as we noted earlier, which is finite (if Q^* is in the language, then $[\![Q^*\phi]\!]_v$ is the same singleton). Finally, if $[\![\phi]\!]_v$ is finite or cofinite for every such v, then $[\![\forall p\phi]\!]_v = [\![\phi]\!]_u \cap [\![\phi]\!]_{u'}$ where u and u' agree with v except for assigning W and $\emptyset$ to p, respectively. Since u and u' also only assign finite or cofinite sets to the propositional variables, $[\![\phi]\!]_u$, $[\![\phi]\!]_{u'}$ and thus $[\![\phi]\!]_u \cap [\![\phi]\!]_{u'}$ are cofinite or finite.

38. Note that the result that $Q^n\top$ expresses a different set of worlds for each n can be derived from K2$^+$ alone, or from K3$^+$ and the existence of $\top$ and closure of existence under Q.
39. The crucial step is showing that if $[\![\phi]\!]_v$ expresses a finite or cofinite set relative to any assignment that assigns finite or cofinite sets, $[\![\forall p\phi]\!]_v$ expresses a finite or cofinite set. However, if the domain of propositions is infinite, $[\![\forall p\phi]\!]_v$ potentially corresponds to an infinite intersection of sets of the form $[\![\phi]\!]_u$: even if the latter are all finite or cofinite, an infinite intersection of them needn't be.
40. More formally, in the model currently under consideration, $[\![\lambda]\!] = \{w \mid w \notin \sigma(w)\}$. This set is finite if and only if all but finitely many worlds $x \in W$, are members of $\sigma(x)$. Thus, all but finitely many singleton sets must be paired with their members according to σ, which means that there are only finitely many worlds left for the rest of the finite/cofinite sets to be paired with, which is impossible. Similarly, it cannot be cofinite because it would mean that all but finitely sets of the form $W \setminus \{x\}$ get mapped to x under σ, leaving only finitely many worlds to get mapped to the finite/cofinite sets.
41. It is relatively easy to tweak the model M so as to get K1 by brute force. Let σ^* be a bijection between $\mathbb{N} \setminus \{0\}$ and finite and cofinite sets given by $\sigma^*(n+1) = \sigma(n)$. For $n > 0$ $|Q|(n) = \{\sigma^*(n)\}$, and $|Q|(0) = \{0\} \cup \{n \mid n \notin \sigma^*(n)\}$. Assuming that the quantifiers range over finite and cofinite sets, we can show that λ is vacuously true at 0, and thus that $[\![\lambda]\!] = \{0\} \cup \{n \mid n \notin \sigma^*(n)\}$, so this model satisfies K1 and K1$^+$. But there is no reason to think that it satisfies K2 or K2$^+$.
42. Suppose that $[\![\phi]\!] = I^+$ where I is a set of nth stage possibilities (a set of n-tuples). Then, by construction, $[\![Q\phi]\!]$ is the set of worlds where $I \in w(n)$. (This is where the requirement that $w(n)$ consists only of propositions not generated at earlier stages comes in.) So the set of $n+1$-stage possibilities that contain I in their $n+1$th component has $[\![Q\phi]\!]$ as its completion, and so this intension belongs to the $n+1$th stage.
43. We shall show that for every formula ϕ, there is some n such that $[\![\phi]\!]_v$ is a completed stage n intension for every choice of v that maps variables to completed stage 1 propositions (and thus a closed sentence will always express a bounded proposition). The inductive hypothesis clearly holds for propositional variables. For each n, the completions of stage n intensions are closed under arbitrary Boolean operations. This means that if $[\![\phi]\!]_v$ and $[\![\psi]\!]_v$ are completed stage n propositions for each v mapping variables to completed sets of proto-worlds, $[\![\phi \wedge \psi]\!]_v$ and $[\![\neg\phi]\!]_v$ are also completed stage n intensions. Similarly, if $[\![\phi]\!]_v$ is a completed stage n intension for each v mapping variables to completed stage 1 propositions, then $[\![\forall p\phi]\!]_v$ is an intersection of completed stage n intensions, and is thus a completed stage n intension. $[\![\Box\phi]\!]_v$ and $[\![\phi \equiv \psi]\!]_v$ are always either $\emptyset$ or W, and thus completions of sets of stage 1 intensions. Finally, if $[\![\phi]\!]_v$ is a completed stage n intension for each v of the required type, $[\![Q\phi]\!]_v$ is a completed stage $n+1$ proposition.
44. One might think that fictional names are a counterexample to the first-order version of this generalisation. Even if Sally thinks that Pegasus flies, there is

nothing Sally thinks flies, or so the objection goes. For his part, Williamson suggests that 'we should distrust attempts to use fictional or mythological names to refute metaphysical or logical theses. Such terms have a confusing variety of uses [...]. The more such uses are studied, the less they seem to threaten classical theses of logic and metaphysics'. (See Williamson 2013, 153). Elsewhere, he suggests in a Fregean spirit that he is conducting his inquiry not in ordinary natural language but in a language free of certain defects that afflict natural languages. In this connection, a non-referring use of 'Pegasus' is cited as one such defect: 'we should not distort our formal language by allowing for such a term'. (See Williamson 2013, 132).

45. For relevant discussion, see Williamson (1988).
46. Note that a language which could form disjunctions which contained a disjunct for every instantiation of a sentence of that very same language would not have a well-founded syntax since the instantiation of the disjunction itself would have to be a disjunct. One could countenance non-well-founded languages like this, but they are subject to their own liar-like paradoxes. For example, consider the sentence beginning with an infinite string of negations of ordinal type ω. Such a sentence would be identical to its own negation and paradox would quickly ensue.
47. See Chapter 6 of Williamson (2013).
48. This may be a bit quick, but it is beyond the scope of our paper to fully explore the implications of subtly different construals of the vicious circle principle.
49. The model we describe is not a regular model in the sense of Section 4 since the extension of Q is not given by a set of sets of points, but by a set of closed formulae of the language. Let $W = \mathbb{N}$, and $R = W \times W$, and let $Dom(n) = \mathcal{P}(\mathbb{N})$. Finally, let σ be a bijection between $\mathcal{L}$ and $\mathbb{N}$. Truth at a world is defined recursively. For the most part, we follow the clauses in Section 4. $[\![Q\phi]\!]_v = \{n \mid \phi$ is closed and $n = \sigma(\phi)\}$. We're not offering this as a model that's metaphysically illuminating, but it does establish the consistency of Comp, $K2^+$ and K2.
50. We have discovered a very similar result in Myhill (1979), deriving a paradox from an unmodalised version of K4.
51. One could assert that no zero-level propositions are materially equivalent to λ by denying the existence of any zero-level propositions. Or one could go for a very radical form of ramification in which the truth functional connectives themselves have a hierarchical structure.
52. Note that sometimes philosophers adopt the convention of calling a schema true if its universal closure is true (where each schematic letter is bound by a quantifier of the requisite type), but that won't help in the relevant setting since it will send us headlong into the first problem of insufficient generality.
53. Note that when this is spelled out fully, an unary operator (for example) is indexed by a pair of numbers: one that signifies the level of the input proposition and the other that signifies the level of the output proposition. Recalling the notation from the introduction, we would represent this by the ramified type $(()/n)/m$.
54. We note in passing that some implementations of Tarski's stratificational approach to the liar adopts this second tack: for example, he endorses the schema $\neg Tr_n(\ulcorner\phi\urcorner)$ when ϕ contains a truth predicate of level n or higher.
55. The most straightforward definition of sharing a grammatical category has to do with substitution preserving grammaticality. Unfortunately, this is not available to someone who takes type mismatch to induce ungrammaticality since, for example, switching names of different types may turn a grammatical

sentence into an ungrammatical one, even though all names belong to the same grammatical category. Here is an alternative definition that is available to such a theorist, at least if they opt for a cumulative approach: two expressions belong to the same grammatical category just in case there is some expression such that both are in the range of its significance. If instead one goes non-cumulative, it's no longer clear how to make good on the vision that one is stratifying within a grammatical category since the notion of grammatical category is no longer clear. Such a view arguably collapses into Fregeanism.

56. As noted in the introduction, on standard ways of setting up the framework, the expressions of the form $\forall p$ can be ordered by the natural numbers because the variable p can be assigned a ramified level – the expression $\forall$ itself is syncategorematic.
57. One typically pursues this idea by ramifying the variables instead of the quantifiers, by insisting that a variable of type τ be only assigned values in a particular domain, and treating quantifiers themselves as type neutral rather like Russellian negation (as noted above). One could also achieve a similar effect by having a single type of variable, but by ramifying the device, we use λ abstract. It is unclear to us what, if anything, is at stake between these variants.
58. This seems very much in line with a solution entertained, albeit in a compressed way, by Kripke to his own puzzle.
59. As argued in a lot more detail.
60. The ordering technique we have in mind is one according to which if 'expresses' expresses$'$ expressing$''$, then expressing$''$ comes immediately below expressing$'$ in the ordering. Ruling out loops and infinite descending sequences together would still guarantee no more than a partial order; further arguments would be needed to secure the kind of well-order that would justify indexing by the natural numbers.
61. Even if one took Kaplan's lessons to heart, this wouldn't make trouble for the models that we employed in Section 4. Note, importantly, that Kaplan makes it definitive of possible world semantics that one treats the object-level propositional quantifiers as ranging over all sets of worlds, but note that the models of Section 4 therefore do not count as 'possible worlds' models in that sense. There is more to be said about the relation between possible world semantics, as Kaplan describes, and what typically goes under the name 'possible worlds model theory', but we are not going to pursue these issues further here.
62. Of course for reasons given earlier, hardcore syntactic constraints on truthfulness and meaningfulness adopted by some ramifiers do not sit well with coarse-grained theories of propositions that are often adopted within possible worlds semantics.
63. This could be achieved by stipulating that all propositional variables have the type $()/0$: this effectively means quantification over propositions doesn't come in levels, but the operators still get ramified types of the form $(()/n)/m$ for $n < m$ and so cannot be iterated. If we are going for a falsity interpretation of level mismatch, we would not require any special syntactic restrictions.
64. Much of what we have to say can be adapted to the view that there is a hierarchy of thinking relations, but no straightforward syntactic scope constraint concerning when attitudinal verbs can be embedded within one another to produce truths.
65. For example, let $W = \{w\}$, $D(w) = \{\emptyset, \{w\}\}$, $|Q_1|_w = \emptyset$, $|Q_2|_w = \{\{w\}\}$.

66. For what it's worth, we do recognise that there may be something a little unstable about a ramification strategy that imposes austere syntactic constraints on attitudinal verbs, but does not ramify the variables or quantifiers. One effect of this combination is that one can have syntactically well-formed instances of the premises of Leibniz's law, such as Qp and $p = Qq$, but ungrammatical instances of the conclusion, such as QQq. But as we see it, there is no similar instability attaching to a view that espouses a hierarchy of relations but imposes no syntactical prohibitions. In other respects, the view might make similar moves to the more traditional ramifier, for example, with the traditional ramifier, it may think that our inclination to the hypothesis 'it's possible to uniquely think that everything you're thinking is false' or 'it's possible for it to uniquely seem as though you're saying that everything it seems as though you're saying is false' is based on a blindness to the underlying structure of our mental states and speech acts.
67. On the interpretation on which type mismatch induces ungrammaticality, this interpretation of K2 is especially natural as we don't in general consider ungrammatical substitution instances of schematic principles.
68. We let the quantifiers range over bounded propositions, and the extension of Q_n is the extension of Q (as defined in Section 4.3) intersected with the n-level propositions. To see that this model works, one first shows that if the highest type in ϕ is n, then ϕ expresses an n-level intension. Thus, every sentence whose type does not exceed n expresses an n-level intension, and so is Q_{n+1} at some world.
69. For our part, we don't see any special appeal to a view that says K2 is obviously logically coherent, but not the conjunction of K2 with a supervenience thesis. Why not say instead that the logical mastermind would pour scorn on both on account of Prior's theorem?
70. We might note in passing that, although not in the Russellian tradition, one might want to sort first-order variables into types with disjoint domains (see McDaniel 2009). One might also envisage cross-cutting domains – a picture encouraged by some of the literature on quantifier variance. Talk of 'domains of quantification' here and in the text is of course a bit of cheat (see footnote 7.1 below).
71. Even this characterisation of how the quantifiers relate to each other is something of a cheat since it pretends to be able to talk all at once about all the levels, yet we have no quantifier that can do this. This kind of consideration motivates Williamson (2003) to think that absolutely general first-order quantification is unavoidable; arguably, these considerations generalise to quantification into sentence position.
72. In this framework – where operators and connectives aren't typed – it's not so obvious whether any type mismatches can occur since a variable cannot be in the scope of another. Thus, the exact treatment of type mismatches doesn't matter so much in this context.
73. Such a manoeuvre surely would not constitute special pleading, for even the Fregean would not countenance 'is a horse is not a concept' as an instance of the first-order generalisation 'nothing is a concept'.
74. This thought turns on the cumulative aspect of the hierarchy, but as noted in footnote 55, it's hard to properly distinguish the Fregean from the non-cumulative ramifier since we wouldn't want the distinction to turn on very superficial syntactic categories.

References

Bacon, Andrew. 2014."Radical Anti-Disquotationalism." *MS*.
Bacon, Andrew. 2011. "Vagueness at Every Order." *MS*.
Bradwardine, Thomas. 2010. *Insolubilia*. Leuven: Peeters.
Church, Alonzo. 1976. "Comparison of Russell's Resolution of the Semantical Antinomies with that of Tarski." *Journal of Symbolic Logic* 41 (4): 747–760.
Davies, Martin. 1981. *Meaning, Quantification, Necessity: Themes in Philosophical Logic*. Routledge & Kegan Paul.
Divers, John. 2006. *Possible Worlds*. Routledge.
Dorr, Cian. 2012. "Propositional Profusion and the Liar." *MS*.
Glanzberg, Michael. 2001. "The Liar in Context." *Philosophical Studies* 103 (3): 217–251.
Harris, J. H. 1982. "What's So Logical about the 'logical' axioms?" *Studia Logica* 41 (2–3): 159–171.
Kaplan, David. 1977. "Demonstratives." In *Themes from Kaplan*, edited by Joseph Almog, John Perry, and Howard Wettstein, 481–563. Oxford University Press.
Kaplan, David. 1995. "A Problem in Possible Worlds Semantics." In *Modality, Morality and Belief: Essays in Honor of Ruth Barcan Marcus*, edited by Walter Sinnott-Armstrong, Diana Raffman, and Nicholas Asher, 41–52. Cambridge University Press.
Kripke, Saul A. 2011. "A Puzzle about Time and Thought." In *Philosophical Troubles. Collected Papers Vol I*, edited by Saul A. Kripke. Oxford University Press.
Lewis, David K. 1986. *On the Plurality of Worlds*. Blackwell Publishers.
McDaniel, Kris. 2009. "Ways of Being." In *Metametaphysics: New Essays on the Foundations of Ontology*, edited by David John Chalmers, David Manley, and Ryan Wasserman, Oxford University Press.
McGee, Vann. 2000. "Everything." In *Between Logic and Intuition: Essays in Honor of Charles Parsons*, edited by Gila Sher and Richard L. Tieszen, 54–78. Cambridge University Press.
Myhill, John. 1979. "A Refutation of an Unjustified Attack on the Axiom of Reducibility." In *Bertrand Russell Memorial Volume*, edited by Bertrand Russell and G. W. Roberts, 81–90. Humanities Press.
Parsons, Charles. 1974. "The Liar Paradox." *Journal of Philosophical Logic* 3 (4): 381–412.
Prior, A. N. 1956. "Modality and Quantification in S5." *Journal of Symbolic Logic* 21 (1): 60–62.

Prior, A. N. 1961. "On a Family of Paradoxes." *Notre Dame Journal of Formal Logic* 2 (1): 16–32.

Prior, A. N. 1977. *Worlds, Times, and Selves*. Duckworth.

Ramsey, Frank Plumpton. 1960. *The Foundations of Mathematics and Other Logical Essays*. Paterson, NJ: Littlefield, Adams.

Rayo, Agustín, and Gabriel Uzquiano. 1999. "Toward a Theory of Second-Order Consequence." *Notre Dame Journal of Formal Logic* 40 (3): 315–325.

Rayo, Agustín, and Timothy Williamson. 2003. "A Completeness Theorem for Unrestricted First-Order Languages." In *Liars and Heaps*, edited by J. C. Beall. Oxford University Press.

Russell, Bertrand. 1908. "Mathematical Logic as Based on the Theory of Types." *American Journal of Mathematics* 30 (3): 222–262.

Sher, Gilar, and Richard L. Tieszen. eds. 2000. *Between Logic and Intuition: Essays in Honor of Charles Parsons*. Cambridge University Press.

Slater, B. H. 1986. "Prior's Analytic." *Analysis* 46 (2): 76–81.

Smith, Nicholas J. J. 2006. "Semantic Regularity and the Liar Paradox." *The Monist* 89 (1): 178–202.

Tarski, Alfred. 1936. "The Concept of Truth in Formalized Languages." In *Logic, Semantics, Metamathematics*, edited by A. Tarski, 152–278. Oxford University Press.

Tucker, Dustin, and Richmond H. Thomason. 2011. "Paradoxes of Intensionality." *Review of Symbolic Logic* 4 (3): 394–411.

Uzquiano, Gabriel. 2015. "A Neglected Resolution of Russell's Paradox of Propositions." *Review of Symbolic Logic* 8 (2): 328–344.

Williamson, Timothy. 1988. "Equivocation and Existence." *Proceedings of the Aristotelian Society* 88 (n/a): 109–127.

Williamson, Timothy. 1998. "Bare Possibilia." *Erkenntnis* 48 (2/3): 257–273.

Williamson, Timothy. 2003. "Everything." *Philosophical Perspectives* 17 (1): 415–465.

Williamson, Timothy. 2013. *Modal Logic as Metaphysics*. Oxford University Press.

Reply to Bacon, Hawthorne and Uzquiano

Timothy Williamson

Wykeham Professor of Logic, University of Oxford, Oxford, UK

In 'Higher-order free logic and the Prior-Kaplan paradox' (Bacon, Hawthorne, and Uzquiano 2016), Andrew Bacon, John Hawthorne and Gabriel Uzquiano (BHU) explore various strategies for avoiding some disturbing results in higher-order logic derived by Prior (1961), Kaplan (1995) and others, under apparently reasonable assumptions. Specifically, the results involve quantification into sentence position. In particular, they are derivable in systems of higher-order modal logic of the sort defended in *Modal Logic as Metaphysics*, so I cannot just say that they are not my problem.

Typical of the results is this:

Prior's Theorem $\quad Q\forall p(Qp \rightarrow \neg p) \rightarrow \exists p(Qp \,\&\, p) \,\&\, \exists p(Qp \,\&\, \neg p)$

Here Q is any sentence operator. For instance, we can read Q as 'TW visibly wrote at 8.45 that', with reference to the morning on which I wrote this paragraph. Under this interpretation, roughly paraphrased in English, Prior's theorem says that if TW visibly wrote at 8.45 that whatever TW visibly wrote at 8.45 is not so then something TW visibly wrote at 8.45 is so and something TW visibly wrote at 8.45 is not so, and consequently TW visibly wrote at least two things at 8.45.

The main problem with Prior's Theorem is *not* that it is beyond the remit of logic to give us this sort of information. As I argue in Williamson 2013, logic is defined by its generality, not by its neutrality or uninformativeness. Strong, informative theories are as valuable in logic as they are in any other science. The main problem with Prior's theorem is much worse and more straightforward. It just looks to be false, under various interpretations including the one above. For, as I can assure the reader, a natural description of what happened at 8.45 on the morning I wrote this paragraph is this: TW visibly wrote that whatever TW visibly wrote at 8.45 is not so; TW visibly wrote nothing else (in the relevant sense). Given the natural description, TW visibly wrote only one thing at 8.45, so the consequent of Prior's theorem is false, while its antecedent is true. Thus Prior's theorem looks to make false predictions about easily observable events.

On the face of it, the problem has nothing special to do with higher-order *modal* logic. Prior's theorem contains no specifically modal operators, and the reading above on which it appears false was specified in non-modal terms. However, BHU develop modal aspects of the problem in several ways.

First, BHU formulate various plausible-looking possibility claims that can be refuted by reasoning like that in the derivation of Prior's theorem. For instance, if you doubt my description of what *did* happen at 8.45 this morning, but concede that it (metaphysically) *could* have happened that way, your concession is incompatible with the necessitation of Prior's theorem, which is standardly derivable in higher-order modal systems, including mine. BHU use some plausible-looking generalized possibility claims as benchmarks in their search for higher-order modal systems that avoid Prior-like results: one can test systems for their consistency with such claims. However, since the original problem arose without appeal to modality, we may reasonably hope for a non-modal explanation of what went wrong. For understanding the root of the problem, modality looks like one moving part too many, an unhelpful complication and distraction.

The second way in which BHU develop a modal aspect of the problem is by treating quantification into sentence position intensionally, modelling it in terms of quantification over sets of worlds, at least in some of the approaches they consider. That is also how *Modal Logic as Metaphysics* treats quantification into sentence position. However, that too is not essential to the underlying problem, as BHU show. The problem arises even if quantification into sentence position is treated extensionally, in effect as quantification over truth-values, since the consequent of Prior's theorem then implies, for instance, that at 8.45 I visibly wrote two very coarse-grained things that differ in truth-value. Equally, the problem arises even if quantification into sentence position is treated hyperintensionally, in effect as quantification over very fine-grained entities such as Fregean thoughts, since it still seems false that at 8.45 I visibly expressed in writing two Fregean thoughts that differ in truth-value. Thus, the underlying problem is robust on the dimension of fineness of grain in the interpretation of quantification into sentence position. The intensional approach is not to blame, though an adequate diagnosis must no doubt take account of the operative level of individuation.

BHU emphasize a third modal theme with reference to *Modal Logic as Metaphysics*. The first part of their paper constructs a variety of ingenious models to investigate 'Fregean' higher-order modal systems in which, as in the book, one avoids a Russellian hierarchy of ramifications within a given grammatical category. The sentential quantifiers are interpreted as ranging over intensions (in effect, sets of worlds). However, by contrast with the book, the quantifiers do not mandatorily range over *all* intensions. Rather, they are restricted to a proper or improper subset of intensions; BHU restrict the term 'proposition' to intensions in that subset. The effect is to invalidate both the rule of universal instantiation and various comprehension principles, thereby undermining Prior's proof of his theorem and most similar proofs. The result is a higher-order *free* modal logic. But although the domain of quantification is restricted, the restriction is not relativized to worlds. BHU are still doing constant domain semantics, as in the book. Thus sentential quantifier analogues of the Barcan formula and its converse and the necessity of existence are still valid. I derive those principles by universal instantiation, but they hold in BHU's models even though universal instantiation does not. As BHU say, this pulls apart two strands of thought that go together very naturally in the argument of *Modal Logic as Metaphysics*: in their terms, 'First that existence doesn't modally come and go, and second, that existence comes cheap'.

A quick way to check the separation of the two strands is by considering models with only one world. In such models, there is no contingency at all, so trivially existence doesn't modally come and go. But the domain of quantification may still be restricted, for instance to truths, so that existence isn't cheap in BHU's sense: in such models, $\forall p\, p$ is true but $p \,\&\, \neg p$ false, so universal instantiation fails, and $\neg(p \,\&\, \neg p)$ is true but $\exists p\, \neg p$ false, so existential introduction fails.

However, 'existence is cheap' is a slightly misleading slogan for the aspect of the theory defended in *Modal Logic as Metaphysics* that BHU have in mind (although I do not think that they misunderstand the theory). Of course, some philosophers may use the word 'existence' for an 'expensive' property, such as concreteness, that some things have and others lack, but that is irrelevant because the book explicitly avoids using the word 'existence' in its theorizing. The real issue must concern *being something*, on an unrestricted reading of that phrase at whatever order is in play. It is also irrelevant that, on my view, any individual is something in all possible circumstances, and likewise for any proposition, for that is just the claim that 'existence doesn't modally come and go', with which BHU are *contrasting* the claim that 'existence is cheap'. Rather, they seem to have in mind the unrestricted validity at all orders of the standard introduction rule for the unrestricted quantifier $\exists$ and the standard elimination (instantiation) rule for the unrestricted quantifier $\forall$, according to the theory. But of course that validity does not mean that the rules can be applied to whatever has the syntactic appearance of an expression of the appropriate category. After all, it is no objection to the standard introduction rule for disjunction ($\vee$) that *A* may be true while $A \vee B$ is not because (we may suppose) although *B* has the syntactic appearance of a sentence, it has never been assigned a meaning, so the disjunction too is semantically defective. Theorizing in any discipline, logic, physics or anything else, is liable to go wrong when one does it in semantically defective terms. To speak for convenience with typical ambiguity: for any grammatical type whatsoever, unless an expression of that type is semantically defective, it has a semantic value of the appropriate kind (perhaps relative to an assignment), and an unrestricted quantifier of the same type ranges over all values of that kind, so the standard quantifier rules preserve truth (when applied to semantically non-defective expressions). None of this prejudges the question of how easy or hard it is for an expression of a given type to be semantically non-defective. With these clarifications, I will for simplicity continue using BHU's phrase 'existence is cheap' for the claim that the standard quantifier rules are truth-preserving.

In BHU's sense, there are models in which existence doesn't modally come and go but isn't cheap. As they also note, the converse is much harder to make sense of: cheap existence modally coming and going. For if it sometimes modally goes, it is not as cheap as all that. More formally, the characteristic theses of higher-order necessitism are derivable from the standard higher-order rule of universal instantiation (the first-order case is trickier).

Although BHU construct various Fregean models in which the existence of propositions isn't cheap, but doesn't modally come and go, and various Priorian theorems are false, in the end they declare themselves pessimistic about the prospects for this strategy in defusing Prior–Kaplan paradoxes. For too many other plausible assumptions have to be given up in blocking all routes to the paradoxical conclusions. Within the Fregean framework, one can derive versions of the theorems from plausible principles about the closure of the class of propositions under various operations (6.1) or the supervenience of all truths on fundamental truths (6.2), or by making only minimal assumptions about the existence of propositions and consequently ruling out more complex but still apparently possible scenarios (6.3). Moreover, without the standard quantifier rules, higher-order quantifiers are much less useful for enhancing the expressive power of the language (6.4 and 6.5). One would have to be very confident indeed that the Priorian theorems have false readings to be willing to give up so much of higher-order logic in order to vindicate the legitimacy of those readings.

In Section 7 of their paper, BHU consider strategies that reject universal instantiation, and so deny that existence is cheap, within a Russellian framework, ramifying either the quantifiers for a given grammatical category, or the attitude verbs (such as 'write' and 'think') used to specify the paradox-generating readings of the operator *Q*, or both. Once again their conclusions are pessimistic: they provisionally judge the costs to outweigh the benefits.

In concluding, BHU write 'our own inclination is to take Prior's results at face value', so they incline to rejecting principles that conflict with them (Section 8). But, as they say, 'There still remains the question of how to implement that discovery'. They suggest that we may still need to postulate ramifications of subtly different relations associated with a given attitude verb to explain why Prior's results *seem* to fail, but without ramifying the quantifiers or postulating failures of universal instantiation. This direction of research is congenial to my own instincts. I have suggested such a non-logical version of indefinite extensibility in dealing with the semantic and set-theoretic paradoxes (Williamson 1998). The guiding abductive methodology is to keep the overall logical framework simple and strong, while explaining the paradoxes by allowing messier proliferation at the level of less general distinctions. That way one prevents the resultant restrictions from undermining the explanatory power of theories in other disciplines (such as the natural sciences), as they do when logical principles themselves are restricted in response to the paradoxes, and so have to be supplemented with numerous ad hoc assumptions in order to recover their full strength in scientific applications (Williamson forthcoming).

That general methodology is applicable to the case BHU discuss, where someone restricts the rule of universal instantiation for quantification into sentence position by adopting some form of free logic instead. Thus by itself $\forall p\, A$ no longer in general yields its instance $A(B/p)$ (subject as always to the normal syntactic qualifications to prevent free variables in B becoming bound when B is substituted for p in A). To derive the instance in free logic, one also needs an auxiliary premise such as $\exists q(B \equiv q)$ (where q does not occur in B and $\equiv$ is the sentential analogue of identity), to ensure that B expresses a proposition in the domain of quantification.

Now suppose that the background logic for a theory about some topic far from Prior-Kaplan paradoxes employs quantification into sentence position. For instance, that might be the most economical way of understanding some talk of 'conditions' or the like in natural science, since (unlike first-order quantification) it avoids ascribing an apparatus of nominalizing and denominalizing devices to mediate between variables in name position and sentences; for a detailed account of how scientific applications of the mathematical theory of dynamical systems are naturally understood in terms of quantification into sentence position see Williamson 2016. Thus, a scientific explanation of some phenomenon may involve deriving it from various theoretical assumptions that include generalizations of the form $\forall p\, A$. In applying those generalizations, the derivation will typically need to instantiate them. But, as we noted, in the free logical framework each such move generally requires postulating an auxiliary non-logical assumption of the form $\exists q(B \equiv q)$. If such an assumption is justified at all, it will be by further auxiliary assumptions, such as that universal instantiation fails only where intentionality is involved and that the present phenomenon does not involve intentionality — odd-looking assumptions for an explanation in physics! Thus, restricting universal instantiation to block Prior-Kaplan paradoxes has unfortunate repercussions in all sorts of cases far from those paradoxes. By contrast, ramifying the attitude verbs is far more local in its effects, since those verbs are absent from most theories in the natural sciences.

Of course, such general methodological remarks alone do not satisfyingly explain away the apparent falsifications of predictions made by Prior's theorem about observable events. I will not attempt to provide any such explanation here. However, I do suggest that an appropriate setting for such an explanation is likely to be an externalist theory of content, on which what, if anything, one has an attitude towards depends on many factors and preconditions that are utterly non-obvious to both oneself and others; they can only be uncovered through painstaking theoretical investigation. Such a setting helps reduce the shock of Prior's results, although we are still very far from completely removing their sting.

References

Bacon, Andrew, John Hawthorne, and Gabriel Uzquiano. 2016: "Higher-order Free Logic and the Prior-Kaplan Paradox." *Canadian Journal of Philosophy* 46 (4–5): 493–541.

Kaplan, David. 1995. "A Problem in Possible Worlds Semantics." In *Modality, Morality, and Belief: Essays in Honor of Ruth Barcan Marcus*, edited by W. Sinnott-Armstrong, D. Raffman, and N. Asher, 41–52. Cambridge: Cambridge University Press.

Prior, A. N. 1961. "On a Family of Paradoxes." *Notre Dame Journal of Formal Logic* 2: 16–32.

Williamson, Timothy. 1998. "Indefinite Extensibility." *Grazer Philosophische Studien* 55: 1–24.

Williamson, Timothy. 2013. *Modal Logic as Metaphysics*. Oxford: Oxford University Press.

Williamson, Timothy. 2016. "Modal Science." *Canadian Journal of Philosophy* 46 (4–5): 453–492.

Williamson, Timothy. forthcoming. "Semantic Paradoxes and Abductive Methodology." In *The Relevance of the Liar*, edited by B. Armour-Garb. Oxford: Oxford University Press.

Williamson on Fine on Prior on the reduction of possibilist discourse

Kit Fine

Department of Philosophy, New York University, New York, NY, USA

ABSTRACT
I attempt to meet some criticisms that Williamson makes of my attempt to carry out Prior's project of reducing possibility discourse to actualist discourse.

Timothy Williamson's *Modal Logic as Metaphysics* (2013; MLM) is a tour de force — comprehensive in its scope, brilliant in its argumentation, and startling in its conclusions. It merits discussion on a wide range of different fronts, but I hope I can be forgiven for focusing on chapter 7 of the book, in which Williamson criticizes my attempt to carry out Prior's project of reducing possibilist discourse to actualist discourse.

My response is in three main parts. I begin by discussing what the reductive project should be. Williamson and I disagree on this question and, although it is not important for the evaluation of my own reductive proposal, it is important for a broader understanding of the metaphysical issues at stake. I then discuss and evaluate Williamson's criticisms of my original reductive proposal. Although I believe that these criticisms can to some extent be met, they point to the need for a more satisfactory and less contentious form of reduction. Finally, I lay out the new proposed reduction; it is based on the idea of finding a general way of extending a reduction of first-order discourse to higher order discourse.

The main technical innovation of the paper is the introduction of the suppositional calculus, which I hope may be of some independent interest quite apart from its connection with the present reductive project. Although I have not attempted to be rigorous, I hope it is clear from my informal remarks how a more rigorous treatment might proceed.[1]

1. What is the problem?

Part of Williamson's aim is to reshape the debate between the actualist and the possiblist. He takes this debate to be 'hopelessly muddled' (23–4) and thinks that 'rather than trying to make sense of the confused dispute between actualism and possibilism, we can more profitably replace it with enquiry into clearer issues in roughly the same territory,' where 'perhaps the deepest of those is the dispute between contingentism and necessitism' (308). I must admit of a fondness for the old 'confused dispute' and something of a distaste for its proposed replacement; and so let me briefly explain why, though the question calls for a much more thorough discussion than I can give here.

As I understand it, the debate between the actualist and the possibilist rests on two distinctions. The first is the distinction between possible and actual objects; and the second is the distinction between what is and is not real. The actualist then claims that only actual objects are real, while the possibilist denies that this is so.[2]

This formulation of the debate is therefore as clear as the two distinctions — or notions — upon which it rests. Williamson seems to find both distinctions obscure (though obscurity in either would be enough to damn the formulation). In §1.6 of MLM, Williamson considers various possible definitions of what it is for an object to be actual and finds them all lacking. But why is a definition required? I would have thought that the notion was perfectly clear from examples. There are various actual people — Socrates, for example, Timothy Williamson or myself — and also various merely possible people — the various children that Wittgenstein might have had, for instance. The actual people are actual objects, the actual and merely possible people possible objects. What could be clearer than that?

It might be thought that an actual person is a person while a merely possible person is not a person but something that is possibly a person and hence just as actual as a person. This is not my own way of looking at the matter nor, I think, the most natural way.[3] A merely possible person *is* a person, just as an actual person is a person; the difference between them lies not in whether or not they are persons but in whether or not they are actual (pace §1.3 of MLM). But no matter. Let us grant that a merely possible person is not a person but only possibly a person. Still, there is a clear intuitive distinction between being a merely possible king, let us say, and a merely possible person. Bonnie Prince Charles was not a king, only possibly a king, but that does not make him a possible object; he is as actual as you or I. Likewise, a possible child of Wittgenstein will not be a person, on the view we are considering, but only possibly a person. But that is surely enough to make him or her a merely possible object; there is no reason, in this particular case, to suppose that the object is also actual.[4]

Williamson briefly discusses the idea of reality or fundamentality in §7.1 of MLM. He there takes the issue between the actualist and possibilist to be one about 'whose quantifiers are more natural or basic' and comments that 'the inexpressibility of the dispute in more robust and tractable terms would be a disappointing outcome. For presumably the ordering in terms of relative naturalness or basicness is supposed to be of metaphysical interest because it is not determined just by the psychology of speakers, but rather by non-linguistic features of reality as a whole. If so, the dispute would be more perspicuous if each side could formulate its view of the disputed features in less extrinsic terms.'

I agree. But my own characterization of what it is for an object to be real *is* stated in such terms (Fine 2009). For it is supposed that we have a conception of how things are in reality. For an object to be real is then for it to figure in reality; and so to be an actualist is to hold that how things are in reality is entirely a matter of how *actual* things are. We might also put the point by saying that we are capable in principle of fully describing reality without reference to what is non-actual, but such appeal to language or to what we do is not strictly required.

Let us turn to Williamson's preferred formulation of the issue in terms of necessitism and contingentism ('N/C,' for short). The necessitist holds, under an absolutely unrestricted use of the quantifier, that necessarily everything necessarily has being (is identical to something), while the contingentist denies that this is so (as explained in chapter 1 of MLM).

Whence my distaste for this formulation of the issue? One reason is that it makes me a necessitist. I am perfectly happy to concede that the unrestricted quantifiers range over all possible objects and so am happy to accept that necessarily everything necessarily has being.[5] But as an actualist, I also believe that there is a separate issue as to whether only actual objects are real — or some such thing, whether or not it is exactly stated in this way. Williamson's formulation of the issue is therefore completely incapable of getting at the sense in which I am an actualist. Far from clarifying the issue, it ignores it.

But it is worse than that. For I believe that *anyone* in their right mind should accept necessitism. Suppose I show you three blades and three handles and ask how many possible knives can be made from the blades and handles? Surely the correct answer is nine; there are nine possible knives that can be made from the given blades and handles. Yet, according to the contingentist, the correct answer is zero (assuming that no one of the blades is ever attached to a handle). This strikes me as absurd — perhaps not as absurd as denying that there are chairs and tables, but absurd all the same.

Of course, the contingentist, along with his fellow anti-realistists, will attempt to fend off the absurdity. He will claim that although it is not strictly and literally true that nine possible knives can be made from the blades, still there is a sense in which it is correct to say that there are nine possible knives. And a number of possible reasons might be given for this: perhaps we engage in fictitious talk; perhaps we fail to use the expression 'there are' as a quantifier; or perhaps we speak metaphorically. But, to my mind, this is simply to put bad linguistics at the service of bad philosophy. For apart from the philosophical prejudice against there being possible objects, there is no good reason to take any of these hypotheses seriously. In a somewhat different connection, Williamson writes 'neither side [of the N/C dispute] is entitled to patronize the other by treating them as linguistically incompetent, unable to express their own views accurately in their own language.' (311); and it seems to me that we should extend a similar degree of respect to the ordinary speakers of our language.

Williamson briefly touches on the question of triviality in his book. He writes

> Now we must consider the charge that there is no live issue for the opposite reason, because necessitism is too obviously true. The idea is that if 'something' covers merely possible coins as well as coins, then trivially this coin is necessarily something, and so on. The idea is fallacious. That 'something' is unrestricted implies that the things it ranges over include whatever merely possible coins there are, *if any* [my emphasis] (MLM, 15).

The idea is indeed fallacious. But the charge of triviality we are making is not based upon this idea but simply on the idea that it is evidently true there are nine possible knives that can be made from the three handles and the three blades. In the same way, it is evidently true, on the basis of observation, that there are chairs and table — not because 'something' covers whatever chairs and tables there are but because observation makes it evident to us, once 'something' is allowed to cover chairs and tables, that there are chairs and tables.

If I am right, then the issue of necessitism vs. contingentism is a non-issue and any right-minded philosopher should accept necessitism without further ado. But quite apart from the question of triviality, it is far from clear to me that Williamson's approach provides any gain in clarity over the more usual approach. It certainly looks that way when we attend to the *formulation* of the issue; for the necessitist and contingentist can state their respective theses within the language of quantified modal logic, while the actualist and possibilist must appeal to the relatively obscure idioms of actuality and reality (or the like). But things look different when we attend, not to the formulation of the various theses, but to their defense.

For the necessitist needs to capture the truth in what the contingentist is saying when he says that it is possible that Socrates (and other 'contingents') do not have being. To this end, he will appeal to some feature *F* and say, instead, that it is possible that Socrates (and the like) are not *F*. The feature Williamson appeals to earlier in the book is being concrete (§1.2) and later he appeals to the related notion of being chunky in providing a possible defense of contingentism (315). But what is it to be concrete? In a footnote (6), Williamson remarks, 'the term "concrete" is used informally throughout this book. For present purposes, we need not decide between various ways of making it precise (being material, being in space, being in time, having causes, having effects).'

But none of these various glosses will serve the general demands made on the notion. Being material or being in space will not do since someone who thinks of Socrates as a Cartesian ego may still of him (or rather, it) as a contingent being. Being in time or having causes or effects will not do either. For consider someone who thinks that each possible world is a contingent being (having being only at that world). Then no possible world will be in space or time or have causes or effects; and, in the case of the empty possible world, it will not even be a material being or have proper parts. It therefore seems as if only the actualist's more rarefied notion of being actual will fit the bill.[6] Indeed, apart from Williamson's hostility to the notion, it is far from clear to me why the necessitist and contingentist should not simply expropriate the actualist's notion of being actual for their own purposes. Certainly, there is nothing in their respective positions that rules it out; and, once this is done, they will be on the same conceptual footing in this regard as their actualist and possibilist counterparts.

Another apparent advantage of the N/C formulation is that it does not require appeal to the seemingly obscure idea of what is fundamental (or real). But the necessitism and possibilist must appeal to an equally obscure idea of cash-value. For the contingentist is under some obligation to find the cash value (or 'truth') in the statements made by the necessitist, and likewise for the necessitist (cf. MLM, 312).

But why is it not easy to discharge this obligation? Why can the contingentist, for example, not take the cash value of a necessitist's statement *A* simply to be given by its 'neutral' consequences (those independent of their differing metaphysical commitments)? We feel that in certain cases this is not enough, that we require a neutral statement or set of neutral statements that is *equivalent* to the given statement, that otherwise something will be lost. But why and when? Williamson argues in the particular case of the statement that 'there are infinitely many possible stars' and of some other statements that something significant *would* be lost if the contingentist could find no neutral equivalent (MLM, 348–349); somehow the contingentist would not be able to make distinctions that they should (or perhaps would like) to make. But we are provided with no clear or principled conception of why or when an equivalent is required. For a 'soft' actualist like myself, by contrast, there is no such problem; for I will be willing to accept whatever the possibilist says is *true* and hence will take myself to be under an obligation to account for its truth. The contingentist may play fast and loose with the necessitist's falsehoods, but the soft actualist cannot play fast and loose with the possibilist's truths; and so even if the actualist's position is more obscure in its formulation, it is much clearer what is required for its defense.

2. The standard first-order reduction

In what follows, I shall depart from Williamson's own exposition in chapter 7 of MLM and construe the issue of reduction in terms of the debate between actualism and possibilism, since that is how I originally conceived of the issue and how I still prefer to think of it. However, it should be relatively easy to transpose what I say to fit the debate between contingentism and necessitism; and sometimes, though not always, I shall indicate what further changes this would entail.

There is a standard 'translational' reduction of first-order possibilist discourse to first-order actualist discourse, described in Fine (1977a). Let us suppose that the formulas of interest to us are drawn from a first-order language *L*, which contains the 'inner' quantifiers $\forall$ and $\exists$, ranging in each possible world over the objects that are actual in that world and which also contains the 'outer' quantifiers Π and Σ, ranging in each possible world over all of the possible objects, i.e. over all of the objects that are actual in some possible world. (Under the N/C construal of the debate, the language *L* under consideration will contain just one type of quantifier which will, in effect, range over all actual objects for the contingentist and all possible objects for the necessitist; and later, when we give the suppositional reduction, we may allow the outer quantifier to range over all individuals whatever, even when they are not possible.)

We shall assume that the first-order quantifiers Π and Σ or $\forall$ and $\exists$ range over worlds as well as over individuals of a more ordinary sort. We shall assume, more specifically, that, at each possible world, the inner domain of quantification will contain that world and that world alone, while the outer domain of quantification will contain every possible world. Thus at each world, the only world taken to be actual is that world itself. There are other ways to achieve the same effect. It could be supposed that at each world only that world *obtains*, for example, or that at each world there is only one world-proposition that is *true*. But however exactly the assumption is formulated, I shall assume that it need not be a point of contention between the actualist and the possibilist (or between the contingentist and the necessitist).

We shall also assume that the language *L* contains a finite stock of predicates and contains, in addition, a predicate *W* for being a world and a predicate *E* for being 'existent' which, in the present case, may be identified with being actual. Thus *W* will be true, at each possible world, of that world alone, while *E* will be true, at each possible world, of the individuals which are actual at that world and hence, in particular, of that world itself (under Williamson's preferred formulation, *E* should be understood to mean not actuality, but chunkiness). It should be noted that the claim $\Box\forall x\, \Box(Ex \equiv \exists y\, x = y)$ (necessarily, every actual object is necessarily such that necessarily it exists iff there is something actual that it is) will hold for the actualist, but that the corresponding claim $\Box\Pi x\Box(Ex \equiv \Sigma y\, x = y)$ will not hold for the possibilist.

In what follows, let us use *w, v, u* and the like as variables for worlds. $\forall w\, Aw$, for example, should be regarded as an abbreviation for $\forall x(Wx \supset Ax)$. It should be evident from the intended interpretation of *W* that the following two 'axioms' will hold:

World Existence: $\Box\exists w\top$ (necessarily, there is a world)

World Identity: $\Box\forall w\forall v\, w = v$ (necessarily, any two worlds are the same).

Call a formula *A* of *L* *actualist* if it does not contain the quantifiers Πx or Σ. Then what the actualist would like is to construct, for each formula *A* of *L*, an actualist formula A^* equivalent to A. 'Equivalent' here could mean a number of things – provably equivalent within a suitable logic for the language L, semantically equivalent under the intended semantics for the language L, equivalent according to our informal understanding of the language L.

I hope it will be obvious that the formula A^* we construct below is informally equivalent to A. It could also be shown to be provably equivalent to A within a suitable formal system and semantically equivalent within a suitable semantics, though this is not something I shall attempt to establish. There are somewhat weaker standards for a successful reduction that one might lay down, not requiring equivalence but some suitable form of supervenience. But since I believe the stronger standards can be met, there will be no need for us to consider the weaker standards.

Given a formula A and world-variable w, let us use $[A]_w$ as an abbreviation for $\Diamond(Ew \wedge A)$ (possibly, w is actual and A is the case) and, in the particular case in which A is the atomic formula $Rx_1x_2 \ldots x_n$, let us use $Rx_1x_2 \ldots x_nw$ in place of $[Rx_1x_2 \ldots x_n]_w$. Intuitively, $[A]_w$ says that A is true at w and $Rx_1x_2 \ldots x_nw$ says that R holds of $x_1, x_2, \ldots x_n$ in w.

The basic idea behind the proposed actualist reduction of possibilist discourse is that every formula ΣxBx in A should be replaced by $\exists w\Diamond\exists x[Bx]_w$ (the actual world is such that possibly some object is such that A is true at w). Thus, the modalized quantifier $\Diamond\exists x$ has the effect of quantifying over all possible objects and the combination $[Bx]_w$ has the effect of taking the evaluation of Bx back to the actual world. Since the idea and its variants are already familiar from the literature, I hope that there is no need to go into further detail.[7]

3. Going higher order

Let us now suppose that L is extended to a higher order language, with quantification over properties and relations of individuals, properties and relations of such properties and relations, and so on all the way through the type-theoretic hierarchy. As will become clear, the exact way in which the language is extended to the higher order will not much matter for our purposes. However, certain modifications to the proposed reduction would have to be made if L was allowed to contain non-logical predicates of higher order in addition to the given stock of first-order predicates.

Let us consider a particular case in which one might wish to give a higher order reduction, the case in which one has a plural possibilist quantifier ΣV (with V ranging over all 'pluralities' of possible individuals).[8] It will not do to follow the original first-order strategy, and replace $\Sigma V\, BV$ with $\exists w\Diamond\exists V\, [BV]_w$, for we want ΣV, in effect, to range over all pluralities of possible individuals, while the modalized quantifier $\Diamond\exists V$ will only range, in effect, over pluralities of *compossible* individuals, i.e. pluralities of individuals for which it is possible that all of them are actual.

To get over the difficulty, one might follow the strategy I have advocated in some previous papers (Fine 1977a; 146–148; 1977b, 161–162; 2005a, 173–174) and replace the plural possiblist quantifier ΣV with a string of modalized quantifiers $\Diamond\exists x_1\Diamond\exists x_2 \ldots$. The formulas $\Sigma V\, BV$ will first be replaced with $\exists w\Diamond\exists x_1\Diamond\exists x_2 \ldots [BV]_w$ and each occurrence of Vx (x is a V) in BV will then be replaced with the corresponding disjunction $x = x_1 \vee x = x_2 \vee \ldots$.[9] I no longer wish to advocate this strategy, but it is worth considering if only because of the interest of the issues that it raises.

Williamson has two main objections against such an account: the first is to the effect that no string of modalized quantifiers $\Diamond\exists x_1\Diamond\exists x_2 \ldots$, if intelligible, could be long enough to say all that needs to be said (353–356); and the second is that no infinitely long string of modalized quantifiers $\Diamond\exists x_1\Diamond\exists x_2 \ldots$ is even intelligible (356–360).[10]

Let us consider these objections in reverse order. As Williamson points out, it is not always clear what a given infinite string of quantifiers and modal operators should mean and so 'the onus is on contingentists to explain what they intend the infinite sequence of modal operators and restricted [my actualist] quantifiers to mean.' He then considers various proposals for fixing the intended meaning and finds them all lacking.

I think that there is a real problem here and that all but one of the proposals that Williamson considers are indeed inadequate. The one exception is the explanation of $\Diamond\exists x_1\Diamond\exists x_2 \ldots$ as something analogous to a branching quantifier, where the branches are given by $\Diamond\exists x_1$, $\Diamond\exists x_2$, … respectively — or, better still, by $\Diamond\exists x_1\Box$, $\Diamond\exists x_2\Box$, …. He says 'the problem [with this solution] is that a finite or infinite set of unordered sentence operators rarely has a natural meaning as an operator in its own right' (359); and so a meaning has to be given. 'However, the only apparent way of extending the available semantic treatments of branching quantifiers to modalized examples … is by treating the modal operators as quantifiers, just as in the possible worlds semantics, which contingentists cannot use to give the intended truth conditions' (360).

I wish to take exception to the last sentence. Indeed, there is something odd in the thought that the contingentist must somehow have recourse to the possible worlds semantics to explain the modalized quantifiers when no such recourse is required for the unmodalized quantifiers. To see why this is an unreasonable demand, consider the case of ordinary syntactic substitution. Given an expression E and some expression F that it contains, we may then substitute some other expression G for (all occurrences of) F in E, thereby obtaining the expression $E[G/F]$. One substitution may be performed *successively*, after another. Thus, we may first substitute G_1 for F_1 in E and then substitute G_2 for F_2 in the result $E\,[G_1/F_1]$ to obtain $E\,[G_1/F_1]\,[G_2/F_2]$. But we may also *simultaneously* substitute G_1 for F_1 and G_2 for F_2 in E to obtain $E\,[G_1/F_1 \mid G_2/F_2]$ (as long as F_1 and F_2 are non-overlapping expressions in E). The result of a successive substitution will not in general be the same as the result of the corresponding simultaneous substitution. Suppose E is the formula $p \vee q$. Then $E\,[q/p][p/q]$ will be $p \vee p$ while $E\,[q/p|p/q]$ will be $q \vee p$.

We can usually give no clear meaning to an infinite successive substitution. If E is an expression containing the infinitely many distinct sentence letters p_1, p_2, p_3, ..., then what meaning is to be given to $E\,[p_1/p_2][p_2/p_3]\ldots$? But infinite simultaneous substitution *will*, in general, make sense. If E is an expression containing the infinitely many non-overlapping expressions F_1, F_2, F_3, ..., then $E\,[G_1/F_1|G_2/F_2|G_3/F_3|\ldots]$ is the well-defined result of simultaneously replacing each of F_1, F_2, F_3, ... with G_1, G_2, G_3, ..., respectively.

Let us now ask: How are we capable of understanding an infinite simultaneous substitution $E\,[G_1/F_1|G_2/F_2|G_3/F_3|\ldots]$? A natural answer suggests itself. Each of the terms G_1/F_1, G_2/F_2, G_3/F_3, ... corresponds to a syntactic operation on expressions, the operation that is performed when F_i is replaced in the given expression E with G_i. Moreover, these operations are in a clear sense independent of one another. If we were to assign a subordinate to execute each of the operations, then each of the subordinates could proceed with his task without any interference from the others. And it is because of the independence in this sense of the different syntactic operations corresponding to the terms G_1/F_1, G_2/F_2, G_3/F_3, ... that we are able to make sense of performing all of these operations simultaneously, i.e. to make sense of the infinite simultaneous substitution $E\,[G_1/F_1|G_2/F_2|G_3/F_3|\ldots]$.

Similarly for any other operations or 'actions' Suppose that there are infinitely many light switches in heaven and God commands that each of the switches be flipped from Off to On (let there be light!). Then since the actions of flipping a switch are independent of one another, there is no difficulty in making sense of all of them being performed simultaneously. Generally, given any independent actions a_1, a_2, a_3, ..., we can make sense of the action $a = (a_1|a_2|a_3|\ldots)$ which is the simultaneous execution of each of them.

What goes for syntactic operations, or for actions in general, also goes for semantic operations. The semantic operations a_1, a_2, a_3, ..., corresponding respectively to the modalized quantifiers $\Diamond\exists x_1\Box$, $\Diamond\exists x_2\Box$, $\Diamond\exists x_3\Box$, ... are independent of one another in the relevant sense (as long, of course, as the variables x_1, x_2, x_3, ... are all distinct from one another). This is a fact that can be recognized by the actualist (or contingentist) as much as by the possibilist (or necessitist) and no more requires appeal to some 'higher order' apparatus than in the analogous case of substitution. But then the actualist is in a perfectly good position to make sense of ($\Diamond\exists x_1\Box$, $\Diamond\exists x_2\Box$, $\Diamond\exists x_3\Box$, ...) as the simultaneous application of the semantic operations signified by the individual prefixes $\Diamond\exists x_1\Box$, $\Diamond\exists x_2\Box$, $\Diamond\exists x_3\Box$, ... I have no definition of independence that might apply across the board to all these different cases, but that is no reason to doubt the intelligibility of the notion or the legitimacy of its application to 'parallel processing' in any given case.[11]

For this reason, I am not inclined to take Williamson's second objection seriously. The first is more serious, but it raises difficult issues in the theory of classes and in the metaphysics of modality about which I am not sure what to say. What we would like is: (i) to have as many variables as possible objects; and (ii) to allow all variables to appear in a single formula, perhaps many times over. Since any object can be considered a variable, (i) is tantamount to the claim (i)′ that there should be as many (actual) objects as there are possible objects; and since any formula can be regarded as a class, (ii) is tantamount to the claim (ii)′ that there should be a class of all actual objects (and also suitable classes of classes of actual objects). There are versions of the theory of classes that allow for (ii)′ and perhaps may serve our purposes.[12] But (i)′ is more dubious and it is not even clear how the actualist might state it, since there will presumably be no actual one–one correspondence between the actual individuals and all possible individuals. Perhaps it is sufficient for the possibilist to accept (i)′, or something analogous, as true. He will then be in a position to recognize the correctness of the actualist's infinitary reduction even if the actualist herself cannot state the assumptions upon which its correctness depends.

But this is all very speculative. Add to it the desirability of providing a finitary reduction (MLM, 353), and we have reason enough to look for an alternative approach.

4. The suppositional approach

Let me explain in broad outline the idea behind the alternative approach before proceeding to greater detail.

There is something which we might call the 'modal pluriverse.' This is constituted by the totality of possible worlds, the totality of possible individuals, and the pattern of instantiation of certain properties and relations in each of the possible worlds. Thus, a simple 'toy' pluriverse might be constituted by three possible worlds *w*, *v*, and *u* and two possible individuals *a* and *b*, with *w* containing no actual individuals, *v* containing two actual individuals, *a* and *b*, both with the property *F*, and *u* containing one actual individual *a*, without the property *F*.

Our interest will not be in all properties or relations whatever, or even in the fundamental properties or relations, but merely in those that can be expressed by means of the atomic predicates of the possibilist sentences whose reduction is in question. Thus, we will view the pluriverse through the 'lens' of the sentences by which it might be described. We shall also suppose that the pluriverse is 'centered,' in that one of the worlds will be picked out as the actual world. A pluriverse will therefore correspond, at least under a set-theoretic representation, to one of the models of Kripke (1963) for the quantified system of modal logic S5.

The possibilist is relatively well-equipped to describe the pluriverse, since he can freely quantify over all possible worlds and all possible individuals. The actualist is not so well equipped; and indeed it is not even clear that he can form a complete and intelligible conception of what the pluriverse is. Our first step is therefore to describe the possibilist's pluriverse from an actualist point of view, i.e. by only making appeal to resources available to the actualist.[13] Now if every possible individual were actual, the expressive edge that the possibilist enjoys over the actualist would disappear. So what we will have the actualist do is describe a hypothetical pluriverse, one which is just like the actual pluriverse but for the fact that every possible individual is actual. Within such a pluriverse, the distinction between what is possible and what is actual will disappear and so it must be reinstated as a further distinction made within the realm of what is actual.

We might imagine that each possible world within the possibilist's pluriverse contains a barrier between what is actual and what is possible — with all the actual objects of the world lying to the right of the barrier and all the merely possible objects of the world lying to its left. The actualist first slides the barrier all the way to the left so that all possible objects lie to its right and he then introduces a dummy barrier to mark where the actual barrier once lay (under the N/C construal of the debate, the new dummy barrier will mark the distinction between what is and is not 'chunky').

This is the first stage in the proposed reduction. We are still not done since, even if we have been successful in the first stage, we will only have provided a *first-order* description of the pluriverse, one describing the first-order behavior of the possible individuals in different possible worlds but not one also describing the higher order behavior of the possible individuals or the behavior of properties or relations or other such entities in the different possible worlds.

At the next stage of the reduction, we appeal to the general idea that the higher order behavior of the pluriverse can be fully determined on the basis of its first-order behavior. Of course, inner quantification over actual properties and relations will in general behave differently from outer quantification over properties and relations, just as quantification over actual individuals in general behaves differently from quantification over possible individuals. But, in the present case, the hypothetical pluriverse is one in which each possible individual is actual and hence one in which each property or relation of the possibilist is actual; and so there will be no discrepancy between inner quantification over these entities in the hypothetical pluriverse and outer quantification over these entities in the actual pluriverse. The actualist is thereby able to simulate possibilist discourse within the actual pluriverse by means of the corresponding actualist discourse within the hypothetical pluriverse.

There is a certain resemblance between my approach and that of Sider (2002)[14], since we both take a possibilist sentence to be true if it is true according to the possibilist's pluriverse. But, as will become apparent, there are two significant differences. First, and most significantly, the way in which we specify the pluriverse is different: I specify it via suppositions while he specifies it directly by means of an infinitary sentence or a modal model. This means that my specification is finitary, not subject to worries over the cardinality of the set of possible objects or set of worlds and complete (no possibilist sentence is left undecided), while his specification is infinitary, subject to worries over cardinality and possibly incomplete (so that some possibilist sentences may be left undecided). Second, we have a different understanding of how the pluriverse determines the truth of a possibilist sentence: he makes use of the ordinary notion of first-order logical consequence (extended to an infinitary language), while I make essential use of a broader notion of logical consequence which allows higher order truths to be determined on the basis of first-order truths. I am thereby able to capture the higher order features of possibilist discourse, which is not something that can be properly accommodated within his own approach.

5. The suppositional calculus

Let us turn to the details of the alternative approach. We begin by developing the suppositional calculus within which the actualist will provide his description of the pluriverse.

The suppositional calculus (SC) is a calculus in which we are able to reason explicitly about the structure of suppositions and what follows from a supposition. It is of the greatest importance for understanding what is distinctive about the calculus to appreciate that a suppositional term, such as 'suppose that *A*,' is taken to signify an act – the intellectual act of supposing that *A*. Given that a suppositional term signifies an act, we may then use the general means available to us for specifying complex acts in terms of simpler acts to specify complex suppositional acts. So, for example, just as I can say 'do α and then do β,' so I can say 'first suppose *A* and then suppose *B*.' Moreover, just as I can use indexed modalities to indicate what will follow from a given act α — where $\Box_\alpha A$ is taken to mean that the truth of *A* will result from the performance of α, so we can use indexed modalities to indicate what will follow from a given suppositional act σ — where $\Box_\sigma A$ is now taken to mean that the truth of *A* will result from the supposition of σ.[15]

Of course, systems of natural deduction — especially when laid out in the style of Fitch — constitute a kind of suppositional calculus. We can see from a derivation within such a system that a given formula will follow from the supposition of other formulas. But such a system does not enable us to explicitly talk or reason about what follows from a supposition. It does not enable us to say, let alone show, for example, that some formula does *not* follow from the supposition of other formulas; and, as we shall see, it provides very limited means for saying what the suppositions might be.

As the previous paragraphs have perhaps made clear, the expression 'supposition' is multiply ambiguous. It may refer to a suppositional term, such as 'suppose *A*,' or to the instruction or act expressed by such a term, or the sentence or proposition (*A* or <*A*>) that is supposed. In case there is need of disambiguation, we can refer to the 'suppositional clause,' the 'suppositional act' and the 'supposed sentence or proposition.'

In what follows, we shall simply describe the language of the suppositional calculus and indicate in an informal manner how it is to be interpreted. In a fuller treatment, we should also provide a semantics and proof theory for the calculus. This is an interesting project in its own right; and I am hopeful, once it carried out, that we be able to provide a more formal vindication of the proposed reduction, under which the possibilist formula to be reduced will be both provably and semantically equivalent to its actualist counterpart.

Although my sole aim in this paper has been to put the calculus at the service of the actualist, it seems plausible that the calculus will have a number of other applications. Within logic, the calculus provides us with a means of generalizing the notion of a sequent, where the suppositional term σ provides us with a more general notion of an antecedent and where a sequent $\sigma \vdash A$ (or $\sigma \vdash A_1, A_2, \ldots, A_n$) will now be expressed by the modal formula $\Box_\sigma A$ (or $\Box_\sigma [A_1, A_2, \ldots, A_n]$); and within the semantics for natural language, the suppositional terms σ provide us with a means of keeping track of the structure and flow of suppositions within a given piece of discourse.

The calculus contains two main kinds of clause: suppositional and indicative. For ease of expression, we call the former 'S-terms' or 'terms' and the latter 'I-formulas' or 'formulas.' We suppose that the initial formulas are 'sourced' from the given language L (so, strictly speaking, the notions of formula and term will be relative to *L*).

Terms and formulas are defined by the following clauses (informal interpretation to follow):

(i) If *A* is an formula then /*A*/is a term;
(ii) \\ is a term;
(iii) if σ and τ are terms, then so is $(\sigma|\tau)$;
(iv) if σ and τ are terms, then so is $(\sigma; \tau)$;
(v) if *A* a formula and σ a term then $A \rightarrow \sigma$ is a term;
(vi) if σ is a term then so is $\forall x\sigma$;
(vii) if σ is a term then so is $\Box\sigma$;
(viii) if *A* is a formula of *L* then *A* is a formula
(ix) formulas are closed under the usual operations for forming formulas (negation, conjunction, actualist quantification, etc.)
(x) if σ is a term and *A* a formula, then $\Box_\sigma A$ is a formula.

Let us make some comments on the various types of construction:

The terms under (i) and (ii) constitute the two basic types of supposition, all other suppositions being formed from them by means of certain basic operations. /*A*/ (read 'suppose that *A*') allows us to suppose the truth of an indicative formula, while the back step constant \\ (read 'revert to the previous supposition') allows us to go back to what (if anything) was previously supposed. Thus, within a context in which one supposition is immediately embedded within another, \\ has the effect of taking us back to the immediately preceding supposition (this corresponds to the operation of moving one step leftward within a Fitch-style system of natural deduction); and when the context is one in which there is no immediately preceding supposition, \\ has the effect of going to the 'empty' suppositional state, in which nothing has been supposed.

(iii) and (iv) provide two ways of forming composite suppositions. $(\sigma|\tau)$ (read 'σ and τ') allows one *jointly* to suppose σ and τ. Thus $(/A/|/B/)$, in particular, tells one jointly to suppose *A* and *B*. $(\sigma; \tau)$ (read 'σ and then τ') allows one *successively* to suppose σ and τ. Thus $(/A/; /B/)$, in particular, tells one successively to suppose *A* and *then* to suppose *B*, thereby embedding the supposition of B within the supposition of *A*. Our understanding is that when one forms an embedded supposition, the previous suppositions remain in force. If one were to suppose, alternatively, that the previous suppositions do not remain in force, then the present conception of an embedded supposition should be defined as $(\sigma|(\sigma; \tau))$. The composite suppositions $(/A/|/B/)$ and $(/A/; /B/)$ are, of course, to be distinguished from the corresponding conjunctive supposition $/(A \wedge B)/$. For in the former cases, the two conjunct propositions are each supposed, either jointly or successively, while, in the latter, case the single-conjunctive proposition is supposed.

(v) allows us to form a conditional supposition $A \rightarrow \sigma$ (read 'if *A* then σ'). Thus, one might only want *Ga* to be supposed if *Fa* is the case; and this would then be expressed as $Fa \rightarrow /Ga/$. A conditional supposition, such as $Fa \rightarrow /Ga/$, should, of course, be distinguished from the supposition of the corresponding conditional $/Fa \rightarrow Ga/$. A conditional supposition only comes into effect if its antecedent is in fact the case, while the supposition of the corresponding conditional will come into effect regardless of the circumstances. We may say, in general, that a supposition is *weak* or *conditioned* if what it takes to be given is dependent on the actual circumstances and that otherwise it is *strict* or *unconditioned*. Thus, the construction $A \rightarrow \sigma$ (along with the construction $\forall x\, \sigma$, to be considered below) provides us with the means of forming conditioned suppositions.

(vi) allows us to form a generalized supposition $\forall x\, \sigma$ (read 'for all *x*, σ'). $\forall x\, \sigma$ requires the joint supposition of σ for each value of the variable *x*. The generalized supposition $\forall x\, /Ax/$ is, of course, to be distinguished from the corresponding general supposition $/\forall x\, A/$. The latter requires the supposition of a general proposition while the former requires the supposition of all of its instances. The generalized construction may be usefully combined with the conditional construction. For example, $\forall x(\text{Man}(x) \rightarrow /\text{Woman}(x)/)$ will require us to suppose, of each man, that he is a woman.

There is an important respect in which generalized suppositions differ from other generalized injunctions. Suppose I state 'climb every mountain!,' something which might be formalized as $\forall(x)(\text{Mountain}(x) \rightarrow \text{Climb}(x))$.

Then in order to execute this injunction, I must, for each mountain *x*, perform the individual act of climbing *x*. But suppose now that I state 'for each mountain, suppose that it is climbed,' something which might be formalized as $\forall x(\text{Mountain}(x) \rightarrow /\text{Climbed}(x)/)$. Then in order to 'execute' or make this supposition, it is not necessary that, for each mountain *x*, I perform the individual act of supposing that *x* is climbed. All I need to do is to state 'for each mountain, suppose that it is climbed.' There is a performative aspect to suppositional statements, which means that nothing beyond the statement of the supposition itself is required to ensure that the supposition is made.

(vii) is perhaps the most contentious of our constructions and allows us to turn a conditioned supposition σ into an unconditioned supposition $\Box\sigma$ (read 'necessarily, σ'). The necessity here should be taken to be metaphysically necessity; and $\Box\sigma$ then strengthens the supposition σ by requiring it to take effect not merely in the actual circumstances but in any counterfactual circumstances that might obtain. Suppose, for example, that σ is the conditional supposition $A \rightarrow /B/$, requiring the supposition of *B* when *A* is the case. $\Box\sigma$ will then require the supposition of *B* when *A* is possibly the case since, in any actual or counterfactual circumstance in which *A* is the case, σ would require the supposition of *B*.

Note that we do not understand $\Box(A \rightarrow /B/)$ to mean that *B* would be supposed if *A* were the case. Indeed, we can see on general grounds that this cannot be so, since we always want σ to tell us to form a supposition. But $\Box(A \rightarrow /B/)$, on the proposed reading, is an indicative sentence and only tells us what *has* been supposed under certain counterfactual circumstances, not what *is to be* supposed. Rather, $\Box(A \rightarrow \sigma)$ should be taken to mean 'necessarily, if ever *A* then form (i.e. actually form) the supposition σ.' The sole use we shall make of $\Box$ is in constructions of the form $\Box\forall x\Box(Ax \rightarrow /Bx/)$.

This construction tells us, in effect, to form the supposition *Bx* for any possible object *x* for which it is possible that *Ax*. Thus, $\Box\forall x\Box$ here serves as a quantifier over possible objects and so $\Box\forall x\Box(Ax \rightarrow /Bx/)$, were we allowed to use the possibilist quantifier Π, might also be expressed in the form $\Pi x(\Diamond Ax \rightarrow /Bx/)$.

Even if the above use of $\Box$ is for some reason rejected, it is still plausible that the actualist should be able to make sense of generalized suppositions of the form $\Pi x(\Diamond Ax \rightarrow /Bx/)$. For he can already make sense of the indicative claim $\Pi x(\Diamond Ax \rightarrow Bx)$ (for every possible object *x* for which it is possible that *Ax* it is the case that *Bx*). But if he can make sense of this claim, then what is to prevent him from making sense of the corresponding suppositional sentence $\Pi x(\Diamond Ax \rightarrow /Bx/)$ (for every possible object *x* for which it is possible that *Ax* suppose that *Bx*), given that the only difference between them lies in the substitution of a supposition for a judgment?

Consider a related case. The actualist can certainly make sense of the claim that every possible act of killing is forbidden ($\Pi x(\Diamond Kx \rightarrow Fx)$, since he can express it in the form $\Box\forall w\Box\forall x(\Diamond Kx \rightarrow Fxw)$.[16] But then surely he can also make sense of the injunction that every possible act of killing is to be forbidden

($\prod x(\Diamond Kx \rightarrow F!x)$) — something someone might bear in mind were they to envisage killing someone! Perhaps this injunction can be stated in corresponding fashion as $\Box\forall x(\Diamond Kx \rightarrow F!x)$, where the injunction $F!x$ automatically brings one back to the actual world. But even if it cannot, there would appear to be no essential difficulty in moving from the intelligibility of the indicative to the intelligibility of the corresponding injunction.

Of course, the actualist may be in no position to make some of the individual suppositions $/Bx/$. But this is equally true of the possibilist when he forms the supposition $\prod x\,\sigma x$ or even of the actualist when he forms the supposition $\forall x\,\sigma x$. Nor would it appear to be required that each of the individual suppositions should somehow be collected together in a single entity or that it should be possible to regard each of them as the supposition of a proposition that actually exists. For if such a constraint is not required for the intelligibility of the indicative sentence $\prod x(\Diamond Ax \rightarrow Bx)$, then why should it be required for the intelligibility of corresponding suppositional sentence $\prod x(\Diamond Ax \rightarrow /Bx/)$?

In what follows, I shall take the liberty of using $\prod x(\Diamond Ax \rightarrow /Bx/)$ as an abbreviation for $\Box\forall x\Box(Ax \rightarrow /Bx/)$, but with the understanding that there may be some alternative way of making the intended meaning of $\prod x(\Diamond Ax \rightarrow /Bx/)$ acceptable to the actualist.

(x) provides us with our sole means, beyond usual apparatus of connectives and quantifiers, for forming new formulas. Given a term σ and a formula A, it allows us to form the 'entailment' $\Box_{\sigma}A$, which may be read 'under the supposition σ, it logically follows that A' and which we may write in the form '$\sigma \vdash A$' to avoid confusion with $\Box A$. Here, σ is the *antecedent* or *suppositional base* and A the *consequent* of the entailment. But note that the antecedent is an S-term, not an indicative sentence, and signifies a complex suppositional act for which the propositions to be supposed may be several rather than single, embedded one within the other, and dependent upon the circumstances. We may abbreviate the formula $\backslash\backslash \vdash A$ to $\vdash^{-1} A$. Thus, $\vdash^{-1} A$ tells us that A follows from what was previously supposed.

The intended notion of logical consequence is logical consequence in a broad sense which goes beyond classical first-order logical consequence in a number of ways. First, it encompasses higher order logic and logic with generalized quantifiers — such as 'for finitely many' or 'uncountably many' — as long as these quantifiers are of a purely logical character, and it also encompasses any other higher order constructions — such as plural quantification or quantification over classes — whose use by the possibilist may be in question. Second, it accommodates modal reasoning. We will take A, for example, to be a logical consequence of $\Box A$. There may, of course, be different opinions as to how higher order modal logic should be formulated. But it is important, if we are not to prejudge the issue between the actualist and possibilist or between the contingentist and necessitist, that the logic should be acceptable to both and that it should, in particular, remain non-commital on the validity of the Barcan Formula (and its converse). Third, we should allow the consequence relation to hold between what, in effect, are singular positions. Quine has inveighed against quantifying into modal contexts, but his concerns may be met in the present case if we take the identity and distinctness of the individuals in the different singular proposition to be an aspect of their logical form (as proposed in Fine 2005, 108–112); and this is how we shall understand the notion in what follows.

6. The suppositional base

We now want to specify some suppositions that will serve to describe the possibilist's pluriverse from an actualist point of view. We will do this by means of a suppositional term of the form $\sigma; \tau$, with one supposition within the other and with each of the component suppositions σ and τ formulated in terms acceptable to the actualist.

So let us specify each of the terms σ and τ in turn. The unembedded supposition σ will be the joint supposition $(\sigma_1|\sigma_2|\sigma_3|\sigma_4|\sigma_5) = (((((\sigma_1|\sigma_2)|\sigma_3)|\sigma_4)|\sigma_5))$ of the following suppositions, for *R* any *n*-place predicate of L[17]:

σ_1: $\Pi x / \Box \exists y\, x = y /$

σ_2: $\Pi x_1 \Pi x_2 \ldots \Pi x_n \Pi w\, (Rx_1x_2\ldots x_nw \rightarrow /Rx_1x_2 \ldots x_nw/)$

σ_3: $\Pi x_1 \Pi x_2 \ldots \Pi x_n \Pi w (\neg Rx_1x_2\ldots x_nw \rightarrow /\neg Rx_1x_2\ldots x_nw/)$

σ_4: $/\Box \exists w Ew \wedge \Box \forall w \forall v (Ew \wedge Ev \supset w = v)/$

σ_5: $\forall w / \Pi x_1 \Pi x_2 \ldots \Pi x_n\, (Rx_1x_2\ldots x_nw \equiv Rx_1x_2 \ldots x_n)/$

The embedded supposition τ is simply:

τ: $/\Pi x (\vdash^{-1} \exists y\, x = y)/$

Let me make some comments on the interpretation and role of these various suppositions:

σ_1 is the sole supposition of σ to be counterfactual, i.e. to be incompatible with what is the case. It tells us to suppose, of each possible object, that necessarily there is some actual thing that it is. Of course, the actualist believes no such thing, but he may still intelligibly hypothesize that this is how things are.

σ_2 tells us to suppose, whenever a given *n*-place predicate *R* holds of some possible individuals in a possible world, that this is the case; and σ_3 tells us to suppose, whenever a given predicate does not hold of some possible individuals in a possible world, that this is not the case. Thus, the behavior of the given properties and relations will be the same in the hypothetical pluriverse as it is in the actual pluriverse.

It is important to bear in mind that the predicate *E* for existence has been taken to be one of the given atomic predicates. Thus, as a special case of σ_2 and σ_3, we have:

σ_2': $\Pi x \Pi w\, (Exw \rightarrow /Exw/)$

σ_3': $\Pi x \Pi w\, (\neg Exw \rightarrow /\neg Exw/)$

These clauses tell us that the behavior of the existence predicate is the same in the hypothetical pluriverse as it is in actual pluriverse. Since we have also, in effect, supposed $\Box\forall x\Box\exists y\, x = y$ (that necessarily each object is necessarily something), we will no longer have the usual connection $\Box\forall x\Box(Ex \equiv \exists y\, x = y)$ between existence and actuality. Existence in the hypothetical pluriverse will be the shadow, so to speak, of actuality in the actual pluriverse.

It should also be noted that there is no need to assume Williamson's 'being constraint' of chapter 4 of MLM (also called the 'falsehood principle' in Fine 1981), according to which an atomic predicate can only be true in each world of the actual objects of that world. For even if a dyadic predicate *R*, say, is true of two non-actual objects *x* and *y* in a given world *w*, σ_2 will still trigger the supposition that this is so.

σ_4 tells us to suppose that possible worlds behave as possible worlds. We can no longer suppose, of course, that World Existence and Identity hold since worlds, like everything else, will necessarily be actual. But we make the corresponding suppositions that necessarily some world exists and necessarily any two existent worlds are the same.

σ_5 tells us to suppose that the actual world behaves as the actual world, with truth of an atomic predication in the actual world being tantamount to truth simpliciter.

Finally, the embedded supposition τ requires us to suppose that each possible object is identical to one whose being follows from the preceding supposition. Since it is only the possible objects *a*, *b*, … from the original actualist pluriverse whose being will follow from the preceding supposition, the present embedded supposition requires, in effect, that each possible object should be identical to one of *a*, *b*, …. It is in this way that we can guarantee that the original ontology of possible objects is exhaustive of the possibilist ontology without using an infinitary sentence.

If the predicate *E* is allowed to apply to higher order entities, then we might also include such suppositions as:

τ': $/\Pi P \Pi w(EPw \rightarrow \vdash^{-1} EPw)/$.

This guarantees that *E* will not have an application to such higher order entities beyond its original application in the actual pluriverse. It should be noted that it is only in the suppositions τ or τ' that we make use of the back space operator \\; and so even if we have reservations about the general use of the operator, it is only its use in the present context that is required to legitimate the present reduction.

We are now in a position to state the reduction. Let *A* be a possibilist statement from the higher order version of the language *L*, possibly containing the outer quantifiers and the existence predicate at each type, but none of the inner quantifiers. Let *A′* be the result of replacing all of the outer quantifiers by the corresponding inner quantifiers. $\Pi x\Box\Sigma y\, x = y$, for example, becomes $\forall x\Box\exists y\, x = y$. Of course, *A′* will not be an acceptable actualist substitute for *A* since *A′* may often be false when *A* is true, as in the previous example. However, the formula *A′* will be useful as an intermediate step in giving the required reduction *A** of *A*.

For given a possibilist formula *A* from the higher order version of the language *L*, the actualist reduction *A** may be taken to be the formula $\sigma; \tau \vdash A'$, to the effect that *A*′ follows from the supposition $\sigma; \tau$.

We may convince ourselves of the adequacy of the reduction as follows. The statement *A* will be true if *A* logically follows from a complete first-order description of the possibilist's first-order pluriverse (in which it is said what objects there are, what worlds there are, which world is actual and how the objects behave in each of the worlds). *A* will follow from such a description iff *A*′ follows from a complete first-order description of the actualist analog of such a pluriverse (in which necessarily each actual object is necessarily actual). But $\sigma; \tau$ constitutes a complete first-order description of the actualist analog of the possibilist pluriverse; and so *A*′ will follow from this complete first-order description of the actualist analog iff *A*′ follows from the supposition ($\sigma; \tau$).

There are two key aspects of the proposed 'demonstration.' First, it is supposed that a complete first-order description of the pluriverse will logically entail whatever holds at the higher level; the first-order is taken, in this sense, to be *determinative* of the higher order. Second, we postulate devices of supposition that enable us to provide what is, in effect, a complete first-order description of the possibilist pluriverse. Of course, we have only demonstrated equivalence of truth-value. But the demonstration is broadly logical in character and so, under any reasonable conception of content, *A* will have the same content as *A**.

It is also very plausible, as I have tried to argue, that the reductive statement *A** will indeed be acceptable to the actualist — nothing in the ideology of *A*′ or in the suppositional apparatus of $\sigma; \tau$ will require appeal to mere possibilia; and so we have what would appear to be an adequate and acceptable account of higher order possibilist discourse in actualist terms.

7. Comparisons, consequences, objections

The present reduction has a number of advantages over my previous reduction. The main advantage is that it is completely finitary and, in particular, requires no appeal to infinitely long conjunctions or disjunctions or to infinitely long strings or branches of quantifiers interlaced with modal operators. But it also has a certain flexibility in its favor. For the general idea of determining higher order truths on the basis of first-order truths will have application to other forms of higher order truth (such as quantification over pluralities or sets) and to areas other than quantified modal logic.

I had always thought that there should not be a separate problem about effecting a reduction of higher order possibilist discourse that the intelligibility of higher order discourse should somehow follow from the intelligibility of first-order discourse. But I could come up with no satisfactory way to articulate the thought. The present account shows how this might be done. For it shows how, once we are able to state how things are at the first-order level, we can appeal to a general notion of logical consequence to say how things are at the higher level.

However, the proposed reduction is not completely painless. The ideology of the actualist's reduction will need to go beyond the ideology of the possibilist's discourse in a number of key respects. For it will need to make reference to worlds or the like, even when the possibilist makes no such reference. Of course, in this respect, the higher order reduction is not so different from the first-order reduction (though for the purposes of the first-order reduction, we can use Vlach's up and down arrows in place of quantifiers over worlds). More significantly – and this *is* a major difference from the first-order case — our actualist will need to employ a concept of logical consequence that is hyper-intensional; consequence will not preserved under the 'substitution' of different necessarily equivalent propositions. And this might be regarded as a reason in itself for rejecting the reduction.

But quite apart from a general prejudice against hyper-intensional notions (which is perhaps no better founded than the earlier prejudice against intensional notions), it is not clear that there is anything especially problematic about our notion of logical consequence. There is, of course, the need to fix the relevant sense of consequence; and it may also be necessary to fix the intended sense of the higher order quantifiers. But once this is done, then it is far from clear that we do not have a notion that is in reasonably good order.

In a recent paper, Fritz and Goodman (2014) raise the specter of 'revenge.' Suppose that the actualist uses some further modality, such as our broad notion of logical necessity, in attempting to reduce possibilist discourse. Then does not the problem arise anew for objects which are possible with respect to the new modality? In our own case, even if we can explain away metaphysical possibilia with the help of a logical modality, how do we explain away the logical possibilia?

There *is* a possible danger of this sort. But it does not necessarily arise, since it cannot simply be taken for granted that the Barcan Formula (or its converse) will be true for any given modality under an unrestricted reading of the quantifier, even if this is so for other modalities. The present case strikes me as a case in point. Let us suppose, for the sake of argument, that there are just five objects *a*, *b*, *c*, *d*, and *e*; and let *a*, *b*, *c*, *d*, and *e* be names for these objects. Then under a reading of $\Box$ as 'it is logically necessary that' and using Πx for the unrestricted quantifier, $\Pi x \Box (x = a \vee x = b \vee \ldots \vee x = e)$ should hold, since every object x will be identical to *a* or *b* or ... or *e*. But surely $\Box\, \Pi x (x = a \vee x = b \vee \ldots \vee x = e)$ will not hold; it will not be a *logical* necessity (something guaranteed by logical form alone) that every object *x* will be identical to *a* or *b* or ... or *e*. This means that there can be no intelligible notion of possible object corresponding to the new modality, since otherwise quantification over such objects would conform to the Barcan formula; and hence the possibility of revenge does not even arise.[18]

Fritz and Goodman have also objected to a counterfactual reduction, in which the possibilist statement *A* is, in effect, replaced with the statement that *A′* would be true if necessarily everything were necessarily to exist. I myself have general misgivings about counterfactual analyses (of necessity, knowledge, cause, and the like). For the use of any counterfactual requires us to take something as 'fixed'; and there is always the danger that the only way to get the counterfactual analysis to come out as correct is to presuppose the very notion to be analyzed in understanding what it is that is to be kept fixed (similar misgivings also arise over fictionalist analyses).

There is a way in which my own reduction is counterfactual in character, since I have the actualist counterfactually supposing that necessarily every actual object is necessarily actual. However, I do not simply rely on this counterfactual supposition in drawing out its consequences. I also specify in complete detail everything else that must obtain, so that it is simply a matter of logic what the consequences should be. Thus, general misgivings over the import or circular character of counterfactual forms of analysis will have no bearing in the present case.

Notes

1. I should like to thank the audiences at the 2014 Prior conference in Oxford and at the 2014 meeting of the Bucharest Society of Analytic Philosophy for helpful comments. I have also benefitted from conversations with Walter Dean and from conversations with and written comments from Peter Fritz and Jeremy Goodman.
2. This is part of my general stance on ontological questions (Fine 2009). However, the formulation in this case might be regarded as over-simple for at least two reasons. One is that the actualist might wish to make a modal claim, to say that *necessarily* only actual objects are real; another is that a philosopher might believe that only actual objects are real, not because he is an actualist, but because he is a Pythagorean, let us say, and believes that only numbers are real and that all numbers are necessarily actual. But I take it that such subtleties in the formulation of the dispute are not here at issue.
3. I provide a defense of my own view in chapter 9 of Fine (2005a), especially §9.
4. At one point (23), Williamson asks 'And why should the alternative to the view that everything does the harder thing [being actual] be a view on which everything *could* do the harder thing [be possibly actual]?' A good question, to which the answer is that this is not the only alternative. One might think, for example, that there are sets of incompossible objects which could not even be possibly actual and, indeed, such a view would naturally follow from a position which took all urelements to be possibly actual and yet took some of the urelements to be incompossible.
5. I also believe that there is no intelligible notion of unrestricted quantification (Fine 2006). But I do not press the point here, since I agree with Williamson (MLM, 15) that we can get at 'the point' behind the N/C dispute without appeal to the notion.
6. Additionally, if one thinks, as I do, that a wooden chair, say, is necessarily wooden and hence necessarily concrete then the appeal to concreteness is doomed from the start.

7. But I should note that, strictly speaking, the reduction applies recursively to all formulas and that care should be taken to avoid clash of variables.
8. In the more general case, the outer quantifiers may also range over impossible objects – over sets of incompossibles, for example.
9. This is not quite Williamson's proposal on 353 of MLM, but the differences are incidental to the purpose at hand. There is also a subtlety concerning empty pluralities which I have, for convenience, ignored.
10. Fritz and Goodman (2014) point out that an especially severe form of this difficulty arises with the quantifier 'most.'
11. The study of parallel processing or 'process algebra' has become a significant branch of computer science in its own right (see Boeten 2005 for a brief survey). I might also note that the case of substitution shows that Williamson's putative tests for independence (idempotence and commutativity, 358) are clearly off the mark.
12. The theory of classes advocated in Fine (2005b) is one such example.
13. This is a place where appropriate adjustments will need to be made under the N/C construal of the debate.
14. A similar approach to Sider's is considered in Fine (1977a), 147 and developed in §3 of Fine (2002). My own approach also avoids the problems raised by Leuenberger (2006) and extended in the formal appendix to Fritz and Goodman (2014).
15. A symbolism of this sort is familiar from dynamic programming logic (Harel 1984), another well-developed branch of computer science.
16. This claim makes most sense if we take it to be essential to an act of killing that it be an act of killing.
17. If only the identity, and not also the distinctness, of individuals is an aspect of logical form, then we should add the supposition $\Pi x \Pi y(\neg x = y \rightarrow /\neg x = y/)$ (suppose, of any two distinct possible objects, that they are distinct).
18. As Jeremy Goodman has pointed out to me, the contingentist also does not think that the Barcan Formula holds for metaphysical necessity and so this may be a respect in which there is a significant difference between the A/P and N/C formulations of the debate.

References

Boeten, J. C. M. 2005. "A Brief History of Process Algebra." *Theoretical Computer Science* 335: 131–146.

Fine, K. 1977a. *Prior on the Construction of Possible Worlds and Instants*. in Prior and Fine 1977b, 116–168'; reprinted as chapter 4 of Fine (2005a).

Fine, K. 1977b. "'Properties, Propositions and Sets." *Journal of Philosophical Logic* 6: 135–191.

Fine, K. 1981. "Model Theory for Modal Logic, Part III: Existence and Predication." *Journal of Philosophical Logic* 10: 293–307.

Fine, K. 2002. *The Problem of Possibilia*, in Loux and Zimmerman (2002), 161–179; reprinted as chapter 6 of Fine (2005a).

Fine, K. 2005a. *Modality and Tense: Philosophical Papers*. Oxford: Oxford University Press.

Fine, K. 2005b. "Class and Membership." *Journal of Philosophy* 102: 547–572.

Fine, K. 2006. "Relatively Unrestricted Quantification." In *Chapter 1 of 'Absolute Generality'*, edited by A. Rayo and G. Uzquiano, 20–44. Oxford: Oxford University Press.

Fine, K. 2009. "The Question of Ontology." In *Metametaphysics: New Essays on the Foundations of Ontology*, edited by D. Chalmers, D. Manley, and R. Wassermann, 157–177. Oxford: Oxford University Press.

Fritz, P., and J. Goodman. 2014. *Counting Incompossibles*, unpublished draft.

Harel, D. 1984. "Dynamic Logic." In *Handbook of Philosophical Logic, Volume II: Extensions of Classical Logic*, edited by D. Gabbay and F. Guenthner, chapter 10, 497–604. Dordrecht: Reidel.

Kripke, S. 1963. "Semantical Considerations on Modal Logic." *Acta Philosophica Fennica* 16: 83–94; reprinted in 'Reference and Modality' (ed. L. Linsky), Oxford: Oxford University Press, 1971.

Leuenberger, S. 2006. "A New Problem of Descriptive Power." *Journal of Philosophy* 103 (3): 145–159.

Loux, M., and D. Zimmerman, eds. 2002. *The Oxford Handbook of Metaphysics*. Oxford: Oxford University Press.

Prior, A. N., and K. Fine. 1977. *Worlds, times and Selves*. London: Duckworth; chapter 6 of Fine (2005a).

Sider, T. 2002. "The Ersatz Pluriverse." *Journal of Philosophy* 99: 279–315.

Williamson, T. 2013. *Modal Logic as Metaphysics*. Oxford: Oxford University Press.

Reply to Fine

Timothy Williamson

Wykeham Professor of Logic, University of Oxford, Oxford, UK

As should be clear, *Modal Logic as Metaphysics* owes a large debt to Kit Fine's massive contributions to technical aspects of modal metaphysics, though it also takes issue with some of his proposals. In 'Williamson on Fine on Prior on the reduction of possibilist discourse', (Fine 2016), he defends much in his earlier proposals while providing it with a new basis. In this reply, I follow his order of discussion.

1. What is the problem?

Fine's work on modal metaphysics is framed by the dispute between actualism and possibilism. *Modal Logic as Metaphysics* finds that dispute confused, and reframes modal metaphysics in terms of the clearer dispute between contingentism and necessitism. The first section of his paper argues that we should still focus on his original dispute rather than my proposed alternative.

Fine does not deny that necessitism and contingentism are clearer doctrines than actualism and possibilism. The necessitist asserts 'Necessarily everything is necessarily something'; the contingentist denies what the necessitist asserts. In Fine's current formulation, the actualist asserts 'Only actual objects are real' (or perhaps 'Necessarily only actual objects are real'); the possibilist denies what the actualist asserts. The necessitist's assertion and the contingentist's denial can be formulated in the language of first-order modal logic, with the quantifiers read absolutely unrestrictedly and the modal operators read metaphysically. By contrast, 'actual' and 'real' in the actualist's assertion and the possibilist's denial are not to be defined in the language of first-order (or higher-order) modal logic. Instead, Fine treats 'actual' as an undefined term, introduced by examples: Wittgenstein is actual, his children are merely possible. He introduces 'real' differently: to be real is to figure in reality, where the term 'reality' is itself undefined, though to some extent explained by a sketch of the role it is meant to play in metaphysical theorizing (Fine 2009). Fine himself contrasts 'the relatively obscure idioms of actuality and reality (or the like)' with the language of quantified modal logic.

Of course, not even absolutely unrestricted quantification and metaphysical modality are above challenge. Fine himself rejects the former, though not the latter. However, he agrees with me that relevantly similar doctrines can be defined using even weaker resources. For instance, reinterpreting the quantifier as restricted to possibly concrete objects and the modal operator as restricted to physical possibilities preserves the most salient contrasts between contingentism and necessitism. Indeed, Fine's formulations of actualism and possibilism themselves seem to require a quantifier unrestricted at least within reality, since 'Only actual objects are real' is equivalent to 'Everything real is an actual object', where a restriction on 'everything' that excluded some real things would undermine the actualist's intention.

Another apparent difference between the two pairs of doctrines is that whereas contingentism is the contradictory of necessitism, possibilism may be the contrary rather than the contradictory of actualism. For Fine allows that there may be *impossible* objects, in the sense of objects that are not even possibly actual. Consider the formally consistent metaphysical theory that some real things are necessarily actual, some real things are necessarily non-actual, and every real thing is either necessarily actual or necessarily non-actual. That theory is inconsistent with actualism, but it also seems to be inconsistent with possibilism, on any natural understanding of that term. Thus a more general dispute may set actualism against its contradictory rather than against possibilism. However, that does not matter for present purposes.

Fine regards the dispute between necessitism and contingentism as clear enough to be getting on with. His principal objection to focusing on it is simply that he takes it to be obvious which side is right: necessitism. Indeed, he writes: '*anyone* in their right mind should accept necessitism' and 'the issue of necessitism vs. contingentism is a non-issue and any right-minded philosopher should accept necessitism without further ado'. To back up these claims, Fine gives an example of counting the possible knives that could but never will be made from three given blades and three given handles (I use a similar example in favour of necessitism in Williamson 2000). He takes it as just evidently true that there are nine, as the necessitist says, not zero, as the contingentist says; for the contingentist to postulate a non-literal sense of the question to which they can regard the answer 'nine' as correct 'is simply to put bad linguistics at the service of bad philosophy'. Naturally, I feel some sympathy with Fine here, and he is under no obligation to spend more of his time on contingentism. But his impatient dismissal does not make the issue go away.

Fine does not quite say that all those in their right minds accept necessitism, just that they *should* do so; but what if they don't? For instance, Robert Stalnaker presumably counts as in his right mind. He rejects necessitism and develops a contingentist alternative with a clear understanding of what is at issue (Stalnaker 2012). I regard him as wrong, but not as confused. One might hope to do better than just telling Stalnaker to 'accept necessitism without further ado'. After all, many contingentists regard themselves as in a position to dismiss necessitism without further ado, for example on the grounds that it is just evidently true that the table is only contingently something. Conducted thus, the dispute between necessitism and contingentism faces deadlock. In logic, metaphysics and any other serious inquiry, we need methods that enable us to work through clashing senses of the obvious with some hope in the long run of an adequate resolution. Granted, most of what seems obvious to us is indeed true, but a methodology that crashes every time something false seems obvious is insufficiently robust for our purposes. *Modal Logic as Metaphysics* applies and recommends an abductive methodology for first principles of logic; it enlarges the evidence base on which we choose between them so that we can finesse deadlock on a few individual data points. That resembles what happens in natural science: if we cannot agree whether the outcome of a particular experiment refutes a theory, we can test the theory in other ways.

Whatever Fine's individual preferences, those interested in modal metaphysics have collectively to engage with the issue between necessitism and contingentism. Moreover, since it concerns a central choice-point in how we reason about modal matters, with ramifying consequences in many directions, it has to take a fundamental role in modal metaphysics. In comparison, Fine's issue between actualism and possibilism looks secondary, because he in effect sets up the options as two types of necessitism. For he explains the term 'actual' through a contrast between two kinds of thing: those like Wittgenstein (the actual things) and those like Wittgenstein's (merely possible) children (the non-actual things). But it is the necessitist who postulates things like Wittgenstein's merely possible children. From a contingentist perspective, there are no such things, so no contrast with which to explain 'actual'.

Once we understand the dispute between actualism and possibilism as a dispute between rival forms of necessitism, is it clear enough to be productive? Can we fruitfully investigate whether only actual objects are real? I do not regard the sentence 'Only actual objects are real', used in Fine's way, as meaningless. It does not take much to avoid utter meaninglessness in a natural language, and Fine's explanations of 'actual' and 'real' achieve that much. But I am sceptical as to how much more than that they achieve, especially in the case of 'real'. It is not that theorists are obliged to define their new theoretical vocabulary in more basic terms; I agree with Fine that it would be unfair to demand such a reductive definition of 'real'. Sometimes a new theory introduces a valuable distinction we could not previously make. But the theory needs to be a good one, to constrain the use of the term enough to be productive. In the case of 'real', I see no such theory. There are links with other vague phrases of the sort that hand-waving metaphysicians like to gesture with, such as 'consist in', 'nothing more than' and 'account for', which are used in characterizing the relation between the unreal and the real, but such links do not amount to a serious theory. In this respect, 'real' (in Fine's sense) compares poorly with 'necessary' (in the metaphysical sense): our current theories of necessity are multiply constrained by logic, common sense and natural science (Williamson 2007, 2013, 2016) to a much greater extent than are our current theories of reality. If that sounds strange, that is because we tend to hear 'reality' in something like its ordinary sense, in which the primary contrast is with 'appearance', 'fiction' and the like. *Of course* natural science tells us lots about what *is* as opposed to merely *appearing* to be, but that is not the contrast Fine intends with his special metaphysical sense of 'real'. For in the latter sense, something can *be*, not just appear to be, yet not *really* be. That is why 'reality' talk in this special metaphysical sense is so much less constrained than it sounds.

I will not insist here on scepticism about the metaphysical sense of 'real'. As elsewhere, the proper test is an abductive one. If 'real' in that sense is latching onto something profoundly important, then we can expect it sooner or later to figure in powerful theories whose success it would be curmudgeonly to dispute. They might be like theories of natural or social science, only more abstract; they might be more like theories of logic, only more concrete; for present purposes it does not matter which. If such a theory does emerge, it will go far towards vindicating the importance of the dispute between actualism and possibilism as Fine understands it. At the time of writing, I see no such theory on the horizon.

I now turn to the question of mappings between contingentist and necessitist discourse (Williamson 2013, chapter 7), and between actualist and possibilist discourse. In early discussions of actualism and possibilism, such mappings were presented as *translations* between the actualist's language and the possibilist's language. The background assumption was that in effect the actualist and the possibilist spoke different languages, albeit homophonic ones: they assigned different meanings to the quantifiers, and correspondingly different meanings to the sentences in which they occurred, and were speaking past each other when they appeared to disagree. The standard of correctness for the translations was that a sentence in the actualist's language should be mapped to a sentence in the possibilist's language with the same meaning, or at least the same coarse-grained truth condition, and *vice versa*. *Modal Logic as Metaphysics* argues against such a conception of the debate between contingentism and necessitism. My contingentist and my necessitist use a common language with shared meanings: in particular, they both use absolutely unrestricted quantifiers, as is the norm in metaphysics. They know what they are doing; there is no misunderstanding between them. Their apparent theoretical disagreement is just what it seems. No translation is needed, and no mapping from a sentence expressing contingentism to a sentence compatible with necessitism or *vice versa* is an adequate translation. As far as I understand the debate between actualism and possibilism in Fine's sense, it conforms to a parallel argument. The actualist and the possibilist assign the same meanings to 'actual', 'real', and the quantifiers, but may still disagree as to what is real. I will not labour these points here, because Fine does not argue that either debate depends on meaning variance between the two sides.

In some cases, necessitist discourse cannot be usefully mapped into contingentist discourse, or *vice versa*. For instance, an idiosyncratic necessitist might claim that even if a river were not spatiotemporally located, it would still be ugly or beautiful in ways that do not supervene on anything else. Few contingentists would wish to capture those alleged distinctions in their own terms. Conversely, an idiosyncratic contingentist might claim that a number exists or not contingently in ways that do not supervene on anything else. Few necessitists would wish to capture those alleged distinctions in their own terms. However, those are unusual cases. More typically, each side regards the other's discourse not as capricious but as marking genuine distinctions, despite misdescribing them as a result of seeing them through the distorting lens of false metaphysical assumptions — like someone whose literal use of the sentences 'The gods are angry' and 'The gods are not angry' marks but misdescribes the distinction between situations in which there is thunder and situations in which there is no thunder. Consequently, each side wishes to capture in its own terms the genuine distinctions the other's discourse marks, purifying them of the misdescriptions.

To make this idea more precise, *Modal Logic as Metaphysics* sketched a neutral sublanguage of the necessitist and contingentist's shared language of quantified modal logic. Its sentences are neutral in the sense that they do not raise any of the issues in dispute between the necessitist and the contingentist. To start with contingentist discourse, the contingentist's metaphysical theory is CON. The contingentist *marks a distinction* by the sentences A and B if and only if they are not equivalent given CON, in other words, the biconditional $A \leftrightarrow B$ is not a logical consequence of CON. To capture this distinction in neutral terms, we want to map A and B to neutral sentences $(A)^{con}$ and $(B)^{con}$, respectively, which are not logically equivalent. By contraposition, if $(A)^{con}$ and $(B)^{con}$ are logically equivalent, then $A \leftrightarrow B$ should be a logical consequence of CON:

(F1) If $\vDash (A)^{con} \leftrightarrow (B)^{con}$ then CON $\vDash A \leftrightarrow B$

If B is already neutral, then $(B)^{con}$ should just be equivalent to B, for there is no purifying to be done. Since $(A)^{con}$ is neutral, this implies:

(F2) $\vDash ((A)^{con})^{con} \leftrightarrow (A)^{con}$

(Compare proposition 1.3 at Williamson 2013, 371.) But from F1 and F2 we can deduce F3 by substituting $(A)^{con}$ for A and A for B in F1:

(F3) CON $\vDash (A)^{con} \leftrightarrow A$

(Compare proposition 1.8 at Williamson 2013, 372.) Conversely, we can derive F1 from two instances of F3. Thus, for each sentence of the language, the mapping should give us a neutral sentence equivalent to it within the contingentist's metaphysical theory.

There is a parallel line of thought for necessitist discourse. The necessitist's metaphysical theory is NEC. The necessitist *marks a distinction* by the sentences

A and B if and only if the biconditional $A \leftrightarrow B$ is not a logical consequence of NEC. To capture this distinction in neutral terms, we want to map A and B to neutral sentences $(A)^{nec}$ and $(B)^{nec}$, respectively, which are not logically equivalent; hence:

(F4) If $\vDash (A)^{nec} \leftrightarrow (B)^{nec}$ then NEC $\vDash A \leftrightarrow B$

If B is already neutral, then $(B)^{nec}$ should just be equivalent to B, for there is no purifying to be done. Since $(A)^{nec}$ is neutral, this implies:

(F5) $\vDash ((A)^{nec})^{nec} \leftrightarrow (A)^{nec}$

But from F4 and F5 we can deduce F6 by substituting $(A)^{nec}$ for A and A for B in F4:

(F6) NEC $\vDash (A)^{nec} \leftrightarrow A$

(Compare proposition 1.21 at Williamson 2013, 373.) Conversely, we can derive F4 from two instances of F6. Thus, for each sentence of the language, the mapping should give us a neutral sentence equivalent to it within the necessitist's metaphysical theory.

Since CON and NEC are mutually inconsistent but individually consistent, inevitably some biconditionals will be theorems of NEC but not of CON, and others will be theorems of CON but not of NEC. In such cases, the side for which $A \leftrightarrow B$ is a theorem will regard the other side as misdescribing the distinction they mark by applying A but withholding B or *vice versa*.

The derivations of F3 and F6 address Fine's reasonable question as to why each side should want a neutral *equivalent* of each sentence A within the other's theory, rather than simply looking at all the neutral *consequences* of A within the theory. Equivalents in the sense of F3 and F6 are needed to capture the *distinctions* the other side is marking but misdescribing. Without F3, F1 would not hold (since F2 is unproblematic), and without F6, F4 would not hold (since F5 is unproblematic). If two sentences with the same neutral consequences were not equivalent within the metaphysical theory, no distinction they marked could be captured in neutral terms.

Of course, as Fine indicates, we should aim to get more precise about where the pressure on each side is coming from to regard the other side's discourse as marking genuine distinctions. It does not come simply from the bare nature of the metaphysical doctrines of necessitism and contingentism, for we have already seen that some versions of each put no such pressure on the other side. Instead, the pressure should come from more specific successfully working theories whose core of modal metaphysics is CON or NEC — somewhat as the pressure on nominalists to capture distinctions made in platonist mathematical terms comes from successful applications of mathematics in the natural and social sciences. *Modal Logic as Metaphysics* does not attempt to provide the details, but elsewhere I argue that applications of probability theory and dynamical systems theory in the natural sciences implicitly take for granted a necessitist modal logic (Williamson 2016). It is also plausible that much contingentist discourse marks genuine distinctions, even if it misdescribes them.

Since the pressure to capture the distinctions marked by the other side's discourse comes from various specific necessitist and contingentist theories, we should not expect the neutral zone to be constant irrespective of the specifics of those theories. What is in dispute between one contingentist and one necessitist need not be in dispute between another contingentist and another necessitist. I allow for this by giving only a broad, deliberately schematic characterization of the technical term ('chunky') used to demarcate the neutral zone (Williamson 2013, 314). The book provides a template that can be applied to many specific versions of contingentism and necessitism. As Fine points out, various specific characterizations of the chunky contemplated in the book will not fit all versions of contingentism, but that was only to be expected. Only when contingentists starts fleshing out the barebones doctrine of contingentism in reasonable ways is there good reason to view them as marking genuine distinctions. This dependence on the version of contingentism at issue also means that his offer of his more rarefied sense of 'actual' in place of 'chunky' is unlikely to do the job, precisely because it does not depend on the specific versions at issue. In any case, since his introduction of his sense of 'actual' presupposes necessitism, it seems unsuitable for demarcating the neutral ground between a contingentist and a necessitist.

There are indeed complicated and somewhat elusive problems as to why, when and how the contingentist and the necessitist need to capture each other's distinctions. But such obscurities do less harm if confined to one part of the attempt to answer the overall question (contingentism or necessitism?), as they are in *Modal Logic as Metaphysics*, rather than infecting the question itself, thereby rendering any attempt to answer it problematic. They do not counterbalance the obscurity of the issue between actualism and possibilism.

Fine suggests that no problem like those just considered arises for the attempt to assess actualism and possibilism. He explains only very briefly why not, and what he says is puzzling. He writes:

> For a 'soft' actualist like myself, by contrast, there is no such problem; for I will be willing to accept whatever the possibilist says as *true* and hence will take myself to be under an obligation to account for its truth.

This cannot be right in general, for Fine's actualist 'claims that only actual objects are real, while the possibilist denies that this is so'. When the possibilist says 'Some non-actual objects are real', the actualist Fine is *not* willing to accept 'Some non-actual objects are real' as true, on pain of contradiction (his actualist and possibilist say the same thing when they use such a sentence). Perhaps he means only that he will be willing to accept as true whatever the possibilist says *not involving 'real' and similar terms*. But then there is still a problem about the sense, if any, in which the actualist can capture the distinctions marked in the possibilist's use of sentences like 'There are infinitely many real possible stars'. Is there meant to be some advantage in confining the differences between the actualist and the possibilist to sentences involving special metaphysical vocabulary such as 'real'?

There may be a dilemma here. First, suppose that there is no pressure on the actualist or possibilist to capture the distinctions marked in the other's use of 'real'. That would confirm the earlier suspicion that the use of the term is dangerously under-constrained. Alternatively, suppose that there is such pressure. Then the dispute between actualism and possibilism raises complicated and somewhat elusive problems as to why, when and how the actualist and the possibilist need to capture each other's distinctions, similar to the problems raised by the dispute between contingentism and necessitism.

For all these reasons, I continue to regard 'Contingentism or necessitism?' as a much better question than 'Actualism or possibilism?'

Fine formulates the rest of his paper in terms of actualism and possibilism rather than contingentism and necessitism. As he says, it is relatively ease to transpose what he says to fit the latter debate too, given the structural similarities. I will work with his formulations of his arguments here.

2. Fine's standard reduction

In Sections 2 and 3 of his paper, Fine sketches his original reduction of possibilist to actualist discourse (Fine 1977), and responds to two criticisms in *Modal Logic as Metaphysics* of an analogous attempt to map necessitist discourse to a neutral equivalent in the sense of F6, which are equally relevant to his reduction. He describes the reduction as 'translational' only in scare quotes.

My criticisms apply once the possibilist discourse goes beyond the limitations of a first-order language, in particular by the addition of a plural quantifier or other higher-order quantifiers. The initial difficulty for reducing the possibilist plural quantifier ΣV is that the modalized actualist plural quantifier $\Diamond\exists V$ is not an adequate substitute, because the former only requires some objects that individually can each be actual, while the latter requires some objects that jointly can all be actual *together*. Instead, Fine uses an infinite string of modal operators and actualist singular quantifiers $\Diamond\exists x_1 \Diamond\exists x_2 \ldots$ (for the other required adjustments of the formula see Fine 1977, 2016).

Modal Logic as Metaphysics develops two objections to the adequacy of such strings as substitutes for the original plural quantifiers. The first is that the number of variables in the string will impose an (infinite) upper bound on the size of the pluralities taken into consideration, whereas the original plural quantifier imposes no such upper bound (353–356). Fine is unsure what to say about the issues raised by this objection; he sketches but does not endorse one line of response. Partly for this reason, he no longer wishes to rely on such infinite strings in his reduction. I will nevertheless consider his more confident response to my second objection, because it is interesting in its own right.

The second objection is to the very meaningfulness of infinite embeddings of operators. An infinite sequence of meaningful operators is not in general meaningful: the simplest example is an ω-sequence of negations ¬¬¬ … Fine agrees. However, he argues that we *can* make sense of applying infinitely many operations simultaneously, providing that they are all *independent* of each other. He says that he has no definition of 'independent', but he explains what he has in mind with examples. In one case, each operation is the syntactic substitution in a formula of a given expression for each occurrence of a given atomic expression of the same type; if no two substitutions are for the same atomic expression, the operations are all mutually independent and it makes sense to apply infinitely many of them simultaneously to an infinitary formula. In Fine's other example, each operation is the flipping of a light switch; if no two operations are for the same switch (and presumably no two switches are for the same light!), they are all mutually independent and it makes sense to apply infinitely many of them simultaneously to infinitely many light switches. Fine claims that the modalized quantifiers $\Diamond\exists x_1$, $\Diamond\exists x_2$, … — or, as he prefers, $\Diamond\exists x_1\Box$, $\Diamond\exists x_2\Box$, … — all perform mutually independent operations (given that the variables x_1, x_2, … are all mutually distinct), so that we can understand the infinite string as applying all those operations simultaneously to the semantic value supplied by the complement formula.

I have no objection to simultaneously making infinitely many substitutions or flipping infinitely many switches. The question is whether simultaneously applying infinitely many modalized quantifiers is relevantly similar to Fine's unproblematic examples. One obvious feature of the latter is that the operations are all *local*. He requires all the substitutions to be on non-overlapping constituents of the formula operated on, for otherwise things go wrong. Similarly, the flippings must all be of separate switches controlling separate lights. More generally, if a system has infinitely many non-overlapping constituents, it makes sense to perform simultaneous operations on each of these constituents. But do $\Diamond\exists x_1\Box$, $\Diamond\exists x_2\Box$, … perform local semantic operations in that sense? Each of them operates on a whole formula, just as other sentential operators do. When they are applied simultaneously to the same formula, their fields of operation clash. It is not enough that the variables x_1, x_2, … are all mutually distinct, for the simultaneous application of infinitely many quantifiers 'no x_1', 'no x_2', … is just as problematic as the simultaneous application of infinitely many negations. Thus Fine's examples are not similar enough to the case at issue to be of much help.

Consider the two modal operators $\Diamond$ and $\Box$. What are we supposed to get when we apply them 'simultaneously' to a formula A? Do we get something equivalent to $\Diamond\Box$A (which is equivalent in Fine's background modal logic S5 to $\Box$A), or to $\Box\Diamond$A (which is equivalent in S5 to $\Diamond$A), or something else entirely? Since there is no natural answer, $\Diamond$ and $\Box$ presumably count as not independent of each other. Similarly, what are we supposed to get when we apply $\Diamond$ and $\exists x_1$ 'simultaneously' to A? Do we get something equivalent to $\Diamond\exists x_1 A$, or to $\exists x_1 \Diamond A$ (which is not equivalent to the former, since the Barcan and converse Barcan formulas presumably fail for Fine's actualist quantifier), or something else entirely? Again, there is no natural answer, so $\Diamond$ and $\exists x_1$ presumably count as not independent of each other. For similar reasons, $\Box$ and $\exists x_1$ presumably count as not independent of each other. Thus the operators composing $\Diamond\exists x_1\Box$ are *not* independent of the operators composing $\Diamond\exists x_2\Box$. Why then should we believe that $\Diamond\exists x_1\Box$ is independent of $\Diamond\exists x_2\Box$? That would require some 'cancelling out' of the numerous dependencies between their constituents. That such cancelling out occurs in this particular combination (but not most others) is not obvious; it would require a proof, for which we have no hint. Thus the appeal to independence fails to do the work Fine requires of it.

Of course, one could interpret the simultaneous application of $\Diamond\exists x_1\Box$, $\Diamond\exists x_2\Box$, ... as consisting in the simultaneous application of each occurrence of $\Box$, followed by the simultaneous application of the occurrences of $\exists x_1$, $\exists x_2$, ..., followed by the simultaneous application of each occurrence of $\Diamond$. Naturally understood, that is equivalent to the application of $\Diamond\exists x_1, x_2, \ldots\Box$. But, even if we forget the previous problem about cardinality, that is at best tantamount to the modalized actualist plural quantifier $\Diamond\exists V\Box$, which is a compossibility reading of just the sort that gave rise to the original problem.

If the meaning of the infinite sequence of modal operators and actualist quantifiers seems clear, it is because one finds the meaning of the possibilist plural quantifier to be reduced clear in its own right: but that is of no aid to the actualist reduction. Such an infinite sequence is equally unhelpful to the attempt to find a neutral equivalent of higher-order necessitist discourse in the sense of F6. I stand by both the objections in *Modal Logic as Metaphysics*.

3. Fine's suppositional reduction

In the second half of his paper, Sections 4–7, Fine presents an ingenious new reduction of possibilist to actualist discourse, to replace his old one. It is finitary, thereby avoiding the problems just discussed. It is also robust, because it treats the higher-order facts as supervening on the first-order facts in a way that does not depend on how exactly the higher-order facts are formulated. If the reduction works, there will be a corresponding way of giving a neutral equivalent of higher-order necessitist discourse.

Fine's new reduction uses an elaborate *suppositional calculus*, in which one can explicitly formulate complex suppositions and talk about what does or does not follow from them. The general idea is this. Let A be a formula in possibilist terms and A′ be the result of replacing all the possibilist quantifiers in A by corresponding actualist quantifiers. Then A is equivalent to a formula of the suppositional calculus saying that A′ follows from a certain complex supposition $\sigma; \tau$ in actualist terms. Very roughly, $\sigma; \tau$ is the supposition that things are just as they actually are on a possibilist description, except that everything is necessarily actual (even though actualists do not believe that, they can still suppose it). Fine's account has many subtle details, for which the reader should consult his paper.

The suppositional calculus is of great interest in its own right. As Fine suggests, it may well have many fruitful applications not touching the issues between actualism and possibilism and between contingentism and necessitism. My only concern here is with the legitimacy of Fine's present application of it to those issues. I will argue that the supposition $\sigma; \tau$ cannot be understood in actualist terms, and therefore that Fine has not succeeded in reducing possibilist discourse to actualist discourse. A similar objection applies to the corresponding attempt to use the suppositional calculus to give a neutral equivalent of necessitist discourse.

To see the problem, we can examine one constituent σ_1 of $\sigma; \tau$. Fine writes it thus:

$$(\sigma_1) \quad \Pi x / \Box \exists y(x = y) /$$

This needs careful unpacking. The quantifier Πx is a *possibilist* first-order universal quantifier. However, Fine uses it officially as an abbreviation of $\Box\forall x\Box$, where $\forall x$ is an actualist first-order universal quantifier, though with the proviso that there may be some other way of formulating what σ_1 means in actualist terms. Anyway, for clarity we may rewrite σ_1 thus:

$$(\sigma_1^A) \quad \Box\forall x\Box / \Box\exists y(x = y) /$$

Here $\exists y$ is an actualist first-order existential quantifier. Thus we may read $\Box\exists y(x = y)$ as 'necessarily x is something actual'. Putting a formula between occurrences of / forms an instruction to suppose its content. Thus we may read $/\Box\exists y(x = y)/$ as: 'suppose that necessarily x is something actual!' Consequently, Πx in σ_1 and $\Box\forall x\Box$ in σ_1^A are applied to *instructions*. As Fine explains, the intended effect is (and needs to be) not to form an instruction to suppose a universally quantified content but rather to suppose each instance. Thus the intended effect of σ_1^A and σ_1^A is intended to be: 'For every possible object, suppose that necessarily it is something actual!' Can one do that?

Two initial worries are that one is being required to make too many suppositions simultaneously, and that some of those required suppositions are about objects to which one has no means of referring. However, those worries depend on a more psychologized view of supposing than Fine intends. They would be equally relevant to less metaphysically fraught instructions such as 'For every grain of sand, suppose that it is on a beach'. As Fine explains, he is treating 'suppose' as a performative. If one felicitously says 'For every grain of sand, suppose that it is on a beach', then for every grain of sand one *thereby* supposes that it is on a beach. There is no need for many phenomenologically separate acts of supposing, one for each grain of sand. So far, Fine is within his rights.

The deeper problem is this. Possible objects, in the possibilist's sense, include some non-actual, merely possible objects. For each of those, σ_1 and σ_1^A require one to suppose something about it. But to suppose something about a non-actual object is to go beyond actualist terms. Since those suppositions are integral to Fine's plan, possibilist discourse has not been reduced to genuinely actualist discourse.

The difficulty is not confined to the component σ_1 of the overall supposition $\sigma; \tau$. It also arises for Fine's components σ_2 and σ_3. Their details do not matter here, but they involve the application of possibilist universal quantifiers to conditional suppositions, where to make the conditional supposition $A \rightarrow /B/$ is not to (unconditionally) suppose a conditional $A \rightarrow B$, but rather if A holds to suppose B and otherwise to suppose nothing. The closest Fine comes to addressing the difficulty is in discussing the application of the possibilist quantifier to such conditional suppositions, but if what he says works for σ_2 and σ_3, it will also work for σ_1, since an unconditional supposition is equivalent to a conditional supposition with a tautologous antecedent.

Fine begins his discussion of the matter by noting that one can make sense of an indicative sentence such as $\Pi x(\Diamond A(x) \rightarrow B(x))$ in actualist terms (where $A(x)$ and $B(x)$ are not themselves problematic). His original reduction establishes that, since it works for first-order modal languages and the possibilist quantifier Πx is first-order. So far so good. But if the actualist can make sense of $\Pi x(\Diamond A(x) \rightarrow B(x))$, Fine asks:

> Then what is to prevent him from making sense of the corresponding suppositional sentence $\Pi x(\Diamond A(x) \rightarrow /B(x)/)$ (for every possible object x for which it is possible that $A(x)$ suppose that $B(x)$), given that the only difference between them lies in the substitution of a supposition for a judgement?

He is surely right that the difference between the act of supposition and the act of judgement does not matter here, for the task is to capture the *content* of the act in actualist terms, irrespective of which act it is. But what stands to $\Pi x(\Diamond A(x) \rightarrow /B(x)/)$ as judgement stands to supposition is *not* the plain $\Pi x(\Diamond A(x) \rightarrow B(x))$. For, if there are many non-actual objects x for which it is possible that $A(x)$, then $\Pi x\ (\Diamond A(x) \rightarrow /B(x)/)$ expresses a complex act comprising many acts of supposing that $B(x)$ for various non-actual objects x, whereas $\Pi x(\Diamond A(x) \rightarrow B(x))$ does *not* express a complex act comprising many acts of judging that $B(x)$ for various non-actual objects x. The latter complex act goes outside the realm of the actual in exactly the same way as the former does. The proper notation for it would be something like $\Pi x(\Diamond A(x) \rightarrow \#B(x)\#)$, where $\#B(x)\#$ expresses the act of judging that $B(x)$ just as $/B(x)/$ expresses the act of supposing that $B(x)$. Of course, one can use Fine's original reduct of $\Pi x(\Diamond A(x) \rightarrow B(x))$ to make a general judgement in actualist terms, but we have been given no reason why doing so should involve many acts of judging that $B(x)$ for various non-actual objects x.

To take a simpler example, the actualist can make the general judgement that $\Box\forall x\Box\ \neg(C(x)\ \&\ \neg C(x))$ without making many individual judgements about non-actual objects. The rest of Fine's discussion of the issue remains focused on the structural analogy between supposition and judgement, and so misses the heart of the problem.

Of course, one can formulate general propositions in actualist terms about what is (actually) to be supposed: for instance, that necessarily for every actual object *x* necessarily if it is possible that A(*x*) then it is actually to be supposed that B(*x*). But that does not mean that the contents of the suppositions so described are themselves expressible in actualist terms. The actual may contain talk about the non-actual. Since we have been given no further reduction of B(*x*) for non-actual *x* to actualist discourse, to make the supposition /B(*x*)/ is to step outside actualist discourse. But making Fine's overall supposition $\sigma;\tau$ involves making the constituent suppositions σ_1, σ_2 and σ_3, which in turn involves making constituent suppositions of the form /B(*x*)/ for non-actual *x*. Thus making Fine's overall supposition $\sigma;\tau$ involves stepping outside actualist discourse. Fine reduces a sentence A in possibilist discourse to the claim that the sentence A′ in actualist discourse follows from the supposition $\sigma;\tau$. But since making that supposition involves stepping outside actualist discourse, we have not been given a reduction of possibilist to actualist discourse in any useful sense.

The same problem arises when Fine's strategy is applied to the search for neutral equivalents of necessitist discourse. The contents of the relevant suppositions or judgements about individual non-actual objects fall outside the neutral zone, and so cannot be legitimately used. The neutral zone may contain talk about matters outside the neutral zone. The strategy does not give us neutral equivalents of necessitist discourse in any useful sense.

References

Fine, Kit. 1977. "Prior on the Construction of Possible Worlds and Instants." In Arthur Prior and Kit Fine *Worlds, Times and Selves*, 116–168. London: Duckworth.

Fine, Kit. 2009. "The Question of Ontology." In *Metametaphysics: New Essays on the Foundations of Ontology*, edited by David Chalmers, David Manley, and Ryan Wasserman, 157–177. Oxford: Oxford University Press.

Fine, Kit. 2016. "Williamson on Fine on Prior on the Reduction of Possibilist Discourse." *Canadian Journal of Philosophy* 46 (4–5): 548–570.

Stalnaker, Robert. 2012. *Mere Possibilities: Metaphysical Foundations of Modal Semantics*. Princeton, NJ: Princeton University Press.

Williamson, Timothy. 2000. "Existence and Contingency." *Proceedings of the Aristotelian Society* 100: 117–139.

Williamson, Timothy. 2007. *The Philosophy of Philosophy*. Oxford: Blackwell.

Williamson, Timothy. 2013. *Modal Logic as Metaphysics*. Oxford: Oxford University Press.

Williamson, Timothy. 2016. "Modal Science." *Canadian Journal of Philosophy* 46 (4–5): 453–492.

First-order modal logic in the necessary framework of objects

Peter Fritz

Department of Philosophy, Classics, History of Art and Ideas, University of Oslo, Oslo, Norway

ABSTRACT

I consider the first-order modal logic which counts as valid those sentences which are true on every interpretation of the non-logical constants. Based on the assumptions that it is necessary what individuals there are and that it is necessary which propositions are necessary, Timothy Williamson has tentatively suggested an argument for the claim that this logic is determined by a possible world structure consisting of an infinite set of individuals and an infinite set of worlds. He notes that only the cardinalities of these sets matters, and that not all pairs of infinite sets determine the same logic. I use so-called two-cardinal theorems from model theory to investigate the space of logics and consequence relations determined by pairs of infinite sets, and show how to eliminate the assumption that worlds are individuals from Williamson's argument.

1. Introduction

Modal logic is a branch of formal logic with a wide variety of applications in philosophy, computer science, linguistics, economics, and other fields. In *Modal Logic as Metaphysics* (2013), Timothy Williamson applies first-order modal logic as well as some of its higher order extensions to metaphysics. One of the primary roles which modal logic serves in this book is as a regimented language in which various metaphysical theses are formulated. This use is best illustrated with an example.

The overall aim of Williamson's book is to defend necessitism, the view, roughly, that it is necessary what there is. More explicitly, one might describe it as the view that nothing could have failed to be anything, and that this is necessarily so, i.e. as the view that necessarily, everything is necessarily something. For various reasons, such as to rule out ambiguities, it can be helpful to formulate such views in a formal language. Williamson formulates necessitism in first-order modal logic as follows:

(**NNE**) $\Box\forall x\Box\exists y\, x = y$

The role of the formula NNE in Williamson's discussion is not just as an aid to understanding the natural language statement of necessitism given before, e.g. by indicating how to resolve potential scope ambiguities. Rather, NNE on its own is intended to be a statement of the view of necessitism. So, in addition to having formal properties such as being derivable in certain calculi but not others, and being valid on certain model theories but not others, NNE also needs to be endowed with meaning just as the English sentences of this article are endowed with meaning. Williamson's use of NNE therefore presupposes an *intended interpretation* of the formal language of first-order modal logic. In keeping with the variety of uses to which modal logics can be put, this need not be distinguished among various other possible interpretations of the language; it must only be isolated for the particular discussion in Williamson's book.

Williamson specifies his intended interpretation of first-order modal logic by saying that he intends quantifiers to be read unrestrictedly, and modal operators as expressing metaphysical necessity and possibility; he assumes that the intended interpretation of the identity connective and the Boolean operators is clear. This stipulation is of course not uncontroversial, as the intelligibility of both unrestricted quantification and the metaphysical interpretation of modal expressions have been questioned. Even more controversially, Williamson extends his intended interpretation to higher order modal logic by claiming that there is a way of interpreting higher order quantifiers analogously to the unrestricted interpretation of first-order quantifiers. This interpretation is neither to be understood in terms of our informal talk of properties and relations nor in terms of any set-theoretic model theory. For the purposes of this article, it will be assumed that with these stipulations, Williamson has succeeded in singling out an intended interpretation for first- and higher order modal logic; for further discussion, see Williamson (2003). In the following, talk of *properties* and *relations* will sometimes be used as informal glosses of what should properly be formulated in higher order logic; the term *individual* will be used to indicate first-order quantification. Thus, sets will be understood to be individuals. This is worth noting since in impure set theory, the term *individual* is sometimes used for non-sets (also called *urelemente*).

The intended interpretation described so far does not provide any interpretation of non-logical constants. So, while closed formulas without non-logical constants, such as NNE, are given an intended interpretation, closed formulas with non-logical constants are only partially interpreted. However, such formulas at least receive an intended interpretation relative to an interpretation of the non-logical constants occurring in it. (As noted below, quantification over such interpretations can be understood in terms of higher order quantification.)

Consequently, the intended interpretation also determines which formulas are true on every interpretation of the non-logical constants (and the intended interpretation of the logical constants). Williamson calls these formulas *metaphysically universal*. Let Λ_{MU} be the set of metaphysically universal formulas of first-order modal logic. (The formal language will be specified more precisely below). Analogously, let $\vDash_{\mathrm{MU}}$ be the relation which relates a set of closed formulas of first-order modal logic to a single such formula just in case the latter is true on any interpretation of non-logical constants on which all members of the former are true; of course, this closely corresponds what Tarski (2002 [1936]) calls the relation of *following logically*. Whatever one calls Λ_{MU} and $\vDash_{\mathrm{MU}}$, Williamson (2013, 92) observes that their investigation has to be carried out.

Williamson (2013, 145) tentatively suggests an argument for the claim that Λ_{MU} is determined by a certain kind of Kripke frame for first-order modal logic consisting of two infinite sets. In the usual terminology, one of these sets is the *domain of quantification* and the other the *set of worlds*. The frames have no varying domain function since Williamson holds that it is a necessary matter what individuals there are; similarly, they have no accessibility relation since he holds that it is a necessary matter whether a given proposition is necessary. (Williamson's endorsement of the latter is in fact rather tentative, but since this won't be criticized in the following, there is no harm in exploring a position which might be somewhat stronger than Williamson's own.) Such a frame determines a set of closed formulas of first-order modal logic as the set of such formulas which are true in every world of every model based on it, on the usual definitions of model and truth (defined more explicitly below). It is easy to see that only the cardinalities of the two sets making up a given frame matter for which set of formulas it determines. Thus, any two infinite cardinals κ, λ can be understood as directly determining a set of closed formulas of first-order modal logic; call it $\Lambda_{\kappa,\lambda}$. Here, κ indicates the cardinality of the domain of quantification and λ is the cardinality of the set of worlds; we might therefore read '$\Lambda_{\kappa,\lambda}$' as 'the logic of κ individuals and λ worlds.' Williamson's argument aims to establish that $\Lambda_{\mathrm{MU}} = \Lambda_{\kappa,\lambda}$ for some infinite cardinals κ and λ. The analogous definition of a consequence relation $\vDash_{\kappa,\lambda}$ raises the question whether $\vDash_{\mathrm{MU}} = \vDash_{\kappa,\lambda}$ for some infinite cardinals κ and λ.

As Williamson (2013, 144) notes, one might conjecture that Λ_{MU} is uniquely determined by the claim that $\Lambda_{\mathrm{MU}} = \Lambda_{\kappa,\lambda}$ for some infinite cardinals κ and λ, using some modal analog of the Löwenheim–Skolem theorem. And indeed, an analog of the downward Löwenheim–Skolem theorem was already proven in Kripke (1959, 7), from which it follows that $\Lambda_{\aleph_0,\aleph_0} \subseteq \Lambda_{\kappa,\lambda}$. However, Kripke's result does not establish the converse containment. In fact, as Williamson points out, this is not the case, as $\forall x \Diamond \forall y (Uy \leftrightarrow x = y)$ (which might be read as saying that every individual could be the unique U) can only be verified by a model based on a frame in which there are at least as many worlds as individuals, and so that the negation of this formula is a member of $\Lambda_{\kappa,\lambda}$ just in case $\kappa > \lambda$. On the one hand, this shows that $\Lambda_{\kappa,\lambda}$ and $\vDash_{\kappa,\lambda}$ depend on the choice of κ and λ; on the other hand, it is clear by cardinality considerations that many pairs of infinite cardinals κ, λ determine the same sets $\Lambda_{\kappa,\lambda}$ and $\vDash_{\kappa,\lambda}$.

How exactly do $\Lambda_{\kappa,\lambda}$ and $\vDash_{\kappa,\lambda}$ depend on κ and λ? This is the question to be investigated in Section 2. It turns out that $\Lambda_{\kappa,\lambda}$ and $\vDash_{\kappa,\lambda}$ are uniquely determined by the condition that $\lambda \geq \kappa$, but sensitive to the cardinalities of λ and κ if $\lambda < \kappa$, with the different pairs of infinite cardinals determining infinitely many different logics. Furthermore, in some instances, whether two such pairs determine the same logic depends on set-theoretic principles which are likely not to be settled by first-order ZFC. This section is of independent interest; as will become clear, it could be reformulated in terms of extensions of the standard Löwenheim–Skolem theorem to first-order modal logic on the present model theory. Section 3 returns to Williamson's argument that Λ_{MU} is given by a pair of infinite cardinals. Some difficulties in evaluating one of the argument's premises will be raised, and it will be noted that the argument relies on worlds being individuals. The main contribution of this section is to show how to dispense with this last assumption, independently of any particular way of understanding possible worlds.

Metaphysical universality is a simple and natural way of understanding logical validity and consequence. In the case of non-modal first-order logic, there are well-developed arguments which uniquely determine the set of metaphysically universal sentences, namely those derivable in a standard calculus extended by an axiom schema which says that there are at least n individuals, for each $n < \omega$ (Friedman 1999; Williamson 2000a; Rayo and Williamson 2003); model-theoretically, this is the set of sentences valid on any fixed infinite set, or the class of all infinite sets. Williamson's necessary framework of objects (Williamson 2000b; 2013, Section 8.1), according to which it is necessary what individuals there are as well as necessary which propositions are necessary, is the strongest and simplest non-trivial theory of modal metaphysics available. The results below show that even on this assumption, the addition of necessity to first-order logic turns determining the set of metaphysically universal sentences into an extremely difficult issue; doing so involves not only settling difficult questions in metaphysics, concerning the relative numbers of worlds and individuals, but potentially also difficult questions in set theory or higher order logic. In general, such questions only become harder if the necessary framework of objects is rejected (although particular metaphysical assumptions might of course settle the matter). On the present way of understanding logical validity and consequence, first-order modal logic is therefore far less well understood than first-order non-modal logic.

2. Set-theoretic models

To specify the formal languages under consideration more precisely, fix a countably infinite set V of individual variables and a relational signature σ containing a countable infinity of relation symbols for each finite arity. Let $\mathcal{L}_\Box$ be a standard first-order modal language with identity based on V and σ. As explained above, it will be assumed that there is an intended interpretation of this language, on which quantifiers are interpreted unrestrictedly, and $\Box$ as expressing metaphysical necessity. Let an *$\mathcal{L}_\Box$-theory* be a set of $\mathcal{L}_\Box$-sentences (closed $\mathcal{L}_\Box$-formulas), and an *$\mathcal{L}_\Box$-sequent* a tuple $\langle\Gamma,\varphi\rangle$ consisting of an $\mathcal{L}_\Box$-theory Γ and an $\mathcal{L}_\Box$-sentence φ. Call an $\mathcal{L}_\Box$-sentence *metaphysically universal* if it is true on every interpretation assigning an n-ary relation to each n-ary relation symbol in σ; call an $\mathcal{L}_\Box$-sequent $\langle\Gamma,\varphi\rangle$ *metaphysically universal* if on every such interpretation, all elements of Γ are true only if φ is true. Let Λ_{MU} be the set of metaphysically universal $\mathcal{L}_\Box$-sentences and $\models_{\mathsf{MU}}$ the set of metaphysically universal $\mathcal{L}_\Box$-sequents.

The notion of an interpretation of the relation symbols of σ appealed to in the definition of metaphysical universality is informal, but – it will be assumed in the following – nonetheless rigorous. The definition can be formalized in a higher order metalanguage, along the lines of Williamson (2013, Section 5.7), and this can be carried over to the extension of $\mathcal{L}_\Box$ by second-order quantifiers, interpreting such quantifiers homophonically. Since the relevant metalanguage is always of a higher order than the object language under consideration, the assumption that the resulting truth predicates are transparent does not conflict with Tarksi's undefinability theorem.

Assume the truth of first-order ZFC (interpreting quantifiers as implicitly restricted to the pure sets) and the von Neumann construction of ordinals, on which every ordinal is the set of ordinals smaller than it. Let a *frame* be a pair $\mathfrak{F} = \langle D, W\rangle$ of non-empty (pure) sets; informally, D and W will be called the sets of 'individuals' and 'worlds,' respectively. Let a *Kripke model based on* $\mathfrak{F}$ be a tuple $\mathfrak{M} = \langle D, W, i, @\rangle$, where i is a function mapping each n-ary relation symbol in σ to a function mapping each $w \in W$ to a set $i(R)(w)$ of n-tuples on D, and $@ \in W$; informally, i and $@$ will be called the 'interpretation function' and the 'actual world', respectively. As usual, the interpretation of unary relations will be simplified by identifying any individual with its singleton sequence. Truth of an $\mathcal{L}_\Box$-formula φ in $\mathfrak{M}$ is defined as usual relative to an element $w \in W$ and an assignment function $a : V \to D$, written $\mathfrak{M}, w, a \models \varphi$. (Note that frames do not contain an accessibility relation, so $\Box\varphi$ is true in a world if φ is true in every world; similarly, there is no relativization of the domain of first-order quantifiers to worlds.) From this, truth of a sentence φ in $\mathfrak{M}$ is derived as $\mathfrak{M}, @, a \models \varphi$ for all assignments a, and written $\mathfrak{M} \models \varphi$. A *Kripke model of* an $\mathcal{L}_\Box$-theory Γ is a Kripke model in which all members of Γ are true. Γ is *satisfiable* on a frame $\mathfrak{F}$ if there is a Kripke model of Γ based on $\mathfrak{F}$. Here and in the following, any notion – such as being satisfiable – is extended from theories to sentences, treating a sentence as its singleton.

A frame $\mathfrak{F}$ naturally determines a logic (a set of sentences) and a consequence relation (a set of sequents), by counting a sentence as valid if it is true in all models based on $\mathfrak{F}$, and a sequent as valid if its conclusion is true in all models based on $\mathfrak{F}$ in which all of its premises are true. Any frames with the same numbers of individuals and worlds will determine the same logic and the same consequence relation, so define, for any cardinals κ and λ, $\models_{\kappa,\lambda}$ to be the set of $\mathcal{L}_\Box$-sequents $\langle\Gamma,\varphi\rangle$ such that $\mathfrak{M} \models \varphi$ for all Kripke models $\mathfrak{M}$ of Γ on $\langle\kappa,\lambda\rangle$, and define $\Lambda_{\kappa,\lambda}$ to be the set of $\mathcal{L}_\Box$-sentences φ such that $\emptyset \models_{\kappa,\lambda} \varphi$. (As usual, cardinals are taken to be initial ordinals.) Thus $\Lambda_{\kappa,\lambda}$ and $\models_{\kappa,\lambda}$ can informally be thought of as the logic and consequence relation of frames with κ individuals and λ worlds.

2.1. Reduction

To investigate the space of logics and consequence relations $\Lambda_{\kappa,\lambda}$ and $\models_{\kappa,\lambda}$ for infinite cardinals κ and λ, membership in $\Lambda_{\kappa,\lambda}$ and $\models_{\kappa,\lambda}$ will first be reduced to the so-called two-cardinal problem in model theory. As the two-cardinal problem has been investigated extensively, this relatively simple reduction immediately gives rise to a number of corollaries which would be extremely difficult to prove directly. Such corollaries will then be used to lay out the basic features of the space of logics and consequence relations determined by pairs of infinite cardinals, including results on compactness and recursive enumerability.

Let $\mathcal{L}$ be the language of (non-modal) first-order logic with identity based on σ and V. Let a *model* be a tuple $\mathfrak{A} = \langle D, e\rangle$ where D is a non-empty set and e is a function which maps every n-ary relation symbol R of σ to a set $e(R)$ of n-tuples on D; informally, D and e will be called the set of 'individuals' and the 'interpretation function,' respectively. Truth of an $\mathcal{L}$-formula φ in $\mathfrak{A}$ is defined as usual relative to an assignment function $a : V \to D$, written $\mathfrak{A}, a \models \varphi$. From this, truth of a sentence φ in $\mathfrak{A}$ is derived as $\mathfrak{A}, a \models \varphi$ for all assignments a, and written $\mathfrak{A} \models \varphi$. An $\mathcal{L}$-*theory* is a set of sentences of $\mathcal{L}$; an $\mathcal{L}$-*sequent* is a tuple $\langle\Gamma,\varphi\rangle$ consisting of an $\mathcal{L}$-theory Γ and an $\mathcal{L}$-sentence φ. A *model of* an $\mathcal{L}$-theory Γ is a model in which all members of Γ are true.

To state the two-cardinal problem, fix a unary relation symbol U of σ, and say that an $\mathcal{L}$-theory Γ *admits* $\langle\kappa,\lambda\rangle$, for cardinals κ and λ, if there is a model $\mathfrak{A} = \langle D, e\rangle$ of Γ such that $|D| = \kappa$ and $|e(U)| = \lambda$. The two-cardinal problem is to determine under which conditions an $\mathcal{L}$-theory admitting a given pair of infinite cardinals admits another given pair of infinite cardinals. One might think of this as the problem of extending the standard Löwenheim–Skolem theorem to the case of two cardinals, one for the domain and one for the interpretation of U; this was in fact how Robert Vaught introduced the problem

(Morley and Vaught 1962, Section 6). Let $T_{\kappa,\lambda}$ be the set of $\mathcal{L}$-theories admitting $\langle\kappa,\lambda\rangle$, and $T_{\kappa,\leq\lambda} = \bigcup_{\aleph_0\leq\lambda'\leq\lambda} T_{\kappa,\lambda'}$.

From now on, let λ, κ, λ', κ', etc. always be infinite cardinals. The following two elementary facts for any $\mathcal{L}$-theory Γ will be required:

Fact (i). If Γ admits $\langle\kappa,\lambda\rangle$ and $\lambda \leq \kappa' \leq \kappa$, then Γ admits $\langle\kappa',\lambda\rangle$.

Fact (ii). If Γ admits $\langle\kappa,\lambda\rangle$ then Γ admits $\langle\kappa',\kappa'\rangle$.

They can be derived from the Löwenheim–Skolem theorem for first-order logic; proofs can be found in Chang and Keisler (1990, Proposition 3.2.7 (i) and (ii)).

Membership in $\Lambda_{\kappa,\lambda}$ and $\vDash_{\kappa,\lambda}$ is straightforwardly reduced to satisfiability on $\langle\kappa,\lambda\rangle$:

Lemma 1: *Let $\langle\Gamma,\varphi\rangle$ be an $\mathcal{L}_\Box$-sequent.*

(i) $\Gamma \vDash_{\kappa,\lambda} \varphi$ *if and only if* $\Gamma \cup \{\neg\varphi\}$ *is not satisfiable on* $\langle\kappa,\lambda\rangle$.
(ii) $\varphi \in \Lambda_{\kappa,\lambda}$ *if and only if* $\neg\varphi$ *is not satisfiable on* $\langle\kappa,\lambda\rangle$.

Proof: Immediate. □

This lemma makes the connection to extensions of the Löwenheim–Skolem theorem precise. It should be noted that the particular class of Kripke models used here matters. E.g. if an accessibility relation and variable individual domains were introduced, it would be straightforward to prove that the same $\mathcal{L}_\Box$-theories can be satisfied on all pairs of infinite cardinals, since one could always add an unconnected world with an arbitrary number of new individuals. Of course, there are also more interesting questions one could pose in such a setting; see, e.g. Bowen (1979, Chapter 5). Such variant classes of models won't be considered in the following. Similarly, it is also essential to the results established below exactly which resources are available in $\mathcal{L}_\Box$. E.g. adding further modal operators, such as explicit quantifiers over worlds, would substantially change the dependence of $\Lambda_{\kappa,\lambda}$ and $\vDash_{\kappa,\lambda}$ on κ and λ.

The main results to be established in this section are Theorems 3 and 6. The first reduces an $\mathcal{L}_\Box$-theory being satisfiable on $\langle\kappa,\lambda\rangle$ to a corresponding $\mathcal{L}$-theory being a member of $T_{\kappa+\lambda,\leq\lambda}$, and the second reduces an $\mathcal{L}$-theory being a member of $T_{\kappa+\lambda,\leq\lambda}$ to a corresponding $\mathcal{L}_\Box$-theory being satisfiable on $\langle\kappa,\lambda\rangle$. With various two-cardinal results from model theory, these two theorems will entail many useful corollaries, which will be derived in the two subsequent sections. To prove Theorems 3 and 6, observe first that since in any Kripke model, adding any number of copies of one of its worlds doesn't change which $\mathcal{L}_\Box$-sentences are true in it, increasing the number of worlds in a frame doesn't destroy satisfiability:

Lemma 2: *If $\lambda \leq \lambda'$ then any $\mathcal{L}_\Box$-theory satisfiable on $\langle\kappa,\lambda\rangle$ is satisfiable on $\langle\kappa,\lambda'\rangle$.*

Proof: Assume $\lambda \leq \lambda'$ and let Γ be an $\mathcal{L}_\Box$-theory satisfiable on $\langle \kappa, \lambda \rangle$. Let $\mathfrak{M} = \langle \kappa, \lambda, i, @ \rangle$ be a Kripke model of Γ. Define $\mathfrak{M}' = \langle \kappa, \lambda', i', @ \rangle$ such that for all relation symbols R in σ and $w < \lambda'$,

$$i'(R)(w) := \begin{cases} i(R)(w) & \text{if } w < \lambda \\ i(R)(@) & \text{otherwise} \end{cases}$$

It is routine to show that for all $\mathcal{L}_\Box$-sentences φ, $\mathfrak{M} \models \varphi$ if and only if $\mathfrak{M}' \models \varphi$, and so that $\mathfrak{M}'$ is a Kripke model of Γ. □

The syntactic mapping from $\mathcal{L}_\Box$-theories to $\mathcal{L}$-theories used in Theorem 3 is a version of the so-called standard translation, which translates necessity as explicit universal quantification (informally, over worlds). To do so, the arity of each relation symbol is increased by one, and each atomic predication is translated by adding a special variable ϑ to the arguments (which can informally be understood as recording the world of evaluation). Necessity is translated using a universal quantifier binding ϑ, restricted to a new relation symbol U (informally, the worlds). Quantifiers are translated by restricting them to the negation of U (informally, the individuals). Using this recursive translation, which will be called ρ, the present version of the standard translation, called τ, is derived in a second step using a new unary relation symbol A (which informally can be understood as identifying the actual world): $\tau(\varphi)$ is the sentence which says that there is a unique U which is A, and letting ϑ be this element, $\rho(\varphi)$ is true.

To define this procedure more precisely, fix a unary relation symbol A of σ which is distinct from U, and $\vartheta \in V$. Let ρ be an injection which maps every n-ary relation symbol of σ to an $n+1$-ary relation symbol of σ, and every element of V to an element of V, such that none of U, A, and ϑ are in its image. Extend ρ to a mapping from $\mathcal{L}_\Box$-formulas to $\mathcal{L}$-formulas using the following recursion clauses:

$$\begin{aligned}
\rho(Rx_1 \ldots x_n) &:= \rho(R)\rho(x_1)\ldots\rho(x_n)\vartheta \\
\rho(x = y) &:= \rho(x) = \rho(y) \\
\rho(\neg\varphi) &:= \neg\rho(\varphi) \\
\rho(\varphi \wedge \psi) &:= \rho(\varphi) \wedge \rho(\psi) \\
\rho(\Box\varphi) &:= \forall\vartheta(U\vartheta \rightarrow \rho(\varphi)) \\
\rho(\forall x\varphi) &:= \forall\rho(x)(\neg U\rho(x) \rightarrow \rho(\varphi))
\end{aligned}$$

For any $\mathcal{L}_\Box$-sentence φ, define

$$\tau(\varphi) := \exists x(Ux \wedge \forall y(Ay \leftrightarrow x = y)) \wedge \forall\vartheta(A\vartheta \rightarrow \rho(\varphi))$$

and extend τ in the obvious way to any $\mathcal{L}_\Box$-theory Γ:

$$\tau(\Gamma) := \{\tau(\varphi) : \varphi \in \Gamma\}.$$

To prove Theorem 3, two model transformations corresponding to the syntactic mapping τ will be used, transforming a Kripke model into a corresponding model, and *vice versa*:

Theorem 3: *For any $\mathcal{L}_\Box$-theory Γ, Γ is satisfiable on $\langle\kappa,\lambda\rangle$ if and only if $\tau(\Gamma) \in T_{\kappa+\lambda,\leq\lambda}$.*

Proof: Let $\Gamma \neq \emptyset$; otherwise, the claim is trivial. Assume first that Γ is satisfiable on $\langle\kappa,\lambda\rangle$, witnessed by $\mathfrak{M} = \langle\kappa,\lambda,i,@\rangle$. Writing $X \uplus Y$ for the disjoint union of sets X and Y, let $\mathfrak{A} = \langle\kappa \uplus \lambda, e\rangle$ be a model such that $e(U) = \lambda$, $e(A) = \{@\}$, and $e(\rho(R)) = \{\langle o_1,\ldots,o_n,w\rangle : \langle o_1,\ldots,o_n\rangle \in i(R)(w), w < \lambda\}$ for any n-ary relation symbol R in σ. A routine induction on the complexity of formulas establishes that for all $\mathcal{L}_\Box$-sentences φ, $\mathfrak{M} \models \varphi$ if and only if $\mathfrak{A} \models \tau(\varphi)$. So $\tau(\Gamma)$ admits $\langle\kappa+\lambda,\lambda\rangle$ and thus is a member of $T_{\kappa+\lambda,\leq\lambda}$.

Assume now that $\tau(\Gamma) \in T_{\kappa+\lambda,\leq\lambda}$. Then for some $\lambda' \leq \lambda$, $\tau(\Gamma)$ admits $\langle\kappa+\lambda,\lambda'\rangle$. By Fact (i), $\tau(\Gamma)$ admits $\langle\kappa+\lambda',\lambda'\rangle$. Let $\mathfrak{A} = \langle\kappa \uplus \lambda', e\rangle$ be a witness such that $e(U) = \lambda'$. Let $\mathfrak{M} = \langle\kappa,\lambda',i,@\rangle$ be a Kripke model such that for every n-ary relation symbols R in σ and $w < \lambda'$, $i(R)(w) = \{\langle o_1,\ldots,o_n\rangle : \langle o_1,\ldots,o_n,w\rangle \in e(\rho(R))\}$, and @ is such that $e(A) = \{@\}$ (recall that Γ is non-empty). Again, a routine induction on the complexity of formulas establishes that for all $\mathcal{L}_\Box$-sentences φ, $\mathfrak{M} \models \varphi$ if and only if $\mathfrak{A} \models \tau(\varphi)$. So Γ is satisfiable on $\langle\kappa,\lambda'\rangle$, and thus by Lemma 2, Γ is satisfiable on $\langle\kappa,\lambda\rangle$. □

For Theorem 6, a complementary mapping from $\mathcal{L}$-sentences to $\mathcal{L}_\Box$-sentences will be used. The idea behind this translation is to add two constraints: on the one hand, for there to be at least as many worlds as *U*s, and on the other hand, for there to be infinitely many *U*s. The first can be achieved by stating that each *U* is possibly the unique *U*. To achieve the second, a new relation symbol *R* will be used, stating that it is a serial, transitive, and asymmetric on *U*.

More precisely, let π be an injection which maps every n-ary relation symbol of σ to an n-ary relation symbol of σ distinct from *R*, and which maps *U* to itself. Extend π to a function from $\mathcal{L}$-formulas to $\mathcal{L}$-formulas which only replaces a relation symbol *F* by $\pi(F)$. For any $\mathcal{L}$-sentence φ, define

$$\mu(\varphi) := \quad \pi(\varphi) \wedge \delta \wedge \iota,$$

where δ and ι are the following $\mathcal{L}_\Box$-sentences:

$$\begin{aligned} \delta :=& \quad \forall x(Ux \to \Diamond\forall y(Uy \leftrightarrow x = y)) \\ \iota :=& \quad \forall x(Ux \to \exists y(Uy \wedge Rxy))\wedge \\ & \quad \forall x\forall y\forall z((Ux \wedge Uy \wedge Uz \wedge Rxy \wedge Ryz) \to Rxz)\wedge \\ & \quad \forall x\forall y((Ux \wedge Uy \wedge Rxy) \to \neg Ryx) \end{aligned}$$

As with τ, extend μ in the obvious way to any $\mathcal{L}$-theory Γ:

$$\mu(\Gamma) := \quad \{\mu(\varphi) : \varphi \in \Gamma\}.$$

To prove Theorem 6, two auxiliary lemmas are established:

Lemma 4: *For any $\mathcal{L}$-theory Γ, if Γ admits $\langle\kappa,\lambda\rangle$ then $\mu(\Gamma)$ is satisfiable on $\langle\kappa,\lambda\rangle$.*

Proof: Assume that Γ admits $\langle\kappa,\lambda\rangle$, witnessed by a model $\mathfrak{A} = \langle\kappa, e\rangle$. Let $\mathfrak{M} = \langle\kappa, \lambda+1, i, \lambda\rangle$ (using ordinal addition) be a Kripke model such that for all $w < \lambda$, $i(U)(w) = \{\langle w\rangle\}$, $i(R)(\lambda) = \{\langle u,v\rangle : u < v < \lambda\}$, and, for any relation symbol F in σ, $i(\pi(F))(\lambda) = e(F)$. It is routine to show that for any $\mathcal{L}$-sentence φ, $\mathfrak{A} \vDash \varphi$ if and only if $\mathfrak{M} \vDash \mu(\varphi)$, and so that $\mathfrak{M}$ is a Kripke model of $\mu(\Gamma)$. Since $|\lambda+1| = \lambda$, $\mu(\Gamma)$ is satisfiable on $\langle\kappa,\lambda\rangle$. □

Lemma 5: *For any $\mathcal{L}$-theory Γ, if $\mu(\Gamma)$ is satisfiable on $\langle\kappa,\lambda\rangle$ then Γ admits $\langle\kappa,\lambda'\rangle$ for some $\lambda' \leq \lambda$.*

Proof: Let $\Gamma \neq \emptyset$; otherwise, the claim is trivial. Let $\mathfrak{M} = \langle\kappa,\lambda,i,@\rangle$ be a Kripke model witnessing that $\mu(\Gamma)$ is satisfiable on $\langle\kappa,\lambda\rangle$. Define a model $\mathfrak{A} = \langle\kappa, e\rangle$ by letting $e(F) = i(\pi(F))(@)$ for any relation symbol F in σ. $\mathfrak{A}$ is a model of Γ. Furthermore, by ι, $\aleph_0 \leq |e(U)|$, and by δ, $|e(U)| \leq \lambda$ (recall that $\Gamma \neq \emptyset$). □

Theorem 6: *For any $\mathcal{L}$-theory Γ, $\Gamma \in T_{\kappa+\lambda,\leq\lambda}$ if and only if $\mu(\Gamma)$ is satisfiable on $\langle\kappa,\lambda\rangle$.*

Proof: Assume first that $\Gamma \in T_{\kappa+\lambda,\leq\lambda}$. Then there is a $\lambda' \leq \lambda$ such that Γ admits $\langle\kappa+\lambda,\lambda'\rangle$. If $\lambda > \kappa$, then note that by Fact(ii), Γ admits $\langle\kappa,\kappa\rangle$, and so by Lemma 4, $\mu(\Gamma)$ is satisfiable on $\langle\kappa,\kappa\rangle$, and so by Lemma 2, $\mu(\Gamma)$ is satisfiable on $\langle\kappa,\lambda\rangle$. If $\lambda \leq \kappa$, then $\kappa+\lambda = \kappa$, so Γ admits $\langle\kappa,\lambda'\rangle$. So by Lemma 4, $\mu(\Gamma)$ is satisfiable on $\langle\kappa,\lambda'\rangle$ and thus by Lemma 2, $\mu(\Gamma)$ is satisfiable on $\langle\kappa,\lambda\rangle$.

Assume now that $\mu(\Gamma)$ is satisfiable on $\langle\kappa,\lambda\rangle$. Then by Lemma 5, Γ admits $\langle\kappa,\lambda'\rangle$ for some $\lambda' \leq \lambda$. If $\lambda > \kappa$, then note that by Fact (ii), Γ admits $\langle\lambda,\lambda\rangle$, and since then $\kappa+\lambda = \lambda$, Γ admits $\langle\kappa+\lambda,\lambda\rangle$ and thus $\Gamma \in T_{\kappa+\lambda,\leq\lambda}$. If $\lambda \leq \kappa$, then $\kappa+\lambda = \kappa$, so Γ admits $\langle\kappa+\lambda,\lambda'\rangle$, and thus $\Gamma \in T_{\kappa+\lambda,\leq\lambda}$. □

2.2. Logics and consequence relations

A helpful way of mapping out the logics and consequence relations determined by pairs of infinite cardinals, which also lends itself to reformulations in terms of analogs of the Löwenheim–Skolem theorem, is to establish conditions under which one is contained in another. Using Theorems 3 and 6, this can be reduced to the two-cardinal problem:

Theorem 7:

(i) $\vDash_{\kappa',\lambda'} \subseteq \vDash_{\kappa,\lambda}$ *if and only if* $T_{\kappa+\lambda,\leq\lambda} \subseteq T_{\kappa'+\lambda',\leq\lambda'}$.

(ii) $\Lambda_{\kappa',\lambda'} \subseteq \Lambda_{\kappa,\lambda}$ *if and only if for every $\mathcal{L}$-sentence φ, if $\{\varphi\} \in T_{\kappa+\lambda,\leq\lambda}$ then $\{\varphi\} \in T_{\kappa'+\lambda',\leq\lambda'}$*

Proof: (i) If $\vDash_{\kappa',\lambda'} \not\subseteq \vDash_{\kappa,\lambda}$, then by Lemma 1, there is an $\mathcal{L}_\Box$-theory Γ which is satisfiable on $\langle\kappa,\lambda\rangle$ but not satisfiable on $\langle\kappa',\lambda'\rangle$. By Theorem 3, $\tau(\Gamma) \in T_{\kappa+\lambda,\leq\lambda}$ and $\tau(\Gamma) \notin T_{\kappa'+\lambda',\leq\lambda'}$, so $T_{\kappa+\lambda,\leq\lambda} \not\subseteq T_{\kappa'+\lambda',\leq\lambda'}$.

If $T_{\kappa+\lambda,\leq\lambda} \nsubseteq T_{\kappa'+\lambda',\leq\lambda'}$, then there is an $\mathcal{L}$-theory Γ such that $\Gamma \in T_{\kappa+\lambda,\leq\lambda}$ and $\Gamma \notin T_{\kappa'+\lambda',\leq\lambda'}$. By Theorem 6, $\mu(\Gamma)$ is satisfiable on $\langle\kappa,\lambda\rangle$ but not satisfiable on $\langle\kappa',\lambda'\rangle$, so by Lemma 1, $\models_{\kappa',\lambda'} \nsubseteq \models_{\kappa,\lambda}$.

(ii) Analogous. □

As a first corollary, it follows immediately that a frame in which there are at least as many worlds as individuals determines the unique weakest logic and consequence relation determined by any pair of infinite cardinals:

Corollary 8: *If* $\kappa \leq \lambda$ *then* $\models_{\kappa,\lambda} \subseteq \models_{\kappa',\lambda'}$ *(and so* $\Lambda_{\kappa,\lambda} \subseteq \Lambda_{\kappa',\lambda'}$*).*

Proof: Assume $\kappa \leq \lambda$. Fact (ii) entails that for any $\mathcal{L}$-theory Γ, if $\Gamma \in T_{\kappa'+\lambda',\leq\lambda'}$ then Γ admits $\langle\lambda,\lambda\rangle$, and therefore $\Gamma \in T_{\kappa+\lambda,\leq\lambda}$. So $T_{\kappa'+\lambda',\leq\lambda'} \subseteq T_{\kappa+\lambda,\leq\lambda}$ and therefore, with Theorem 7, $\models_{\kappa,\lambda} \subseteq \models_{\kappa',\lambda'}$. □

A second corollary follows from the so-called infinite gap two-cardinal theorem, due to Vaught (1965a) (see also Chang and Keisler (1990, Theorem 7.2.6)), which entails that if a theory admits $\langle\kappa,\lambda\rangle$ which are sufficiently far apart, it admits $\langle\kappa',\lambda'\rangle$. For this and the next few corollaries, two precise notions of distance among cardinals are required; one is in terms of successor cardinals, and one in terms of the powerset operation. So define, for any ordinal α, $\aleph_\alpha(\kappa)$ and $\beth_\alpha(\kappa)$ as follows: $\aleph_0(\kappa) = \beth_0(\kappa) = \kappa$; $\aleph_{\alpha+1}(\kappa) = \aleph_\alpha(\kappa)^+$ and $\beth_{\alpha+1}(\kappa) = 2^{\beth_\alpha(\kappa)}$; and, if α is a limit ordinal, $\aleph_\alpha(\kappa) = \bigcup_{\beta<\alpha}\aleph_\beta(\kappa)$ and $\beth_\alpha(\kappa) = \bigcup_{\beta<\alpha}\beth_\beta(\kappa)$. The relevant consequence of the infinite gap two-cardinal theorem can now formulated as saying that an $\mathcal{L}$-theory admitting $\langle\beth_\omega(\lambda'),\lambda'\rangle$ also admits $\langle\kappa,\lambda\rangle$ (for $\lambda \leq \kappa$). It follows that any frame in which there are sufficiently more individuals than worlds determines the unique strongest logic and consequence relation determined by any pair of infinite cardinals:

Corollary 9: *If* $\kappa \geq \beth_\omega(\lambda)$*, then* $\models_{\kappa',\lambda'} \subseteq \models_{\kappa,\lambda}$ *(and so* $\Lambda_{\kappa',\lambda'} \subseteq \Lambda_{\kappa,\lambda}$*).*

Proof: Let $\kappa \geq \beth_\omega(\lambda)$, and consider any $\Gamma \in T_{\kappa+\lambda,\leq\lambda}$. Then there is a $\lambda^* \leq \lambda$ such that Γ admits $\langle\kappa+\lambda,\lambda^*\rangle$. By Fact (i), Γ admits $\langle\beth_\omega(\lambda^*),\lambda^*\rangle$ (as $\lambda^* \leq \lambda$, $\lambda^* \leq \beth_\omega(\lambda^*) \leq \beth_\omega(\lambda) \leq \kappa+\lambda$). By the infinite gap two-cardinal theorem, Γ admits $\langle\kappa'+\lambda',\lambda'\rangle$, and therefore $\Gamma \in T_{\kappa'+\lambda',\leq\lambda'}$. Hence $T_{\kappa+\lambda,\leq\lambda} \subseteq T_{\kappa'+\lambda',\leq\lambda'}$ and therefore by Theorem 7, $\models_{\kappa',\lambda'} \subseteq \models_{\kappa,\lambda}$. □

An immediate consequence of these two corollaries is that if $\kappa \leq \lambda$, then $\models_{\kappa,\lambda} = \models_{\aleph_0,\aleph_0}$, and if $\kappa \geq \beth_\omega(\lambda)$, then $\models_{\kappa,\lambda} = \models_{\beth_\omega,\aleph_0}$, and analogously for the corresponding logics. To have non-arbitrary ways of referring to these, call them $\models_{\leq}/\Lambda_{\leq}$ and $\models_{\gg}/\Lambda_{\gg}$, respectively.

What the first two corollaries leave open are $\models_{\kappa,\lambda}$ and $\Lambda_{\kappa,\lambda}$ where $\lambda < \kappa < \beth_\omega(\lambda)$; call these 'intermediate.' The following two corollaries show that in this intermediate region, the comparative cardinalities matter; furthermore, in case the generalized continuum hypothesis (abbreviated GCH, the claim that $\aleph_\alpha = \beth_\alpha$ for all ordinals α) fails, they show that both the comparison in terms of successor cardinals and the comparison in terms of the powerset operation matter.

Corollary 10: *For each* $n < \omega$, $\Lambda_{\beth_n(\lambda)^+,\lambda} \nsubseteq \Lambda_{\beth_n(\kappa),\kappa}$ *(and so* $\vDash_{\beth_n(\lambda)^+,\lambda} \nsubseteq \vDash_{\beth_n(\kappa),\kappa}$*).*

Proof: There is an $\mathcal{L}$-sentence β_n which admits, for every λ', $\langle \beth_n(\lambda'), \lambda' \rangle$ but not $\langle \beth_n(\lambda')^+, \lambda' \rangle$ (such a sentence is specified in the proof of Chang and Keisler (1990, Proposition 3.2.7 (iii), attributed to Raphael Robinson, unpublished)). So β_n admits $\langle \beth_n(\kappa), \kappa \rangle$ and thus $\{\beta_n\} \in T_{\beth_n(\kappa),\leq\kappa}$. Assume for contradiction that $\{\beta_n\} \in T_{\beth_n(\lambda)^+,\leq\lambda}$. Then there is a $\lambda' \leq \lambda$ such that β_n admits $\langle \beth_n(\lambda)^+, \lambda' \rangle$, and so by Fact (i), β_n admits $\langle \beth_n(\lambda')^+, \lambda' \rangle$ (since $\lambda' \leq \lambda$, $\lambda' \leq \beth_n(\lambda')^+ \leq \beth_n(\lambda)^+$). $\lightning$, so $\{\beta_n\} \notin T_{\beth_n(\lambda)^+,\leq\lambda}$. Thus $\{\beta_n\} \in T_{\beth_n(\kappa),\leq\kappa}$ and $\{\beta_n\} \notin T_{\beth_n(\lambda)^+,\leq\lambda}$. With Theorem 6, it follows that $\mu(\beta_n)$ is satisfiable on $\langle \beth_n(\kappa), \kappa \rangle$ but not $\langle \beth_n(\lambda)^+, \lambda \rangle$, and so that $\neg\mu(\beta_n)$ witnesses $\Lambda_{\beth_n(\lambda)^+,\lambda} \nsubseteq \Lambda_{\beth_n(\kappa),\kappa}$. □

While (Chang and Keisler 1990, Proposition 3.2.7 (iv), attributed to Michael Morley, unpublished) record the existence of an $\mathcal{L}$-theory which admits, for every λ, $\langle \aleph_n(\lambda), \lambda \rangle$ but not $\langle \aleph_n(\lambda)^+, \lambda \rangle$, the construction of such a theory is left as an exercise to the reader. To ensure that the corresponding analog to Corollary 10 holds for logics, not just consequence relations, the following lemma shows that there is a *finite* such theory.

Lemma 11: *For each* $n < \omega$*, there is an* $\mathcal{L}$*-sentence* α_n *which admits, for every* λ, $\langle \aleph_n(\lambda), \lambda \rangle$ *but not* $\langle \aleph_n(\lambda)^+, \lambda \rangle$.

Proof: The following illustrates the proof for $n = 1$; the general case is obtained by iterating the construction. Let R be a binary and B a ternary relation symbol of σ. Let φ be the following $\mathcal{L}$-sentence, which requires R to be a linear order and B to induce, for every x, an injection from the elements R-related to x to U:

$$\begin{aligned}\alpha_1 := \quad & \forall x \forall y (((Rxy \wedge Ryx) \rightarrow x = y) \wedge \\ & \forall z ((Rxy \wedge Ryz) \rightarrow Rxz) \wedge \\ & (Rxy \vee Ryx)) \wedge \\ & \forall x (\forall y (Ryx \rightarrow \exists z (Uz \wedge \forall z' (Bxyz' \leftrightarrow z = z'))) \wedge \\ & \forall y \forall y' \forall z ((Bxyz \wedge Bxy'z) \rightarrow y = y'))\end{aligned}$$

α_1 admits $\langle \lambda^+, \lambda \rangle$: Interpreting R as the standard order on ordinals. For each $a < \lambda^+$, choose an injection $i_a : a \rightarrow \lambda$ (which exists since $|a| \leq \lambda$), and interpret B to relate abc just in case $i_a(b) = c$.

α_1 does not admit $\langle \lambda^{++}, \lambda \rangle$: Assume R is interpreted as a linear order, and let a be an element which is above λ^+ elements in this order (such an element must exist for cardinality reasons). Thus there can be no injection from these elements to λ, and so no interpretation verifying α_1. □

Corollary 12: *For each* $n < \omega$, $\Lambda_{\aleph_n(\lambda)^+,\lambda} \nsubseteq \Lambda_{\aleph_n(\kappa),\kappa}$ *(and so* $\vDash_{\aleph_n(\lambda)^+,\lambda} \nsubseteq \vDash_{\aleph_n(\kappa),\kappa}$*).*

Proof: Analogous to the proof of Corollary 10, using Lemma 11. □

In this proof, $\neg\mu(\alpha_1)$ witnesses $\Lambda_{\lambda^{++},\lambda} \nsubseteq \Lambda_{\kappa^+,\kappa}$. What does this sentence look like? Apart from relabeling R in α_1, μ conjoins α_1 with two $\mathcal{L}_\Box$-sentences δ and ι which can informally be understood as requiring there to be infinitely many Us and there to be a world corresponding to each U. The most straightforward way of understanding how $\neg\mu(\alpha_1)$ imposes a constraint on the relation between cardinalities of individuals and worlds is in two steps, one imposing a constraint on the relation between cardinalities of individuals and Us and one imposing a constraint on the relation between cardinalities of Us and worlds. This line of thought applies as well to the witnesses used in the proof of Corollary 10 above and the proofs of several further corollaries below.

Especially without the GCH, the last two corollaries show that the intermediate region exhibits interesting structure. The following two corollaries impose two basic containment properties on the whole space, and so in particular on this intermediate region. The first simply applies the earlier observation that satisfiability can't be destroyed by increasing the number of worlds; the second notes that it can also not be destroyed by reducing the number of individuals:

Corollary 13: *If $\lambda' \leq \lambda$ then $\vDash_{\kappa,\lambda} \subseteq \vDash_{\kappa,\lambda'}$ (and so $\Lambda_{\kappa,\lambda} \subseteq \Lambda_{\kappa,\lambda'}$).*

Proof: By Lemmas 1 and 2. □

Corollary 14: *If $\kappa' \leq \kappa$ then $\vDash_{\kappa',\lambda} \subseteq \vDash_{\kappa,\lambda}$ (and so $\Lambda_{\kappa',\lambda} \subseteq \Lambda_{\kappa,\lambda}$).*

Proof: Let $\kappa' \leq \kappa$. Then $\kappa' + \lambda \leq \kappa + \lambda$, so, by Fact (i), for all $\lambda' \leq \lambda$, $T_{\kappa+\lambda,\lambda'} \subseteq T_{\kappa'+\lambda,\lambda'}$. So $T_{\kappa+\lambda,\leq\lambda} \subseteq T_{\kappa'+\lambda,\leq\lambda}$, and therefore, by Theorem 7, $\vDash_{\kappa',\lambda} \subseteq \vDash_{\kappa,\lambda}$. □

The following diagram summarizes the positive corollaries obtained so far for logics and consequence relations determined by a pair of cardinals in the ℶ-series:

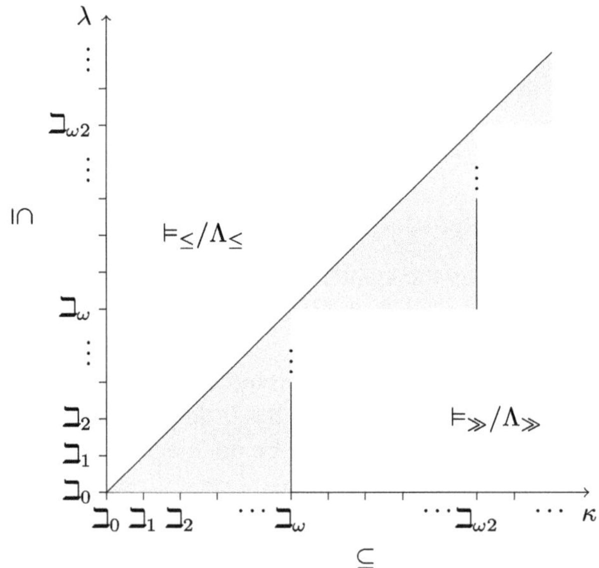

The two curves in the diagram divide the space of logics and consequence relations it represents into the region of $\vDash_{\leq}$, which includes those on the diagonal line, the shaded intermediate region, and the region of $\vDash_{\gg}$, which includes those on the vertical lines. ⊆ marks the containment properties along both axes established in the last two corollaries. It should be noted that the diagram might be misleading in two respects: First, it only records pairs of cardinals in the $\beth$-series, so unless the GCH is true, it does not represent all pairs of infinite cardinals. Second, although the shaded area and lines suggest continuity, the represented space is obviously discrete. Not represented in the diagram is Corollary 10, which entails that any two represented consequence relations in the intermediate region are distinct if they, as one might say, differ in how far they are from the diagonal.

Assuming the GCH, every intermediate consequence relation is identical to $\vDash_{\aleph_n(\kappa),\kappa}$ for some κ and $n < \omega$. One might conjecture that in this case, at least the choice of κ does not matter. But it turns out that this conjecture is unlikely to be settled by ZFC, even assuming the GCH: On the one hand, the conjecture follows from $V = L$, the axiom of constructibility, which entails the GCH. On the other hand, given the consistency of a very mild large cardinal hypothesis, ZFC + GCH is consistent with the negation of the conjecture.

For the first of these observation, the conjecture will be derived from the so-called gap n conjecture, which says that every $\mathcal{L}$-theory admitting $\langle\aleph_n(\kappa), \kappa\rangle$ also admits $\langle\aleph_n(\lambda), \lambda\rangle$. According to Chang and Keisler (1990, 524), that this holds for all $n < \omega$ was shown to follow from $V = L$ in unpublished work by Ronald Jensen. (Details of the gap 1 case can be found in Jensen, R. Björn (1972); details of the gap 2 case and a sketch of the gap n case can be found in Devlin (1984, Chapter VIII).)

Corollary 15: *If $V = L$, then for all $n < \omega$, $\vDash_{\aleph_n(\lambda),\lambda} = \vDash_{\aleph_n(\kappa),\kappa}$ (and so $\Lambda_{\aleph_n(\lambda),\lambda} = \Lambda_{\aleph_n(\kappa),\kappa}$).*

Proof: Assume $V = L$ and $\Gamma \in T_{\aleph_n(\kappa),\leq\kappa}$. Then Γ admits $\langle\aleph_n(\kappa), \kappa'\rangle$ for some $\kappa' \leq \kappa$, and thus $\langle\aleph_n(\kappa'), \kappa'\rangle$ by Fact (i). So by the gap n conjecture, Γ admits $\langle\aleph_n(\lambda), \lambda\rangle$ and therefore $\Gamma \in T_{\aleph_n(\lambda),\leq\lambda}$. So $T_{\aleph_n(\kappa),\leq\kappa} \subseteq T_{\aleph_n(\lambda),\leq\lambda}$, and thus, with Theorem 7, $\vDash_{\aleph_n(\lambda),\lambda} \subseteq \vDash_{\aleph_n(\kappa),\kappa}$. By symmetry, $\vDash_{\aleph_n(\lambda),\lambda} = \vDash_{\aleph_n(\kappa),\kappa}$. □

Assuming $V = L$, the logics and consequence relations therefore form $\omega+1$-chains, linearly ordered by the proper subset relation, and determined by the following sequence of pairs of cardinals:

$$\langle\aleph_0, \aleph_0\rangle, \langle\aleph_1, \aleph_0\rangle, \langle\aleph_2, \aleph_0\rangle, \ldots, \langle\aleph_\omega, \aleph_0\rangle$$

The second observation is based on a consistency proof of the failure of the gap 2 conjecture due to Jack Silver. This proof uses an $\mathcal{L}$-sentence k which admits $\langle\kappa^{++}, \kappa\rangle$ if and only if a certain claim $K_2(\kappa)$ holds. ($K_2(\kappa)$ states the existence of a so-called Kurepa tree satisfying certain properties in terms of κ. A sketch of the construction of k can be found in Vaught (1965b), see also the discussion in Chang and Keisler (1990, 527).) Using μ and α_2 (as in Lemma 11), this can be turned into a sentence which is a member of $\Lambda_{\kappa^{++},\kappa}$ if and only if $K_2(\kappa)$ fails:

Lemma 16: $\neg\mu(k \wedge \alpha_2) \in \Lambda_{\kappa^{++},\kappa}$ *if and only if* $\neg K_2(\kappa)$.

Proof: $\neg\mu(k\wedge\alpha_2) \in \Lambda_{\kappa^{++},\kappa}$ if and only if $\mu(k\wedge\alpha_2)$ is not satisfiable on $\langle\kappa^{++},\kappa\rangle$. By Theorem 6, this is the case if and only if $\{k,\alpha_2\} \notin T_{\kappa^{++},\leq\kappa}$. Since α_2 does not admit $\langle\kappa^{++},\kappa'\rangle$ for any $\kappa' < \kappa$, this is equivalent to the claim that k does not admit $\langle\kappa^{++},\kappa\rangle$, which, as noted, is the case if and only if $\neg K_2(\kappa)$. □

On the assumption that ZFC is consistent with the existence of two uncountable inaccessible cardinals, Silver (1971) has proven that ZFC + GCH is consistent with $K_2(\aleph_1) \wedge \neg K_2(\aleph_0)$ as well as consistent with $K_2(\aleph_0) \wedge \neg K_2(\aleph_1)$. Thus on this assumption, ZFC + GCH neither settles whether $\neg\mu(k \wedge \alpha_2)$ is a member of $\Lambda_{\aleph_2,\aleph_0}$, nor whether it is a member of $\Lambda_{\aleph_3,\aleph_1}$, nor whether these logics are the same:

Corollary 17: *If* ZFC *is consistent with the existence of two uncountable inaccessible cardinals, then* ZFC + GCH *is consistent with (i) as well as consistent with (ii):*

(i) $\Lambda_{\aleph_2,\aleph_0} \not\subseteq \Lambda_{\aleph_3,\aleph_1}$ *(and so* $\models_{\aleph_2,\aleph_0} \not\subseteq \models_{\aleph_3,\aleph_1}$*)*
(ii) $\Lambda_{\aleph_3,\aleph_1} \not\subseteq \Lambda_{\aleph_2,\aleph_0}$ *(and so* $\models_{\aleph_3,\aleph_1} \not\subseteq \models_{\aleph_2,\aleph_0}$*)*

Proof: Immediate by Lemma 16 and the result mentioned above. □

In the same way, an analogous negative result for $\langle\aleph_3,\aleph_1\rangle$ and $\langle\aleph_7,\aleph_5\rangle$ can be derived, and the independence of $\neg\mu(k \wedge \alpha_2) \in \Lambda_{\aleph_2,\aleph_0}$ from ZFC (rather than ZFC + GCH) can be obtained from the slightly weaker assumption that ZFC is consistent with the existence of *one* uncountable inaccessible cardinal. (See Chang and Keisler (1990, 527–528), who attribute the relevant results to Silver and Robert Solovay.)

Some particular positive results have been obtained in the case of the gap 1 conjecture, which lead to some particular positive corollaries for logics and consequence relations determined by pairs of infinite cardinals $\langle\kappa^+,\kappa\rangle$, although negative results are also available for such pairs. The first positive corollary is based on what is known as Vaught's Two-Cardinal Theorem:

Corollary 18: *If* $\kappa > \lambda$, *then* $\models_{\aleph_1,\aleph_0} \subseteq \models_{\kappa,\lambda}$ *(and so* $\Lambda_{\aleph_1,\aleph_0} \subseteq \Lambda_{\kappa,\lambda}$*).*

Proof: As proven in Morley and Vaught (1962, Theorem 6.2) (see also Chang and Keisler (1990, Theorem 3.2.9)), for all $\lambda' \leq \lambda$, $T_{\kappa,\lambda'} \subseteq T_{\aleph_1,\aleph_0}$, so $T_{\kappa+\lambda,\leq\lambda} \subseteq T_{\aleph_1,\aleph_0}$, and thus by Theorem 7, $\models_{\aleph_1,\aleph_0} \subseteq \models_{\kappa,\lambda}$. □

Using what is known as Chang's Two-Cardinal Theorem, this first positive corollary leads to a second one:

Corollary 19: *If the* GCH *holds and* κ *is regular, then* $\models_{\kappa^+,\kappa} = \models_{\aleph_1,\aleph_0}$ *(and so* $\Lambda_{\kappa^+,\kappa} = \Lambda_{\aleph_1,\aleph_0}$*).*

Proof: As proven in Chang (1965) (see also Chang and Keisler (1990, Theorem 7.2.7)), $T_{\aleph_1,\aleph_0} \subseteq T_{\kappa^+,\kappa}$, so $T_{\aleph_1,\aleph_0} \subseteq T_{\kappa^+,\leq\kappa}$, and thus by Theorem 7, $\models_{\kappa^+,\kappa} \subseteq \models_{\aleph_1,\aleph_0}$. With Corollary 18, the claim follows. □

The relevant negative results depend on large cardinal assumptions which are stronger than those used in the gap 2 case. Similar to the use of Kurepa trees in the gap 2 case, they use so-called special κ^+-Aronszajn trees, whose existence will be written as $A(\kappa^+)$. Again, there is an $\mathcal{L}$-sentence a which admits $\langle\kappa^+, \kappa\rangle$ if and only if $A(\kappa^+)$ (a construction is given in Chang and Keisler (1990, 7.2.11); the observation that such a sentence exists is attributed to Frederick Rowbottom and Silver, independently and unpublished).

Lemma 20: $\neg\mu(a \wedge \alpha_1) \in \Lambda_{\kappa^+,\kappa}$ *if and only if* $\neg A(\kappa^+)$.

Proof: Analogous to Lemma 16. □

While $A(\aleph_1)$ holds, the consistency of ZFC with the existence of a Mahlo cardinal entails the consistency of ZFC with $\neg A(\aleph_2)$, and the consistency of ZFC with the existence of a supercompact cardinal entails the consistency of ZFC + GCH with $\neg A(\aleph_{\omega+1})$ (see Chang and Keisler (1990, 7.2.10. (i), 7.2.12. (i) & (iii)), who attribute these results to Nachman Aronszajn, Mitchell (1972) and Shelah (1979)). Thus:

Corollary 21: *(i) If* ZFC *is consistent with the existence of a Mahlo cardinal, then* ZFC *is consistent with* $\Lambda_{\aleph_2,\aleph_1} \not\subseteq \Lambda_{\aleph_1,\aleph_0}$ *(and so* $\models_{\aleph_2,\aleph_1} \not\subseteq \models_{\aleph_1,\aleph_0}$*).*

(ii) If ZFC *is consistent with the existence of a supercompact cardinal, then* ZFC + GCH *is consistent with* $\Lambda_{\aleph_{\omega+1},\aleph_\omega} \not\subseteq \Lambda_{\aleph_1,\aleph_0}$ *(and so* $\models_{\aleph_{\omega+1},\aleph_\omega} \not\subseteq \models_{\aleph_1,\aleph_0}$*).*

Proof: Immediate by Lemma 20 and the results mentioned above. □

2.3. *Axiomatizations*

If it could be argued that $\models_{\mathrm{MU}}$ is given by a particular pair $\langle\kappa, \lambda\rangle$, it would be interesting to specify it using a standard axiomatic calculus. This requires (i) $\models_{\kappa,\lambda}$ to be compact, (ii) $\models_{\kappa,\lambda}$ to satisfy the deduction/detachment property, and (iii) $\Lambda_{\kappa,\lambda}$ to be recursively enumerable. (ii) is easily seen to hold in general:

Proposition 22: $\Gamma \cup \{\varphi\} \models_{\kappa,\lambda} \psi$ *if and only if* $\Gamma \models_{\kappa,\lambda} \varphi \to \psi$.

Proof: Routine. □

This raises the question which pairs of infinite cardinals determine compact consequence relations, and which determine recursively enumerable logics. For those pairs that satisfy both conditions, it is also interesting to consider whether there are any natural axiomatizations of the logics they determine. The next result shows that all of this is easily done in the case of $\models_{\leq}$. For concreteness, the construction will be based on the axiomatization described in Hughes and Cresswell (1996, Chapter 13), which will be called LPC + S5. Its theorems are the same as those of the system which was presented in a more compressed manner in Kripke (1959), and there proven to be sound and complete with respect to the class of all frames.

Theorem 23: *Let* LPC$^\infty$+ S5 *be the result of adding the following axiom schema to* LPC + S5*:*

$$E_n := \quad \exists x_0 \ldots \exists x_{n-1} \bigwedge_{i<j<n} x_i \neq x_j$$

Let $\vdash$ *be the set of sequents* $\langle \Gamma, \varphi \rangle$ *such that* $\bigwedge_{\gamma \in \Gamma_0} \gamma \rightarrow \varphi$ *is derivable in* LPC$^\infty$+ S5 *for some finite* $\Gamma_0 \subseteq \Gamma$. $\vdash = \vDash_{\leq}$.

Proof: If $\Gamma \nvdash \varphi$, then $\Gamma \cup \{E_n : n < \omega\} \nvdash \varphi$, so there is no finite $\Delta \subseteq \Gamma \cup \{E_n : n < \omega\}$ such that $\bigwedge_{\delta \in \Delta} \delta \rightarrow \varphi$ is derivable in LPC$^\infty$+ S5, and so no such sentence is derivable in LPC + S5 either. Thus by the adequacy of LPC + S5 (Hughes and Cresswell 1996, Chapter 14), $\Gamma \cup \{E_n : n < \omega\} \cup \{\neg\varphi\}$ is satisfiable on a frame, which must contain infinitely many individuals. By Corollaries 13 and 8, this can be assumed to be $\langle \aleph_0, \aleph_0 \rangle$, so $\Gamma \nvDash_{\leq} \varphi$.

Since the axioms of LPC + S5 are valid on all frames, and the instances of E_n are valid on $\langle \aleph_0, \aleph_0 \rangle$, the axioms of LPC$^\infty$+ S5 are valid on $\langle \aleph_0, \aleph_0 \rangle$. The validities on $\langle \aleph_0, \aleph_0 \rangle$ are easily seen to be closed under the rules of LPC + S5 (necessitation, modus ponens, and a version of universal generalization). Thus all sentences provable in LPC$^\infty$+ S5 are valid on $\langle \aleph_0, \aleph_0 \rangle$. Now, if $\Gamma \vdash \varphi$, then $\bigwedge_{\gamma \in \Gamma_0} \gamma \rightarrow \varphi$ is derivable in LPC$^\infty$+ S5 for some finite $\Gamma_0 \subseteq \Gamma$. Consider any model of Γ based on $\langle \aleph_0, \aleph_0 \rangle$; this must be a model of Γ_0, and so a model of φ, hence $\Gamma \vDash_{\leq} \varphi$. □

In case $\kappa > \lambda$, the issues are more subtle. Theorems 3 and 6 allow us to derive a number of results on compactness and recursive enumerability. Starting with compactness, they yield the following theorem, which reduces the compactness of $\vDash_{\kappa,\lambda}$ to a corresponding property for $T_{\kappa+\lambda,\leq\lambda}$:

Theorem 24: *The following are equivalent:*

(i) $\vDash_{\kappa,\lambda}$ *is compact.*
(ii) *If every finite subset of an* $\mathcal{L}$*-theory* Γ *is a member of* $T_{\kappa+\lambda,\leq\lambda}$ *then* $\Gamma \in T_{\kappa+\lambda,\leq\lambda}$.

Proof: Assume first that (ii) is not the case, witnessed by an $\mathcal{L}$-theory Γ. Then by Theorem 6, $\mu(\Gamma)$ is not satisfiable on $\langle \kappa, \lambda \rangle$ although all of its finite subsets are; thus $\vDash_{\kappa,\lambda}$ is not compact.

Assume now that $\vDash_{\kappa,\lambda}$ is not compact, so there is an $\mathcal{L}_\Box$-theory Γ which is not satisfiable on $\langle \kappa, \lambda \rangle$ although all of its finite subsets are. By Theorem 3, all finite subsets of $\tau(\Gamma)$ are members of $T_{\kappa+\lambda,\leq\lambda}$ but $\tau(\Gamma) \notin T_{\kappa+\lambda,\leq\lambda}$; thus (ii) is not the case. □

For $\vDash_{\gg}$ and $\vDash_{\aleph_1,\aleph_0}$, results are available from which compactness follows:

Corollary 25: $\vDash_{\gg}$ *and* $\vDash_{\aleph_1,\aleph_0}$ *are compact.*

Proof: As shown in Vaught (1965a), if every finite subset of an $\mathcal{L}$-theory Γ admits $\langle \beth_\omega, \aleph_0 \rangle$ then Γ admits $\langle \beth_\omega, \aleph_0 \rangle$; see also Chang and Keisler (1990, Theorem 7.2.6, Exercise 7.2.12). As shown in Morley and Vaught (1962), if every finite subset of an $\mathcal{L}$-theory Γ admits $\langle\aleph_1,\aleph_0\rangle$ then Γ admits $\langle\aleph_1,\aleph_0\rangle$, see also Chang and Keisler (1990, Exercise 3.2.13). The claim follows by Theorem 24. □

No corresponding compactness results seem easily available for the remaining intermediate consequence relations. Indeed, Shelah (2005) contains, for each $n > 1$, a consistency result for the existence of an $\mathcal{L}$-theory which does not admit $\langle\aleph_n,\aleph_0\rangle$ although all of its finite subsets admit it. With Theorem 24, a corresponding consistency result for the incompactness of $\models_{\aleph_n,\aleph_0}$, for $n > 1$, follows.

The question of recursive enumerability can also be reduced to one which has been explored in model theory, using the standard notion of consistency in first-order non-modal logic:

Theorem 26: *If there is a recursive $\mathcal{L}$-theory Γ such that for any $\mathcal{L}$-sentence φ, $\Gamma \cup \{\varphi\}$ is consistent if and only if $\{\varphi\} \in T_{\kappa+\lambda,\leq\lambda}$, then $\Lambda_{\kappa,\lambda}$ is recursively enumerable.*

Proof: Assume there is such a theory Γ, and consider any $\mathcal{L}_\Box$-sentence φ. $\varphi \in \Lambda_{\kappa,\lambda}$ iff $\neg\varphi$ is not satisfiable on $\langle\kappa,\lambda\rangle$. By Theorem 3, this is the case iff $\{\tau(\neg\varphi)\} \notin T_{\kappa+\lambda,\leq\lambda}$. By assumption, this is the case iff $\Gamma\cup\{\tau(\neg\varphi)\}$ is inconsistent, which is the case iff $\tau(\varphi)$ is derivable from Γ in a standard calculus for first-order logic. This yields a method for recursively enumerating $\Lambda_{\kappa,\lambda}$. □

Corollary 27: *$\Lambda_{\gg}$ and $\Lambda_{\aleph_1,\aleph_0}$ are recursively enumerable.*

Proof: For $\models_{\gg}$, let $\kappa = \beth_\omega$ and $\lambda = \aleph_0$. The existence of a set satisfying the condition on Γ in Theorem 26 follows from the corresponding two-cardinal theorem by Vaught (1965a); explicit specifications of such sets are given in Barwise (1977) and Schmerl (1977). In the case of $\models_{\aleph_1,\aleph_0}$, the existence of a set satisfying the condition on Γ in Theorem 26 for $\kappa = \aleph_1$ and $\lambda = \aleph_0$ was noted in Vaught (1964) based on results in Fuhrken (1964); an explicit specification of such a set is given in Keisler H. Jerome (1966). In each case, the recursive enumerability of the relevant set of theorems follows by Theorem 26. □

As the preceding corollaries indicate, $\models_{\gg}$ and $\models_{\aleph_1,\aleph_0}$ stand out as compact consequence relations whose corresponding logics are easily shown to be recursively enumerable. It would be interesting to develop axiomatizations for them as well. Indeed, these two stand out also in that any of the other logics and consequence relations are likely to be extremely difficult to axiomatize; recall that previous corollaries exhibited various potential failures of ZFC (or even ZFC + GCH) to decide membership in logics such as $\Lambda_{\aleph_2,\aleph_0}$, $\Lambda_{\aleph_2,\aleph_1}$ and $\Lambda_{\aleph_{\omega+1},\aleph_\omega}$.

LPC$^\infty$ + S5 is an infinitary axiomatization of $\Lambda_{\leq}$ in the sense that it is obtained from LPC + S5 by adding an axiom schema with infinitely many instances. It turns out that this is an essential feature, in the sense that there is no finite set of axioms which added to LPC + S5 yields an axiomatization of $\Lambda_{\leq}$ (even if a rule of substitution is added to LPC + S5, which Hughes and Cresswell (1996, Chapter 13) do not include). Moreover, this is the case for any logic $\Lambda_{\kappa,\lambda}$; in this sense, none of the logics under consideration here is finitely axiomatizable.

Without setting up the relevant variant of LPC + S5 in detail, here is a sketch of the proof: Consider a model $\mathfrak{M}$ based on $\langle \kappa, \lambda \rangle$ which interprets every n-ary relation symbol in σ as the function mapping every element of λ to κ^n, the set of all n-tuples on κ. For any $i < \omega$, define $\mathfrak{M}_i$ to be such a model based on $\langle i, \lambda \rangle$, i.e. one which interprets every n-ary relation symbol in σ as the function mapping every element of λ to i^n. Using a back and forth system, it can be shown that any $\mathcal{L}_\Box$-sentence of quantifier rank up to i is true in $\mathfrak{M}$ if and only if it is true in $\mathfrak{M}_i$. (The relevant definitions can be adapted from Fritz (2013).) Considering any finite set $\Gamma \subseteq \Lambda_{\kappa,\lambda}$, let i be the maximum of the quantifier ranks of its members. Since $\mathfrak{M}$ verifies all members of Γ, so does $\mathfrak{M}_i$. All axioms LPC + S5 are true in $\mathfrak{M}_i$ (under any assignment function), and the $\mathcal{L}_\Box$-formulas true in $\mathfrak{M}_i$ (under any assignment function) are closed under the rules of LPC + S5. But E_{i+1}, a sentence saying that there are at least $i + 1$ individuals, is not true in $\mathfrak{M}_i$. Thus E_{i+1}, which is a member of $\Lambda_{\kappa,\lambda}$, is not derivable in the axiomatization resulting from adding the members of Γ to LPC + S5.

3. Metaphysical universality

Recall that the previous section developed the theory of logics and consequence relations determined by pairs of infinite cardinals on the assumption of first-order ZFC, a theory of *pure* sets. Since there are non-sets, there is no pure set containing all individuals, and so no intended frame, i.e. no frame $\langle D, W \rangle$ such that D is the set of individuals and W is the set of worlds. (Even in a variant model theory based on one of the usual extensions of ZFC to impure sets, there is no intended frame, since there is no universal set.) An argument is therefore needed to show that Λ_{MU} and $\vDash_{MU}$ are determined by some pair of infinite cardinals. Williamson (2013, 117–118, 145) tentatively suggests such an argument, conditional on Premise 1, stated below.

The next section summarizes this argument, albeit somewhat loosely. The argument relies on two premises, which will be discussed in subsequent sections. On the one hand, some difficulties in evaluating the second premise of the argument will be raised. On the other hand, the first premise relies on the assumption that worlds are individuals; it will be shown how to eliminate this assumption.

3.1. Williamson's argument

Let $\mathcal{L}^2$ be the extension of $\mathcal{L}$ by an infinite set of relational variables for each arity, as well as quantifiers binding them; let $\mathcal{L}^2_\Box$ be the corresponding extension of $\mathcal{L}_\Box$. For any $\mathcal{L}_\Box$-formula φ, let $\forall\varphi$ be the result of uniformly replacing each relation symbol in φ by a distinct relational variable of appropriate arity, and prefixing the resulting formula by a string of universal quantifiers binding all of these variables; the result is an $\mathcal{L}^2_\Box$-formula. Unless explicitly noted otherwise, first- and second-order quantifiers will be interpreted unrestrictedly, and $\Box$ as expressing metaphysical necessity, as above. Thus, for any $\mathcal{L}_\Box$-sentence φ, $\forall\varphi$ is a fully interpreted $\mathcal{L}^2_\Box$-sentence. With this, the definition of Λ_{MU} can be summarized as follows:

$$\Lambda_{\mathrm{MU}} = \{\varphi \in \mathcal{L}_\Box : \forall\varphi \text{ is true}\}$$

Recall that this truth predicate is assumed to be transparent, as third- and in general higher order quantifiers are assumed to be available in the metalanguage.

As in Section 2.1, let ρ be an injection which maps every n-ary relation symbol of σ to an $n+1$-ary relation symbol of σ, and every element of V to an element of V, such that neither U nor ϑ are in its image; this was extended to a version of the standard translation from $\mathcal{L}_\Box$-formulas to $\mathcal{L}$-formulas. Instead of this extension of ρ, the following will use a variant extension ζ which differs from the above extension of ρ only in that first-order quantifiers are not restricted to the negation of U. Thus, the most important clauses are:

$$\begin{aligned}
\zeta(Rx_1 \ldots x_n) &:= \rho(R)\rho(x_1)\ldots\rho(x_n)\vartheta\\
\zeta(\Box\varphi) &:= \forall\vartheta(U\vartheta \rightarrow \zeta(\varphi))\\
\zeta(\forall x\varphi) &:= \forall\rho(x)\zeta(\varphi)
\end{aligned}$$

Extend the notation $\forall\varphi$ introduced above by writing $\forall[U]\varphi$ for the result of the same procedure, exempting U from being replaced by a variable and bound by a universal quantifier. Thus for any $\mathcal{L}_\Box$-sentence φ, $\forall[U]\zeta(\varphi)$ is an $\mathcal{L}^2$-formula in which all relation symbols apart from U are interpreted, and no variable apart from ϑ is free.

> **Premise 1:** For any $\mathcal{L}_\Box$-sentence φ, $\forall\varphi$ is true iff $\forall[U]\zeta(\varphi)$ is true, interpreting U as being a world and ϑ as the actual world.

Note that in the version of the standard translation used in this premise, the necessity operator is translated as a universal quantifier restricted to U, and first-order quantifiers are unchanged. In neither case is the relevant quantifier relativized to a condition formulated in terms of the variable ϑ representing the world of evaluation, so Premise 1 is where the assumptions of the necessity of what there is and what is necessary comes into the argument.

To state the second premise of the argument, say that an $\mathcal{L}^2$-theory is *satisfiable on the intended interpretation* if it is true on some interpretation of the non-logical constants, keeping the unrestricted interpretation of first- and second-order quantifiers. Say that such a such a theory is *satisfiable on a set* if it has a model in the standard set-theoretic semantics, according to which first-order quantifiers are interpreted as ranging over the members of a given set A, n-ary second-order quantifiers are interpreted as ranging over sets of n-tuples on A, and non-logical constants are interpreted accordingly.

Premise 2: Every $\mathcal{L}^2$-theory satisfiable on the intended interpretation is satisfiable on a set.

Using these two premises, the argument runs as follows: By definition of Λ_{MU}, all members of $\{\forall\varphi : \varphi \in \Lambda_{\text{MU}}\} \cup \{\neg\forall\varphi : \varphi \notin \Lambda_{\text{MU}}\}$ are true, so by Premise 1, all members of the $\mathcal{L}^2$-theory $\Gamma = \{\forall[U]\zeta(\varphi) : \varphi \in \Lambda_{\text{MU}}\} \cup \{\neg\forall[U]\zeta(\varphi) : \varphi \notin \Lambda_{\text{MU}}\}$ are true, interpreting U as being a world and ϑ as the actual world. Thus Γ is satisfiable on the intended interpretation, and so by Premise 2 satisfiable on a set. In particular, there is a model $\mathfrak{A} = \langle D, e\rangle$ and assignment function a such that D is some cardinal κ, $e(U)$ is a cardinal $\lambda \leq \kappa$, $a(\vartheta) \in \lambda$, and each member of Γ is true relative to $\mathfrak{A}$ and a. To establish $\Lambda_{\text{MU}} = \Lambda_{\kappa,\lambda}$, it now suffices to prove the following claim, which is routine: for any $\mathcal{L}_\Box$-sentence φ, $\mathfrak{M}, a \vDash \forall[U]\zeta(\varphi)$ iff φ is valid on $\langle\kappa, \lambda\rangle$.

It should be noted that Williamson's argument only applies to Λ_{MU}. To turn it into an argument for the claim that $\vDash_{\text{MU}} = \vDash_{\kappa,\lambda}$ for some cardinals κ and λ, one could strengthen Premise 2 to apply to sentences of an extension of $\mathcal{L}^2$ which allows for suitably infinitary conjunctions. Although the following focuses on Williamson's original argument concerning Λ_{MU}, much of it can be adapted to the case of $\vDash_{\text{MU}}$.

3.2. *Premise 2: Kreisel's principle*

A version of Premise 2 is discussed in Kreisel (1967), and investigated formally in Shapiro (1987), who calls it *Kreisel's principle*. Shapiro formalizes Kreisel's principle in the context of second-order ZFC. Williamson notes that Shapiro proves it to have a standard model satisfying second-order ZFC, assuming that there are $\aleph_2$ inaccessible cardinals. However, such a semantic consistency proof does not establish the truth of the premise. Williamson also notes that Shapiro derives in second-order ZFC that Kreisel's principle is equivalent to a certain reflection principle in second-order set theory. Such reflection principles have played an important role in modern set theory, and it has been suggested that adopting them can be justified on the basis of the iterative conception of set. However, some variants of such reflection principles have turned out to be inconsistent, and consequently, the status of the (as far as we know) consistent reflection principles is disputed, see, e.g. Koellner (2009) for a recent overview.

Furthermore, it is not even clear that Shapiro's results apply to Premise 2 as used in Williamson's argument, since Shapiro (1987, 311) interprets all quantifiers to be restricted to pure sets, whereas Williamson appeals to an unrestricted reading of quantifiers. It is conceivable that Kreisel's principle holds on pure sets without holding in general, i.e. that every second-order theory satisfiable on the intended interpretation restricted to pure sets is satisfiable on a pure set even though there is some second-order theory satisfiable on the unrestricted intended interpretation which is not satisfiable on any pure (or even impure) set. Further assumptions of (impure) set theory are needed to show that this is not the case. Alternatively, one might directly endorse the relevant unrestricted version of the Kreisel Principle, which might turn out to be equivalent to a suitable reflection principle in impure set theory. Although reflection principles are usually discussed in the context of pure set theory, Burgess (2004) introduces an impure set theory containing such a principle as one of its axioms.

3.3. Premise 1: worlds in higher order logic

Premise 1 captures the assumption that on a suitable way of understanding possible worlds, necessity is equivalent to being true in all possible worlds. This won't be questioned in the following. But Premise 1 also contains the assumption that possible worlds can be taken to be individuals, since ζ replaces $\Box$ by a restricted universal *first-order* quantifier. Maybe they can be so understood according to the modal realism of Lewis (1986), but Williamson (2013, 17) explicitly rejects this view.

In line with a view which is perhaps more popular than modal realism (see, e.g. Stalnaker (1976) and Fine (1977)), Williamson's treatment of possible world semantics in Williamson (2013, Section 3.3) identifies possible worlds with possible propositions which are maximal in the sense of strictly entailing each proposition or its negation. Furthermore, in Williamson (2013, Chapter 5), propositional quantification is treated as nullary second-order quantification, i.e. quantification into sentence position. It would therefore be natural for Williamson to treat quantification over worlds as restricted quantification into sentence position. He does not explicitly adopt this position in Williamson (2013, Section 5.7), where he gives a homophonic interpretation of higher order modal logic. Rather, he leaves it open how to understand worlds by introducing a primitive type for worlds.

In the following, an argument will be sketched which shows that no matter how worlds are understood, Premise 1 can be adapted without loss of plausibility. For generality, $\mathcal{L}^2$ will be extended by a type w for worlds, the interpretation of which will be left open. E.g. quantifiers binding variables of type w might be interpreted as propositional quantifiers restricted to propositions which are maximal in the sense described above. To define the syntax of this extended language $\mathcal{L}^{2w}$, let the type of a relational symbol be given by a finite sequence of *e*s and *w*s, indicating that the symbol combines with a corresponding string of individual and world variables to form an atomic predication. E.g. $\langle e, e, w\rangle$ is the type of a relation symbol R which can be combined with two first-order variables x and y and a world variable χ to form an atomic predication $Rxy\chi$. Based on a countably infinite supply of individual variables, world variables, as well as variables and constants of each relational type, which can be combined in the way just illustrated to form atomic predications, recursively define the formulas of $\mathcal{L}^{2w}$ using Boolean operations as well as quantifiers binding first-order, world, and relational variables. However worlds are understood, worlds χ and ν are taken to be identical if they are indistinguishable in the sense that $\forall X(X\chi \leftrightarrow X\nu)$, where X is a variable of type $\langle w \rangle$.

To adapt Premise 1 to $\mathcal{L}^{2w}$, let ξ be a version of the standard translation which differs from ζ in using a world variable θ in place of the first-order variable ϑ, and translating $\Box$ using an unrestricted universal world-quantifier binding θ:

$$\xi(\Box\varphi) := \quad \forall\theta\xi(\varphi)$$

Of course, each n-ary relation symbol must now also be translated using an $n+1$-ary relation symbol of type $\langle e, \ldots, e, w \rangle$. The required premise is then:

> **Premise 1a:** For any $\mathcal{L}_\Box$-sentence φ, $\forall\varphi$ is true iff $\forall\xi(\varphi)$ is true, interpreting θ as the actual world.

Grant the truth of this premise for the sake of the argument – how to substantiate it will depend on how one understands worlds. $\forall\xi(\varphi)$ is in general a sentence of $\mathcal{L}^{2w}$ rather than $\mathcal{L}^2$, so the original argument using Premise 2 (Kreisel's principle) must be modified. The new argument distinguishes two cases: there being at least as many individuals as worlds, and there being more worlds than individuals.

Since set theory need not be applicable to worlds, how should this talk of 'at least as many' and 'more' be understood? There being at least as many individuals as worlds should be understood as there being an injective functional relation between worlds and individuals, and there being more words than individuals should be understood as there being no such relation. Here, quantification over relations between worlds and individuals should be understood as quantification binding variables of type $\langle w, e \rangle$, i.e. variables which combine with an expression of type w (denoting a world) and an expression of type e (denoting an individual) to form a sentence. For readability, the argument is written in English using phrases such as 'relation between worlds and individuals,' rather than a formal higher order language, in which it would be expressed more properly.

> **Case 1:** There are at least as many individuals as worlds; i.e. there is an injection i from worlds to individuals. For this case, the argument relies on the following premise:
>
> **Premise 1b:** If there is an injection i from worlds to individuals, then for any $\mathcal{L}_\Box$-sentence φ, $\forall\xi(\varphi)$ is true, interpreting θ as the actual world, iff $\forall[U]\zeta(\varphi)$ is true, interpreting U as the image of i and ϑ as the image of the actual world under i.

This premise is motivated by the kind of unrestricted comprehension principle endorsed in Williamson (2013, 227): with it, any sequence of relations witnessing the falsity of $\forall\xi(\varphi)$ can be turned, via *i* into a counterexample of $\forall[U]\zeta(\varphi)$, and *vice versa*. Premises 1a and 1b entail:

> For any $\mathcal{L}_\Box$-sentence φ, $\forall\varphi$ is true iff $\forall[U]\zeta(\varphi)$ is true, interpreting *U* as the image of *i* and ϑ as the image of the actual world under *i*.

This claim can take the place of Premise 1 in the original argument.

> **Case 2:** There are more worlds than individuals. (The following would also apply to the case in which there are as many worlds as individuals, but this is already covered by Case 1.) In this case, it will be shown that $\Lambda_\leq = \Lambda_{\mathrm{MU}}$, first showing $\subseteq$ and then $\supseteq$.

Assume first that $\varphi \in \Lambda_\leq$. Then by Theorem 23, φ is derivable in LPC$^\infty$+ S5. With Premise 1c it follows that $\varphi \in \Lambda_{\mathrm{MU}}$, as required:

> **Premise 1c:** For any $\mathcal{L}_\Box$-sentence φ, if φ is derivable in LPC$^\infty$+ S5, then $\forall\varphi$ is true.

A natural way of substantiating this premise is by an inductive argument over length of derivation in (a suitably modified version of) LPC$^\infty$+ S5 along the lines of the argument in Williamson (2013, Section 3.3) for the propositional case.

For the other direction, it is helpful to establish an auxiliary claim for any given $\mathcal{L}_\Box$-sentence φ. Assume that $\forall[U]\zeta(\varphi)$ is true, interpreting *U* as being an individual (and ϑ as some particular individual – it does not matter which). On this interpretation, $\neg\forall[U]\zeta(\neg\eta)$ is true, where $\eta = \forall x\Diamond\forall y(Iy \leftrightarrow x = y)$ (interpreting $\rho(I)$ as identity). By Premise 2, $\{\forall[U]\zeta(\varphi), \neg\forall[U]\zeta(\neg\eta)\}$ is therefore satisfiable on a cardinal κ, interpreting *U* as a cardinal $\lambda \leq \kappa$. Thus φ is valid on $\langle\kappa, \lambda\rangle$, but $\neg\eta$ is not. Since $\neg\eta$ is not valid on $\langle\kappa, \lambda\rangle$, $\kappa = \lambda$, and therefore $\varphi \in \Lambda_\leq$.

Contraposing this line of reasoning, if $\varphi \notin \Lambda_\leq$, then $\forall[U]\zeta(\varphi)$ is false, interpreting *U* as being an individual and ϑ as some individual. With Premise 1d it follows that $\forall\xi(\varphi)$ is false, interpreting θ as the actual world, and so, via Premise 1a, that $\varphi \notin \Lambda_{\mathrm{MU}}$, as required:

> **Premise 1d:** If there are more worlds than individuals, then for any $\mathcal{L}_\Box$-sentence φ, if $\forall[U]\zeta(\varphi)$ is false, interpreting *U* as being an individual and ϑ as some individual, then $\forall\xi(\varphi)$ is false, interpreting θ as the actual world.

This premise can be motivated by adapting the proof idea of Lemma 2: Since there are more worlds than individuals, there is a surjection *s* from worlds to individuals, mapping the actual world to the individual used to interpret ϑ. As in the argument for Premise 1b, given a sequence of relations which witnesses the falsity of $\forall[U]\zeta(\varphi)$, the unrestricted comprehension principle allows us to construct, via s, a sequence of relations witnessing the falsity of $\forall\xi(\varphi)$, interpreting θ as the actual world.

Acknowledgements

Thanks to Jeremy Goodman and Timothy Williamson for comments on a draft, Øystein Linnebo, Beau Mount and Sam Roberts for discussions of Kreisel's principle, and audiences in Cambridge, Oxford, Paris and Montreal – especially my commentators, Michael Gabbay in Cambridge and Beau Mount in Oxford – for questions and comments.

Funding

This research was supported by an AHRC/Scatcherd European Scholarship.

References

Barwise, Jon. 1977. "Some Eastern Two Cardinal Theorems." In *Logic, Foundations of Mathematics, and Computability Theory*, edited by Robert E. Butts and Jaakko Hintikka, 11–31. Dordrecht: D. Reidel.

Bowen, Kenneth A. 1979. *Model Theory for Modal Logic: Kripke Models for Modal Predicate Calculi*. Dordrecht: D. Reidel.

Burgess, John P. 2004. "E Pluribus Unum: Plural Logic and Set Theory." *Philosophia Mathematica* 12: 193–221.

Chang, C. C. 1965. "A Note on the Two Cardinal Problem." *Proceedings of the American Mathematical Society* 16: 1148–1155.

Chang, C. C., and H. Jerome Keisler. 1990. *Model Theory*. 3rd ed. Amsterdam: North-Holland.

Devlin, Keith J. 1984. *Constructibility*. Berlin: Springer-Verlag.

Fine, Kit. 1977. *Postscript to Worlds, Times and Selves* (with A. N. Prior). London: Duckworth.

Friedman, Harvey M. 1999. *A Complete Theory of Everything: Satisfiability in the Universal Domain*. http://www.math.ohio-state.edu/~friedman/.

Fritz, Peter. 2013. "Modal Ontology and Generalized Quantifiers." *Journal of Philosophical Logic* 42: 643–678.

Fuhrken, G. 1964. "Skolem-type Normal Forms for First-order Languages with a Generalized Quantifier." *Fundamenta Mathematicae* 54: 291–302.

Hughes, G. E., and M. J. Cresswell. 1996. *A New Introduction to Modal Logic*. London: Routledge.

Jensen, R. Björn 1972. "The Fine Structure of the Constructible Hierarchy." *Annals of Mathematical Logic* 4: 229–308.

Keisler, H. Jerome 1966. "First Order Properties of Pairs of Cardinals." *Bulletin of the American Mathematical Society* 72: 141–144.

Koellner, Peter. 2009. "On Reflection Principles." *Annals of Pure and Applied Logic* 157: 206–219.

Kreisel, Georg. 1967. "Informal Rigour and Completeness Proofs." In *Problems in the Philosophy of Mathematics*, edited by Imre Lakatos, 138–186. Amsterdam: North-Holland.

Kripke, Saul A. 1959. "A Completeness Theorem in Modal Logic." *Journal of Symbolic Logic* 24: 1–14.

Lewis, David. 1986. *On the Plurality of Worlds*. Oxford: Basil Blackwell.

Mitchell, William. 1972. "Aronszajn Trees and the Independence of the Transfer Property." *Annals of Mathematical Logic* 5: 21–46.

Morley, Michael, and Robert Vaught. 1962. "Homogeneous Universal Models." *Mathematica Scandinavica* 11: 37–57.

Rayo, Agustín, and Timothy Williamson. 2003. "A Completeness Theorem for Unrestricted First-order Languages." In *Liars and Heaps: New Essays on Paradox*, edited by J. C. Beall, 331–356. Oxford: Oxford University Press.

Schmerl, James H. 1977. "An Axiomatization for a Class of Two-cardinal Models." *The Journal of Symbolic Logic* 42: 174–178.

Shapiro, Stewart. 1987. "Principles of Reflection and Second-order Logic." *Journal of Philosophical Logic* 16: 309–333.

Shelah, Saharon. 1979. "On Successors of Singular Cardinals." In *Logic Colloquium 78*, edited by Maurice Boffa, Dirk van Dalen, and Kenneth McAloon, 357–380. Amsterdam: North-Holland.

Shelah, Saharon. 2005. "The Pair $(\aleph_n, \aleph_0)$ May Fail $\aleph_0$-compactness." In *Logic Colloquium '01*, edited by Matthias Baaz, Sy-David Friedman, and Jan Krajíček, 402–433. Boca Raton, FL: A K Peters/CRC Press.

Silver, Jack. 1971. "The Independence of Kurepa's Conjecture and Two-cardinal Conjectures in Model Theory." In *Axiomatic Set Theory. Proceedings of Symposia in Pure Mathematics, Volume XIII, Part I*, edited by Dana S. Scott, 383–390. Providence, RI: American Mathematical Society.

Stalnaker, Robert. 1976. "Possible Worlds." *Noûs* 10: 65–75.

Tarski, Alfred. 2002 [1936]. "On the Concept of Following Logically." *History and Philosophy of Logic* 23 :155–196. Originally published in Polish and German in 1936.

Vaught, R. L. 1964. "The Completeness of Logic with the Added Quantifier 'There are Uncountably Many'." *Fundamenta Mathematicae* 54: 303–304.

Vaught, R. L. 1965a. "A Löwenheim–Skolem Theorem for Cardinals for Apart." In *The Theory of Models*, edited by J. W. Addison, L. Henkin, and A. Tarski, 390–401. Amsterdam: North-Holland.

Vaught, R. L. 1965b. "The Löwenheim–Skolem Theorem." In *Methodology Logic and Philosophy of Science*, edited by Yehoshua Bar-Hillel, 81–89. Amsterdam: North-Holland.

Williamson, Timothy. 2000a. "Existence and Contingency." *Proceedings of the Aristotelian Society* 100: 321–343.

Williamson, Timothy. 2000b. "The Necessary Framework of Objects." *Topoi* 19: 201–208.

Williamson, Timothy. 2003. "Everything." *Philosophical Perspectives* 17: 415–465.

Williamson, Timothy. 2013. *Modal Logic as Metaphysics*. Oxford: Oxford University Press.

Reply to Fritz

Timothy Williamson

Wykeham Professor of Logic, University of Oxford, Oxford, UK

In 'First-order modal logic in the necessary framework of objects' (Fritz 2016), Peter Fritz starts from some remarks in *Modal Logic as Metaphysics* about first-order necessitist modal logic and develops a far more systematic theory. The paper is a technical tour de force. One of the main aims of my book was to open up new areas for investigation (not just debate). In that respect, Peter Fritz's work, along with that of Jeremy Goodman and others, is vindicating it more rapidly and extensively than I had dared hope.

As a simple and plausible working hypothesis, *Modal Logic as Metaphysics* assumes that the propositional modal logic of metaphysical modality is S5. In terms of Kripke frames for propositional modal logic, one can, therefore, drop the accessibility relation, in effect treating all worlds as mutually accessible. The logic of such a frame depends only on the number of worlds, irrespective of which is designated as the actual world. If the number is infinite, the logic of the frame is S5 itself. If the number is finite, the logic of the frame is a proper extension of S5; the extra theorems are implausibly restrictive on the metaphysical reading of the modal operators, so we can dismiss that case (Williamson 2013, 111).

When we move to first-order necessitist modal logic, we inhabit the frames simply by adding a single domain of individuals. The extra formulas validated if the domain is finite are again implausibly restrictive on the metaphysically intended reading of the language, so we can dismiss that case too. Thus, the structure of the frames of interest for first-order necessitist modal logic is fixed by two infinite cardinals, the number of individuals (Fritz's κ) and the number of worlds (Fritz's λ). However, by contrast with the propositional case, that they are infinite does not fix the logic of the frame. The logic also depends on their relative cardinality. I give an example of a first-order modal formula valid on such frames with more individuals than worlds but invalid on those with no more individuals than worlds (Williamson 2013, 145). Section 2 of Fritz's paper massively extends that observation, by reducing the issues to well-studied questions in non-modal model theory. Restricting attention to frames of the relevant sort and their logics, he shows that any frame with no more individuals than worlds determines the unique weakest logic, while any frame with enough more individuals than worlds determines the unique strongest logic. If there are only 'slightly' more individuals than worlds, the logic is very sensitive to how many more, and to delicate set-theoretic issues involving Cantor's generalized continuum hypothesis; there are many intermediate logics. In general, increasing the number of worlds and decreasing the number of individuals never strengthen the logic; decreasing the number of worlds and increasing the number of individuals never weaken it. Fritz provides a helpful map of all the cases, and a wealth of information about the logics. He axiomatizes the weakest logic simply by adding as axioms 'There are at least *n* individuals' for all natural numbers *n* to a standard axiomatization of necessitist S5. He also shows that the strongest logic is recursively enumerable, so in principle recursively axiomatizable, and notes that it would be interesting to develop an axiomatization for it. Indeed: ideally, one wants a perspicuous schema for the additional axioms that somehow clarifies their metaphysical significance.

One upshot of Fritz's technical results is to show that, already in first-order modal logic, necessitism comes in many varieties, even when finite restrictions on the number of individuals or the number of worlds are excluded. Given the methodology of Williamson 2013, these logics are all genuine rivals of each other, for if one of them has all and only the metaphysically universal formulas of the language as theorems, then the others do not. The next step is to find robust considerations that favour one of them over the rest, or at least some of them over others. Of course, there is a tempting argument by the plural analogue of Cantor's theorem that there must be more individuals than worlds, because for any worlds there is an individual proposition true in all and only those worlds. There is also a tempting argument by the plural analogue of Cantor's theorem that there must be more worlds than individuals, because for any individuals there is a world in which all and only those individuals are thought about at the last moment of time. The challenge is to develop more stable and reliable ways of resolving the issues that overcome such conflicting temptations.

The frames in Fritz's Section 2 are all set-sized, so none of them really provides the metaphysically intended interpretation of the first-order modal language, with absolutely unrestricted quantifiers. Thus it is by no means automatic that the metaphysically universal formulas are all and only those valid on a frame with κ individuals and λ worlds, for some fixed set-sized κ and λ. I sketch an argument to bridge the gap (Williamson 2013, 117–118, 145). That is the topic of Section 3 of Fritz's paper. My argument relies on a sort of reflection principle suggested by Kreisel, which says that every theory in a standard non-modal second-order language satisfiable on the intended interpretation is satisfiable on a set (Premise 2, on Fritz's analysis). It also relies on an equivalence between the metaphysical universality of a formula of the first-order modal language, expressed by its universal generalization in a corresponding second-order modal language, and its 'translation' into a non-modal second-order language with quantification over worlds, envisaged as a special case of propositions (Premise 1 on Fritz's analysis).

Concerning Premise 1, Fritz points out that it would be in the higher-order spirit of *Modal Logic as Metaphysics* to replace the first-order quantification over worlds (propositions, treated as individuals) by quantification into sentence position, along the lines of chapter 5. He shows how to carry out this modification, while neither making special assumptions about worlds nor losing philosophical plausibility. I welcome this refinement of the argument.

I supported Premise 2, Kreisel's Principle, by appeal to a consistency proof for it by Stewart Shapiro (for a universe of pure sets). As Fritz points out, such a result falls some way short of proving the Principle *true*. I did not suggest otherwise. In the book, I merely say that from a contemporary set-theoretic perspective, Kreisel's Principle 'looks quite plausible' (Williamson 2013, 118). At the very least, Shapiro's result shows that a refutation of Kreisel's Principle would in effect have to refute many other widely held assumptions about set theory too.

In the long run, we might hope to approach the problem from the other side, by arguing more directly for constraints on first-order necessitist modal logic, without the detour through set-sized frames, and then use Fritz's results to show that any logic meeting those constraints is the logic of some set-sized frame. It is not yet clear whether that hope is realistic.

References

Fritz, Peter. 2016. "First-order Modal Logic in the Necessary Framework of Objects." *Canadian Journal of Philosophy* 46 (4–5): 585–610.

Williamson, Timothy. 2013. *Modal Logic as Metaphysics*. Oxford: Oxford University Press.

Williamson on necessitism

Jeremy Goodman

Department of Philosophy, NYU, New York, NY, USA

ABSTRACT
I critically discuss some of the main arguments of *Modal Logic as Metaphysics*, present a different way of thinking about the issues raised by those arguments, and briefly discuss some broader issues about the role of higher-order logic in metaphysics.

This paper is a study of some of the main arguments of *Modal Logic as Metaphysics* (hereafter *MLaM*).[1] In Section 1, I will discuss Williamson's argument in Chapter 6.2 that contingentists should be higher-order contingentists. In Section 2, I will consider Williamson's argument for necessitism in Chapter 7. In Section 3, I will present a different way of thinking about the issues raised by that argument. In Section 4, I will consider some broader issues about the role of higher-order logic in metaphysics.

It would be hard to overstate the influence of *MLaM* on my own philosophical development, as a source of both ideas and inspiration. Although my discussion will mostly be critical, it takes place against a backdrop of broadly shared methodological assumptions and priorities. I hope it conveys some sense of the richness and subtlety of the underlying philosophical issues, and of Williamson's contribution to our understanding of them.

1. From contingentism to higher-order contingentism

In Chapter 6.2 of *MLaM*, Williamson argues that contingentists – those who think that $\neg\Box\forall x\Box\exists y(y = x)$ – should be higher-order contingentists. In particular, they should think that it is a contingent matter which properties there are: $\neg\Box\forall F\Box\exists G(F \approx G)$, where $\approx$ abbreviates some higher-order analogue of the identity predicate, taking monadic predicate variables rather than individual variables as arguments. (Following Williamson, I will use 'property'-talk to pronounce in English claims that are properly formalized using quantification into monadic predicate positions.) There are a number of options for how to interpret $\approx$: these include necessary co-extensiveness ($\Box\forall x(Fx \leftrightarrow Gx)$), necessary co-intensiveness ($\Box\forall x\Box(Fx \leftrightarrow Gx)$), higher-order indiscernibility ($\forall O(OF \leftrightarrow OG)$), and taking the notion as primitive (i.e. taking as axiomatic that it is reflexive and obeys some appropriate version of Leibniz's law). Williamson discusses these issues in Chapter 6.1, and I will return to them later. For now, I will follow Williamson and treat $\approx$ as shorthand for talk of necessary co-extensiveness, and hence treat the claim that it is a contingent matter which properties there are as shorthand for $\neg\Box\forall F\Box\exists G\Box\forall x(Fx \leftrightarrow Gx)$. In this section I want to raise some questions about Williamson's reasons for thinking that contingentists should accept this further claim.

1.1. Symmetry

With respect to the combination of contingentism and higher-order necessitism, Williamson writes that 'The onus is on the metaphysician who postulates such logical differences between orders to justify the asymmetry in treatment' and that 'The default preference is for a uniform metaphysics, on which being is contingent at all orders or none' (274). Is this true?

It often happens in theorizing about inductively defined classes (like the natural numbers) that base cases of the induction (e.g. 0) fail to satisfy important generalizations that are satisfied by the remaining members of the class. Of particular interest here is the hierarchy of syntactic types of expressions of ML_P, the higher-order formal language in which Williamson formulates competing theories of modal metaphysics. As he explains in Chapter 5.5, types are defined inductively as follows: e (the type of individual variables) is the only basic type, every finite sequence of types is a type, and nothing else is a type. Intuitively, an expression of type $\langle t_1, \ldots, t_n\rangle$ is a predicate that takes n arguments, respectively of types $t_1 \ldots, t_n$, in that order. (In this broad sense of 'predicate' formulas are 0-place predicates of type $\langle\rangle$, dyadic sentential operators like 'and' are 2-place predicates of type $\langle\langle\rangle, \langle\rangle\rangle$, etc. We sometimes indicate the type of an expression with a superscript on its first occurrence in a formula.) For every type of expression, there is a corresponding necessitist thesis: in the case of type e, it is $\Box\forall x^e\Box\exists y^e(y = x)$; in the case of a predicative type $\langle t_1, \ldots, t_n\rangle$, it is $\Box\forall X^{\langle t_1,\ldots,t_n\rangle}\Box\exists Y^{\langle t_1,\ldots,t_n\rangle}\Box\forall x^{t_1} \ldots \forall x^{t_n}(Xx_1 \ldots x_n \leftrightarrow Yx_1 \ldots x_n)$.

Williamson claims that considerations of symmetry in theory choice militate in favor of accepting all such theses or none. I don't want to claim that there is no theoretical pressure in this direction. But I do think that, taking into account only purely formal considerations, a split decision as regards type-e necessitism, on the one hand, and necessitism for all predicative types $\langle t_1, \ldots, t_n\rangle$, on the other hand, is not unprincipled. For it regularly happens in theorizing with(/about) higher-order languages like ML_P that special provision has to be made for the case of individuals(/expressions of type *e*); this shows up both in axiomatics (for example, as above in the characterization of the correspondence between syntactic types and necessitist theses) and when doing model theory. These precedents suggest that, while there may be strong theoretical pressures to accept $\langle t_1, \ldots, t_n\rangle$-necessitism if and only if one accepts $\langle t'_1, \ldots, t'_m\rangle$-necessitism, considerations of symmetry 'between orders' provide much less reason to expect that *e*-necessitism and $\langle t_1, \ldots, t_n\rangle$-necessitism stand or fall together.

1.2. Parity

A different argument for the conclusion that contingentists should be higher-order contingentists is premised on the claim that the considerations in favor of contingentism correspond to parallel considerations in favor of higher-order contingentism. But is there such parity of motivation?

A first thing to note is that, while many philosophers think that contingentism is obviously true, fewer think that higher-order contingentism is obviously true. So insofar as one is motivated to be a contingentist by an appearance of obviousness, there does not seem to be parity in motivation as regards higher-order contingentism. Having said this, I agree with Williamson that the question of necessitism versus contingentism is not at all obvious, and that it needs to be decided instead on broadly theoretical grounds. So let us set aside the asymmetry of obviousness and consider more theoretical arguments.

Like Williamson, I think that the best theoretical argument against necessitism is that it makes the modal 'float free' of the non-modal in some objectionable way. (Williamson discusses such arguments in Chapter 8.2.) Insofar as such considerations motivate contingentism, do they likewise motivate higher-order contingentism?

Here, I think the answer depends on exactly how the objection to necessitism is understood. One option is to understand the issue in terms of supervenience – for example, in terms of the thesis that all qualitative properties of and relations between individuals supervene on a distinguished class of 'fundamental' or 'perfectly natural' properties and relations. This supervenience thesis is in tension with necessitism given certain essentialist assumptions. Suppose I had never been born and my computer had never been made. According to the necessitist, there would still have been such a thing as me and such a thing as my computer. Presumably neither of us would instantiate any fundamental properties or stand in any fundamental relations – or at least, none that would differentiate us. Yet, given popular essentialist assumptions, we would differ in the (seemingly qualitative) respect of possibly being a person: I would have this property but my computer would not. So there would be things (me and my computer) that differed qualitatively without differing in any fundamental respect, in violation of the supervenience of the qualitative on the fundamental.

A contingentist can respond by claiming that in the relevant counterfactual circumstances there would have been no such thing as me and no such thing as my computer, thereby blocking the alleged counterexample to the supervenience thesis. On its face, this response does not require also positing any higher-order contingency (for example, it does not require claiming that there would have been no such property as being identical to me). So it seems that, insofar as one's contingentism is motivated by a desire to maintain the supervenience of the qualitative on the fundamental, there are not parallel considerations in favor of higher-order contingentism.

One might reply that the supervenience of the qualitative on the fundamental ought to be understood more broadly so as to concern not only the qualitative properties of individuals but also the qualitative properties of properties of individuals. This stronger thesis might then support a parallel argument for higher-order contingentism, for if in the relevant counterfactual circumstances there still was the property of being me and the property of being my computer, then these properties would seem to differ in the qualitative respect of being possibly instantiated by a person. Arguably, though, they would not be differentiated in any fundamental respects, although it is admittedly much less clear how the operative notion of fundamentality applies to properties of and relations among properties of individuals, especially when such talk is ultimately to be understood in higher-order terms.[2] I am not sure to what extent the motivations for the supervenience thesis concerning properties of individuals generalize to the supervenience thesis concerning properties of properties of individuals, but at any rate Williamson's discussion of the supervenience of the qualitative on the fundamental in Chapter 8.2 is formulated only in terms of properties of individuals.

There are also non-supervenience-theoretic ways of articulating the idea that necessitism makes the modal float free of the non-modal. For example, some contingentists will be motivated by the idea that every truth should be grounded in a 'non-modal' truth (whatever that means). This grounding thesis appears to be threatened by higher-order necessitism. For had there been no such thing as me, there might then seem to be no non-modal ground for the truth that I am possibly a person. Higher-order contingentists can hold onto the grounding thesis by claiming that, in the relevant circumstances, there would have been no such proposition as the proposition that I am possibly a person. (I use 'proposition'-talk as shorthand for claims appropriately formalized using quantification into sentence position, again following Williamson.) This in turn constitutes an argument for (first-order) contingentism, given the assumption that, necessarily, for any individual there is the proposition that it is possibly a person.

On this way of thinking about the modal 'floating free' of the non-modal, the contingentist solution does lead to (and indeed essentially involves) a form of higher-order contingentism. However, at least when the challenge is framed in terms of such propositions as that I am possibly a person, it requires a hyperintensional theory of grounding. This is because that proposition is a necessary truth, and so is necessarily equivalent to the presumably non-modal necessary truth that all chairs are chairs. Since Williamson is rather hostile to hyperintensional metaphysics (more on which later), by his lights the supervenience-theoretic formulation of the 'floating free' worry is a (comparatively) firmer foundation for contingentism than the ground-theoretic worry. And as I argued above, the supervenience thesis that Williamson considers leads to contingentism but does not in any straightforward way lead to higher-order contingentism.

To be clear, in suggesting that the best arguments for contingentism do not correspond to parallel arguments for higher-order contingentism, I am not saying that there are no good arguments from contingentism to higher-order contingentism. There are other ways in which the two positions might be related. For example, one might have the view that all properties and propositions can in some sense be built out of, or pinned down in terms of, the individuals that there are together with qualitative properties of and relations between those individuals. Contingency in which individuals there are might then induce corresponding contingency in which properties and propositions there are, given plausible auxiliary assumptions. Such views about the metaphysics of properties and propositions are certainly worth taking seriously, as are the versions of higher-order contingentism they suggest when combined with (first-order) contingentism.[3] So higher-order contingentism is certainly one principled version of contingentism. But it needn't be the only principled version of contingentism, since the principles motivating contingentism as such do not inevitably lead to higher-order contingentism either by themselves or together with uncontentious metaphysical assumptions.

Before moving on, I'd like to mention one wrinkle about the connection between essentialism and the modal 'floating free' of the non-modal. In Chapter 8.2, Williamson expresses sympathy for an anti-essentialist response, according to which the same qualitative possibilities are open to all individuals – my computer could have been a person, and I could have been a computer, for example. Although I think this form of anti-essentialism is an interesting view in its own right, it is arguably not enough to answer the 'floating free' worry in full generality. This is because those, like Mackie (2006), who are inclined toward such anti-essentialist theses nevertheless want to maintain that possibilities in which I am a computer are far removed from actuality. And it seems plausible that, if a possibility w is far removed from actuality, and another possibility v is nearby actuality, then in v actuality is nearer than w. Now consider some nearby possibility v in which I was never born and my computer never made. (If you don't think there are any such possibilities, modify the example accordingly.) Provided that every world in which I am a computer is far removed from actuality, and assuming that the relevant notion of comparative modal distance can be understood in terms of counterfactuals in the familiar way, it follows that in the world where I am never born and my computer is never made I bear the following asymmetric (and plausibly qualitative) relation to my computer: $\lambda xy(\Diamond(Px \leftrightarrow \neg Py) \wedge ((Px \leftrightarrow \neg Py) \mathrel{\Box\!\!\rightarrow} Px))$, where P expresses personhood and $\Box\!\!\rightarrow$ expresses the counterfactual conditional. The failure of this modal relation to supervene on the fundamental is therefore not blocked by mere anti-essentialism. Of course, the anti-essentialist could go further, and say that, among the contingently non-concrete individuals, all qualitative possibilities are equally counterfactually nearby. But this is a much more radical claim than the one Williamson explicitly considers.[4]

1.3. 'Modal tracking'

Williamson also gives a more concrete argument that contingentists should be higher-order contingentists. The argument is that, if contingentism is true, then for any ordinary material thing x (like you or me) it is a contingent matter whether there is a haecceity of x, where a property F is a *haecceity of x* just in case it is necessarily co-extensive with being x ($\Box\forall y(Fy \leftrightarrow y = x)$). Let us turn to that argument.

Williamson writes:

> Informally write '*X* tracks *y*' for '*X* is a haecceity of *y* and *X* cannot be a haecceity of anything other than *y*,' with '*X* is a haecceity of *y*' defined as previously. [...] [M]y haecceity necessarily tracks me. Even if I had never been, [given higher-order necessitism] there would still have been a property tracking me (and only me). But how can it lock onto me in my absence? In those circumstances, what makes me rather than something else its target? (269)

This passage is puzzling. Williamson appears to be arguing as follows: if he is a contingent being and higher-order necessitism is true, then his haecceity could track him in his absence; but his haecceity could not track him in his absence; so if contingentism is true (and he is a contingent being), we should reject higher-order necessitism. The problem with this argument is that, given the definition of 'tracks' (which, incidentally, seems gratuitous, since the defined notion is equivalent to the notion 'is a haecceity of' in terms of which it is defined), it is uncontroversial that Williamson's haecceity necessarily tracks him: even if there were no such thing as his haecceity, it would still be a haecceity of him, since having it would still be necessarily co-extensive with being him. So while one could treat Williamson's rhetorical question as material for an argument that he is a necessary being (and indeed he considers related arguments in Chapter 6.5), it does not make much sense as a premise in an argument for the contingent being of his haecceity in a dialectical context where we are granting for the sake of argument that he is a contingent being.

Maybe this is just a slip. After all, Williamson could instead have written 'How could *there have been* something that in my absence locked onto me?'. But if it is a slip, it is a telling one, since the ease of making it suggests that we are mistaking a general puzzling feature of contingentism for a particular puzzling feature of the combination of contingentism and higher-order necessitism.

1.3.1. Modal tracking by accident

Williamson is careful not to claim that, if contingentism is true, then no individual could have had a haecceity in circumstances in which there was no such thing as that individual. Consider a knife *k* made from handle *h* and blade *b*. Suppose *h* and *b* were manufactured but never joined, and so *k* was never made. According to most contingentists, this would be a circumstance in which there was no such thing as *k*. Nevertheless, it might still be a circumstance in which there would be a haecceity of *k*. For in those circumstances, there would be the property of being something that in those circumstances would have been a knife made from *h* and *b* had the two been joined in the appropriate way. And given not unreasonable assumptions, this property would be a haecceity of *k*.

Williamson argues that this point does nothing to suggest that every material object, or even *k*, *necessarily* has a haecceity. For suppose *b* and *h* had never been made, perhaps because the history of the cosmos went quite differently and the solar system never formed. How could we single out *k* in those circumstances in terms of its possible relations to the material things there would then have been? Or consider some elementary particle α. Had there not been that particle (which contingentists will normally agree is possible), how could we single out α in such circumstance in terms of its possible relations to the material things there would then have been?

If we believed only in the material objects recognized by common sense, I agree that the task looks hopeless. However, many metaphysicians believe that there are many more material objects than those recognized by common sense. They start with the idea that a statue and the matter from which it is made, despite coinciding, are different objects, since they differ as regards the possible circumstances in which they would exist. They then conclude that, on pain of arbitrariness, we should say that any eligible modal profile is the modal profile of some possible material object. We can formulate this principle of plenitude in higher-order modal terms without begging any questions about higher-order contingentism: we simply say that, necessarily, for every possibly instantiated property *X* that necessarily is instantiated only by material objects, it is possible that there be a material object *x* that necessarily fuses the things that are *X* (i.e. necessarily, everything that is *X* is part of *x* and every part of *x* shares a part with something that is *X*).[5] Given the not outlandish assumptions that, necessarily, (1) there is at least one material object, and (2) any two material objects that necessarily coincide (in the sense of sharing parts with the same things) are qualitatively discernible, the principle of plenitude allows us to argue that, necessarily, every material object necessarily has a haecceity, where we assume only that there is a property corresponding to any condition specified in terms of individuals there are and their qualitative properties and relations.[6] Assuming that, necessarily, every non-material object (numbers, sets, etc.) can be singled out in terms of its qualitative properties and relations to material objects, this entails that, necessarily, everything necessarily has a haecceity.

There are other independently motivated and contingentist-friendly metaphysical principles that have this conclusion as an unintended consequence. For example, if we accept the principle of conditional excluded middle for counterfactuals (roughly, the principle that negating a non-vacuous counterfactual is equivalent to negating its consequent), we can argue that necessarily everything necessarily has a haecceity from the premise that, necessarily, any two contingent beings are possibly qualitatively discernible.[7] In general, it strikes me as a wide open question whether contingentists should be higher-order contingentists, turning on many subtle question about superficially unrelated matters.[8]

2. The 'cash-value' argument

The most challenging part of *Modal Logic as Metaphysics* is the extended argument for necessitism in Chapter 7. In this Section, I want to raise some worries about that argument.

2.1. *The formal result*

First, let me informally describe the formal results underlying the argument. I will make some simplifications, which I will note along the way.

Williamson has us consider two theories: *chunky-style contingentism* (CSC) and *chunky-style necessitism* (CSN). CSC consists of the claim that, necessarily, everything is 'chunky' – we will consider later how this predicate ought to be interpreted. CSN is the combination of necessitism, the claim that everything could be chunky, and a claim, for each primitive predicate of the language, to the effect that it is chunkiness-entailing in all of its arguments ($\Box\forall x_1 \ldots \forall x_n(Rx_1 \ldots x_n \rightarrow (Cx_1 \wedge \ldots \wedge Cx_n))$).

Williamson then considers two modal languages that differ only in that, while one contains only individual quantifiers, the other contains plural quantifiers too.[9] It does *not* contain higher-order quantifiers in the sense of 'higher-order' that has been operative throughout this paper – i.e. for no types $t_1, \ldots, t_n$ in the type hierarchy defined earlier does it contain quantified variables of type

$\langle t_1, \ldots, t_n \rangle$. This point is worth emphasizing (and Williamson does emphasize it), since his notational choice to use 'Xx' to regiment 'x is one of X' might suggest that 'X' is a variable of type $\langle e \rangle$, when in fact it is a plural variable that does not fall under any type in the aforementioned type hierarchy.

Williamson next defines a consequence relation $\vDash$ in the familiar way, using variable-domain Kripke models. Plural quantifiers range over sets of individuals in the domain of the world of evaluation. An individual(/tuple of individuals) is in the extension of a monadic(/polyadic) predicate at a world only if it is(/they are all) in the domain of that world. In this way, we validate the *being constraint* ($\Box\forall x_1 \ldots \Box\forall x_n \Box(Rx_1 \ldots x_n \rightarrow \exists y(y = x_i))$, for $1 \leq i \leq n$): informally, all atomic predications entail the being of the individuals of which they are predications. Note, however, that the imposition of the being constraint is not essential to the overall argumentative strategy – parallel theorems can be established in a setting where the constraint is relaxed provided we modify CSC, CSN, the definition of a model, and the definition of 'neutrality' (to be explain presently) in corresponding ways; see Fritz (2013).

Lastly, Williamson defines a *neutral* formula to be any formula logically equivalent to a formula in the image of a recursively defined mapping $(\cdot)^{con}$ from formulas to formulas. Informally, the effect of this mapping is to restrict all individual quantifiers to chunky things, all plural quantifiers to pluralities of chunky things, and to conjoin all atomic predications with claims to the effect that the individuals involved are chunky. The idea is that neutral formulas are those that are concerned only with 'the realm of the chunky'.

Say that A is *equivalent to B given Γ* just in case $\Gamma \vDash A \leftrightarrow B$. Williamson's main results are as follows. In the language with only singular quantification, every formula is equivalent to a neutral formula given CSC and equivalent to a neutral formula given CSN. By contrast, in the language enriched with plural quantification, although every formula is equivalent to a neutral formula given CSC, some formulas are not equivalent to any neutral formula given CSN.

2.2. Philosophical motivation

Williamson's theorem does not wear its philosophical significance on its sleeve. He explains its relevance roughly as follows.

Imagine two metaphysicians, Kit and Bob. It is common knowledge between them that Kit is a necessitist and Bob is a contingentist. Despite this metaphysical disagreement, they are generally trusting of each other. Kit often finds that, when Bob asserts something incompatible with his necessitism, Kit is able to find a kernel of truth in Bob's utterance – intuitively, he is able to factor out Bob's contingentism from his utterance and obtain something that both of them could in principle agree on. For example, when Bob says that there could have been no such thing as the Eiffel tower (which is inconsistent with necessitism), Kit – being trusting of Bob *modulo* his contingentism – comes to believe that there could have been no such concrete object as the Eiffel tower. This claim has the advantage of being consistent with both of their respective metaphysical visions and equivalent to what Bob said given Bob's general views about contingent being. Conversely, suppose Kit says that something non-concrete could have been a child of Wittgenstein. This is inconsistent with Bob's brand of contingentism. Nevertheless, Bob seems to be able to factor out the necessitism from Kit's claim resulting in something that both of them could in principle agree on – for example, the claim that it is possible that something be Wittgenstein's child. This claim is consistent with both of their respective metaphysical visions and equivalent to what Kit said given Kit's brand of necessitism.

There seems to be something systematic going on here. Necessitists can learn from contingentists and vice versa, by extracting 'neutral cash value' from each other's utterances: sentences that are neutral, in the sense of being orthogonal to their background metaphysical disagreement, and cash value, in the sense that the claim extracted is as good as the statement from which it was extracted to the person who made the latter statement and so (setting metaphysical debates aside) can serve equally well their goal of exchanging information.

The theories CSC and CSN are meant to formalize the respective metaphysical positions of interlocutors like Kit and Bob, and the notion of a neutral formula is means to formalize the informal notion of neutrality just described – namely, the notion of being orthogonal to their background metaphysical disagreement. The idea is that Kit and Bob's disagreement permits agreement on a wide range of matters – in particular, on matters regarding only 'chunky' individuals. But chunkiness is not merely characterized in terms of what Kit and Bob can agree on. It also plays an important role in their respective metaphysical visions. In particular, Bob thinks that everything is chunky – it is this together with a recognition of the contingency of chunkiness that leads him to be a contingentist in the first place. Kit, by contrast, thinks that everything could be chunky. This is motivated by his concern only for matters chunky – his necessitism and recognition of the contingency of chunkiness forces him to countenance things that are contingently non-chunky, but leaves him free to deny that any things are necessarily non-chunky.

Williamson's theorem shows that, insofar as their discourse can be regimented in the formal languages he considers, and their theoretical commitments faithfully captured respectively by CSC and CSN, Kit will always be able to extract a neutral cash value from Bob's utterances: sentences only about the chunky, and so orthogonal to their metaphysical dispute, and yet equivalent given Bob's commitments to the original utterances. (Such equivalents are delivered by the mapping $(\cdot)^{con}$.) And Bob can do the same for Kit's utterances of sentences that can be formalized in the plural-quantification-free fragment of their common language. But some of Kit's sentences involving plural quantification can be shown to have no neutral equivalents for him. Kit's assertion that Jack and Jill stand in the ancestral of the relation of being possible nemeses is an example, where the ancestral (i.e. transitive closure) of a condition is defined using plural quantification in the familiar way.

This limitative result is supposed to be pretty embarrassing for Bob – indeed, embarrassing enough to make him reconsider his commitment to CSC. And reflection on Bob's embarrassing predicament is meant to suggest that all of us have strong reason to reject contingentism. The embarrassment is supposed to consist in the fact that, by Bob's lights, he can't quite make sense of what Kit is up to – of what picture Kit has about the possibilities for chunky things that is behind some of his utterances. And that seems absurd: Bob can understand perfectly well how Kit is thinking. If his metaphysical scruples prevent him from being able to articulate that picture of modal reality, then so much the worse for those scruples.

2.3. *Worries about the mode of argument*

This is not a familiar mode of argument, and I don't think it is a good one.

The most obvious objection is the most straightforward: whether one can find neutral equivalents of one's interlocutor's utterances is in general not a good test of the tenability of one's position. For one thing, it is usually not possible for either party. Consider a dispute between a proponent of string theory and a proponent of loop quantum gravity, or between a Christian and a Muslim, or about pretty much anything. In all such disputes, when the dispute is in clear view and the people involved are intelligent and communicating in good faith, it will be possible for them to learn from each other in ways that to a large extent admit of systematic generalizations and that are naturally described as attempts to find the kernel of truth in what the other person says. But there won't be an algorithm that meets a standard anything like the one to which Williamson is holding the chunky-style contingentist. Perhaps many theoretical disagreements can be sidestepped most of the time, but it is hard to think of any interesting theoretical disagreements that can be sidestepped all of the time, even by one party.

Note also how fragile the possibility of neutral cash value extraction is in the case at hand. Consider a parallel dispute between Kit* and Bob*, who respectively accept C*SN and C*SC, which differ from CSN and CSC in that they concern not chunkiness but a closely related property chunkiness*.[10] Neither Kit nor Bob accepts C*SN or C*SC, and neither Kit* nor Bob* accepts CSN or CSC. So Kit will not be able to always produce neutral equivalents of Bob*'s utterances. What does Williamson's mode of argument predict about this case? It can't be that everyone ought to feel pressure to give up their view. Moreover, it seems like, in whatever pre-theoretically obvious sense Kit can learn from Bob, Kit can learn from Bob* too. In general, the possibility of 'neutral cash value extraction' is neither a good test of theoretical adequacy nor a good test of the extent to which interlocutors can in some pre-theoretically recognizable sense learn from each other's discourse. This is true even for disputes only slightly removed from the one Williamson has us imagine.

Here is another way to dramatize the point. Williamson notes that, on many salient interpretations of 'chunky,' real necessitists are not chunky-style necessitists, since they believe in things that most contingentists think there could not even possibly be – for example, they believe that there is a non-empty set of all of Wittgenstein's possible children. Suppose Kit told Bob that the set of Wittgenstein's possible children was his favorite set, despite the impossibility of all of its members being chunky together. On pain of inconsistency, Kit must not accept CSN, since his commitments are incompatible with the atomic predicates 'is a member of' and 'is the favorite set of' both being chunkiness-entailing in the way that CSN requires. His background necessitist theory is then presumably something weaker than CSN, and this fact will prevent Bob from being able to find a neutral equivalent of Kit's utterance, since Kit's background theory will make fewer sentences equivalent than CSN does. In this connection, Williamson says that in confining his attention to the chunky-style necessitist he is making things easier for the contingentist. This is true in the sense that it makes it easier for them to rise to the challenge he has set for them. But it also suggests that he has not set them the right challenge. For there doesn't seem to be any sense in which Bob ought to be embarrassed by some inability to make sense of what Kit was getting at here.[11]

A different worry about the argument concerns the notion of neutrality. Consider some sentence *A* (such as the aforementioned ancestral-involving one) that is not equivalent given CSN to any neutral sentence. Williamson's theorem establishes that there is such a sentence. But this is not yet to establish that there is no sentence equivalent to *A* given CSN that is intuitively not at issue in Kit and Bob's metaphysical dispute. That would require an argument that all such intuitively not-at-issue sentences are neutral in the technical sense. Williamson treats this claim as a working hypothesis, but never argues for it at any length. *Prima facie*, there would seem to be counterexamples. For example, it is common ground between Kit and Bob that there are no dragons – this certainly looks like something independent of their metaphysical dispute by appeal to which Bob ought to be able to at least try to find common ground with Kit. But '$\neg\exists xDx$' is not neutral in Williamson's sense, since it is not equivalent to any sentence whose quantifiers are explicitly restricted by the predicate 'is chunky.' Of course, this point doesn't suggest any sentence equivalent to *A* given CSN that, though not neutral, is intuitively orthogonal to their metaphysical disagreement. But it highlights one further respect in which the formal result is some distance from the pre-theoretical phenomenon it is intended to characterize.

2.4. Worries about 'chunkiness'

So far I have left the notion of chunkiness schematic, as Williamson does to a large extent. But it cannot be left completely schematic. For one thing, if we allow that the sorts of hypothetical interlocutors we are interested in might differ in their interpretation of 'chunky', then Williamson faces the challenge of explaining why Kit isn't in as bad shape with respect to Bob* as Bob is in with respect to Kit.

Williamson does in fact elaborate on the sort of thing he has in mind. At some points, he characterizes being chunky as being grounded in the concrete, and at other times as being not contingently non-concrete. But although these notions might coincide according to proponents of CSC and CSN, they do not coincide according to Williamson: insofar as he is willing to operate with the relevant notion of grounding (about which he sometimes voices reservations) he thinks that impure sets of possibly concrete but not possibly co-concrete objects are neither grounded in the concrete nor contingently non-concrete. Maybe according to Williamson we should imagine Kit and Bob as accepting CSN/CSC on both interpretations of 'chunky,' and also believing – falsely, according to Williamson – that necessarily everything is, necessarily, grounded in the concrete just in case it is not contingently non-concrete, but he does not say so explicitly.[12]

A different strategy for isolating the intended notion of chunkiness is in terms of its role in motivating contingentism. Williamson writes:

> Why do contingentists reject the necessitist postulation of contingently non-concrete objects, such as merely possible people? As just noted, the reason is often not any objection to the non-concrete as such. It may rather be an objection to the *contingently* non-concrete. (314, emphasis original)

More generally, the suggestion is that chunkiness is whatever condition is both uncontroversially contingent and, by contingentists' lights, necessarily equivalent to being identical to something.

One problem with this suggestion is that not all contingentists are motivated to be contingentists by this sort of reasoning – as Williamson puts it, not all contingentists are chunky-style contingentists. Williamson writes as if most familiar versions of contingentism are chunky-style, so that his discussion of CSC applies to most actual contingentists. This strikes me as both a sociological and a dialectical mistake. As for the sociology, I think that most contingentists – at least, those who are motivated by general theoretical considerations – are motivated by some version of the aforementioned worry that necessitism makes the modal float objectionably free of the non-modal. This is an argument for contingentism of a very different shape from the one suggested in the passage just quoted.[13]

The dialectical error is more important. Of non-chunky style forms of contingentism, Williamson writes:

> The challenge to proponents of other forms of contingentism is to show rigorously how their favored form overcomes the limitations of CSC explained in this Chapter. (315)

This seems to me to get the situation backwards, and in so doing to highlight one of the most puzzling features of Williamson's setup. In that setup, the necessitist is judged on the basis of how well he can come up with neutral equivalents of the contingentist's claims, and the contingentist is judged on the basis of how well he can come up with neutral equivalents of the necessitist's claims. But the relevant notion is equivalence by the lights of the *speaker*, not by the lights of the hearer. Bob's commitment to CSC, or any other version of contingentism, has no bearing whatsoever on which sentences of Kit's he can find equivalents of. That depends only on Kit's commitments. On the other hand, Bob's commitment to CSC is crucial for Kit: If Bob didn't have that commitment, then Kit would be unable to come up with neutral formulas equivalent by Bob's light to various of his claims. So insofar as we need to take such contingentists seriously, Williamson should see this as a challenge for necessitists, not for contingentists. And we should take such contingentists seriously. After all, it is not as though supervenience-motivated contingentists are any harder for necessitists to learn from than are their hypothetical chunky-style cousins.

(One might think that this misses the point that what counts as neutral in a dispute depends on the respective theories of the disputants. In reply: that kind of dependence only amplifies the problem. For the less the contigentist is committed to, the more is up for grabs, so if anything *fewer* sentences should count as neutral, in which case it will be harder still for the necessitist to find neutral equivalents of contingentists' discourse. Of course, maybe this more demanding notion of neutrality would also make it harder for the contingentist to extract neutral equivalents of necessitists' discourse, but if so it's a wash – we would be left in the normal situation in theoretical disputes where for no side is every claim equivalent by its lights to something that is not in dispute. And that would be enough to undermine Williamson's argument, since then there wouldn't be the relevant sort of asymmetry between necessitism and contingentism.)

Later on, Williamson himself makes an exactly parallel point in the other direction. He is responding to the suggestion that contingentists might try to get around his argument by endorsing some version of Plantinga's theory of individual essences. He writes:

> [The strategy] does not fit the dialectical context. [...] [T]he present challenge is to find neutral equivalents for plurally quantified sentences given a necessitist theory. The Plantinga-style postulates (45)–(47) had to be added to the necessitist [theory], not to the contingentist [theory]. Thus, the question is not whether contingentists have good reason to postulate (45)–(47), but whether necessitists have. Perhaps a few necessitists will like (45)–(47). But they are not needed to

> solve any problem for necessitists, for they have no special difficulty in making sense of quantified modal logic. (352)

Williamson is entirely correct that, in the present dialectical context, the contingentist's metaphysical commitments regarding individual essences are completely irrelevant. But this is an indication that something has gone wrong with the dialectic.

What has gone wrong is that, in the case of the Plantinga-style contingentist, incompatible dialectical constraints are being imposed on the notion of chunkiness. On the one hand, chunkiness has to be a sufficiently undemanding property that it can encompass everything the contingentist believes in, since CSC says that, necessarily, everything is chunky. On the other hand, chunkiness has to be a sufficiently demanding property that to restrict one's attention to the chunky is to restrict one's attention to an arena isolated from questions of necessitism versus contingentism. For at least a certain Plantinga-style contingentist in a dispute with a necessitist skeptical of individual essences, no notion plays both roles, because the question of whether there are individual essences is, for him, intimately tied up with the question of necessitism – essences are not neutral ground, despite being countenanced by the contingentist. (For example, following Jager (1982), the imagined contingentist might postulate such essences in order to make sense of possible-worlds model-theory for modal logic, which is something necessitism would have allowed him to do by other means.)

2.5. *Adding higher-order quantification*

Williamson argues at length that the existence of formulas not equivalent to any neutral formula given CSN is robust with respect to various ways of enriching the modal language with infinitary conjunction and infinitary quantification. But he does not consider whether the result is robust with respect to enriching the language with higher-order quantifiers – i.e. by allowing for quantified variables of predicative types of the type-hierarchy described earlier. This might seem like a strange omission, given the focus on such quantification in the previous two chapters of the book. But the reason for the omission is clear enough. In the case of plural quantification, there is a fairly obvious way of using variable-domain Kripke models to define a consequence relation for the expanded language, and likewise an obvious way of extending the notion of a neutral formula to the expanded language (i.e. of adding the appropriate clause for plurally quantified formulas to the definition of the mapping $(\cdot)^{con}$). By contrast, in the case of higher-order quantification, neither issue is straightforward.

Suppose we were to resolve the two issues as follows. We extend the consequence relation to the higher-order language by considering the class of intensional models that validate the *unrestricted comprehension* schema

$\exists X^{\langle t_1,\ldots,t_n\rangle}\Box\forall x_1^{t_1}\ldots x_n^{t_n}(Xx_1\ldots x_n \leftrightarrow A)$, and hence validate higher-order necessitism. (Following Williamson, we count the result of prefixing an instance of a schema with any string of universal quantifiers and necessity operators as itself an instance of the schema; we also require that X not occur free in A.) As for the definition of neutrality, we have a trivial clause for higher-order quantifiers, letting $(\forall X^{\langle t_1,\ldots,t_n\rangle}A)^{\text{con}} := \forall X(A)^{\text{con}}$. In such a setting, every formula A *will* be equivalent to a neutral formula given CSN. (Informally, claims involving quantification over modally rigid properties of necessarily existing haecceities will count as neutral equivalents of necessitists' claims involving plural quantification.) In Williamson's dialectical setting, this is the sense in which higher-order necessitism might afford the contingentist a way of making sense of necessitist's discourse. What matters is not that higher-order necessitism is something the contingentist happens to accept, but rather that it forms part of the background of their dispute with the necessitist, in a way that is reflected both in the definition of the relevant consequence relation and in the definition of a neutral formula.

By contrast, we might define the consequence of relation and the notion of neutrality in a more complicated way that would be friendly to higher-order contingentism. Recall the view, mentioned earlier, that there are all and only those propositions, properties and relations whose identities can be 'pinned down' in terms of qualitative properties of and relations among individuals there are. Fine (1977b) and Fritz and Goodman (forthcoming-b) show how to define classes of intensional models that are naturally seen as corresponding to this general vision, and Fritz and Goodman also show how these ideas might be encoded in an object-language comprehension principle.[14] Using these ideas, we might consider a weaker, higher-order-contingentism-friendly consequence relation that corresponds to the broader class of models alluded to. We might also adopt a non-trivial clause for higher-order quantification in the characterization of a neutral formula, by requiring that the quantifiers be restricted to entities of the relevant type that are pinned down in terms of the qualitative properties of and relations between the chunky individuals there are, in the sense formalized by the comprehension principle alluded to above. In light of recent work by Peter Fritz, it seems safe to conjecture that, relative to this consequence relation and definition of neutrality, *not* all formulas will have neutral equivalents given CSN.[15]

We can also consider mixing and matching these options. That is, we could consider a higher-order-contingentism-friendly consequence relation but not impose any restriction on the higher-order quantifiers in the definition of a neutral formula, or consider a higher-order-necessitist consequence relation but require neutral formulas to be equivalent to ones all of whose higher-order quantifiers are restricted to entities that can be pinned down in terms of qualitative properties of and relations among chunky individuals. It seems fairly safe to conjecture that, on either split decision, there will be formulas not in-the-operative-sense-equivalent to any in-the-operative-sense-neutral formula given CSN.

So there is some hope that by 'going higher-order necessitist' the contingentist might be able to answer Williamson's challenge in the form he poses it. But their higher-order necessitism must be extremely thoroughgoing: it must form part of the background logic that characterizes their dispute with the necessitist. (Provided this condition is met, it seems reasonable not to impose any restriction on the higher-order quantifiers in the definition of a neutral formula, despite the formal possibility of such a split decision.) This point is relevant to views like those discussed earlier, according to which higher-order necessitism falls out as an unintended consequence of some substantive metaphysical thesis like a certain plenitudinous modal mereology. In a debate between such a contingentists and a necessitist, it seems inappropriate to describe higher-order necessitsim as something to which they are each committed as a matter of 'logic,' although admittedly the issue is somewhat murky.

By contrast, consider the character Williamson dubs the 'ultra-minimalist contingentist,' who 'lay[s] down [the unrestricted] comprehension principle [as something] attractive simply [...] for its general theoretical virtues, and derive[s] the non-contingency of higher-order being as a corollary, but uphold[s] the contingency of first-order being as common sense. Such an ultra-minimalist contingentist abjures any attempt to explain or justify the asymmetry [as regards necessitism at different orders] on deeper metaphysical grounds, as likely just to lead to trouble.' (275,276) Unlike the plenitude-motivated higher-order necessitist, the ultra-minimalist contingentist accepts unrestricted comprehension in much the same spirit as the necessitist does. So it does seem appropriate, in considering a dispute between such a contingentist and a necessitist, to treat instances of unrestricted comprehension as theorems of the background logic characterizing their disagreement. Of course, Williamson has other arguments against ultra-minimalist contingentism, which we have considered already. But as far as the argument of Chapter 7 is concerned, it seem to me that – even if we set aside all of the general worries about the argument – the ultra-minimalist contingentist comes out unscathed, because he can for any given formula produce a higher-order formula that is both in the dialectically relevant sense neutral and in the dialectically relevant sense logically equivalent to the original formula given CSN. If that is right, then even considered on its own terms Williamson's argument fails to touch all forms of chunky-style contingentism.

2.5.1. Adding primitive higher-order identity

I want to briefly mention another way in which Williamson's argument is not robust with respect to a certain way of enriching the background language with higher-order devices. Suppose that for every type t we enrich our language with a primitive identity predicate of type $\langle t, t \rangle$. Suppose further that the dialectical situation is such that we may treat as logical axioms these predicates' reflexivity ($a^t =^{\langle t,t\rangle} a$) and obeying Leibniz's law ($a^t =^{\langle t,t\rangle} b^t \rightarrow A \leftrightarrow A[b/a]$, provided b is free for a in A). (The schematic letters 'a' and 'b' can be instantiated with simple or complex expressions of the relevant type, and as before we count any result of prefixing an instance of a schema with a string of necessity operators and universal quantifiers as itself an instance of the schema.) In order to allow the contingentist to be a higher-order contingentist, we do not require the identity predicates or lambda-terms to satisfy the 'being constraint' (explained in Section 2.1 above), although we may still insist on the being constraint for all primitive predicates other than the identity predicates. Now consider the consequence relation corresponding to the class of models obtained by modifying the clauses in Fritz and Goodman (forthcoming-b) in the natural way so as to interpret lambda-terms and to validate the axioms governing the primitive identity predicates. It turns out that, with respect to this consequence relation, every formula *is* equivalent given CSN to a neutral formula (even if in the definition of neutrality we require that quantifiers be restricted in the way discussed above).[16] Fritz and Goodman (fothcoming-a, [Section 3.6]) discuss the philosophical tenability of this version of higher-order contingentism in a related dialectical context, to be explained presently.

3. A different dialectical strategy

In the last section, I gave a somewhat negative assessment of Williamson's argument in Chapter 7 of *MLaM*. But I do think there is a good (though not decisive) argument for necessitism in the vicinity, which draws on a similar combination of observations about the intelligibility of superficially 'necessitist' discourse and formal results about the undefinability of various classes of variable-domain Kripke-models. That argument is developed in detail in Fritz and Goodman (fothcoming-a). Here, I will simply sketch the argument, highlighting some ways in which its dialectical structure differs from Williamson's.

Williamson challenges contingentists to make sense of the discourse of a certain imaginary character, the chunky-style necessitists. As I see the situation, the genuinely pressing challenge for contingentists can be posed without considering any such character. This is because contingentists themselves, as ordinary English speakers, feel the temptation to talk in superficially necessitist ways – to make such speeches as 'there are n possible knives that could be made from a given spare handle and n spare blades' and 'most possible people will never be born.' Of course, mindful of their contingentism, if pressed they will not accept these sentences (or at least their most syntactically faithful renderings in a formal language) as literally true on a metaphysical interpretation of their modalities and an unrestricted interpretation of their quantifiers. Nevertheless, it is implausible that we fail to systematically cognize and communicate sensible claims about modal reality through the use of these sentences (even if this use is characterized by some sort of hedging). The question, then, is what we are up to.

On its face, this seems like a more pressing challenge than Williamson's. When asked to find the kernel of truth in the discourse of someone you disagree with, there is always the option of saying that there is no way of extracting from their utterance something equivalent to it by their lights and reasonable by your lights. But this option is unavailable if the challenge is that of self-understanding. Contingentists ought to recognize that they themselves have a systematic way of using quantificational expressions to communicate claims about, as it were, 'merely possible' people. They therefore face a challenge to come up with a systematic paraphrase of such 'modalized' uses of quantifiers: to produce sentences in which the quantifiers are interpreted unrestrictedly that literally express the propositions they have in mind when they feel the pull of superficially necessitist ways of speaking.

Note that, unlike certain debates in the literature on actualism and possibilism (which Williamson rightly criticizes), this way of thinking about the challenge does not assume that necessitists and contingentists speak different languages between which we are for some reasons compelled to find a translation. For one thing, the protagonists 'the necessitist' and 'the contingentist' are nowhere to be found – everyone, regardless of their commitments, faces the challenges of making sense of the kind of 'modalized quantification' to which I have alluded (and of which Fritz and Goodman give many examples); necessitism simply offers the most straightforward answer to this challenge. Of course, if contingentism is true, then these uses of sentences are either strictly speaking false, or not properly formalized in the way the surface syntax suggests, or their quantifiers are not being given an unrestricted interpretation and their modal operators the relevant metaphysical interpretation.[17] So unlike in Williamson's setup, the discourse that contingentists are being challenged to make sense of is not discourse that is taken by all of those who are inclined to engage in it – namely, not by the contingentists themselves – to be literally true on the unrestricted/metaphysical interpretation, at least not when formalized in the obvious way. (I suspect that it was a desire to avoid considering such discourse that lead Williamson to construct the dialectic that he did.) But this difference between Fritz and Goodman's argument and Williamson's does not constitute a slide back into the old actualism/possibilism debate.

Indeed, in summarizing his own argument for necessitism in Chapter 7 of *MLaM*, Williamson writes as if the challenge he really has in mind is something much like the one I have just sketched – namely, to isolate the propositions obviously in the vicinity of certain uses of quantified modal sentences on the assumption that, because of certain contingentist assumptions, they cannot be the propositions literally expressed by those sentences on a flat-footed interpretation. Here is the relevant passage:

> In short, the necessitist can draw more distinctions than the contingentist can. Every distinction the contingentist can draw can be drawn in neutral terms, so the necessitist can draw it too. The converse fails. The necessitist can draw distinctions the contingentist cannot, because they cannot be drawn in neutral terms. That would not matter if those extra distinctions were bogus. But the contingentist cannot plausibly dismiss them like that, because they are too intimately related to distinctions the latter is committed to regarding as genuine. Thus necessitism provides a clearer view than contingentism of modal reality. (364)

This passage is hard to square with Williamson official dialectical setup, because that setup is concerned with literal unhedged discourse in a common formal language, and anything the necessitist can say in that language the contingentist can say too. The passage seems more in keeping with the way of framing the argument for necessitism that I have been recommending: certain ways of speaking that have a necessitist flavor seem to correspond to sensible ideas; yet these ideas are such that, if contingentism is true, then they are not expressible in the relevant formal languages (because we can prove that associated classes of models are not definable in those languages); so we should reject contingentism.

One advantage of this argumentative strategy is that it dispenses completely with all talk of 'chunkiness.' The restriction to neutral formulas falls away too, since the issue is not framed in terms of any actual or hypothetical dispute. (Of course, there are plenty of methodological challenges that our argument raises that Williamson's does not; we try to address these as best we can in our paper.)

Despite the many differences between our preferred argumentative strategies, though, many of the relevant issues are parallel. For one thing, the model-theoretic results and proof strategies are similar – see Fritz (2013). For another thing, the question of whether being a higher-order necessitist (or countenancing a family of non-being-constrained primitive higher-order identity predicates) affords the contingentist a solution to the challenge turns on similar considerations about whether we can reasonably restrict our attention only to models in which the relevant metaphysical principles hold.

One technical difference worth highlighting is that, while Williamson frames his challenge in terms of plural quantification, Fritz and Goodman frame it in terms of generalized quantifiers (focussing on the non-first-order-definable generalized quantifiers 'there are uncountably many' and 'most'). An advantage of this way of framing the challenge is that it allows us to see an argument Williamson presents in Section 6.4 against higher-order contingentism as a species of whatever general phenomenon is meant to be dramatized in the case of plural quantification (despite the fact that the arguments are presented as independent). Williamson claims that, for the purposes of 'second-order modal mathematics,' we need to be able to instantiate a certain modalization of the claim that a given condition $\leq$ is a complete order – i.e. the claim that, for every possible property X, being possibly $\leq$ has maximal element among the possible X's – with any condition $\varphi(x)$, even relative to an assigment of incompossible entities to parameters occurring in it chosen so that the open formula $\varphi(x)$ fails to express even a possible property. On its face, this seems like simply a bald assertion that the higher-order contingentist might simply reject as a consequence of his overall view. But that would be a mistake. For the condition of being a complete order is a generalized quantifier with two variables, regarding which we can raise the challenge for contingentists to make sense of the corresponding 'necessitist discourse' or 'modalized use' thereof.[18] The invalidity of the relevant inference by higher-order contingentist's lights (and Fritz and Goodman forthcoming-b argue that they should think it is invalid) shows that simply modalizing the ordinary higher-order definition of a complete order does not achieve the desired effect of expressing the intended structural constraint on the patter of satisfaction of being possibly $\leq$ by all possible individuals, as it were. The appeal to second-order modal mathematics can then be understood simply as a reminder that it seems like such claims make perfect sense (although the paticular example is rather involved). What the relevant formal results show is that, in the case of many non-first-order definable generalized quantifiers, it is not just that certain natural ways of sprinkling modal operators into their ordinary non-modal higher-order definitions fail to express the intended condition on models, but indeed no formula express the intended condition.

4. Higher-order logic as metaphysics

In my view, the greatest service Williamson did in writing *MLaM* has nothing to do with modal metaphysics in particular: it was to forcefully show that higher-order languages are a powerful tool for metaphysical theorizing. I hope and believe that this kind of higher-order metaphysics is the way of the future. I want to close by considering two issues arising in this connection.

4.1. *Grainularity*

Williamson operates throughout the book on the assumption that necessary co-extensiveness suffices for the identity of propositions, properties, and relations, where this is understood as at least entailing their higher-order indiscernibility. But the status of this assumption is not entirely clear. Consider, in particular, the following passage:

> Hyperintensionality arises at the level of thought and linguistic meaning, and should be explained at that level, not at the level of anything like a general theory of properties and relations. For present purposes, a coarser grained intensional standard of individuation is more plausible, and certainly much simpler. (266)

I happen to disagree with the claim that hyperintensionality (in the sense of distinctions among necessarily extensional equivalents) arises only at the level of thought and language, but let's set that aside. What I don't understand is Williamson's suggestion (made throughout the book) that the granularity of reality is something that can be relativized to our purposes. One thing Williamson could mean here is that, for present purposes, he is restricting his higher-order quantifiers to entities that don't draw any hyperintensional distinctions.[19] But if that is how he intends to be understood, it seems to conflict with his insistence that his quantifiers are always to be read as unrestricted (although maybe he is only talking about his first-order quantifiers?).

A different thing that someone might mean by this passage is that, among the abstract objects, there are many families of property-like things, that these families disagree as regards their fine-grainedness, that for different theoretical purposes it is most fruitful to use 'property' to talk about different ones of these families, and that for present purposes it is best to interpret 'property'-talk as being about members of the modally-individuated family of property-like abstracta. But this cannot be what Williamson means, because he does not think that higher-order quantification is or can be analyzed as some sort of tacit first-order quantification over a realm of abstract objects.

The great promise of higher-order logic as metaphysics, from my perspective, is that it lets us ask questions about fineness of grain using no non-logical predicates like 'property' or 'propositions' that might then be claimed to admit of different disambiguations, and from which scholastic metaphysics then ensues. For we can unambiguously interpret the dyadic sentential operator $\lambda p^{\langle\rangle} q^{\langle\rangle} (\forall O^{\langle\langle\rangle,\langle\rangle\rangle} (Op \leftrightarrow Oq))$ as involving absolutely unrestricted and irreducibly higher-order quantification into monadic sentential operator position. In terms of this notion, we can frame competing hypothesis about reality's granularity: i.e. necessary and/or sufficient conditions on p being higher-order indiscernible from q. I believe that such questions are among the deepest and most exciting in all of metaphysics. They are deep because of their widespread ramifications, and exciting because they lend themselves to the kind of abductive investigation aided by formal methods that Williamson advocates in the Afterword of *MLaM* and inspiringly demonstrates throughout the book.

Williamson's rhetoric surrounding these issues unfortunately (and uncharacteristically) gives the impression that questions of granularity are to some extent pragmatic ones. I don't think that Williamson really thinks this, but a naive reader could be forgiven for thinking otherwise.

4.2. A non-modal argument for necessitism

Although I have been critical of some of Williamson's arguments for necessitism, I am a necessitist myself. The argument that most convinces me is not the one I described in Section 3, which I think can be resisted by adopting higher-order necessitism in the right spirit. I am motivated rather by considerations having nothing to do with modality in particular, but instead by independently attractive principles about fineness of grain.

Using only Boolean connectives and higher-order quantification, we can for all types $\langle t_1 \ldots t_n \rangle$ define a higher-order analogue of identity $\approx$ in terms of higher-order indiscernibility:

$$\approx^{\langle\langle t_1 \ldots t_n\rangle,\langle t_1 \ldots t_n\rangle\rangle} := \lambda x^{\langle t_1 \ldots t_n\rangle} y^{\langle t_1 \ldots t_n\rangle} (\forall X^{\langle\langle t_1 \ldots t_n\rangle\rangle} (Xx \leftrightarrow Xy)).$$

We can also define Boolean operations on all such types using lambdas (e.g. $X^{\langle t_1 \ldots t_n\rangle} \wedge Y^{\langle t_1 \ldots t_n\rangle} := \lambda x_1^{t_1} \ldots x_n^{t_n} (Xx_1 \ldots x_n \wedge Yx_1 \ldots x_n)$). This allows us to define, for all types $\langle t_1 \ldots t_n \rangle$, relations like

$$\leq^{\langle\langle t_1 \ldots t_n\rangle,\langle t_1 \ldots t_n\rangle\rangle} := \lambda x^{\langle t_1 \ldots t_n\rangle} y^{\langle t_1 \ldots t_n\rangle} (x \approx (x \wedge y))$$

and

$$\leq'^{\langle\langle t_1 \ldots t_n\rangle,\langle t_1 \ldots t_n\rangle\rangle} := \lambda x^{\langle t_1 \ldots t_n\rangle} y^{\langle t_1 \ldots t_n\rangle} ((x \wedge (y \rightarrow y)) \approx (x \wedge y)).$$

Given a sufficiently coarse-grained view of higher-order reality, such relations will satisfying general principles characteristic of a notion reasonably pronounced 'entails.'[20] We can then abductively investigate general principles governing entailments among properties and relations.

In particular, we can investigate the entailment theoretic properties of quantifiers: that is, how entailment relations among generalizations are related to the entailment relations among the conditions being generalized. In this connection, the following simple and powerful principle is extremely attractive:

ADJUNCTION
$\forall p^{\langle\rangle} \forall X^{\langle t\rangle} : (p \text{ entails}^{\langle\langle\rangle,\langle\rangle\rangle} \forall x^t X^{\langle t\rangle} x) \leftrightarrow (\lambda x(p) \text{ entails}^{\langle\langle t\rangle,\langle t\rangle\rangle} X)$

This principle pins down the meaning of the quantifier up to mutual entailments in terms of entailment relations between properties – no small feat – and it has other theoretical attractions too. Yet combined with other plausible principles governing entailment relations among predications, it allows us to argue that my being something is entailed by the logical truth that everything is something ($\forall x \exists y(y = x)$ entails $\exists y(y = a)$). Here is not the place to go into the details of the argument.[21] I merely want to mention the argument as an example of a certain kind of non-modal argument for necessitism. For on either of the two candidate notions of entailment $\leq$ and $\leq'$, that I am a necessary being follows from the above entailment given completely uncontroversial principles of (the *de dicto* fragment of) quantified modal logic (together with what should be an uncontroversial instance of Leibniz's law for $\approx^{\langle\langle\rangle,\langle\rangle\rangle}$). Modal logic *per se* isn't where the action is here. Grain science is.

This last point is relevant to Sider (forthcoming), who defends contingentism by claiming that the sort of abducive criteria Williamson advocates are inappropriate for theorizing about metaphysical necessity because that notion is metaphysically second-rate. While I disagree with Sider about the place of abduction in modal metaphysics, I also think we can sidestep that issue, since abductive criteria are absolutely relevant for investigating the general principles that we can formulate in pure higher-order logic. Surprisingly, some attractive such principles may by themselves lead us right to necessitism's doorstep.

Notes

1. All page references are to Williamson (2013).
2. See Sider (2011) and Dorr and Hawthorne (2013) for discussion. It is also not entirely clear how to formulate the relevant supervenience theses in a higher-order setting. If we assume that necessarily equivalent propositions are identical (as Williamson does), then there is only one necessary proposition, and so necessity can be defined in purely logical higher-order terms as being a tautology (e.g. $\lambda p(\forall O(Op \leftrightarrow O(p \rightarrow p))))$. And normally, when we formalize the claim that a property (strongly) supervenes on a set of properties and relations, we really mean that it supervenes on the set of properties and relations obtained by closing the original set under certain logical operations. (Sometimes such theses are instead formulated in terms of isomorphisms between possible worlds, but as Williamson points out only necessitists can take such formulations at face value.) In the present higher-order setting, this would then have the effect of making necessity part of every supervenience base, in which case the modal will trivially supervene on the non-modal.
3. See Fine (1977b) and Fritz and Goodman (forthcoming-b) for an exploration of such views in the setting of higher-order modal logic.
4. Consideration of modal truths specified in counterfactual terms, rather than merely in terms of what is necessary or possible, also suggests a way of formulating the ground-theoretic objection in a non-hyperintensional setting, since such truths are not invariably non-contingent.
5. This principle is a consequence of (the modal analogue of) the tensed mereology developed in Hovda (2013); for related ideas, see Hawthorne (2006). Fine (1999) also seems to be committed to such a principle, although since he also takes the relevant notion of parthood to be anti-symmetric, he thereby flirts with paradox; see Goodman (xxx-b) for discussion, where I explore a slightly weakened (but for present purposes strong enough) principle restricted to properties *X* that, necessarily, have extensions not too big to form a set.
6. An object-language argument for this conclusion would be quite involved, but one can give a relatively straightforward model-theoretic argument relative to the class of structures discussed in Fritz and Goodman (forthcoming-b).
7. See Goodman (xxx-a) and Fritz and Goodman (Forthcoming) for discussion.
8. Of course, those who, unlike Williamson, accept hyperintensional theories of properties might claim that, although it could turn out to be true 'by accident' that necessarily everything necessarily has a haecceity in Williamson's sense, purely modal-metaphysical considerations establish that there could not be the property of being *me* without there being such a thing as me.
9. The second language also contains polyadic generalizations of plural quantifiers, which we might think of as plurally quantifying over ordered *n*-tuples of individuals; I will ignore this subtlety in what follows. Williamson also interprets

the plural variables in such a way that they can have empty extension – i.e. there are some things (the 'empty plurality') of which nothing is one.

In addition to Boolean connectives and necessity and possibility operators, both languages also include two operators $\downarrow$ and $\uparrow$ which can be seen as generalizing the more familiar rigidifying 'actually' operator: $\downarrow$ has the effect of making the subformula it embeds to be evaluated at the modal scope of any $\uparrow$ binding it, where an occurrence of $\uparrow$ *binds* an occurrence of $\downarrow$ just in case the latter is in the scope of the former and no occurrence of $\uparrow$ has intermediate scope.

10. For purposes of illustration, we might take chunkiness to be being grounded in the concrete and chunkiness* to be instantiating some fundamental property or standing in some fundamental relation, on the assumption that these are distinct properties.
11. Williamson mentions making special provision for sets, but if what he has in mind is something like the proposal in Fine (1977a), it would solve the problem for the 'is a member of' predicate but not for the 'is the favorite set of' predicate. Williamson also justifies the focus on chunky-style necessitsm by saying:

> [I]t is in the spirit of the preferred form of necessitism explained in Chapter 1 to characterize non-chunky things mainly in modal or temporal terms, through the properties they would or could have if they were chunky, or did have when they were chunky, or will have when they are chunky. Such necessitists may be happy to confine their primitive non-logical predicates to those meeting the constraint [of being chunkiness-entailing]. [...] Thus we envisage a necessitist who asserts [the relevant chunkiness-entailment principle] for every primitive non-logical predicate in the language. (326,327).

In reply, the 'favorite set' example shows that, although it may be natural for necessitists to characterize non-chunky objects '*mainly* in modal or temporal terms', they are not likely to characterize them *only* in such terms, which is what matters in the present context.

12. The interpretation of chunkiness as contingent non-concreteness has the advantage of making the chunky-style necessitist's claim that everything is possibly chunky uncontroversial, since for any condition *F* it is uncontroversial (assuming S5) that necessarily everything is possibly not contingently not *F*. The issue is then whether, so understood, the realm of the chunky does in fact constitute 'neutral ground' in Kit and Bob's metaphysical dispute.
13. One might think that, when formulated in supervenience-theoretic terms, the floating-free argument for contingentism can be seen as a version of CSC with chunkiness interpreted as instantiating some fundamental property or standing in some fundamental relation. But this is not so, since the floating-free objection so construed would then not apply to the conjunction of necessitism with the claim that identity is a fundamental relation, a view to which the objection clearly does apply when construed in terms of supervenience.
14. In particular, they discuss ways of adding a 'qualitative' operator to modify their schema Comp_{FS} so as to formalize the Fine-inspired view they call the 'qualitative generation view.'
15. Of particular relevance are the main result of Fritz (Forthcoming) and the connection Fritz (2013) establishes between such results and the existence of neutral equivalents given CSN, although the latter paper only considers first-order languages.

16. Briefly, here is why. Let $\ulcorner \Pi x^t A \urcorner$ abbreviate the formula $\ulcorner \exists p\, p \wedge \lambda x(p) =^{\langle\langle t\rangle,\langle t\rangle\rangle} \lambda x(p \wedge A) \urcorner$. One feature of the relevant class of models is that, at each world of every model, the proposition true at only that world is in the $\langle\rangle$-domain of that world. And relative to an assignment of this 'world proposition' to p, $\ulcorner p \wedge \lambda x(p) =^{\langle\langle t\rangle,\langle t\rangle\rangle} \lambda x(p \wedge A) \urcorner$ is true at the world of evaluation just in case $\ulcorner \lambda x(A) \urcorner$ has universal extension at that world – not merely in the sense of being satisfied by everything in the t-domain of that world, or even the t-domain of some world, but rather that it be satisfied by any type-t intension definable in the model, even 'impossible' ones in the t-domain of no world. It is this feature that lets Π behave like a 'classical' quantifier, in the sense that the Π-analogue of the unrestricted comprehension schema comes out valid. This in turn allows us to use Π-quantification of variables of type $\langle e\rangle$ to simulate necessitists' plural quantification in a reasonably mechanical way.
17. To say that the quantifiers are not interpreted as 'unrestricted' is *not* to say that they receive a restricted interpretation; see Dorr (2005, 2008).
18. For the relevant notion of a generalized quantifier, (see Westerståhl 2011).
19. See Fritz and Goodman (forthcoming-b, [Section 3.4]) for an explanation of how to make this notion precise.
20. My own idiosyncratic views about granularity make $\leq'$ but not $\leq$ behave in an entailment-like way, but that is a long story.
21. The argument is inspired by Dorr (2014); I develop it in Goodman (2016) [Chapter 5].

Acknowledgements

Thanks to Cian Dorr and Peter Fritz for comments on an earlier version of this paper and to audiences at the 2014 Logical and Modal Space workshop at NYU, the 2015 Williamson on Modality workshop in Montreal, and the 2015 Williamson, Logic and Philosophy conference at Peking University.

References

Dorr, Cian. 2005. "What We Disagree about when We Disagree about Ontology." In *Fictionalism in Metaphysics*, edited by Mark Eli Kalderon, 234–286. Oxford: Oxford University Press.

Dorr, Cian. 2008. "There are No Abstract Objects." In *Contemporary Debates in Metaphysics*, edited by Theodore Sider, John Hawthorne, and Dean W. Zimmerman, 32–64. Malden, MA: Blackwell.

Dorr, Cian. 2014. "Quantifier Variance and the Collapse Theorems." *The Monist* 97 (4): 503–570.

Dorr, Cian, and John Hawthorne. 2013. "Naturalness." In *Oxford Studies in Metaphysics*, Vol. 8 edited by Karen Bennett, and Dean Zimmerman, 234–286. Oxford: Oxford University Press.

Fine, Kit. 1977a. *Postscript to Worlds, Times and Selves* (with A. N. Prior). London: Duckworth.
Fine, Kit. 1977b. "Properties, Propositions and Sets." *Journal of Philosophical Logic* 6 (1): 135–191.
Fine, Kit. 1999. "Things and their Parts." *Midwest Studies in Philosophy* 23: 61–74.
Fritz, Peter. 2013. "Modal Ontology and Generalized Quantifiers." *Journal of Philosophical Logic* 42: 643–678.
Fritz, Peter. Forthcoming. "Higher-Order Contingentism, Part 3: Expressive Limitations." Unpublished.
Fritz, Peter, and Jeremy Goodman. Forthcoming-b "Counting Incompossibles." *Mind*.
Fritz, Peter, and Jeremy Goodman. Forthcoming-c. "Higher-Order Contingentism, Part 1: Closure and Generation." *Journal of Philosophical Logic*.
Fritz, Peter, and Jeremy Goodman. Forthcoming-a. "Counterfactuals and Propositional Contingentism". Unpublished.
Goodman, Jeremy. 2016. "The Case for Necessitism." DPhil thesis, University of Oxford.
Goodman, Jeremy. Forthcoming-a. "Consequences of Conditional Excluded Middle." Unpublished.
Goodman, Jeremy. Forthcoming-b. "Matter and Mereology." Unpublished.
Hawthorne, John. 2006. "Motion and Plenitude." *Metaphysical Essays*, 111–144. Oxford: Oxford University Press.
Hovda, Paul. 2013. "Tensed Mereology." *Journal of Philosophical Logic* 42: 241–283.
Jager, Thomas. 1982. "An Actualistic Semantics for Quantified Modal Logic." *Notre Dame Journal of Formal Logic* 23: 335–349.
Mackie, Penelope. 2006. *How Things could have Been*. Oxford: Oxford University Press.
Sider, Theodore. 2011. *Writing the Book of the World*. Oxford: Oxford University Press.
Sider, Theodore. Forthcoming. "On Williamson and Simplicity in Modal Logic." *Canadian Journal of Philosophy* (online encyclopedia).
Westerståhl, Dag. 2011. "Generalized Quantifiers." In *Stanford Encyclopedia of Philosophy*, edited by Edward N. Zalta, Summer 2011 edition.
Williamson, Timothy. 2013. *Modal Logic as Metaphysics*. Oxford: Oxford University Press.

Reply to Goodman

Timothy Williamson

Wykeham Professor of Logic, University of Oxford, Oxford, UK

In 'Williamson on necessitism' (Goodman 2016), Jeremy Goodman makes a variety of probing objections to my arguments for necessitism in *Modal Logic as Metaphysics*. Had the book done justice to all the considerations he raises, it would have been much longer, something for which not all readers would have been wholeheartedly grateful. That is not remotely to imply that Goodman's considerations are irrelevant or unimportant. The dispute between necessitism and contingentism is genuinely caught up in an intricate nexus of complex philosophical and technical issues, all of which deserve to be untangled at length. As Goodman shows, many diverse but internally coherent theories give distinctive answers to the questions. In writing the book, I did not expect or even hope that it would be the last word on anything. My aim was to make some progress myself and to facilitate much more on the part of others. Goodman has contributed significantly to such progress through his explorations of the space of conceivable logico-metaphysical theories of modal reality.

1. First-order contingentism and higher-order contingentism

Section 1 of Goodman's paper criticizes my arguments in Section 6.2 of the book against combining ordinary first-order contingentism with higher-order necessitism. On the target view, it is contingent what objects there are, in other words $\neg\Box\forall x\, \Box\exists y\, x = y$, but necessary what properties there are, or more correctly $\Box\forall X\, \Box\exists Y\, X \approx Y$, where the variables '$X$' and '$Y$' take monadic predicate position and $\approx$ is the appropriate analogue of identity, and likewise for other higher types.

Goodman agrees with me that, given first-order contingentism, there is *some* defeasible presumption in favour of higher-order contingentism because uniform theories are *ceteris paribus* preferable to non-uniform ones. At least, he writes 'I don't want to claim that there is no theoretical pressure in this direction' (Section 1.1). In the reverse direction, I agree with Goodman that if one is going to be contingentist at some orders and necessitist at others, the best place to make the cut is between the first order and the rest, because in an inductively defined hierarchy it is not unusual for the case governed by the base clause to be special, by contrast with the cases governed by the inductive clauses. Goodman and I also agree that the issue between necessitism and contingentism at any order is a difficult theoretical one, not to be settled by pre-theoretical judgements. He does point out that first-order contingentism is far more often considered obvious than higher-order contingentism, in response to the claim that there is parity in motivation between the views. But that asymmetry is hardly surprising, given the unfamiliarity of higher-order quantification by contrast with first-order quantification. It does not show that the considerations moving people to first-order contingentism have no analogues for higher-order quantification; the analogues may just be psychologically harder to access.

One can get a feel for the oddness of the hybrid view by considering a theorist whose logic is first-order contingentist but higher-order necessitist, and whose metaphysics also permits first-order quantification over properties. The first-order ascription of contingent being to the property of being a tiger is consistent in the logic, while the corresponding second-order ascription of contingent being is not. That looks like an unwarranted asymmetry. A good example of a thoughtful contingentist who takes the case for contingentism about individuals to generalize to properties, relations and propositions is Stalnaker (2012).

Some contingentists are moved by the idea that necessitism makes the modal float free of the non-modal in some objectionable sense. I explore this idea in the book, finding the distinction between the modal and the non-modal obscure but conceding that there may be some sense in which necessitism implies that individuals can differ in qualitative modal respects without differing in qualitative non-modal respects, although I argue that such a failure of supervenience would not be objectionable (2013, 380–391). Goodman considers this issue in Section 1.2 of his paper. It is important for him because he thinks that 'the best theoretical argument against necessitism is that it makes the modal "float free" of the non-modal in some objectionable way'. Thus, if the floating free argument has force when formulated in first-order terms but not when formulated in higher-order terms, there is a crucial difference in motivation between first-order contingentism and higher-order contingentism, which would go some way towards motivating the combination of first-order contingentism with higher-order necessitism. As he points out, my discussion of the supervenience issues is formulated only in terms of properties of (and relations between) individuals. However, that restriction was only for the sake of simplicity, since at that point in the book I was no longer considering the hybrid view. I see no special difficulty in formulating higher-order analogues of the floating free argument. Indeed, Goodman himself sketches how such an analogue might go, although he expresses uncertainty about its force. Just as (apparently) first-order necessitism makes the qualitative modal difference between Goodman and his computer float free of any qualitative non-modal differences between them (there are none) in counterfactual circumstances where neither is concrete, so (apparently) higher-order necessitism makes the second-order qualitative modal difference between being Goodman and being his computer float free of any second-order qualitative non-modal difference between them (there are none) in the same counterfactual circumstances. Although the problem does not depend on an intensional view of the higher-orders such as I take in the book, it is exacerbated by that view, which makes the higher-orders modal in their very nature, and thereby undermines the requisite higher-order non-modal differences. Thus what Goodman regards as the best theoretical argument against first-order necessitism does have a natural and arguably at least equally forceful analogue against higher-order necessitism. Although he considers several further variations on these themes, nothing he says warrants his claim that 'the best arguments for contingentism do not correspond to parallel arguments for higher-order contingentism'. On the contrary, the parallelism stands up well, thereby removing a supposed motivation for the hybrid view.

In the book, I briefly discuss a form of anti-essentialism as a response to the floating free argument that some necessitists may wish to make (2013, 389). Goodman says that '[i]n chapter 8.2 Williamson expresses sympathy for' it, but I do not; I merely mention it as a theoretical option. He contends, plausibly, that it will not suffice to block the floating free argument once counterfactual conditionals are included in the domain of the modal. His point reinforces the case for parity between first-order necessitism and higher-order necessitism, since necessitist comprehension principles in higher-order logic rule out the relevant higher-order analogues of anti-essentialism. For instance, there is a property that necessarily everything has and a property that necessarily nothing has.[1]

The central arguments in the book for the parity of first-order and higher-order contingentism turn on the connection between an object and its haecceity, the property of being that object (which there necessarily is, given higher-order necessitism). If there is no such object, how can there be a property necessarily related to it? In Section 1.3 of his paper, Goodman makes much of a passage in which I ask rhetorically about my haecceity in circumstances where there is no such thing as me 'how can it lock onto me in my absence?' (2013, 269). He argues that the challenge fails to address higher-order necessitism, because my haecceity would have locked onto me in my absence on the relevant modal reading ($\Box\forall x\,(Xx \leftrightarrow x = \mathrm{TW})$) given first-order contingentism, whether or not there was such a property in those circumstances, and so irrespective of the difference between higher-order necessitism and higher-order contingentism. He writes:

> Maybe this is just a slip. After all, Williamson could instead have written 'How could *there have been* something that in my absence locked onto me?' But if it is a slip, it is a telling one, since the ease of making it suggests that we are mistaking a general puzzling feature of contingentism for a particular puzzling feature of the combination of contingentism and higher-order necessitism.

There was no slip. My point was just that higher contingentism can mitigate the general puzzling feature of first-order contingentism by denying that there would have been such a property in the relevant circumstances, while higher-order necessitism cannot. Goodman is hardly in a position to repudiate such a move, since he treats an analogous move at first order as effective. For, given an apparent violation of the supervenience of the qualitative on the fundamental, when he and his computer differ in their absence with respect to the qualitative property of possibly being a person, despite supposedly not differing with respect to any fundamental property, he writes in Section 1.2:

> A contingentist can respond by claiming that in the relevant counterfactual circumstances there would have been no such thing as me and no such thing as my computer, thereby blocking the alleged counterexample to the supervenience thesis.

But in the relevant counterfactual circumstances Goodman and his (actual) computer would still have had the modal qualitative difference on the relevant modal reading ($\neg(\Diamond Px \leftrightarrow \Diamond Py)$) with no corresponding fundamental (or non-modal qualitative) difference, whether or not there would have been such things in those circumstances. Goodman's point is presumably that first-order contingentism can mitigate the general puzzling feature of that failure of supervenience by denying that there would have been such things in the relevant circumstances. Fair enough, but a structurally analogous point is being expressed in the passage of which he makes such heavy weather.

In the book, I note that in some cases a contingentist can explain how actual materials uniquely characterize a merely possible object, and thereby make it plausible from a contingentist perspective to claim actual being for its haecceity, without reliance on higher-order necessitism. For instance, parts from which it could be made may have actual being. However, I argued that extending that strategy to a general case for higher-order necessitism from a first-order contingentist starting point would require highly implausible metaphysical assumptions (270–271). In Section 1.3.1 of his paper, Goodman suggests that I underestimated the range of metaphysical resources available to some first-order contingentists in pursuing such a strategy. He gives two ingenious examples, which I will discuss in turn.

Goodman's first example is of a first-order contingentist who endorses the principle of plenitude that 'any eligible "modal profile" is the modal profile of some possible material object'; Goodman provides a more precise unpacking of this idea. He continues:

> Given the not outlandish assumptions that, necessarily, (1) there is at least one material object, and (2) any two material objects that necessarily coincide (in the sense of sharing parts with the same things) are qualitatively discernible, the principle of plenitude allows us to argue that, necessarily, every material object necessarily has a haecceity, where we assume only that there is a property corresponding to any condition specified in terms of individuals there are and their qualitative properties and relations.

He then sketches an extension of the strategy to non-material things. One general concern about this strategy is that its reliance on modal considerations to individuate coincident material objects may undermine what Goodman regards as the best motivation for first-order contingentism: the argument that first-order necessitism makes the modal float objectionably free of the non-modal. A more specific objection to the strategy is that the assumption that necessarily there is at least one material object is, contrary to Goodman, unreasonable. Why could there not have been mere empty space-time? Or, if empty space-time itself counts as a material object, why could there not have been no space-time at all? Both scenarios seem to be metaphysically possible.

Goodman's other example depends on the principle of conditional excluded middle (CEM), that for any antecedent *A* and consequent *C*, either *A* counterfactually implies *C* or *A* counterfactually implies ¬*C*, which is valid on Stalnaker's semantics for counterfactuals but not on Lewis's. On that assumption, 'we can argue that necessarily everything necessarily has a haecceity from the premise that, necessarily, any two contingent beings are possibly qualitatively discernible'. In recent years, support has grown for CEM as a thesis about counterfactual conditionals in natural language. Now one point on which both its proponents and its opponents agree is that its defence requires a story about tiebreaking. For there are surely cases where a possible world *w* is equally similar to two possible worlds w_a and w_b, by whatever standard of similarity is relevant to counterfactuals, so w_a and w_b are tied with respect to *w*. Let *A* be true in just the worlds w_a and w_b, and *C* be true in just w_a. Thus, if *A* counterfactually implies *C* in *w*, then the tie is broken in favour of w_a, while if *A* counterfactually implies ¬*C* in *w*, then the tie is broken in favour of w_b. Hence, if CEM holds in *w*, the tie is broken one way or the other. Some putative readings of the counterfactual break the tie in favour of w_a, other putative readings break it in favour of w_b. Normally, proponents of CEM take either a supervaluationist or an epistemicist view of tiebreaking. On the supervaluationist view, it is indeterminate which reading is correct, but determinate that exactly one reading is correct (in an appropriately minimalist sense of 'correct'). On the epistemicist view, it is unknowable which reading is correct, but knowable that exactly one reading is correct. Such tiebreaking is crucial to the supposed role of counterfactuals in capturing haecceities.

A toy model will clarify the position. Suppose that the world w_a contains only the material object *a*, while the world w_b contains only the material object *b*, where by normal standards *a* in w_a is qualitatively identical to *b* in w_b; the world *w* contains nothing. Suppose that the reading of the counterfactual conditional > breaks the tie in favour of w_a. Then, from the standpoint of any world, this predicate singles out *a*, in the sense of applying uniquely to *a*:

$$\lambda x(\neg\exists y\, y = y > (\exists y\, y = y > \exists y\, y = x))$$

For the outer antecedent ($\neg\exists y\, y = y$) takes the evaluation from any world to w, from which the inner antecedent ($\exists y\, y = y$) takes it to w_a by hypothesis. On the other hand, if the reading of > breaks the tie in favour of w_b, then that predicate singles out b. Under these conditions, for the reading of > to break the tie between w_a and w_b just is in effect for it to single out one of a and b. Clearly, CEM by itself does nothing to explain how the tie is broken; it merely implies that somehow or other it is broken. Since a and b are not in w, the mystery of how w can contain a haecceity of a or of b has merely been transferred to the mystery of how w can contain a reading of > that singles out one of a and b. In such cases, CEM may provide evidence for higher-order necessitism, but not in any way that provides a plausible metaphysical story to make sense of its combination with first-order contingentism, which is the challenge issued in the book.

Clearly, there is much more to be explored in combinations of first-order contingentism with higher-order necessitism. By comparison with the versions I consider in the book, Goodman's speculative hypotheses (which he is in any case disinclined to endorse) are more elaborate, but not thereby more plausible.

2. Capturing distinctions

Section 2 of Goodman's paper criticizes the central argument of chapter 7 of my book. Section 3 sketches what he describes as 'a good (though not decisive) argument for necessitism in the vicinity, which draws on a similar combination of observations about the intelligibility of superficially "necessitist" discourse and formal results about the undefinability of various classes of variable-domain Kripke-models'. Peter Fritz and he have developed the alternative argument in detail elsewhere (Fritz and Goodman, Forthcoming). The issues are complicated; some are explained below. My general view is that although the differences between the two arguments are not trivial, Goodman has exaggerated their significance.

An entry point is my summary of the argument in chapter 7 (2013, 364):

> In short, the necessitist can draw more distinctions than the contingentist can. Every distinction the contingentist can draw can be drawn in neutral terms, so the necessitist can draw it too. The converse fails. The necessitist can draw distinctions the contingentist cannot, because they cannot be drawn in neutral terms. That would not matter if those extra distinctions were bogus. But the contingentist cannot plausibly dismiss them like that, because they are too intimately related to distinctions the latter is committed to regarding as genuine. Thus necessitism provides a clearer view than contingentism of modal reality.

Having quoted this summary, Goodman comments in Section 3:

> This passage is hard to square with Williamson's official dialectical setup, because that setup is concerned with literal unhedged discourse in a common formal language, and anything the necessitist can say in that language the contingentist can say too. The passage seems more in keeping with the way of framing the argument for necessitism that I have been recommending [...]

The alleged discrepancy with my official dialectical set-up comes about because I hold, as does Goodman, that the necessitist and the contingentist share a common language: what words literally mean in the necessitist's mouth is just what they literally mean in the contingentist's mouth. Thus, if what distinctions one can draw depends only on the literal meanings of one's words, the distinctions the necessitist can draw are just the distinctions the contingentist can draw. However, that is not how I used talk of drawing distinctions in the book. I wrote of theorists sometimes seeing their rivals 'as tracking genuine distinctions which they misdescribe through relying on false theoretical assumptions' (Williamson 2013, 312). In that sense, someone who takes 'Thor' to name a flesh and blood Viking god may use the sentences 'Thor is angry' and 'Thor is not angry' literally and unhedged, yet thereby draw or track, but misdescribe, the genuine distinction between situations where thunder is on the way and situations where thunder is not on the way (2013, 309). What distinctions one is drawing depends not only on one's *language*, taken literally and unhedged, but on one's *theory*. If one's theory entails that (in all relevant situations) Thor is angry if and only if thunder is on the way, one can use 'Thor' sentences literally and thereby draw the distinction about thunder. If one's theory has no such entailment, one cannot.

Goodman sketches his preferred form of argument thus:

> [C]ertain ways of speaking that have a necessitist flavor seem to correspond to sensible ideas; yet these ideas are such that, if contingentism is true, then they are not expressible in the relevant formal languages (because we can prove that associated classes of models are not definable in those languages); so we should reject contingentism.

The phrase 'necessitist flavor' alludes to the role of theory in determining what 'ideas' the 'ways of speaking' (non-literally) correspond to. Goodman elaborates slightly on this earlier in Section 3:

> contingentists themselves, as ordinary English speakers, feel the temptation to talk in superficially necessitist ways — to make such speeches as 'there are *n* possible knives that could be made from a given spare handle and *n* spare blades' and 'most possible people will never be born'. Of course, mindful of their contingentism, if pressed they will not accept these sentences (or at least their most syntactically faithful renderings in a formal language) as literally true [...]

But what I find missing in both Goodman's present paper and his one with Fritz (Forthcoming) is any account of what *kind* of necessitist theory contingentists are tempted to superficially talk along with, and how it corresponds to the *kind* of model relevant to issues of definability and undefinability.

Let me be more specific. In Fritz and Goodman's detailed arguments, they study definability with respect to variable domain Kripke models in which contingentists' (first-order) quantifiers are modelled in their literal use by quantifiers over the world-dependent domains of individuals but in their 'superficially necessitist' non-literal use by distinct quantifiers over the world-independent union of all the world-dependent domains of individuals (Fritz and Goodman, Forthcoming). Although that is what one might have expected, it constitutes a specific theoretical choice. From a technical perspective, one can just as well model superficially necessitist non-literal uses of contingentists' (first-order) quantifiers by quantifiers over the world-independent *intersection* of all the world-dependent domains of individuals, or over the domain of individuals associated with the designated actual world of the model, held constant irrespective of the world of evaluation, or over a constant domain that is not required to have any particular relation to the world-dependent domains of individuals. All of those theoretical choices and many more model the 'superficially necessitist' use with a constant domain semantics that validates necessitism. However, they differ radically from each other over the definability of the 'necessitist' quantifiers in terms of the 'contingentist' ones: some make it trivial, some make it impossible. Which class of models is relevant depends on the specific nature of contingentists' 'superficially necessitist' non-literal use of their quantifiers, which in turn depends on what kind of necessitist theory contingentists are simulating. Each of those classes of models is a natural modelling choice for some imaginable and at least superficially necessitist way of speaking. Although Fritz and Goodman may in fact have made the most natural modelling choice for the kind of superficially necessitist way in which contingentists are actually most tempted to speak, they do very little to make the connection.

The gap in Fritz and Goodman's argument is not only at the level of model theory and logic. For suppose that contingentists' 'superficially necessitist' non-literal use of quantifiers involved all sorts of non-logical deviations from their literal use, for example, in what material objects there were supposed to be. Then their definability and undefinability results might again be beside the point.

My postulation in the book of a neutral zone between the necessitist and the contingentist was a way of controlling for such problems. Although Fritz and Goodman may seem to achieve greater generality by doing without one, what is really going on is a failure to articulate the conditions for the appropriateness of the modelling choice on which their argument depends. Their arguments look tidier than mine because they have swept more of the messy non-formal issues under the carpet.

Goodman is especially hostile to my setting up of the argument in terms of communication between a necessitist and a contingentist. He sees himself as posing a much more urgent challenge to contingentists by asking how they justify to themselves their habit of speaking in at least superficially necessitist terms. It becomes their problem when internalized to them. But this difference between his way of setting up the issues and mine may be more presentational than substantive. Many deep philosophical debates can be conducted either out loud between two single-minded opponents or silently in the divided mind of one philosopher. In epistemology, for instance, we need no real-life committed sceptic to debate with, because we can feel the pull of sceptical arguments within ourselves. Similarly, it makes little difference whether the 'superficially necessitist' speeches for which contingentists feel some sympathy are ones they are tempted to make themselves or ones they hear necessitists make. The challenge to explain the apparent value of such speeches is the same. It is not as though the implicit non-literal structure of such speeches becomes transparent to one as soon as they are made within the privacy of one's own head. Indeed, the apparatus of neutrality is a handy way of tracking what is *not* at stake when contingentists wrestle with their own inner temptations.

Goodman sometimes writes as though my argument assumed a demand on any theorist to find neutral equivalents for any of their opponents' utterances, but I agree with Goodman that such a demand would be quite unreasonable; I did not make it. The passage quoted above from p. 364 signals a tight restriction on the demand when I say 'the contingentist cannot plausibly dismiss [the necessitist's distinctions] like that, because they are too intimately related to distinctions the [contingentist] is committed to regarding as genuine'. Of course, the argument will not worry fanatical contingentists who feel no compunction in dismissing the necessitist's distinctions, but then Fritz and Goodman's argument will not worry fanatical contingentists who feel no temptation to speak in superficially necessitist ways. Perhaps such contingentists *ought* to be worried, but if so the reason has to do with the intimate relations between the distinctions they accept and those they reject. Given the restricted nature of my argument, it is unclear what problem Goodman's case in Section 2.3 of the slightly diverging necessitists Kit and Kit* and the slightly diverging contingentists Bob and Bob* pose for it.

Similar comments apply to Goodman's example of a necessitist's talk about his favourite set, with putative reference to a set that is not even possible by the contingentist's lights. Such talk goes beyond the dialectical set-up of the book, on the plausible assumption that it involves an ordinary atomic expression ('favourite') whose extension goes beyond the neutral zone. The contingentist is under no obligation to find a neutral equivalent for such talk in the necessitist's theory. For similar reasons, in Fritz and Goodman's framework, the contingentist is under no obligation to paraphrase such talk in terms that are not 'superficially necessitist'. In these respects, my argument and his are far more similar in dialectical situation than he suggests.

Goodman also has the more general concern that chapter 7's dialectical framing of the issues perversely makes the assessment of each side depend on what the *other* side says, through the challenge to find neutral equivalents within the other side's theory. But my argument aims in the first instance to compare against each other a pair of theories, one necessitist and one contingentist, but in all other respects as like each other as can be. Across such comparisons, neither side varies independently of the other. Thus, neither side drops out as somehow irrelevant to its own assessment. Of course, there are many versions of necessitism, and many versions of contingentism. Moreover, the best version of necessitism and the best version of contingentism may not form a neatly contrasted pair: the auxiliary assumptions most suited to necessitism may differ from the auxiliary assumptions most suited to contingentism. In the end, more holistic comparisons will be needed. Goodman's explorations of the space of contingentist options contribute to that task. But we cannot do everything at once. The more limited and schematic comparisons I undertook make a better starting point.

One schematic aspect of my treatment is the use of the predicate 'chunky' to demarcate the neutral ground. In order to achieve greater generality, I explicitly refrained from choosing between various readings, such as 'not contingently non-concrete', 'grounded in the concrete' and others (2013, 314). Of course, a problem then arises in comparing a version of necessitism that requires one reading of 'chunky' with a version of contingentism that requires a different reading, not to mention versions of necessitism and contingentism that cannot be formulated in terms of a 'chunky' predicate at all. One might be tempted to think that the predicate is more trouble than it is worth. Fritz and Goodman do without it. However, as I have already noted, that is because they prefer not to address the messy theoretical variations behind their summary references to 'paraphrase' and 'implicitly necessitist ways' of talking.

Goodman raises a number of other points in Sections 2 and 3 of his paper. Several concern combinations of first-order contingentism with higher-order necessitism, which I have said something about in Section 1 of this response. I will take up just two more of his points from that part of his paper.

The first point arises in Section 2.3, where Goodman lays down another challenge to my use of neutrality. He notes that my results establish that there is a sentence *A* of the relevant language not equivalent given CSN (the relevant necessitist theory) to any neutral sentence. He then writes:

> But this is not yet to establish that there is no sentence equivalent to *A* given CSN that is intuitively not at issue in Kit [the necessitist] and Bob [the contingentist]'s metaphysical dispute. That would require an argument that all such intuitively not-at-issue sentences are neutral in the technical sense.

Goodman then suggests that $\neg\exists x\, Dx$ ('There are no dragons') is a counterexample. He is right that it is intuitively not at issue, because Kit and Bob agree on it for reasons unconnected with modal metaphysics, and that it is not neutral in the technical sense, because it is not *logically* equivalent to a sentence appropriately restricted by the monadic predicate *C* ('... is chunky'). However, Goodman is assuming that to get from (1) to (3) I need the auxiliary premise (2) (where 'neutral' is understood in the technical sense and 'not-at-issue' is understood intuitively):

(1) No neutral sentence is equivalent given CSN to *A*.
(2) Every not-at-issue sentence is neutral.

(3) No not-at-issue sentence is equivalent given CSN to *A*.

But Goodman's assumption that my argument requires (2) is false. I can get from (1) to (3) by the weaker auxiliary premise (2*):

(2*) Every not-at-issue sentence is equivalent given CSN to a neutral sentence.

For suppose that (1) and (2*) hold but (3) fails. Then some not-at-issue sentence *B* is equivalent given CSN to *A*. By (2*), *B* is equivalent given CSN to a neutral sentence *B**. But equivalence given CSN is symmetric and transitive, so *B** is equivalent given CSN to *A*. Since *B** is neutral, that contradicts (1). Thus (1) and (2*) suffice for (3). Moreover, although $\neg\exists x\, Dx$ is a counterexample to (2), it is not a counterexample to (2*), because $\neg\exists x\, Dx$ is equivalent given CSN to the neutral sentence $\neg\exists x\, (Cx \,\&\, Dx)$ ('There are no chunky dragons'). For CSN has the theorem $\forall x\, (Dx \rightarrow Cx)$ ('All dragons are chunky') if *D* is an ordinary atomic predicate, or even if it is replaced by a complex description of a dragon in partly positive terms. In line with its not-at-issue status, $\neg\exists x\, Dx$ is also equivalent to the neutral sentence $\neg\exists x\, (Cx \,\&\, Dx)$ given the relevant contingentist theory CSC, which also has the theorem $\forall x\, (Dx \rightarrow Cx)$. Goodman's sentence $\neg\exists x\, Dx$ presents no difficulty for my argument.

The other point arises at the end of Section 3 of Goodman's paper, where he discusses a technical contrast between my approach in chapter 7 and Fritz and Goodman's approach: I use plural quantifiers while they use generalized singular quantifiers. The connections they establish linking the necessitism–contingentism issue with the theory of generalized quantifiers are clearly illuminating. In particular, Goodman revisits my argument in Section 6.4 of the book that a second-order contingentist comprehension principle is too weak for the purposes of what I call 'second-order modal mathematics' because it does not permit the modalized version of the mathematical notion of a complete order to work as it should (2013, 285–288). He proposes that the example is best reinterpreted as another case of inexpressibility, like those in chapter 7 and in Fritz and Goodman's work: necessitist claims about modalized complete orders 'make perfect sense' but no formula within the envisaged contingentist framework can 'express the intended condition on models'. It is interesting that the example instantiates that phenomenon. But that is no reason to lose sight of the point I originally made with it. That point can be understood in terms of the unpacking of the implicitly modal content of ordinary scientific applications of non-modal mathematics to phase spaces, illustrated by the use of dynamical systems as models for a formal language with modal and temporal operators and quantification into sentence position (Williamson 2016a). The same method can easily be applied to phase spaces that permit other forms of quantification, such as quantification over individuals and their properties, as in the definition of a complete order. Ordinary non-modal mathematical reasoning implicitly relies on unrestricted quantification over all subsets of a given set domain, which turns out to validate unrestricted higher-order modal comprehension principles once the implicitly modal content of the applications to phase space is unpacked. Thus scientists' reliance on the free use of ordinary non-modal mathematical reasoning in applications to phase spaces implicitly relies on comprehension principles stronger than those higher-order contingentism permits. That is a point about deductive strength, not expressive power.

3. Intensionality

One controversial feature of the higher-order modal logic in the book is that it is intensional, rather than hyperintensional. It treats necessary coextensiveness as the analogue of identity for higher types. Goodman prefers a mildly hyperintensional higher-order modal logic, but his main concern in Section 4.1 of his paper is methodological. Why don't I treat intensionality as a contentious assumption in need of sustained theoretical defence?

Goodman is right that my attitude to the choice of a logic is not pragmatic. The choice raises serious questions of truth and falsity, just like theory choice in any other science. In particular, I do not regard intensionality as an artefact of a convenient restriction of the higher-order quantifiers. For example, the idea that the sentence A & ($A \vee B$) expresses a different state of affairs from the sentence A, because they are about different subject matters, strikes me as an obviously wrong-headed projection of differences in modes of representation onto what is represented. However, that does not mean that I oppose the development of hyperintensional metaphysical theories, including hyperintensional higher-order modal logics with an objective reading of the modal operators. Rather, developing such theories is a good way to test the hypothesis of intensionality. In a more partisan spirit, I anticipate the pleasures of watching hyperintensional theories getting entangled in their own ramifications, or bogged down in their own muddy distinctions.

However, one could be much more sympathetic to hyperintensionality in metaphysics than I am and still think that, methodologically, intensionality is a better place to start. For we are in the early days of higher-order modal logic as a framework for metaphysics, and loading ourselves with the massive extra complexity of a hyperintensional theory may well drastically slow progress. Even if, strictly speaking, hyperintensional logics are more accurate than intensional logics, we may do better by working with a less accurate but tractable theory than with a more accurate but intractable one. That attitude corresponds to the spirit of model-building in science (compare Williamson 2016b). Some philosophers like to load all the complications they can think of into a theory from the start, but that attitude is not conducive to progress. It is often easier to get results within a simple framework and then adapt them to a more complex framework than it is to achieve the same result working from the start in the complex framework. The results of higher-order intensional logic will not all be wasted if we later move to higher-order hyperintensional logic.

A second point is that we can simulate hyperintensionality within an intensional framework, just as we can simulate intensionality within the extensional framework of Kripke models. Carnap did the former with his definition of intensional isomorphism, which enabled him to define a hyperintensional criterion of linguistic synonymy within his intensional framework (1947, 56–64). For instance, the sentences A & ($A \vee B$) and A are non-synonymous in the sense of not being intensionally isomorphic, because they differ in semantic structure, even though they have the same intension. Thus, we can still get the effect of hyperintensional distinctions even though they do not enter the theory at ground level.

A third point is that in constructing formal languages, we have some scope for making the symbols mean what we want them to mean. Thus, different theorists may use what looks like the same formula in different senses. Consequently, not every apparent disagreement is a real disagreement. Although that is not what is going on with the necessitism–contingentism dispute, which is deeply rooted in a shared natural language, it may be a live possibility with respect to some more artificial features of formal languages for higher-order modal logic. Goodman rightly points out that I deny that apparent disagreements over higher-order matters can be resolved by saying that the two sides are talking about different abstract objects, since I deny that higher-order quantification involves tacit quantification over objects of any kind, but it is not out of the question that some higher-order analogue of that picture fits some apparent disagreements. Such equivocation is not excluded from physics, so it can hardly be excluded from metaphysics, even on the uncompromisingly realist view of metaphysics to which I am sympathetic.

For all those reasons, I regard the intensional framework of the book as an appropriate starting point for the metaphysics of higher-order modal logic, on a metaphysical reading of the modal operators.

In the last section of his paper, Goodman gives his own preferred argument for necessitism, involving a characterization of the first-order universal quantifier in terms of not too hyperintensional, not specifically modal higher-order logic. There are many roads to the truth.

Note

1. Of course, the correct statement of the higher-order essentialist claims involves quantification into predicate position; for corresponding essentialist claims involving quantification into sentence position, see the discussion of the modal logic of dynamical systems in Williamson (2016a). Incidentally, for present purposes we must understand anti-essentialism as denying the relevant intensional claims that something *necessarily* has the given property, not just the stronger corresponding hyperintensional claims that it *essentially* has the property, since the latter denials without the former are too weak to engage the floating free argument, which targets modal differences.

References

Carnap, Rudolf. 1947. *Meaning and Necessity: A Study in Semantics and Modal Logic*. Chicago, IL: University of Chicago Press.
Fritz, Peter, and Jeremy Goodman. Forthcoming. "Counting Incompossibles." *Mind*.
Goodman, Jeremy. 2016. "Williamson on Necessitism." *Canadian Journal of Philosophy* 46 (4–5): 614–640.
Stalnaker, Robert. 2012. *Mere Possibilities: Metaphysical Foundations of Modal Semantics*. Princeton, NJ: Princeton University Press.
Williamson, Timothy. 2013. *Modal Logic as Metaphysics*. Oxford: Oxford University Press.
Williamson, Timothy. 2016a. "Modal Science." *Canadian Journal of Philosophy* 46 (4–5): 453–492.
Williamson, Timothy. 2016b. "Model-building in Philosophy." In *Philosophy's Future: The Problem of Philosophical Progress*, edited by Russell Blackford and Damien Broderick. Oxford: Wiley.

Plurals and modals

Øystein Linnebo

Department of Philosophy IFIKK, University of Oslo, Oslo, Norway

ABSTRACT
Consider one of several things. Is the one thing *necessarily* one of the several? This key question in the modal logic of plurals is clarified. Some defenses of an affirmative answer are developed and compared. Various remarks are made about the broader philosophical significance of the question.

1. Introduction

Consider some things, and choose any one of them. Is the chosen thing *necessarily* one of the things from which it was chosen? It is frequently assumed that it is (at least on the assumption that all of the things in question still exist).[1] If some things did not include our chosen thing, then these things would simply not be the things with which we started, that is, the things from which we made a choice. If the things from which we chose exist at all, then *necessarily* (subject to the mentioned existential assumption) they include the chosen thing. Likewise, if some other thing is *not* one of things from which we chose, then this too is a matter of (conditional) necessity. These two frequently assumed thoughts can be formalized as follows:[2]

$$x \prec yy \rightarrow \Box(Ex \wedge Eyy \rightarrow x \prec yy) \qquad (\mathrm{Rgd}^+)$$

$$x \not\prec yy \rightarrow \Box(Ex \wedge Eyy \rightarrow x \not\prec yy) \qquad (\mathrm{Rgd}^-)$$

Let their conjunction be the claim that pluralities are *rigid*, which we abbreviate as (Rgd).[3]

The rigidity principles can be regarded as strong extensionality principles, which state that pluralities are tracked extensionally across possible worlds. The ordinary principle of extensionality states that if every one of *these things* is one of *those things* and vice versa, then *these things* are the very same things as *those things*. (We shall return to the question of how best to formalize this principle.) This is widely assumed to provide a criterion of identity for pluralities; and like criteria of identity in general, the principle is widely thought to hold of necessity. Even so, the principle provides no information about how pluralities are tracked across possible worlds. The rigidity principles fill this gap. They tell us that necessarily any given things have their members by necessity. A plurality is therefore not allowed to vary in its membership across possible worlds. Any variation in membership would result in our talking about some other things, not the things in question.

It is important not to misunderstand the rigidity claims. Assume Sophia is one of the philosophers. Does it follow that she is *necessarily* one of the philosophers? (For simplicity, I suppress the assumption that the objects in question still exist.) If so, this would imply the wildly implausible claim that necessarily Sophia is a philosopher. Fortunately, Kripke long ago taught us how to respond. Let *pp* be the things such that anything is one of them if and only it is in fact a philosopher. What is necessarily the case is that Sophia is one of *pp*. But it is not necessary that *pp* are all and only the philosophers. Sophia might have become a psychologist, not a philosopher. Then, she would not have been included in the ranks of the philosophers, although she would still have been one of *pp*.

To appreciate how the rigidity claims for pluralities have real bite, even after this pre-Kripkean confusion is cleared up, it is useful to contrast pluralities with what I shall call *groups*, such as teams, clubs, committees, and the like.[4] Consider a philosophy department. The department might have had other members than it in fact has: Sophus might have been hired instead of Sophia. Someone who *is* a member of the department might not have been so; and someone who *is not* a member, might nevertheless have been one. To ensure ourselves that there is a genuine difference between the modal profile of pluralities and groups, not merely the result of some undetected use of non-rigid designators, we can run the same argument using a proper name or a variable to pick out Sophia's department. A similar contrast obtains between plurals and predicates, which are clearly not rigid, as is illustrated by the following examples:

(1) Timothy Williamson is a philosopher but he might not have been one.
(2) Bill Clinton is not a philosopher but he might have been one.

In the light of the existence of non-rigid groups, we may want to reconsider the received view that pluralities are rigid. Indeed, some apparent counterexamples to the rigidity of pluralities are found in natural language. A nice example is the following ad which I once saw for a gym:

(3) Join, and become one of us!

The plural pronoun 'us' is naturally taken to stand for a plurality. But when so interpreted, the message presupposes that it is possible to become a member of a plurality of which one is not already a member.[5]

How are we to respond to these apparent counterexamples to the rigidity of pluralities? Interesting though they are, the examples are inconclusive. Plural pronouns ordinarily stand for pluralities. But the apparent counterexamples can be explained away if we allow that a plural pronoun can sometimes stand for a group or function as a covert description. Either way, non-rigid would be unsurprisingly. Consider, for instance, a bohemian parent who upon seeing some particularly smug business school students tells her daughter:

(4) I'm glad you're not one of them.

It is natural to understand the parent as expressing joy that her offspring is not (in some salient respect) *like* the students in question rather than pleasure with a fact about plural non-membership. Thus, (4) poses no more of a challenge to the rigidity of pluralities than the following sentence poses to the necessity of identity:

(5) I'm glad you're not him.

To make progress, we need more systematic considerations for or against the rigidity of pluralities.

The basic thought on which I shall base my defense of plural rigidity is that a plurality is nothing over and above its members. More precisely, a plurality is fully specified when we have circumscribed the things that are its members. By contrast, a group is something over and above its members, as it can have a non-trivial membership criterion, which allows a non-trivial tracking of the group across possible worlds. A group is therefore not fully specified merely by circumscribing its members; we additionally need to specify (at the very least) its membership criterion. Since a plurality is nothing over and above its members, there is no material available that might underwrite a non-trivial tracking across possible worlds. All we have to go on are the members. So the only way to track a plurality is the trivial one, which ensures plural rigidity. On this picture, pluralities manifest extensionality in its purest and strongest form.[6]

In what follows, I shall attempt to clarify and develop these ideas. I shall be particularly interested in the extent to which broadly logic considerations support plural rigidity. The result will be a disentangling and clarification of several strains of the basic thought. We shall find that plural rigidity figures at the heart of a network of ideas having to do with the extensionality of pluralities. Since the ideas in this network are true of core uses of our plural resources in ordinary language and thought, I commend them as an explication of these resources. However, because I accept the existence of non-rigid groups, it would not trouble me particularly if my view had to be regarded as partially a matter of stipulation. Should it turn out that non-rigid groups can sometimes serve as an interpretation of plurals, I would restrict myself to 'strict pluralities,' which, I would claim, are correctly described by the mentioned network of interlocking ideas.

2. Why plural rigidity matters

The question whether pluralities are rigid might seem abstruse and of limited interest. In fact, the question has vast repercussions for a number of debates in philosophical logic, metaphysics, and the philosophy of mathematics.

One example is the debate about what forms of higher order logic there are. Can plural logic be replaced by monadic second-order logic or even reduced to it? Or is some reduction in the opposite direction possible? If pluralities are rigid, then the two forms of logic have different modal profiles: as we have seen, predicates are non-rigid.[7] This makes either kind of reduction implausible and suggests instead that the two kinds of logic should be taken at face value and allowed to co-exist without one swallowing the other.

A related example concerns the semantics of predication. Provided we have ordered pairs at our disposal, it is technically possible to use plurals for this purpose: we can take the semantic value of a predicate to be the plurality of tuples of which the predicate is true. However, this semantics seems contrived. In addition to the lack of homophonicity even on the intended interpretation, the need for *ad hoc* tricks to handle predicates that are true of nothing, and the artificial reliance on tuples, there is the aforementioned mismatch of modal profiles.[8]

Next, the rigidity of pluralities is a pillar of one of Williamson's main arguments for *necessitism*, the metaphysical view that necessarily everything necessarily exists.[9] The denial of this view is *contingentism*. When we go on to consider arguments for the rigidity of plurals, it will be important to keep in mind whether the argument is intended to be given in a necessitist setting (which is always easier) or a contingentist setting (which requires greater care).

Finally, the question of the rigidity of pluralities plays an essential role in an approach to mathematics and the phenomenon of indefinite extensibility that I have developed in recent work and canvass in the final section of this article.[10]

3. A Kripkean plausibility argument for the rigidity of sets

It will be useful to begin our investigation of the rigidity of pluralities by reminding ourselves of Kripke's celebrated argument for the necessity of identity and distinctness, and by observing some striking consequences that this argument has for the metaphysics of sets. Kripke's argument turns on Leibniz's law:

$$x = y \rightarrow (\phi(x) \leftrightarrow \phi(y)) \qquad \text{(Leibniz)}$$

Given the assumption $\Box(x = x)$, the law entails $x = y \rightarrow \Box(x = y)$. Moreover, given the Brouwerian axiom

$$\phi \rightarrow \Box\Diamond\phi \tag{B}$$

we can now also derive the necessity of distinctness: $x \neq y \rightarrow \Box(x \neq y)$.[11]

However, a contingentist may object to the assumption of $\Box(x = x)$. After all, in a negative free logic, '$x = x$' can be used as an existence predicate. If so, then what is assumed is the necessary existence of x! The problem is easily circumvented. The contingentist will have no problem with the assumption that x satisfies the predicate '$\Box(x = x \rightarrow x = \ldots)$'. Applying (Leibniz), this enables us to derive formulations of the necessity of identity and distinctness that are acceptable to the contingentist:[12]

$$x = y \rightarrow \Box(x = x \rightarrow x = y) \qquad (\Box =)$$

$$x \neq y \rightarrow \Box(x \neq y) \qquad (\Box \neq)$$

As before, the derivation of the latter from the former relies on (B).

As Kripke realized, Leibniz's law has important metaphysical consequences. The case of sets provides a nice illustration. Consider the principle of extensionality:

$$\forall u(u \in x \leftrightarrow u \in y) \leftrightarrow x = y \qquad \text{(Ext)}$$

Leibniz's law reveals a respect in which this is quite a strong principle. Let x and y be coextensional sets. By (Ext), x and y are identical. Observe now that x satisfies the predicate

$$\Box\forall u(u \in \ldots \leftrightarrow u \in x).$$

So by Leibniz's law, y too satisfies this predicate. Thus, the principle of extensionality logically entails that two coextensional sets are necessarily coextensional.[13]

This is an important consequence. Assume that sets are like groups in that they are tracked across possible worlds in some intensional way. It would then be a complete mystery why actual coextensionality should suffice for identity and therefore also for necessary coextensionality. Consider the analogous claim concerning groups: If two groups in fact have the same members, they are identical and therefore necessarily have the same members. Since membership in a group is contingent and thus is subject to 'drift' as we consider alternative possible worlds, this claim is wildly implausible. Once membership drift is permitted, there is no guarantee that groups whose members in fact coincide will not drift apart in some other possible world. Yet in the case of sets, the principle of extensionality and the necessity of identity entail that there can be no such drifting apart but that any drift would have to be parallel drift. The only explanation of this prohibition on drifting apart is that there can be no drift in the membership of a set at all; in other words, that sets are rigid.

Although the argument is not deductively valid, it is hard to resist. The only reason to accept the principle of extensionality – and the important consequence noted above – is that a set, unlike a group, is fully specified by its elements. Thus, when tracking a set across possible worlds, there is nothing other than the elements to go on. This ensures that the tracking is rigid.[14] By contrast, when tracking a group, there *is* more than the members to go on. But precisely because there is more to go on, we cannot accept a principle of extensionality for groups. Having the same members does nothing to ensure that two groups coincide in whatever additional factor it is that enables the non-rigid tracking. These considerations give rise to a dilemma that applies not only to sets but to any other notion of collection: either we have to give up the principle of extensionality, or else we have to accept the rigidity principles as well.[15] There is no stable middle ground. Kripke famously taught us that there can be no 'soft identity theory' in the philosophy of mind, according to which mental states are identical with physical ones but only contingently so, only the 'hard identity theory' committed to necessary identity. Our present conclusion is analogous. There can be no 'soft extensionalism' concerning sets or other kinds of collection, only 'hard extensionalism' that incorporates the rigidity claims and the idea of transworld extensionality that they embody.[16]

I shall now consider two objections. The first one is based on a mereological analogue of our plausibility argument concerning sets. Assume that x and y share all their parts; that is $\forall z(z \leq x \leftrightarrow z \leq y)$, where $\leq$ indicates parthood. Provided that parthood is reflexive and anti-symmetric, it follows that x and y are identical and thus also that they necessarily share all their parts. Yet this seems compatible with parthood being non-rigid! Does this undermine my plausibility argument concerning sets?[17] I think not. The crux of my plausibility argument is the claim that any reason to accept the principle of extensionality is also a reason to accept rigidity of elementhood. By contrast, there is reason to accept the two mereological principles which is *not* also a reason to accept rigidity of parthood. I shall have to content myself with conveying the intuitive idea. In order to make sense of contingent parthood, it is useful to think of objects as involving both matter and form.[18] For instance, a molecule that is part of me might not have been so because my form permits me to be tracked to other possible worlds on the basis of something other than just my matter. On this hylomorphic conception, it is natural to take parthood to be sensitive to both matter and form, and mutual parthood, to ensure identity not only of matter but also of form – and hence identity of the objects in question. This points to an explanation of the two mereological principle that is fully compatible with non-rigid parthood.[19]

The second objection takes its departure from the well-known fact that Leibniz's law needs to be restricted. Assume Nikita is the shortest spy. Of course, necessarily the shortest spy is the shortest spy. But it does not follow that necessarily Nikita is the shortest spy. It is often proposed that the Leibniz's law be restricted to *rigid designators* – defined as terms that refer to the same object at every world at which they refer at all – thus excluding terms like 'the shortest spy.' Ordinarily, this restriction works well. But when reasoning about sets or other kinds of collection, it threatens to undermine our Kripkean dilemma. To understand this threat, we need to distinguish between two completely different notions of rigidity. Until this paragraph, we have been concerned exclusively with a metaphysical notion of rigidity. Sets and other kinds of collection are said to be rigid if their membership is a matter of necessity, in the precise sense laid down by the kind of rigidity claims stated above. But as we have just seen, there is also the semantic notion of a rigid designator.

The problem is that it can be hard to disentangle the two kinds of rigidity. Assume that a term t refers at w_1 to a collection comprising a and b, where a and b are all and only the Fs at w_1. Assume that t refers at w_2 to the singleton collection of a, where a is the one and only F at w_2. Is t a rigid designator? Obviously, the question cannot be answered until we have been told how to track the relevant kind of collection from world to world. If the collections are tracked extensionally, then we are considering different collections, and accordingly the designator is non-rigid. But if the collections are tracked intensionally in terms of their membership criterion, then we may well be considering one and the same object, namely the collection of Fs, in which case the designator is rigid after all. The threat to our Kripkean dilemma is now apparent. To show that our use of Leibniz's law is permissible, we must first show that the terms in question are rigid designators. This involves showing that they refer to the same set across possible worlds. But this presupposes that we already know how to track sets across possible worlds! As we have seen, this is a matter of answering the question of metaphysical rigidity. Our Kripkean argument therefore appears powerless to answer the question of metaphysical rigidity as the permissibility of its appeal to Leibniz's law presupposes that an answer has already been given.

Fortunately, the threat can be avoided by reformulating the restriction on Leibniz's law. Say that a term is *purely referential* if its semantic contribution to linguistic contexts in which it occurs is exhausted by its referent; or, as Quine (1960) put, if it 'is used purely to specify its object, for the rest of the sentence to say something about' (177). Rather than restrict Leibniz's law to rigid designators, we can restrict the law to terms that are purely referential. After all, such terms serve merely to stand for, or pick out, their referents. Assume $t_1 = t_2$ is a true identity involving such terms. Then, of course, $\phi(t_1) \leftrightarrow \phi(t_2)$ is true as well, as this merely says of the referent that it is ϕ iff it is ϕ. When Leibniz's law is restricted to purely referential terms rather than to rigid designators, the problem we have discussed dissipates. The only terms involved in our argument are variables. And variables are purely referential: what a variable contributes to a linguistic context in which it occurs is nothing but its value. Thus, our Kripke-inspired argument for metaphysical rigidity goes through after all.

There is a more general lesson here as well. The problem of disentangling the semantic and metaphysical kinds of rigidity points to an unfortunate feature of the notion of a rigid designator: it runs together two kinds of consideration that are best kept apart. First, there is the semantic question of whether a term is purely referential. Then, there is the metaphysical question of how its referent is to be tracked from one possible world to another. It is true that every purely referential term is a rigid designator. But our discussion shows that we get a cleaner separation of the metaphysical and semantic questions by focusing on the notion of pure reference rather than rigid designation.[20]

4. A plausibility argument for plural rigidity

I would now like to extend the argument from the previous section to the case of pluralities. Kripke started with Leibniz's law. Is there an analogue of the law in the case of pluralities? The answer will depend on whether the relation of identity is defined between pluralities. It is far from obvious that it is. My proposal is that our extension of the Kripkean argument should use as its starting point the following indiscernibility principle for pluralities:

$$xx \equiv yy \rightarrow (\phi(xx) \leftrightarrow \phi(yy)) \qquad \text{(Indisc)}$$

where $xx \equiv yy$ abbreviates $\forall u(u \prec xx \leftrightarrow u \prec yy)$. If identity is defined on pluralities, then (Indisc) is merely the result of 'telescoping' the principle of extensionality for pluralities (which can then straightforwardly be expressed) and the plural analogue of Leibniz's law. And even if identity is not defined on pluralities, this 'telescoping' is still expressible. Either way, (Indisc) incorporates a plural analogue of Leibniz's law.

Two concerns arise. Firstly, as we have seen, the ordinary singular version of Leibniz's law needs to be restricted. Analogous considerations apply in the plural case. Fortunately, it is easy to see that (Indisc) is suitably restricted. Since plural variables are purely referential just as much as singular ones are (only in a plural way), (Indisc) is entirely legitimate. In particular, it presupposes no prior answer to the question of the rigidity of pluralities and can thus safely be employed in an argument for this rigidity thesis.

Secondly, is (Indisc) acceptable from a contingentist point of view? To assess this, we need to be more explicit about what semantics we adopt. It is natural to use a semantics on which $x \prec xx$ is true at a world w relative to an assignment σ iff all the things that σ assigns to xx exist at w, and $\sigma(x)$ is one of them.

Furthermore, this semantics makes it natural to adopt a negative free logic.[21] The rules of universal elimination must then be formulated so as to make existential assumptions explicit; for instance, from $\forall x\, \phi(x)$, we can infer $Et \rightarrow \phi(t)$, and likewise for the universal plural quantifier. (We shall shortly have more to say about the plural existence predicate.) Given these choices, it is easy to verify that (Indisc) remains a valid principle even in a contingentist setting.[22]

We are now ready to develop our plausibility argument for the rigidity of pluralities. The next step is to derive from (Indisc) an analogue of the necessity of identity, which for obvious reasons I call *covariation*:

$$xx \equiv yy \rightarrow \Box(xx \equiv yy) \qquad \text{(Cov)}$$

Given (B), we can derive the necessity of $\not\equiv$ as well.

We now come to the heart of the argument. Recall the case of sets. While ($\Box =$) is formally compatible with the non-rigidity of sets, it is far more plausible with rigidity. Precisely the same goes for (Cov) and pluralities. If plural membership was contingent – like membership in a group – why should actual coextensionality ensure necessary coextensionality? There would be nothing to prevent two actually coinciding pluralities from 'drifting apart' as we consider alternative possible worlds. The only explanation of the prohibition on such divergence is that plural membership is not subject to membership drift at all. So we establish the same dilemma as in the case of sets. There can be no 'soft extensionalism' concerning pluralities: either we need to give up the ordinary extensionality principle encapsulated in (Indisc), or else we have to accept the full transworld extensionality associated with the plural rigidity principles. Just as in the case of sets, the former horn is deeply unattractive, as it comes close to just changing the subject. So we conclude that plural rigidity is highly plausible.

It is worth noting that the real locomotive of the argument is (Cov), which is strictly weaker than (Indisc) with which the argument officially began. The covariance principle gives us precisely what its name suggests, namely that two overlapping pluralities necessarily covary in their membership. (Indisc) states that all properties of pluralities supervene on membership. To see that the latter principle goes beyond the former, consider a department whose statutes decree that all and only tenured faculty are to be members of the Hiring Committee and of the Graduate Admissions Committee.[23] Then, the two committees necessarily covary in membership. Nevertheless, the two committees have different powers, namely to hire new faculty and admit graduate students, respectively.

To be even more specific about the relation between (Indisc) and (Cov), I claim that the former 'factorizes' into the latter and the claim that the properties of a plurality supervene on what we may call its *modal membership profile*:

$$\Box(xx \equiv yy) \rightarrow (\phi(xx) \leftrightarrow \phi(yy)) \qquad \text{(Sup)}$$

We see this as follows. Clearly, (Cov) and (Sup) entail (Indisc), which in turn entails each of the former two principles. Moreover, (Cov) and (Sup) are logically independent and encapsulate different philosophical ideas, namely covariation in membership and supervenience of properties on modal membership profile, respectively. A more comprehensive 'factorization' of the ideas associated with the extensionality of pluralities will be offered in Section 6.

5. Formal arguments for plural rigidity

We now have what I regard as a fairly convincing plausibility argument for plural rigidity. But as far as formal arguments are concerned, a gap remains. Our best formal result so far is (Cov), which states that coextensive pluralities are necessarily coextensive. The desired rigidity claims state that a plurality has the same members at any world at which it exists. I would now like to investigate what it takes to formally bridge this remaining gap.

5.1. *Plural existence*

Since we are now aiming for formal rigor, the time has come to be entirely precise about the existential assumptions that are involved in the rigidity claims. This requires a plural existence predicate which we can use to say of some things *xx* that they exist. As we have seen, the existence of a single object *x* can be expressed simply as $x = x$. But what about the plural existence predicate?

One option is to define plural existence distributively in terms of singular existence; that is, to define *Exx* as $\forall x(x \prec xx \rightarrow x = x)$. But this is unsuccessful. For a contingentist, the initial quantifier ranges only over objects that exist at the relevant world, which renders the quantified claim trivially true for any plurality *xx* whatsoever. Another natural but unsuccessful option is to define *Exx*, by direct analogy with its singular cousin, as $xx \equiv xx$. This too is easily seen to trivialize, for exactly the same reason as our previous attempt.

One safe option is simply to adopt a primitive collective plural existence predicate *Exx*, which we stipulate to be satisfied by some things at a world just in case all these things exist at the world. Another option is available as well, provided we lay down the plausible and widely adopted axiom that every plurality is non-empty:

$$\forall xx \exists y(y \prec xx) \qquad \text{(NE)}$$

We can then define *Exx* as $\exists y(y \prec xx)$. Given (NE) and our semantics, this definition is easily seen to work. We adopt the latter option as it is more economical.

In the ensuing discussion, we also need a principle to the effect that any plurality is ontologically dependent on each of its members:

$$x \prec xx \rightarrow \Box(Exx \rightarrow Ex) \qquad \text{(Dep)}$$

So we adopt this as an axiom.[24]

We shall now consider some formal arguments for plural rigidity. Each argument will first be developed from a necessitist point of view, as this is simpler. We shall then use our plural existence predicate to reformulate the argument so as to work in a contingentist setting.

5.2. The argument from uniform adjunction

The first argument relies on an operation $+$ of adjoining one object to a plurality. It is reasonable to assume that, necessarily, to be one of these things and that thing is to be one of these things or to be identical with that thing. We call this principle uniform adjunction:

$$\Box\forall x(x \prec xx + a \leftrightarrow x \prec xx \vee x = a) \qquad \text{(UniAdj)}$$

We now argue as follows. Assume $a \prec xx$. Then, by (UniAdj), we have $xx \equiv xx + a$. So by (Cov), we have

$$\Box(xx \equiv xx + a). \qquad (1)$$

Next, we observe that (UNIADJ) also entails

$$\Box(a \prec xx + a). \qquad (2)$$

From (1) and (2), some simple modal logic ensures our desired conclusion that $\Box(a \prec xx)$.

Gabriel Uzquiano has raised a legitimate concern about the argument.[25] Is it permissible to assume that '$xx + a$' is a rigid designator? It can certainly be disputed that the term is purely referential. Fortunately, there is an answer, which forces us to make explicit an assumption that has so far only been implicit. Our argument assumes that uniform adjunctions exist. We can express this as the closure of the following plural comprehension principle, which asserts the existence of uniform adjunctions:

$$\exists yy\Box\forall u(u \prec yy \leftrightarrow u \prec xx \vee u = z) \qquad (3)$$

This principle is very weak. Indeed, it is something that even an opponent of plural rigidity should assent to, as the principle retains its plausibility even when the plural variables are allowed to range over groups.[26]

As before, assume $a \prec xx$. By (3), let yy be the uniform adjunction of a to xx. From this point onward, the argument proceeds exactly as before, only with yy in the role previously played by $xx + a$. Notice that this argument makes no appeal to (Indisc) other than its single instance, (Cov). In this respect, the argument is like the plausibility argument from Section 4.

Let us now try to adapt the argument to a contingentist setting. Then, uniform adjunction requires the following, more guarded formulation:

$$\Box\big(Exx \wedge Ea \rightarrow \forall x(x \prec xx + a \leftrightarrow x \prec xx \vee x = a)\big) \qquad \text{(UniAdj-c)}$$

Our comprehension principle too must make existential assumptions explicit:

$$\exists yy\Box\big(Exx \wedge Ez \rightarrow \forall u(u \prec yy \leftrightarrow u \prec xx \vee u = z)\big) \qquad (5)$$

Thankfully, it is straightforward to verify that our first target, (Rgd^{+}), follows, and using the Brouwerian axiom B, so does (Rgd^{-}). The argument relies on our dependence axiom, (Dep).

5.3. *The argument from partial rigidification*

Another argument is proposed in Williamson (2010, 699–700). The argument requires that, for any plurality xx, there be a plurality yy that is a partial rigidification of xx in the sense that the two pluralities have the same members but it is impossible for yy to lose any of its members. To be precise, we assume the following plural comprehension axiom:

$$\exists yy(xx \equiv yy \wedge \forall x(x \prec yy \rightarrow \Box x \prec yy)) \qquad (6)$$

We can now argue as follows. Assume $a \prec xx$. Let yy be the partial rigidification of xx. Thus, we have $\Box a \prec yy$. By (Cov), we also have $\Box(xx \equiv yy)$. The latter two claims entail $\Box a \prec xx$, as desired.

Let us now try to develop the argument from a contingentist point of view. As usual, the comprehension axiom needs to be formulated with greater care:

$$\exists yy(xx \equiv yy \wedge \forall x(x \prec yy \rightarrow \Box(Eyy \rightarrow x \prec yy))) \qquad (7)$$

Applying the same strategy as before, we get $\Box(Eyy \rightarrow a \prec yy)$. We also have $\Box(xx \equiv yy)$. But now a problem arises. What we have proved does not entail our target claim $\Box(Exx \rightarrow a \prec xx)$ unless we can assume that the existence of xx ensures the existence of yy. But I submit that a theorist who doubts the rigidity of pluralities has reason to challenge this assumption. To see this, recall that such a theorist regards some pluralities as much like groups. For instance, xx might be the Hiring Committee, whose members happen to be a, b, and c. Then, the partial rigidification yy of xx is ontologically dependent on a, b, and c, whereas *xx* need not be subject to this ontological dependence. To clinch the argument, our theorist would have to add a third conjunct to the conjunction involved in (7), namely $Exx \rightarrow Eyy$. But this modification of (7) would be very strong and problematic from the point of view of anyone not antecedently committed to the rigidity of pluralities.

How does Williamson's argument compare with the one from uniform adjunction? Both rely on only a single instance of (Indisc), namely (Cov). But the arguments differ concerning the plural comprehension axioms that they assume: partial rigidification and uniform adjunction, respectively. From a necessitist point of view, both comprehension axioms are very plausible, and as far as I can see, plausible to roughly the same extent. From a contingentist point of view, on the other hand, the comprehension axiom that is required to complete Williamson's argument is significantly less plausible than that required by the argument from uniform adjunction. Thus, the contingentist has reason to favor the latter.

5.4. The argument from uniform traversability

The final argument for plural rigidity that we shall consider is due to Rumfitt (2005). As before, we first give a simple version of the argument that is acceptable from a necessitist point of view, and then consider how the argument can be adapted by a contingentist.

A finite plurality can be *traversed*, in the sense that its members can be exhaustively listed. Assume, for instance, that *xx* is the plurality whose members are *a*, *b*, and *c*. Then, we can also assert that this traversability is uniform, in the sense that it holds by necessity:[27]

$$\Box\forall x(x \prec xx \leftrightarrow x = a \vee x = b \vee x = c)$$

What about infinite pluralities? A straightforward generalization is available if we allow infinitary disjunctions and assume that every object a has a name $\bar{a}$:

$$\Box\forall x(x \prec xx \leftrightarrow \bigvee_{a \prec xx} x = \bar{a}) \qquad \text{(UniTrav)}$$

We now argue as follows. Assume $u \prec xx$. Then, we can find a such that $u = \bar{a}$. By the necessity of identity, we have $\Box(u = \bar{a})$. This entails the necessitation of $\bigvee_{a \prec xx} x = \bar{a}$. Some simple modal logic now ensures our target $\Box(u \prec xx)$.

Let us now consider matters from a contingentist point of view. (UniTrav) must then be reformulated so as to make all existential presuppositions explicit. We do this as follows. Given any *x*, we can name all of its members and use this to state that, provided that *xx* still exist, to be one of *xx* is just to be identical with one of the aforementioned members. In symbols:

$$\Box(Exx \rightarrow \forall x(x \prec xx \leftrightarrow \bigvee_{a \prec xx} x = \bar{a})) \qquad \text{(UniTrav-c)}$$

As far as I can see, this modified principle is as plausible, given contingentism, as the original principle is, given necessitism. It is therefore satisfying to be able to verify that the original argument for rigidity goes through much as before. (The argument relies on (Dep).)

How does this formal argument for plural rigidity compare with the previous two? In my view, the argument has limited explanatory value as its premise of universal traversability is little other than an infinitary restatement of our target claim that a plurality is fixed in its membership as we shift our attention from one possible world to another. In order to substantiate this view, I shall now show just how strong the mentioned premise is and how substantial the ideological resources used in the argument are.

The clearest way to appreciate the strength of universal traversability is by observing that it entails all the premises of the previous two arguments. For example, let us verify that its necessitist version, (UniTrav), entails covariation. So assume $xx \equiv yy$. Then, we can find a bunch of names $\bar{a}$ such that (UniTrav) holds, and another bunch $\bar{b}$ such that the analogous claim holds concerning *yy* and the $\bar{b}$'s. Each of the *a*'s is identical with one of the *b*'s and vice versa. Since each of the identities holds of necessity, so too does their conjunction. Thus, the two disjunctions $\bigvee_{a \prec xx} x = \bar{a}$ and $\bigvee_{b \prec yy} x = \bar{b}$ are necessarily coextensional. Thus, via (UniTrav) and its analogue concerned with *yy* and the *b*'s, so too are *xx* and *yy*.[28] But clearly, the converse entailment does not hold. For covariation is formally compatible with membership drift, so long as two coextensional pluralities are never allowed to drift apart. It can also be verified that uniform traversability entails the comprehension principles associated with uniform adjunction and partial rigidification.

As for ideology, the argument obviously relies on infinitary resources. It is useful to separate these resources clearly from the modal claims that they are used to express. We can do so by considering what we may call *traversability*, which is (UniTrav) without the initial necessity operator. Even this plain version of traversability has some strong consequences. It makes available what Bernays (1935) calls 'quasi-combinatorial' reasoning; that is, reasoning with infinite totalities as if they were finite and subject to combinatorial operations such as formation of arbitrary subsets and effective traversal. Let me mention two important examples.

Firstly, traversability ensures the permissibility of impredicative plural comprehension axioms of the following form:

$$\exists x(\phi(x) \wedge x \prec xx) \rightarrow \exists yy \forall u\big(u \prec yy \leftrightarrow \phi(u) \wedge u \prec xx\big) \qquad (8)$$

To see this, we first find a bunch of names $\bar{a}$ which provide a traversal of *xx*. We would like another bunch of names $\bar{b}$ which provide a traversal of just those members of xx that satisfy ϕ. This is easy: just go through the former bunch and delete every item that names a non-ϕ. The resulting sub-traversal yields a quantifier free – and thus fully predicative – definition of the desired plurality. The upshot is that traversability functions like a principle of reducibility, in the sense of Russell and Whitehead's famous axiom of reducibility. This principle of reducibility becomes particularly far-researching if there is an all-encompassing or universal plurality, as is standardly assumed. We then get a justification of the full impredicative comprehension scheme.

Secondly, when we work in an intuitionistic theory, traversability ensures the availability of classical quantification restricted to any plurality. Assume that a formula $\psi(x)$ (which may have further free variables) has the following decidability property:

$$\forall x(\psi(x) \vee \neg\psi(x))$$

In effect, this means that the property $\lambda x.\psi(x)$ behaves classically. Then, tranversability ensures that quantification restricted to xx behaves classically as well, in the precise sense that we have the following decidability property:

$$(\forall x \prec xx)\psi(x) \vee \neg(\forall x \prec xx)\psi(x)$$

To see this, observe that by traversability, this restricted quantification reduces to a conjunction of its instances, each of which has been assumed to behave classically.

All these considerations support my view that the argument from universal traversability has little suasive force. Its official premise is very strong and close to the desired target. In addition, the argument relies on very strong ideological resources. Despite these complaints, however, the argument serves to highlight a very important aspect of the extensionality of pluralities, namely their traversability

6. Three 'factors' of the extensionality of pluralities

Let us return to the basic thought from the introduction, namely that a plurality is 'nothing over and above' the circumscribed lot of objects that it comprises. The intervening discussion has disentangled several strains of this thought, whose implications and relations of non-deductive support have been explored and can, for a necessitist, be summarized as follows:

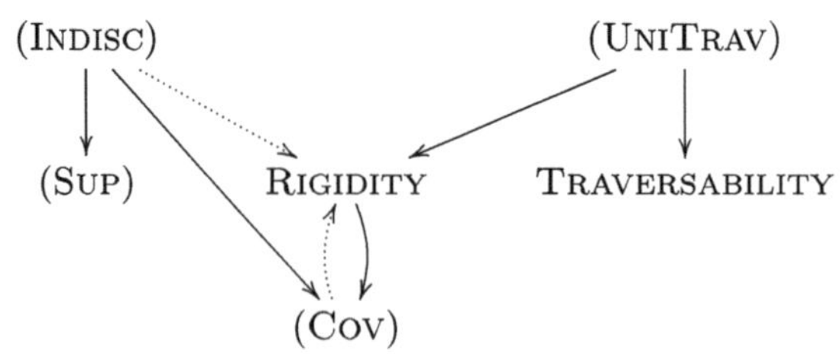

Solid arrows represent one-way implications. Dotted arrows represent non-deductive support, but which can be transformed into implications by adding suitable comprehension axioms, as discussed in Sections 5.2 and 5.3. Formal theses are in parentheses, as usual. Rigidity abbreviates the conjunction of the rigidity claims (Rgd^{+}) and (Rgd^{-}), and, in a contingentist setting, also the dependence claim (Dep). A contingentist can use the same diagram except that (UniTrav) must be replaced with (UniTrav-C), whose left diagonal implication then only yields the two rigidity claims, not (Dep).[29]

My next claim is that the items on the diagram's second floor yield a 'factorization' of all the strains of the extensionality of pluralities that are represented in the above diagram.[30] We begin by observing that each item represents a simple and natural idea.

(a) The properties of any plurality supervene on its modal membership profile, as expressed by (Sup).

(b) A plurality has a rigid membership profile: it has the very same members at any possible world at which it exists.

(c) A plurality is traversable, thus ensuring the permissibility of quasi-combinatorial reasoning applied to the plurality.

Next, we observe that the three 'factors' entail each of the strains of the extensionality of pluralities. It suffices to verify that the items on the top floor of the diagram are entailed by those on the second. For (Indisc), this is implicit in work already done (see the previous footnote). And it can be verified that (UniTrav) (or its contingentist cousin) is entailed by Rigidity and Traversability.

What remains is to verify that the three 'factors' are logically independent of one another. To see that property supervenience, (Sup), does not follow from the other two aspects of extensionality, consider again the case of committees. Imagine an oligarchic department where three senior academics *a*, *b*, and *c* have written into the department statutes that they, and they alone, are to be on the Hiring Committee and the Graduate Admissions Committee. Both committees have a rigid membership profile and are clearly traversable. Yet the two committees are not subject to property supervenience as different powers of decision are vested in them.

Next, to show that a rigid membership profile does not follow from the other two aspects, consider the case of properties, understood as objects that are individuated by the necessary coextensionality of their defining concepts or conditions, and tracked across possible worlds in terms of this concept or condition. Thus understood, properties exemplify the second aspect of extensionality: all the characteristics of any given property are shared by any necessarily coextensive property. However, properties can be subject to contingent membership (or, perhaps better, contingent application), including on domains that are traversable. And as we have seen, the traversability of a domain ensures the traversability of any property on this domain.[31]

Finally, we observe that traversability is not a formal consequence of the other two aspects of extensionality. The principles that explicate these other two aspects do not ensure the availability of the infinitary resources needed for traversability. As Bernays observed, traversability is based on an extrapolation from the finite into the infinite. How far are we willing to extrapolate? The first two aspects of the extensionality of pluralities do not, by themselves, provide any answer to this question.[32]

7. The status of plural comprehension

I wish to end with some remarks about the status of the plural comprehension axioms. Many philosophers appear to regard such axioms as utterly trivial and insubstantial.[33] Provided that a condition is well defined and has at least one instance, *of course* the condition can be used to define a plurality of all and only its instances. This view seems to me misguided and a result of an excessive concern with ontology at the expense of all other questions. Because plural logic is thought to incur no additional ontological commitments over and above those already incurred by the singular quantifiers, the plural comprehension axioms are assumed to be free of the only kind of commitment that one cares about. One of the main upshots of this article is that, irrespective of the question of ontological commitment, pluralities are governed by strong extensionality principles whose satisfaction is a non-trivial matter. Since the plural comprehension axioms make claims about how the plural variables can be interpreted, and since each of these interpretations is governed by non-trivial extensionality principles, these axioms too should be regarded as non-trivial.

To elaborate, let us consider the three 'factors' of the extensionality of pluralities. We begin with the two easier cases. It is not hard to see that traversability is a non-trivial assumption. To say that plural comprehension is permissible on a condition ϕ is to say that we may reason quasi-combinatorially about all the ϕ's. A number of disputes in the foundations of mathematics testify to the non-triviality of this assumption.[34] Nor is property supervenience a trivial matter. Consider the following:

(6) The Hiring Committee met yesterday. They decided to make an offer to Sophia.

Is it permissible to apply the rule of plural existential generalization to 'they'? The answer must be 'no'. Generalizing in this way would ascribe the property of making a job offer to the members of the committee *considered as a mere plurality*, where in reality, the property can only be ascribed to *the committee as such*. For it is only the committee, not the plurality of its members, that has the power to make job offers. Indeed, the property ascribed in the second sentence of (6) fails to supervene on modal membership profile. For as our earlier examples show, two committees can share the same modal membership profile while differing in the powers that are vested in them.[35]

The remaining 'factor' of extensionality is a rigid membership profile. I have argued elsewhere that this is a highly non-trivial matter, and in fact that many concepts and conditions lack an extension with a rigid membership profile.[36] Let me attempt to summarize the central idea of the view. It is useful to begin with a non-mathematical example. Assume you detest web pages that link to themselves and wish to create an inventory of all web pages that are innocent of this bad habit. That is, you wish to create a web page that links to all and only the web pages that do not link to themselves. Can your wish be fulfilled? The answer depends on how the above characterization of your wish is analyzed. Should the scope of the crucial plural description – 'the web pages that do not link to themselves' – be narrow or wide? Depending on its scope, your wish can be analyzed in either of the following two ways:

(N) You wish to design a web page *y* such that, for every web page *x*, *y* links to *x* iff *x* does not link to itself.

(W) There are some web pages *xx* such that, for every web page *x*, *x* is one of *xx* just in case *x* does not link to itself, and you wish to design a web page *y* that links to all and only *xx*.

On the narrow scope reading (N), your wish is flatly incoherent and on a par with the wish to ensure the existence of a Russellian barber:

(B) You wish there to be a barber *y* such that, for all *x*, *y* shaves *x* iff *x* does not shave himself.

But on the wide scope reading (W), I claim, there is no theoretical obstacle to the fulfillment of your wish.

The relevant difference has to do with how the target web pages are specified. On (N), the target is specified intensionally by means of the condition '*x* does not link to itself.' If you attempt to fulfill your wish by creating a new web page *y*, then *y* will be in the scope of the quantifier 'for every web page *x*' and thus be subjected to the mentioned condition, with fatal result. By contrast, on the wide scope reading (W), the target web pages are specified extensionally by means of the plurality *xx*. This makes it unproblematic to envisage – and indeed bring about – an alternative situation in which the associated wish is fulfilled: you simply create a web page *y* that links to all and only *xx*. Of course, *y* has to be a *new* web page, in the sense that it did not exist in the original situation; otherwise it would with fatal consequence fall under the quantifier 'for every web page *x*' that figures in the description of *xx* employed in (W).

It is important to notice the essential role played by the rigidity of pluralities in the argument. Although the plurality *xx* is described, in the original situation, by means of a potentially dangerous condition, there is no requirement that the plurality should remain so described in alternative situations. Like any other plurality, *xx* are tracked to an alternative situation in terms of the members, not in terms of any description that these members happen to satisfy.

Assume now instead that you care about pure sets, not web pages. You have no wishes concerning the latter. But you wish to define a pure set that has as elements all and only those pure sets that are not elements of themselves. It should by now be obvious that this sentence is subject to a scope ambiguity that parallels that of our example concerning web pages.

(N′) You wish to define a pure set *y* such that, for every pure set *x*, *x* is an element of *y* iff *x* is not an element of itself.

(W′) There are some pure sets *xx* such that, for every pure set *x*, *x* is one of *xx* just in case *x* is not an element of itself, and you wish to define a pure set *y* whose elements are all and only *xx*.

On the narrow scope reading, your wish is flatly incoherent. The wide scope reading is more interesting. Assume, as is customary, that standard plural comprehension holds, such that there are indeed some pure sets *xx* that include all and only the non-self-membered pure sets. On this assumption, there is no logical or mathematical obstacle to the fulfillment of your wish. We can make good mathematical sense of the envisaged pure set; for we know exactly what its elements are. Given this, it would run counter to the spirit of modern mathematics to deny that this is a definition in good mathematical standing.

Now we have a problem, however. Unlike web pages, pure sets exist of metaphysical necessity, if at all. The pure set *y* was not created through its definition but existed all along. This means that *y* is in the range of the quantifier 'for every pure set *x*' that figures in the description of *xx* employed in (W′). Once again, the result is fatal. What to do? Our only option, I argue, is to reject the assumption that there are some pure sets *xx* that include all and only the non-self-membered pure sets. That is, we must curtail plural comprehension. There are certain conditions which – despite having a sharp intension – lack an extension with a rigid membership profile. Such conditions cannot figure in plural comprehension axioms.

It would be wrong to think that the view is all negative, however. It is true that plural comprehension must be restricted. But this restriction opens the door to a version of naive set comprehension – every plurality can be used to define a set – which in turn is used in (what I find) an elegant and illuminating approach to set theory.[37]

Summing up, this paper has argued that pluralities are rigid and in fact that this is just one member of a small cluster of strong extensionality principles that govern pluralities. Although these principles can be split into three independent 'factors' (of which plural rigidity is one), they go naturally together as a package. Since the principles are far from trivial, nor are the plural comprehension axioms, which assert the existence of pluralities that are governed by the principles. This realization has an important consequence concerning the applications of the rigidity of pluralities that were outlined in Section 2. It takes us beyond the first three applications, which Williamson accepts, to the final one, canvassed just now, which Williamson is likely to reject.[38]

Notes

1. See Rumfitt (2005, Section VII), Uzquiano (2011), Williamson (2003, 456–457), Williamson (2010, 699–700), and Williamson (2013). The view is challenged in Hewitt (2012).
2. Here and in what follows, a displayed open formula will be short for the necessitation of its universal closure. It is important to realize that this convention differs from another one in discussions of modal logic, which lets a formula be short for its *modal closure*, where this is defined as the result of prefixing the formula with any string of universal quantifiers and necessity operators.
3. Here and in what follows, I use the word 'plurality' as a convenient shorthand to convey claims whose proper expression eschews this word in favor of some plural construction.
4. See e.g. Landman (1989) and Uzquiano (2004).
5. A similar example is attributed to Dorothy Edgington in Rumfitt (2005). Further examples are found in Hewitt (2012).
6. Sets – understood as on the iterative conception – come close but have the additional and complicating factor that their members are 'bound together' into a single object.
7. What about reducing plural logic to monadic second-order logic by translating a plural quantifier by means of a monadic second-order quantifier restricted to concepts that are non-empty and rigid? Technically, this should work. But philosophically, the proposal seems strained and lacking in motivation.
8. See Williamson (2003, Section IX).
9. See Williamson (2010).
10. See Linnebo (2010) and Linnebo (2013). While this approach draws inspiration from Parsons (1983) and to some extent also Putnam (1967) and Hellman (1989), these earlier views do not rely in the same way on the rigidity of pluralities.
11. In fact, as Williamson (1996) has pointed out, ($\Box \neq$) can also be derived without use of the Brouwerian axiom by invoking suitable principles of actuality.
12. Since the necessitation of (Leibniz) ensures $\Box(x = y \rightarrow x = x)$, the existential presupposition '$x = x$', present in ($\Box =$), would be redundant in ($\Box \neq$). For were $x \neq y$ to fail, the mentioned presupposition would anyway be satised. (Thanks to Williamson for this observation.)
13. Here, and in the paragraphs that follow, I leave implicit the proviso that the sets still exist.
14. This is 'the basic thought' from the end of Section 1.
15. For sets, the former option is unattractive. As Boolos (1971, 229–230) reminds us, if ever there was an example of an analytic truth, then the extensionality of sets is one.
16. This is significant for Fregean and neo-Fregean approaches to collections (or extensions, or *Wertverläufe*). These approaches regard extensionality as a

criterion of identity and are thus committed to soft extensionalism. But they also view a collection as in some way 'obtained from' its defining (Fregean) concept and are thus potentially on collision course with hard extensionalism. See Parsons (2012) for a discussion of Frege's concept of extension.

17. Thanks to Jeremy Goodman for articulating this objection.
18. Abstract objects would be a limiting case where the material contribution is nil.
19. The proper analogue of the set theoretic principle of extensionality would be the principle that sameness of *material* parts ensures identity. Now the analogy that seemed to cause trouble for my plausibility argument does work: any reason to accept this principle is also a reason to accept rigidity of material parthood. Of course, anyone attracted to non-rigid parts should respond to this observation by denying that sameness of material parts ensures identity.
20. See Stalnaker (1997) and essays 1–3 of Fine (2005) for some closely related considerations.
21. Notice that this enables us to drop the existential assumptions *Ex* and *Eyy* from (Rgd^-) on p. 1.
22. Here it is essential to observe that (Indisc) is short for the necessitation of its universal closure; cf. footnote 2. Had (Indisc) instead been short for its modal closure, we could have derived $\Box(xx \equiv yy \rightarrow \Box(xx \equiv yy))$, which has contingentist S5 countermodels. (Let *xx* and *yy* be distinct pluralities at w_1, neither of which exists at w_2. Then $xx \equiv yy \rightarrow \Box(xx \equiv yy)$ holds at w_2. Since the antecedent is vacuously true at w_2, so is the consequent. But this contradicts the distinctness of *xx* and *yy* at w_1.) Thanks to Tim Williamson for questions that prompted this clarification.
23. I am assuming that the statutes are partially constitutive of the committees, in the sense that, were one to change the statutes, the original committees would cease to exist and be replaced by new ones. If necessary, this persistence condition for the committees can be written into the statutes.
24. Notice that this enables us to drop the existential assumptions *Ex* from (Rgd^+) on p. 1.
25. A closely related concern is expressed in Uzquiano (2014).
26. Again, it is instructive to consider the mereological analogue, which in the case of (3) is the principle that for any objects x and y there is a sum. A natural formalization of this principle would be:

$$\exists z \Box (x \leq z \wedge y \leq z \wedge \forall u (x \leq u \wedge y \leq u) \rightarrow z \leq u) \tag{4}$$

Assume a molecule m is part of a cat c. For convenience, write $m + c$ for their sum. We easily derive $c \leq m + c$ and $m + c \leq c$, and hence by anti-symmetry, $c = m + c$. Since $m + c$ necessarily has m as part, by Leibniz's law, so does c. However, unlike its plural analogue, this argument involves controversial steps. If $x \leq y$ requires merely that the matter of x be included in that of y, then anti-symmetry can plausibly be denied. On the other hand, if $x \leq y$ is sensitive also to the formal aspects of x and y, then the relevant instance of (4) can plausibly be denied. True, there is a sum $m + c$ which necessarily has both m and c as parts. But this sum has a formal aspect which falsifies the third conjunct of the relevant instance of (4). Although $m \leq c \wedge c \leq c$, we do not have $m + c \leq c$: for this sum has a formal aspect, manifested in its modal profile, that goes beyond anything found in c.

27. In fact, as Jeremy Goodman observed, if a singleton plurality is uniformly traversed by its sole member, then Uniform Adjunction allows us to *prove* that any finite plurality is uniformly traversed by its members.
28. In fact, the entailment of covariance goes through on the contingentist version as well.
29. If desired, one can tweak (UniTrav-C) so as to ensure that (Dep) too follows, namely by adding the following as a third (and perfectly sensible) conjunct: $(Exx \leftrightarrow \bigwedge_{a \prec xx} E\bar{a})$.
30. Recall the claim from p. 662 that (Indisc) can be 'factorized' into (Sup) and (Cov).
31. Of course, *uniform* traversability is another matter, as demonstrated by the argument from Section 5.4.
32. There is a more indirect connection, however. In mathematics, the prevailing view has come to be that quasi-combinatorial reasoning should be extrapolated 'as far as possible.' How far is that? Given (Sup), Rigidity opens for the possibility of extrapolating such reasoning very far, namely to any plurality. The reason is that such reasoning ensures traversability, and thus by (Sup) also uniform traversability, whence by the argument developed in Section 5.4, also Rigidity. Thus, without Rigidity, it would not be permissible to extend quasi-combinatorial reasoning to any plurality.
33. Consider e.g. the breezy arguments for the status of the axioms of plural as 'genuine logical truths' found in Boolos (1985, 342) (corresponding to Boolos (1998, 167)) and Hossack (2000, 422).
34. See Feferman (2005) for a survey of debates concerning the legitimacy of impredicative reasoning in mathematics.
35. Our example from Section 6 of the Hiring Committee and the Graduate Admissions Committee will do.
36. See Linnebo (2010).
37. See Linnebo (2013). This approach uses a non-metaphysical modality to identify the legitimate forms of plural comprehension. This alternative modality allows us to represent the stages of 'the process of set formation'. Plural comprehension is permissible on any condition whose instances can be exhausted by one of these stages.
38. I am grateful for comments from Salvatore Florio, Jeremy Goodman, Simon Hewitt, Timothy Williamson, Gabriel Uzquiano, as well as participants at a London research seminar and a Montreal workshop where this work was presented. Much of the research was undertaken while benefiting from an ERC Starting Grant.

References

Bernays, P. 1935. "On Platonism in Mathematics". 'Sur le platonism dans les mathématiques', *L'enseignement Mathematique*, **34**, pp. 52–69. Reprinted in (Benacerraf and Putnam, 1983).

Benacerraf, P., and Putnam, H. editors 1983. *Philosophy of Mathematics: Selected Readings*, Cambridge: Cambridge University Press. Second edition

Boolos, G. 1971. "The Iterative Conception of Set". *Journal of Philosophy* 68: 215–232. Reprinted in (Benacerraf and Putnam, 1983).

Boolos, G. 1985. "Reading the Begriffsschrift." *Mind* 94 (375): 331–344.

Boolos, G. 1998. *Logic, Logic, and Logic*. Cambridge, MA: Harvard University Press.

Feferman, S. 2005. "Predicativity." In *Oxford Handbook of the Philosophy of Mathematics and Logic*, edited by S. Shapiro, 590–624. Oxford: Oxford University Press.

Fine, K. 2005. *Modality and Tense*. Oxford: Oxford University Press.

Hellman, G. 1989. *Mathematics without Numbers*. Oxford: Clarendon.

Hewitt, S. T. 2012. "Modalising Plurals." *Journal of Philosophical Logic* 41 (5): 853–875.

Hossack, K. 2000. "Plurals and Complexes." *British Journal for the Philosophy of Science* 51 (3): 411–443.

Landman, F. 1989. "Groups, I." *Linguistics and Philosophy* 12 (5): 559–605.

Linnebo, Ø. 2010. "Pluralities and Sets." *Journal of Philosophy* 107 (3): 144–164.

Linnebo, Ø. 2013. "The Potential Hierarchy of Sets." *Review of Symbolic Logic* 6 (2): 205–228.

Parsons, C. 1983. "Sets and Modality". In his *Mathematics in Philosophy*, edited by C. Parsons, 298–341. Cornell, NY: Cornell University Press.

Parsons, C. 2012. "Some Remarks on Frege's Conception of Extension." In *From Kant to Husserl: Selected Essays*, edited by C. Parsons, Cambridge, MA: Harvard University Press.

Putnam, H. 1967. "Mathematics without Foundations." *Journal of Philosophy* LXIV (1):5–22. Reprinted in (Benacerraf and Putnam, 1983).

Quine, W. V. 1960. *Word and Object*. Cambridge, MA: The MIT Press.

Rumfitt, I. 2005. "Plural Terms: Another Variety of Reference." In *Thought, Reference and Experience*, edited by J. L. Bermudez, 84–123. Oxford: Clarendon.

Stalnaker, R. 1997. "Reference and Necessity." In *Blackwell Companion to the Philosophy of Language*, edited by B. Hale and C. Wright, 534–554. Oxford: Blackwell.

Uzquiano, G. 2004. "The Supreme Court and the Supreme Court Justices: A Metaphysical Puzzle." *Noûs* 38 (1): 135–153.

Uzquiano, G. 2011. "Plural Quantification and Modality." *Proceedings of the Aristotelian Society* 111 (2pt2):219–250.

Uzquiano, G. 2014. "Mereology and Modality". In *Mereology and Location*, edited by Kleinschmidt, S., 33–56. Oxford University Press.

Williamson, T. 1996. "The Necessity and Determinacy of Distinctness." In *Essays for David Wiggins: Identity, Truth and Value*, edited by S. Lovibond and S. Williams, 1–17. Oxford: Blackwell.

Williamson, T. 2003. "Everything." In *Philosophical Perspectives 17: Language and Philosophical Linguistics*, edited by J. Hawthorne and D. Zimmerman. Boston: Blackwell.

Williamson, T. 2010. "Necessitism, Contingentism, and Plural Quantification." *Mind* 119 (475): 657–748.

Williamson, T. 2013. *Modal Logic as Metaphysics*. Oxford: Oxford University Press.

Reply to Linnebo

Timothy Williamson

Wykeham Professor of Logic, University of Oxford, Oxford, UK

In 'Plurals and modals' (Linnebo 2016), Øystein Linnebo argues for the metaphysical principle of *plural rigidity*. The strong necessitist version of this principle simply says that whether a given thing is one of some given things is non-contingent. The weaker contingentist version restricts the principle to those circumstances in which there are that thing and those things. I too have argued for the rigidity of pluralities (Williamson 2003, 456–457; 2010, 698–700; 2013, 245–252). Linnebo strengthens the case and draws a detailed, helpful map of the derivability relations between various principles in the vicinity. In most respects our conceptions of the metaphysics of pluralities are similar. This reply addresses the few points of disagreement.

I follow Linnebo in the convenient standard practice of using the singular noun 'plurality', although in a more rigorous formulation of the arguments it would be replaced by a plural term. Similarly, I will sometimes use the noun 'property' in places where it would be replaced by a monadic predicate in a more rigorous formulation.

1. From plural extensionality to plural rigidity

Linnebo and I agree that the natural strategy for deriving the rigidity of pluralities is to start with their extensionality and then find attractive auxiliary assumptions to bridge the gap from extensionality to rigidity. The extensionality of pluralities means that coextensiveness is their analogue of identity: if every one of these is one of those and every one of those is one of these, then these just *are* those. In particular, therefore, coextensive pluralities are necessarily coextensive. By itself, however, that does not exclude a scenario in which pluralities shift their membership across possibilities. For instance, consider a toy model with just two possible worlds, w_1 and w_2, two objects, o_1 and o_2 (both of which are in both worlds), and three pluralities: pp_1, which comprise both objects in both worlds, pp_2, which comprise just o_1 in w_1 and just o_2 in w_2, and pp_3, which comprise just o_2 in w_1 and just o_1 in w_2. In both worlds in that model, extensionality holds but rigidity fails. We can also get that result in models where pluralities vary in the (non-zero) number of their members. Different implementations of the overall argumentative strategy exclude such scenarios in different ways.

Linnebo does not discuss the detailed treatment of both necessitist and contingentist plural modal logics in *Modal Logic as Metaphysics* (241–254). Instead, in Section 5.3 of his paper, he assesses a brief argument for plural rigidity in an earlier paper of mine (Williamson 2010, 700). The latter argument is stated in a necessitist setting. The idea is that for any plurality, there is a coextensive rigid plurality. For instance, let *these* be the committee members, and suppose that just Tom, Dick and Harry are on the committee. It is not contingent whether something is one of Tom, Dick and Harry. But *these* just are Tom, Dick and Harry. Therefore, by extensionality, it is not contingent whether something is one of *these*. Linnebo concedes that this argument for plural rigidity is as plausible as his in the necessitist setting, but objects that it does not generalize properly to the contingentist setting. On such generalizations of the argument to a contingentist setting, in the 2010 paper I say only 'Contingentists will presumably want to qualify these principles to take account of contingency in what there is, but that is not our immediate concern'.

Here is Linnebo's objection. Both his argument and mine start from extensionality, in the form of an indiscernibility principle:

(INDISC) $xx \equiv yy \rightarrow (\varphi(xx) \leftrightarrow \varphi(yy))$

Here, $xx \equiv yy$ abbreviates $\forall u\ (u \prec xx \leftrightarrow u \prec yy)$. Informally, if these are all and only those, then these have a property just in case those do, though Linnebo interprets such a displayed open formula as short for the necessitation of its universal closure.

From (INDISC) we derive the necessity of coextensiveness, the plural analogue of the necessity of identity:

(COV) $xx \equiv yy \rightarrow \Box(xx \equiv yy)$

For the contingentist, the principle that every plurality is coextensive with a rigid plurality takes the restricted form of (7) (in Linnebo's numbering):

(7) $\exists yy(xx \equiv yy \wedge \forall x(x \prec yy \rightarrow \Box(Eyy \rightarrow x \prec yy)))$

Informally, some things *yy* are all and only these *xx*, and if something is one of those *yy*, then necessarily if there are those *yy* then it is one of those *yy*. Two points are worth noting about (7), which is Linnebo's formulation, not mine. First, he has a principle (DEP) that makes the conjunct *Ex* redundant in the more cautious variant of (7) with *Ex* & *Eyy* in place of *Eyy*. Second, the rigidity of *yy* also requires that if some *x* is *not* one of *yy* then necessarily *x* is not one of *yy*, but we can ignore that complication because the same considerations apply to the negative and positive halves of rigidity. In any case, Linnebo makes no objection to (7). The restricted positive rigidity claim to be derived is that if something is one of these *xx*, then necessarily if there are these *xx* then it is one of these *xx*:

$$(\mathrm{RGD}^{+}) \qquad a \prec xx \rightarrow \Box(Exx \rightarrow a \prec xx)$$

As Linnebo notes, what (COV) and (7) enable us to derive from $a \prec xx$ is $\Box(Eyy \rightarrow a \prec yy)$, which has *yy* where the consequent of the target claim (RGD^{+}) has *xx*. No such problem affects the necessitist version of the argument, since it needs no such plural being condition as *Eyy*. We could fill the gap in the argument by adding the extra conjunct $\Box(Exx \rightarrow Eyy)$ to (7), but Linnebo objects that this further assumption would be 'very strong and problematic from the point of view of anyone not antecedently committed to the rigidity of pluralities'.

Fortunately, no extra assumption is required. For from (INDISC) we can derive (COVE) (since $\Box(Exx \leftrightarrow Exx)$ is trivial):

$$(\mathrm{COVE}) \qquad xx \equiv yy \rightarrow \Box(Exx \leftrightarrow Eyy)$$

(COVE) fills the gap in the argument to (RGD^{+}). Deriving (COVE) from (INDISC) involves substituting one plural variable for another in a modal context, but only in the same way as is done in deriving (COV) from (INDISC). Linnebo's defence of the latter derivation and its use in the argument for rigidity applies equally to the derivation of (COVE) from (INDISC) and its use in an argument for rigidity. If his objection to my argument for rigidity worked, then so would a corresponding objection to his argument for rigidity. In effect, the objection would be that (INDISC) itself is 'very strong and problematic from the point of view of anyone not antecedently committed to the rigidity of pluralities'. But (COVE) is no more problematic from such a point of view than (COV), on which Linnebo was happy to rely in his argument for rigidity.

According to Linnebo, 'a theorist who doubts the rigidity of pluralities [...] regards some pluralities as much like groups'. He points out that such a theorist is likely to reject the assumption $\Box(Exx \rightarrow Eyy)$, noting '*xx* might be the Hiring Committee, whose members happen to be *a*, *b*, and *c*. Then the partial rigidification *yy* of *xx* is ontologically dependent on *a*, *b*, and *c*, whereas *xx* need not be subject to this ontological dependence'. But anyone who makes that objection to (COVE) will have a corresponding objection to (COV) too. If there could have been the Hiring Committee even though there was no plurality *a*, *b* and *c*, then the Hiring Committee could have failed to be coextensive with the plurality *a*, *b* and *c*. Any theorist antecedently committed to regarding some pluralities as individuated by properties such as membership of a given committee would be mad to accept (INDISC) in addition.

No arguments such as Linnebo's and mine from plural extensionality to plural rigidity will silence those firmly enough committed to an intensional conception of pluralities. Any mistake can be defended indefinitely by those ruthless enough to say whatever it takes. The value of such arguments is instead for those still open to the attractive idea that coextensiveness is the plural analogue of identity, in showing them how naturally that analogy leads to plural rigidity.

2. Plural comprehension

The concluding part of Linnebo's paper, Section 7, takes a different turn, sketching a rationale for restricting the standard comprehension principle for pluralities. On his view, for some well-defined conditions C, although some things meet C, no things are *the things* that meet C. By contrast, I endorse a strong non-modal comprehension principle for pluralities: for any well-defined condition C, if some things meet C, then some things are the things that meet C (Comp $\prec$, Williamson 2013, 248). Linnebo has developed his view of pluralities in much more detail elsewhere. Here, I will respond only to what he says in the present paper.

On one important point, I emphatically agree with Linnebo: plural comprehension principles are not trivial, logically, metaphysically or epistemologically. Of course, in the standard Quinean sense, ontological commitments are expressed by a first-order singular quantifier. Since the plural quantifiers are irreducible to singular quantifiers, plural comprehension principles do not incur specifically ontological commitments in that sense. However, a theory's plurally quantified claims are just as metaphysically serious as its singularly quantified ones, just as much 'about reality', just as little to be brushed aside as 'mere ideology'. In more methodological terms, they are just as much in the scales when the theory is abductively compared with other theories (Williamson 2013, 260–261). Thus, Linnebo's rejection of strong plural comprehension principles cannot be dismissed on general methodological grounds.

Linnebo's objection to a standard plural comprehension principle is based on Russell's paradox. Given such a principle, there are some pure sets *xx* such that, for every pure set *x*, *x* is one of *xx* just in case *x* is not an element of itself. So 'there is no logical or mathematical obstacle' to defining a pure set *y* whose elements are all and only *xx*:

> We can make good mathematical sense of the envisaged pure set; for we know exactly what its members are. Given this, it would run counter to the spirit of modern mathematics to deny that this is a definition in good mathematical standing.

But:

> [P]ure sets exist of metaphysical necessity, if at all. The pure set *y* was not created through its definition but existed all along. This means that *y* is in the range of the quantifier 'for every pure set *x*' that figures in the description of *xx*[.]

Thus, we cannot escape the inconsistency that *y* is an element of itself just in case *y* is not an element of itself. Linnebo blames the contradiction on the standard plural comprehension principle assumed at the start of the argument, and rejects it. On his view, there are not some pure sets *xx* that include all non-self-membered pure sets and nothing else.

It is hard to see how Linnebo's argumentative strategy could work in plural logic without also working in second-order logic with quantification into predicate position. Thus, if strong plural comprehension principles are to be rejected, so are the corresponding second-order comprehension principles. For plurals play no privileged role in 'modern mathematics'. It is no part of modern mathematical practice that in order to get a set from a property, one must *first* get a plurality from the property, and only then get a set from the plurality. In the sense in which we know what the Russell set's putative members are, we know it just as well by knowing that *x* is a member just in case *x* is a pure set and not a member of itself as we do, less directly, by knowing that *x* is a member just in case *x* is one of *xx* and something is one of *xx* just in case it is a pure set and not a member of itself. Indeed, the plurals look suspiciously like the excess layer of middle management that gets cut out in the interest of efficiency. Of course, Russell's paradox still needs to be blocked, but few working mathematicians worry about how. They simply work in a more or less standard mathematical style that does not in practice court paradox. None of this makes it illegitimate for Linnebo to postulate a deep metaphysical divide between plurals and sets on one side and properties on the other. It is just that the spirit of modern mathematics gives no positive support to his proposal.

Someone might accept the analogy between pluralities and properties, and reject strong comprehension principles for both. However, that is not Linnebo's line. He already hints as much in remarks such as: 'There are certain conditions [such as "*x* is not a member of itself"] which — despite having a sharp intension — lack an extension with a rigid membership profile'. For the claim that '*x* is not a member of itself' has a sharp intension seems to assume a strong comprehension principle for sharp intensions, which are presumably a bit like properties. Moreover, as I emphasize in *Modal Logic as Metaphysics*, without strong comprehension principles, we tend to lose the mathematical advantages of going beyond first-order logic. Thus, if we give them up for plural logic, we need them all the more for second-order logic, non-plurally interpreted.

On the methodological approach of my book, these large issues about the overall shape of quantified modal logic should be settled abductively. Linnebo's restriction of plural comprehension has the compensating advantage that it enables him to maintain an unrestricted principle of set formation from pluralities. However, we speak plurally in many contexts where sets are not at issue. There, Linnebo's restriction of plural comprehension will trip up our plural reasoning, with no compensating advantage. My hunch is that the balance of abductive considerations will favour unrestricted plural comprehension.

References

Linnebo, Øystein. 2016. "Plurals and Modals." *Canadian Journal of Philosophy* 46 (4–5): 655–677.
Williamson, Timothy. 2003. "Everything." *Philosophical Perspectives* 17: 415–465.
Williamson, Timothy. 2010. "Necessitism, Contingentism, and Plural Quantification." *Mind* 119: 657–748.
Williamson, Timothy. 2013. *Modal Logic as Metaphysics*. Oxford: Oxford University Press.

On Williamson and simplicity in modal logic

Theodore Sider

Rutgers University, New Brunswick, NJ, USA

ABSTRACT
According to Timothy Williamson, we should accept the simplest and most powerful second-order modal logic, and as a result accept an ontology of "bare possibilia". This general method for extracting ontology from logic is salutary, but its application in this case depends on a questionable assumption: that modality is a fundamental feature of the world.

1. Necessitism

The central thesis of Timothy Williamson's wonderful book *Modal Logic as Metaphysics* is 'Necessitism'. Put roughly and vividly, Necessitism says that everything necessarily exists. Williamson himself avoids the predicate 'exists', and formulates the view thus:

Necessitism $\forall x \Box \exists y\, y = x$ ('Everything necessarily is something')

Williamson's scruples about 'exists' are reasonable but sometimes require tortured prose, so, choosing beauty over function, I will use the E-word. (But let it be understood that by '*x* exists' I just mean that *x* is identical to something — with quantifiers 'wide open'.)

The considerations favoring Necessitism also lead Williamson to accept the Barcan schema (as well as its converse):

Barcan schema $\Diamond \exists x A \rightarrow \exists x \Diamond A$

So if there could have been a child of Wittgenstein, then there in fact exists something that could have been a child of Wittgenstein.

This thing that could have been a child of Wittgenstein: what is it like? What are its properties? Well, it has the modal property of possibly being a child of

Wittgenstein. And logic demands that it have certain further properties, such as the property of being self-identical, the property of being green if it is green, and so forth. And since Wittgenstein in fact has no children, it has the property of not being a child of Wittgenstein. But it has no other properties beyond these sorts. It is not a human being. It has no physical properties such as mass or spatial location.[1] It is in a sense nonconcrete (which is not to say that it is abstract in any positive sense), a 'bare possibilium'.

Similarly, although Williamson denies that anything could have failed to exist, he allows that an object like me, say, could have failed to be concrete. I could have been a bare possibilium.

Williamson's ontology, then, includes vastly more entities than we ordinarily accept. It is not contingent what entities there are; what is contingent is which entities are concrete.

2. Williamson's argument

Williamson's argument for the Barcan schema and Necessitism is that they are implied by the best modal logic — by the modal logic that best combines various theoretical virtues such as simplicity and strength. An older argument along these lines, versions of which have been defended by Linsky and Zalta (1994, 1996), and by Williamson himself (1998, 2002), claims that the Barcan schema and Necessitism hold in the best first-order quantified modal logic.[2] Contingentists — those who believe that things can exist contingently — have developed quantified modal logics that do not have these consequences. But Necessitists argue that these logics are more complex, or are otherwise inferior to, necessitist quantified modal logic.

What do the contingentist logics look like, and what challenges do they face? Model theoretically, they are based on Kripke models with variable domains. This approach faces the formal challenge of assigning truth values to formulas in worlds where objects denoted in the formula do not exist, and the philosophical challenge of saying how a variable-domain Kripke model can be an *intended* model if necessity and possibility do not reduce to facts about possible worlds (as most Contingentists agree they do not), and if the domains of nonactual worlds in the model contain entities that are not in the domain of its actual world.

Proof-theoretically, the challenges may be brought out by the following argument. Let t name me, and consider the following:

(1) $\vdash t = t$ (reflexivity of identity)
(2) $\vdash \exists y\, y = t$ (1, existential generalization)
(3) $\vdash \Box \exists y\, y = t$ (2, necessitation)

The conclusion says that I necessarily exist. So Contingentists will want to resist the argument. That means rejecting one of the steps. They may, for instance, restrict existential generalization by embracing a free logic, or they may restrict the rule of necessitation. But Necessitists then argue that these restrictions make the logic more complex, or are not independently philosophically motivated, or weaken the logic in unwanted ways.

An important part of Williamson's case for Necessitism is a powerful new development of this style of argument, for second-order rather than first-order quantified modal logic. Nonmodal second-order logic usually includes the following comprehension principle:

Nonmodal comprehension $\exists X \forall x(Xx \leftrightarrow A)$

where any string of universal quantifiers may be prefixed to the result and *A* may be replaced with any formula.[3] When we add modal operators to the language, the question then arises of how to strengthen the comprehension principle. Williamson's preferred strengthening is this:

Modal comprehension $\exists X \Box \forall x(Xx \leftrightarrow A)$

where any combination of universal quantifiers and occurrences of $\Box$ may be prefixed. (Note: the second-order quantifiers in this discussion are not to be given Boolos's plural interpretation; they are 'intensional' second-order quantifiers.)

Modal comprehension implies this:

Haecceities $\forall y \Box \exists X \Box \forall x(Xx \leftrightarrow x=y)$

Haecceities may be paraphrased, somewhat naughtily, thus: 'for every object, *y*, it's necessary that there exists a property that, necessarily, is instantiated by something iff that something is *y*'; or, more succinctly: 'for every object, it's necessary that there is a haecceity of that object'. (The naughtiness is that the irreducibly second-order quantification in Haecceities sounds like first-order quantification over properties in the paraphrases.)

Williamson then argues that Contingentists ought to reject Haecceities; the crux of the argument is this:

> Even if I had never been, by [Haecceities] there would still have been a property tracking me (and only me). But how can it lock onto me in my absence? In those circumstances, what makes me rather than something else its target? (269)

This objection to Haecceities is similar to Kit Fine's (1985) objection to Alvin Plantinga's (1976) individual essences. One might question whether Williamson's objection is as powerful as Fine's, because of a difference between their targets. What Fine objected to was Plantinga's postulation of an *entity*, a property, that would track me (in the sense of necessarily being instantiated by something iff that something is me) even if I didn't exist, whereas Haecceities doesn't say that there would exist any such entity. It doesn't talk about properties at all (despite the naughty paraphrase); the variable *X* in Haecceities is irreducibly second order. Still, the second-order $\exists X$ is meant to be a sort of existential quantification, after all; the strength of Fine's first-order argument arguably carries over to Williamson's second-order argument.

Thus, Contingentists are under pressure to restrict the Modal comprehension principle. But, Williamson then argues that all Contingentist-friendly restrictions would badly restrict second-order modal logic.

One Contingentist strategy would be to revert to the Nonmodal comprehension principle, but now allow modal operators to occur in the prefixes. What a comprehension principle does is imply the existence of a property corresponding to an arbitrarily chosen formula; the envisioned restriction implies that the correspondence is merely 'intra-world' rather than inter-world. And therein lies its fatal weakness, according to Williamson. Second-order quantification (of the type Williamson is discussing) is the analog of first-order quantification over properties in the 'abundant' sense.[4] Thus our second-order modal logic ought to imply, for instance, that properties are closed under conjunction: that for any properties *Y* and *Z*, there exists a conjunction of those properties. A conjunction of *Y* and *Z* is a property that, necessarily, is instantiated by something iff that thing instantiates both *Y* and *Z*; thus we want the following claim:

Conjunction $\forall Y \forall Z \exists X \Box \forall x (Xx \leftrightarrow (Yx \wedge Zx))$

But, note that Conjunction expresses an inter-world, not merely intra-world, correlation, between the property *X* and the chosen formula $Yx \wedge Zx$. *X* must be *necessarily* such as to be instantiated when and only when *Y* and *Z* are instantiated. Conjunction is implied by Williamson's Modal comprehension principle, but not by the weakened principle. The weakened principle merely implies such claims as:

$\forall Y \forall Z \exists X \forall x (Xx \leftrightarrow (Yx \wedge Zx))$

$\Box \forall Y \forall Z \exists X \forall x (Xx \leftrightarrow (Yx \wedge Zx))$

$\Box\Box \forall Y \forall Z \exists X \forall x (Xx \leftrightarrow (Yx \wedge Zx))$

etc.

in which the property *X* whose existence is guaranteed is merely the 'intra-world conjunction of *Y* and *Z*'. The second formula, for instance, says that (to put it in possible worlds talk), in any possible world *w*, for any properties *Y* and *Z* in *w*, there exists in *w* a property *X* that is in *w* instantiated when and only when both *Y* and *Z* are instantiated, but this leaves open that in other worlds *v*, it's not the case that *X* is instantiated when and only when *Y* and *Z* are both instantiated.

Williamson considers other weakenings of Modal comprehension, but in each case he argues that Contingentist-friendly weakenings '[prevent] second-order logic from adequately serving the logical and mathematical purposes for which we need it' (288) — those purposes demanding an abundant conception of properties in a modal context.

How might one resist this argument? One might meet it head-on, either by (i) defending Haecceities against the Williamson/Fine objection, or (ii) conceding that Haecceities is false and seeking out some weakened comprehension principle that does not imply Haecceities but nevertheless is strong enough for the desired applications of second-order logic. I won't pursue either route here. (I don't see how to pursue (ii), and (i) strikes me as being metaphysically ugly.) Instead, I will discuss two ways to resist the argument indirectly. In the next section, I will consider — but reject — the possibility of accepting the argument's conclusion but 'defanging' that conclusion by distinguishing fundamental from nonfundamental existence. Then, in the final section I will argue that if modality is not a fundamental feature of the world, there is no presumption that second-order modal logic be simple and strong; thus we can simply reject the demand for a simple and strong comprehension principle.

3. Fundamental and nonfundamental existence

Some ontologists distinguish between existence and fundamental existence.[5] The distinction is believed to resolve the tension between a pair of conflicting pressures: pressure (which can come from various sources) to increase ontology, and pressure from Occam's razor to contract ontology. The resolution is achieved by increasing ontology while minimizing fundamental ontology, and by claiming that Occam's razor applies only to fundamental ontology.[6] For example, we might bow to inflationary ontological pressure by accepting the existence of shadows, while appeasing Occam by denying that there fundamentally are any such things. Fans of this distinction may be tempted to apply it here, to say that Williamsonian entities exist but don't fundamentally exist. Thus we have our cake and eat it too: denying the fundamental existence of Williamsonian entities appeases Occam, while accepting their existence lets us avoid challenging Williamson's argument head-on; that argument, which concerns existence, not fundamental existence, may simply be accepted.

But consider a restated version of Williamson's argument in which all quantifiers are stipulated to express fundamental existence. The plan was to reject the fundamental existence of Williamsonian entities; thus the restated argument must be rejected. But the plan was also to accept Williamson's original argument. Thus the plan is successful only if the restated argument is weaker than the original argument. But in fact I think that restating the argument in terms of fundamental existence *strengthens* it: as we'll see in the next section, it's natural to think that the case for simplicity in modal logic is strongest if all the concepts involved are fundamental ones.

Furthermore, even setting this aside, facts about what exists nonfundamentally must, like all nonfundamental facts, be somehow based or grounded in fundamental facts. So if Williamsonian entities exist but nonfundamentally so, facts about them must be grounded in facts that do not involve the existence of such entities. But it's unclear whether appropriate grounding facts can be found.

In some cases of nonfundamental existence, it's easy to see what the grounding base might be. Facts about shadows, for instance, might be said to be grounded in certain facts about things and their relationships to light sources and backgrounds. Such facts are fit to ground putative facts about shadows because there is no apparent 'metaphysical gap' between them and the facts about shadows; the existence of my shadow is 'nothing over and above', it's said, my existence and my relations to light sources and backgrounds. Many would also say that there is no metaphysical gap between the existence of a mereological sum and the existence of its parts. There is, however, a metaphysical gap between, for instance, our *evidence* for subatomic particles and the existence of those particles; thus the existence of subatomic particles is not grounded in our evidence for them.

So, can we find a grounding base for the existence of Williamson's nonconcrete entities? Those entities are certainly not composed (in any relevant sense) of more mundane entities. Nor are there more mundane entities that stand to Williamsonian entities as I stand to my shadow, or as holes stand to hole-linings, to take another example. If a grounding story is to be told here, it cannot be parallel to the stories for sums or shadows or holes.

The best bet for a grounding base would seem to be certain modal facts.[7] Suppose a certain knife blade *B* and knife handle *H* are manufactured, but are never joined to make a knife. They could have been so joined: there could have existed a knife made of *B* and *H*. Thus, according to Williamson, there exists a (nonconcrete) object that could have been a knife made of *B* and *H*. Indeed, it's plausible that there is just one such object (Williamson 2000, 335). Here a ground of the existence of this Williamsonian entity suggests itself: the modal fact that there could have existed a knife made of *B* and *H*. (According to this proposal, notice, an *existential* fact, that there exists an entity of such-and-such a type, can be grounded in a fact that is *modal* rather than existential in form: that there could have existed an entity of such-and-such a type. Nonfundamental existence would thus be quite unlike fundamental existence in its underlying metaphysical nature.)

But the case of the possible knife made from *B* and *H* is special in a crucial way: though the possible knife is not concrete, it has concrete 'anchors' in *B* and *H* since it can be uniquely specified as the possible knife that could have been made from those concrete objects. In other cases there are no such concrete anchors. Suppose there are only finitely many subatomic particles: $a_1 \dots a_n$. Then, assuming it is possible that there exists another subatomic particle distinct from $a_1 \dots a_n$, the Barcan schema implies the existence of something that could have been a subatomic particle distinct from $a_1 \dots a_n$; and given other modal principles Williamson accepts — the necessity of distinctness and the 'B' schema from propositional modal logic — this something must in fact be distinct from $a_1 \dots a_n$. Indeed, Williamson will admit the existence of infinitely many such possible subatomic particles, since for any finite number of them $b_1 \dots b_m$, it's possible that there exists a subatomic particle distinct from $a_1 \dots a_n$ and also from $b_1 \dots b_m$. Letting b_i be any one of these nonconcrete possible subatomic particles, it's then hard to see what ground there might be for the existence of b_i, because there is no concrete anchor for b_i as opposed to the other b_js. For instance, the modal fact that there could have existed a subatomic particle distinct from $a_1 \dots a_n$ seems to be an unfit ground for the existence of b_i, since if it were, it would presumably also be the ground for the existence of every other b_j; but surely these facts should all have different grounds.[8]

Although the modal fact that there could have existed a subatomic particle distinct from $a_1 \ldots a_n$ seems unfit to ground *singular* existence-facts about any particular b_i, it seems more fit to ground the *general*[9] fact that there exists some possible subatomic particle or other distinct from $a_1 \ldots a_n$. So I suppose there is room for a position that denies the existence of singular facts about 'unanchored' entities, admitting only general facts about them. Still, it faces some questions. (i) Is it well-placed to accept Williamson's argument that we need his strong comprehension principles; can we get by in second-order modal logic with only these general facts? (ii) How will it resist the argument that if there is a general fact that there exists an *F*, then there must also exist a particular fact of the form *a* is *F*? (iii) Let *a* be something that, in fact, fundamentally exists — a subatomic particle, say. Although *a* exists necessarily, it could have been nonconcrete; and if it had been nonconcrete, there would have been various singular facts about it (it would have existed, been possibly concrete, and so forth). But *a* would then have been unanchored, and wasn't the idea to reject singular facts about unanchored things?[10]

4. The scope of simplicity + strength arguments

My plan for the remainder is to challenge Williamson's argument on methodological grounds. Necessitist second-order logic would indeed be simpler and stronger than the contingentist variety, I grant; but, I will argue, we can nevertheless reject that logic since the presumption that second-order modal logic be simple and strong is weak. However, I believe that Williamson's *general* methodology — his view of how to choose a logic, and of the bearing of this choice on metaphysics and ontology in general — is correct. Williamson lays out his methodology right from the start, in the first paragraph of the book's preface:

> The title of this book may sound to some readers like Good as Evil, or perhaps Cabbages as Kings. If logic and metaphysics appear disjoint, the reason is not just the lingering spell of a logical positivist conception of metaphysics as cognitively meaningless and logic as cognitively meaningful but analytic. Many contemporary philosophers who acknowledge metaphysics as continuous with the rest of science are still inclined to assign logic a more special status. They see it as a neutral referee of disputes between scientific theories, including metaphysical theories, blowing the whistle when the rules are broken, not as a disputing party in its own right. If so, logic says nothing over which there could be such a dispute, on pain of non-neutrality; thus logical theories are quite different in status from scientific theories. This book is written in the contrary conviction that, just as metaphysics is much more like the rest of science than was once thought, so too is logic. Indeed, one role for logic is to supply a central structural core to scientific theories, including metaphysical theories, in essence no more above dispute than any other part of those theories.

I couldn't agree more. And from this point of view, the theoretical pressure to accept a simple yet powerful logic does not differ in kind from the theoretical pressure to accept simple yet powerful laws of nature. David Lewis's *On the Plurality of Worlds* had a massive impact on the way metaphysics was practiced, even on those who did not accept its conclusion, because of the methodology it championed: an approach to ontology in the spirit of the indispensability argument for mathematical entities. Williamson's book will, I predict, have a similar impact, even among those who remain unconvinced of Necessitism, because of its methodology of choosing metaphysical positions using logical considerations.

On, though, to my objection. I grant the analogy between the presumption of simple, powerful laws of nature and the presumption of a simple, powerful logic. But it's not true that there is a presumption of simple and powerful statements governing just any concepts. While such a presumption exists for *some* concepts — concepts of physics, for instance — there is surely no such presumption for the concepts of lying, bachelorhood, or games. It would be misguided to let the simplicity and strength of a proposed 'logic' of lying, bachelorhood, or games carry much weight, if it conflicted with our otherwise reasonable beliefs about those domains. If someone boldly conjectured that all utterances of falsehoods are lies, we would not downgrade the importance of the obvious counterexamples (innocent mistakes, non-assertive utterances ...) by saying 'yes, but the proposed "logic" is *so* simple that we should reconsider whether the alleged counterexamples are really genuine'.[11]

Also, without any restriction on the concepts involved, a presumption in favor of simplicity and strength will give no guidance. As is familiar from Nelson Goodman (1955), if such a presumption using the concepts of blue and green were accompanied by equally strong presumptions using arbitrarily chosen other concepts — such as his grue and bleen — then the presumptions as a whole would yield no guidance for what to believe.

Williamson's argument succeeds only if the concepts of second-order modal logic are like those of physics, and unlike lying, bachelorhood, games, and grue and bleen, in being 'presumptively simple' — in being concepts for which there is indeed a presumption of simple and strong generalizations. I'm going to argue that they are not. In doing so I am going to assume a particular metaphysics, and a particular connection between that metaphysics and epistemology. A similar story could, I hope, be told under other assumptions, but it's best to proceed under some definite assumptions. Roughly put, my claim is that the concepts of second-order modal logic are presumptively simple only if modality is metaphysically *fundamental*; and it isn't.[12]

Now, simplicity arguments are certainly not confined to reasoning about what is fundamental. Simplicity arguments are routinely used by detectives, hunters, and archeologists, not just physicists.[13] A detective prefers the simple hypothesis that the murder was committed by a single person, unless and until the evidence demands a more complex explanation; and the concepts of criminology are of course not fundamental.

But simplicity reasoning in such everyday cases need not be underwritten by any *ultimate* presumption concerning the concepts involved.[14] By 'ultimate' I mean epistemically ultimate; an ultimate presumption is thus a principle or element of epistemology (for instance a prior probability distribution, in a Bayesian setting) that is not derived from any further principle or element of epistemology. Suppose there is an ultimate presumption in favor of simplicity in physics. This alone would underwrite simplicity reasoning in chemistry, for example, because of the manner in which chemical properties are based in physics. (Understand 'basing' however you like.) A presumption of simplicity in physics means that it's rational to assume a priori that the world is likely to be physically simple, that certain sorts of physical complexity are a priori highly unlikely. We assume that it's a priori highly unlikely for there to be a sudden, huge, isolated exception to otherwise simple physical generalizations — all particles in the universe suddenly undergoing discontinuous, drastic, and haphazard changes to their positions, masses, and charges, say. But such an occurrence would also constitute a sudden, huge, and isolated exception to otherwise simple chemical generalizations, and thus would count as chemical complexity as well. More generally: *any* sudden, huge, isolated exception to otherwise simple chemical generalizations would require some such physical complexity or other; thus we assume a priori that these exceptions are unlikely.[15]

It's a reasonable guess that something like this holds more generally. Ordinary appeals to simplicity, such as by detectives, hunters, and archeologists, do not require epistemically ultimate presumptions of simplicity for the concepts they involve. Given a presumption of simplicity in physics, there is no need for a fundamental presumption of simplicity in criminology, hunting, or archeology. When judging that there was probably just a single murderer, a detective has her past experience to work with: murderers have generally acted alone, and when multiple murderers did conspire, they generally left distinctive traces that are in this case absent. I grant that some epistemically ultimate bias toward simplicity is needed to move from this evidence about the past to the conclusion that there is just one murderer at work here in the present, but surely that bias needn't be at the level of criminology: the world would need to be unnecessarily *physically* complex for there to have been two or more murderers. Detectives can even make simplicity-arguments using concepts such as lying, bachelorhood, or games (or even grue) without an ultimate presumption of simplicity for such concepts. The simplest hypothesis available to a detective might be that one of her suspects has lied, but the sort of simplicity ultimately at issue is at the level of physics.

Williamson's argument assumes that there are likely to be simple laws governing the concepts of second-order modal logic. If those concepts had an appropriate basis in physical concepts, then his argument could be underwritten by an ultimate bias in favor of physical simplicity, just like the arguments of detectives, hunters, and archeologists. But this seems very unlikely.

I've been writing as if ordinary simplicity reasoning requires only an ultimate presumption of physical simplicity, but probably that's not quite right; no doubt we need to recognize other 'ultimately simple' concepts — other concepts for which there are ultimate presumptions of simple and strong generalizations.[16] Certain mathematical concepts, and the concepts of first- and perhaps even second-order logic, are likely candidates. But even when the bias toward physical simplicity is thus enhanced, it still seems unable to underwrite Williamson's argument, since it seems unlikely that *modal* concepts have an appropriate basis in physical, mathematical, and (nonmodal) logical concepts.

It's not enough that modal concepts be based in some way or other in the ultimately simple concepts. Grue is based in the physical, but there is no presumption of simple and strong generalizations governing grue. I don't know how to say in general what kind of basing is 'appropriate', in order for presumptions of simplicity about the basis to 'trickle up', generating presumptions of simplicity about the based; but the lesson of grue suggests that disjunctive basing blocks the trickling up. Thus, for example, if my own view (Sider 2011, chapter 12) about how modal concepts are based in nonmodal concepts is correct, we should not expect simple laws governing modal concepts. In my view, for a proposition to be necessary is for it to be true and fall under one of a certain list of types, where the items on the list are heterogeneous and given more or less conventionally.[17] Even given presumptions of simplicity for the concepts in the list, there would be no presumption of simplicity for modality itself, since it's defined disjunctively via the list.

Lewis's 1986 identification of necessity with truth at all concrete worlds might seem less disjunctive and thus more friendly to the presumption. But his counterpart-theoretic reduction of modality *de re* generates hidden complexity in modal logic.[18] And in any case, there is a sense in which Lewis's modal realism is already necessitist, obviating the need for Williamson's argument.[19] (There is also the point that Lewis's modal realism is surely false.)

Perhaps there is some other reduction of necessity on which necessity would be appropriately based in ultimately simple concepts. But I suspect a more likely position for a defender of Williamson's argument is that necessity and possibility are metaphysically fundamental. Modal concepts are, after all, notoriously resistant to reduction; even reductionists like me should admit this. Given such a position, it would be natural to hold that modal concepts are ultimately simple — that there is an ultimate presumption of simplicity for modal concepts, just as there is an ultimate presumption of simplicity for physical concepts. My argument thus rests on rejecting this 'modalist' position.

This issue is large in scope, and I won't pretend to settle it here. All-too-briefly and superficially: my rejection of modalism is based on the conviction that modal concepts do not have the theoretical 'track record of success' that good candidates for being metaphysically fundamental ought to have. In addition to being 'reduction-resistant', physical, logical, and mathematical concepts are essential to theories that have been immensely successful at explaining phenomena. The same cannot be said of modal concepts, in my view.

But my opponent may urge that modal concepts *do* have the required track record: in metaphysics, perhaps; or in ordinary counterfactual reasoning; or, finally, in possible-worlds semantics for natural language. I'll consider these arenas in turn.

Williamson says that 'For quantified modal logic in particular, one of its many roles is to supply a central structural core to theories of modal metaphysics.' (x) Perhaps modal metaphysics has a track record of success that justifies our acceptance of modality at the fundamental level.

There is no question that, in comparison to many other approaches to metaphysics, anyway, modal metaphysics has been a successful research program. The metaphysical boom since the seventies is evidence for this. But many have recently been challenging the hegemony of the modal approach to metaphysics.[20]

Some of these challenges, though, are merely to the completeness of the modal approach. If modality is merely to be *supplemented* by the favored concepts of the challengers — ground, essence, or building — it may still be part of a successful metaphysical research program, and thus have a claim to fundamentality.

Moreover, even if modality is to be replaced rather than supplemented, this would not on its own evade the challenge of Williamson's argument. Suppose Fine convinced us to replace modality with his notion of essence, in our metaphysical endeavors. We would still have the notion of necessity: to be necessary, according to Fine, is to hold in virtue of the essences of all things (1994, 9). This is a direct and nondisjunctive definition of necessity in terms of Fine's essentialist operator 'it is true in virtue of the essences of such-and-such entities that *A*', and so Williamson's argument might retain its force.[21]

But I myself think that modality ought to be replaced in metaphysical inquiry in a more radical way: by a notion of fundamentality from which there is no immediate and nondisjunctive method for recovering the notion of modality (Sider 2011). In my view, the putative explanatory successes of modal metaphysics are either illusory or else may be bettered by broadly fundamentality-theoretic metaphysical explanations.

There is a more basic concern with supporting the fundamentality of modality by its track record within metaphysics: a track record in that arena can only be so impressive. Metaphysics, after all, is a deeply speculative field.

Here I find myself struck again by the parallel between *Modal Logic as Metaphysics* and *On the Plurality of Worlds*. Both Williamson and Lewis draw their ontological conclusions — the former of a domain of contingently nonconcrete entities, the latter of an amodal domain of concrete 'possibilia' — from a premise that the subject matter of modality ought to be theorized in a certain way. For Williamson, the premise is that the logic of possibility and necessity, when combined with second-order quantification theory, should be simple and powerful; for Lewis the premise is that the theory of modality should be simple and powerful but also parsimonious in a certain way, and thus free of primitive modal notions. Lewis famously met incredulous stares, and Williamson probably gets them too. What I suspect is that the incredulous stare comes from a concern with drawing radical ontological conclusions on the basis of promised theoretical successes in modal metaphysics, an area of unclear theoretical value.

Even if modal metaphysics is granted to have a weaker track record than, say, successful scientific theories, it may be argued that it has a strong enough track record to justify *some* confidence in the fundamentality of modal concepts. While this may generate only weak confidence in Williamson's conclusion, that's something he could live with; it is unsurprising, after all, to have weaker confidence in metaphysical conclusions than in scientific conclusions.

However, we must take into account countervailing pressures on our beliefs: there is surely some sort of presumption to minimize ontology. Now, in a discussion of the necessity and permanence of identity, Williamson in fact addresses the question of what to do when ontological parsimony clashes with modal logic:

> For the necessitist or permanentist to mess with the modal or temporal logic of identity in order to avoid ontological inflation would be a lapse of methodological good taste, or good sense, for it means giving more weight to ontology than to the vastly better developed and more successful discipline of logic. (26–27)

While I agree that there is something to Williamson's claim that logic is vastly better developed and more successful than ontology, the truth of the claim depends, I think, on what exactly one means by 'logic' and 'ontology'. Logic as a whole has an undeniably powerful track record — much of it in the sciences and mathematics, in which logical notions and logical inference are indispensable — but the track record of *modal* logic is far less impressive, being mostly within philosophy. And as for ontology: ontology as a whole, inquiry into what there is, includes questions like whether there are black holes, comets, and electrons, and thus has an excellent track record. To discount the importance of parsimony in the present context would require, for the purpose of evaluating track records, simultaneously grouping modal logic with nonmodal logic and grouping philosophical ontology apart from nonphilosophical ontology. Each strikes me as problematic, and the two together even more so.

Let's now leave the arena of metaphysics, and look elsewhere for a track record of success for modal concepts. It may be proposed that our successful practice of reasoning counterfactually, in ordinary, nonphilosophical contexts — stressed by Williamson himself in *The Philosophy of Philosophy* — gives the needed track record. Also there is the semantics of natural language, in which possible worlds have been undeniably valuable. Each of these two nonmetaphysical arenas provides a formidable argument that modal concepts do have a track record of success. Still, there is room for doubt.

Something about the *structure* of possible worlds semantics is undeniably explanatory: it is fruitful to think of thought and discourse using a space of entities that in some sense count as possibilities. But it is less clear that the quality of the explanations in possible worlds semantics rests on the 'possibilities' in question really counting as possibilities in any deep sense. Robert Stalnaker's longstanding attitude toward possible-worlds explanations comes to mind here: he argues that it is not important for there to be a once-and-for-all space of possibilities; we need only a conception of possibilities adequate for the modeling purposes at hand.[22] Similarly, I would argue, though it's clearly fruitful for creatures like us to reason counterfactually, the quality of the practice of counterfactual reasoning is largely insensitive to the nature of the background space of possibilities, since what does the bulk of the explanatory work is the similarity metric on that space, or the contextual restriction on possibilities in the space, depending on one's views about the semantics of counterfactuals.[23] Thus it is unclear whether the concepts of possibility and necessity themselves deserve credit for the successes of possible worlds semantics and counterfactual reasoning.

Taking this line is a little uncomfortable, though. Compare the argument that mathematical entities must be accepted because our best physical theories quantify over them (Putnam 1971; Quine 1951, section 6; 1960, chapter 7). Even if it is possible to paraphrase away the quantification, or to take it instrumentally, that doesn't on its own answer the argument, for the more straightforwardly interpreted theory may be superior, on general theoretical grounds, to the paraphrased or instrumentalist replacements. The Stalnakerian attitude toward the successes of possible-worlds semantics and counterfactual reasoning are similar, and an opponent may make the same complaint.

Notes

1. Whether it has causes and effects depends on what we say about causation by absence.
2. See also Cresswell (1991).
3. This is the monadic case of a more general principle. *X* may not be free in *A*; *A* will in general contain *x* free, and may contain further free variables, which may be bound to the prefixing quantifiers.
4. See Lewis (1986, 59–60), who emphasizes the need to recognize both abundant and 'sparse' properties.
5. See Dorr (2005, 2007), Fine (2009), Schaffer (2009), Sider (2009, 2011, chapters 7–9), Cameron (2010).
6. On the latter claim see Schaffer (2009, 361), Sider (2011, 169, note 10), Bennett (2016, chapter 7).
7. Compare Fine (1977).
8. This general style of kind of argument — requiring different grounds for different facts in certain cases — is pursued by Dasgupta (2014, 573) in a different context.

There is, though, a hard question of exactly when distinct facts should be required to have distinct grounds, since the requirement is not correct in general — most friends of ground would admit, for instance, that the fact that *P* can ground both ∼∼*P* and also ∼∼∼∼*P*. (This hard question, in my view, is symptomatic of a problematic feature of the 'conditional' notion of ground, as opposed to 'biconditional' level-connecting concepts. See Sider (2013a; b).)

9. Well, the fact is partially singular since it involves $a_1 \ldots a_n$, but what's important is that it is general insofar as it involves nonconcreta.
10. The rejection might be restricted to *actually* unanchored things. This would be in the spirit of 'nonserious actualism' (Plantinga 1983; Fine 1985; Salmon 1998), and would arguably not be ad hoc: given *a*'s actual concrete existence, we can single it out and make modal claims about it in particular — even the claim that it could have been nonconcrete. (If *a* had been nonconcrete, what fact would have grounded its existence? Answer: 'The singular modal fact that *a* could have existed fundamentally!' — more of the same nonseriousness.)
11. See Weatherson (2003) for a discussion of a number of related issues.
12. On metaphysical fundamentality see Sider (2011).
13. As Williamson pointed out in response to a draft of this paper.
14. See also Bennett (2016, chapter 7).
15. 'But what if one doesn't know how, or even that, chemistry is based in physics?' — I'm thinking of the epistemology here in an 'externalist' way. The believer needn't compute what she's a priori entitled to believe about chemistry from some initial assumptions about physics. The picture is rather that certain possible worlds are a priori more likely, and that those worlds are in fact physically and hence chemically simple.
16. These concepts might be, but needn't be, metaphysically fundamental, since the ultimacy is epistemic: the presumption of simplicity is underived from other such presumptions. For example, perhaps there are ultimate presumptions of simplicity for certain higher level special-science concepts. This wouldn't require saying that special-science concepts are fundamental, though it would be naturally paired with the idea that special-science lawhood cannot be explained merely by appeal to physical lawhood.
17. Note that my account of necessity is limited in crucial ways — it does not cover iterations of modal operators, for example — and therefore would need to be extended in order to make contact with Williamson's argument, which involves the logic of second-order quantified modal logic. But on any way of extending it, necessity would remain disjunctive in the relevant sense.
18. See Kripke (1972, 45, note 13), Hazen (1979).
19. See Williamson (2013, 16). Williamson's classification of Lewis as a Necessitist might seem to ignore Lewis's (1968) scheme for translating modal formulas. For example, Lewis's translation of $\Diamond\exists xFx \rightarrow \exists x\Diamond Fx$ (an instance of the Barcan schema) into the language of counterpart theory is a false sentence saying that if some L-world (i.e. maximal spatiotemporally interrelated object) contains an *F* then there is an *F* in *this* L-world. Williamson would complain that the translation treats the quantifiers in the modal formula as restricted, whereas Necessitism concerns unrestricted quantification. Though I think that calling Lewis a Necessitist is unnecessarily confrontational (Lewis would insist that *ordinary* modal claims are to be interpreted in his way, and why fight about that?), I do think that the classification is fair insofar as the issue is what Williamson cares about most: access to full second-order modal reasoning. Since Lewis's translation scheme is contingentist it will sometimes interfere with modal second-order reasoning,

but then Lewis will be happy to suspend the translation scheme, as he always did when it caused trouble (Lewis 1986, 12–13).

20. See, for instance, Fine (1994, 2001, 2012), Schaffer (2009), Rosen (2010), Bennett (2016).
21. Note that Fine takes the quantifiers to be 'possibilist' in Fine (1995), his formal development of the logic of essence — 'for the sake of simplicity', he says (244).
22. See, for example, Stalnaker (1986).
23. See Sider (2011, section 12.12) on a related issue. For the two approaches to counterfactuals alluded to in the text, see Stalnaker 1986 and Lewis 1973 on the similarity approach, and Lycan 2001 and von Fintel 2001 on the contextualist approach.

Acknowledgements

Thanks to Karen Bennett and Timothy Williamson.

References

Bennett, Karen. 2016. *Making Things Up*. Oxford: Oxford University Press. Forthcoming.

Cameron, Ross P. 2010. "How to Have a Radically Minimal Ontology." *Philosophical Studies* 151: 249–264.

Chalmers, David J., David Manley, and Ryan Wasserman, eds. 2009. *Metametaphysics*. Oxford: Oxford University Press.

Cresswell, Max. 1991. "In Defence of the Barcan Formula." *Logique Et Analyse* 135 (136): 271–282.

Dasgupta, Shamik. 2014. "The Possibility of Physicalism." *Journal of Philosophy* 111: 557–592.

Dorr, Cian. 2005. "What We Disagree about when We Disagree about Ontology." In *Fictionalism in Metaphysics*, edited by Mark Kalderon, 234–286. Oxford: Oxford University Press.

Dorr, Cian. 2007. "There are No Abstract Objects." In *Contemporary Debates in Metaphysics*, edited by Theodore Sider, John Hawthorne and Dean W. Zimmerman, 32–63. Oxford: Blackwell.

Fine, Kit. 1977. "Postscript." In *Worlds, Times, and Selves*, edited by Kit Fine. London: Duckworth.

Fine, Kit. 1985. "Plantinga on the Reduction of Possibilist Discourse." In *Alvin Plantinga*, edited by J. Tomberlin and Peter van Inwagen, 145–186. Dordrecht: D. Reidel.

Fine, Kit. 1994. "Essence and Modality". In Tomberlin (1994), 1–16.

Fine, Kit. 1995. "The Logic of Essence." *Journal of Philosophical Logic* 24: 241–273.

Fine, Kit. 2001. "The Question of Realism." *Philosophers' Imprint* 1: 1–30.

Fine, Kit. 2009. "The Question of Ontology". In Chalmers et al. (2009), 157–177.

Fine, Kit. 2012. "Guide to Ground." In *Metaphysical Grounding: Understanding the Structure of Reality*, edited by Fabrice Correia and Benjamin Schnieder, 37–80. Cambridge: Cambridge University Press.

von Fintel, Kai. 2001. "Counterfactuals in a Dynamic Context". In *Ken Hale: A Life in Language*, 123–152. Cambridge, MA: MIT Press.
Goodman, Nelson. 1955. "The New Riddle of Induction." In *Fact, Fiction, and Forecast*, edited by Nelson Goodman, 59–83. Cambridge, MA: Harvard University Press.
Hazen, Allen. 1979. "Counterpart-theoretic Semantics for Modal Logic." *Journal of Philosophy* 76: 319–338.
Kripke, Saul. 1972. "Naming and Necessity". In *Semantics of Natural Language*, edited by Donald Davidson and Gilbert Harman, 253–355, 763–769. Dordrecht: D. Reidel.
Lewis, David. 1968. "Counterpart Theory and Quantified Modal Logic." *Journal of Philosophy* 65: 113–126.
Lewis, David. 1973. *Counterfactuals*. Oxford: Blackwell.
Lewis, David. 1986. *On the Plurality of Worlds*. Oxford: Blackwell.
Linsky, Bernard, and Edward N. Zalta. 1994. "In Defense of the Simplest Quantified Modal Logic." In Tomberlin (1994), 431–458.
Linsky, Bernard, and Edward N. Zalta. 1996. "In Defense of the Contingently Nonconcrete." *Philosophical Studies* 84: 283–294.
Lycan, William. 2001. *Real Conditionals*. Oxford: Oxford University Press.
Plantinga, Alvin. 1976. "Actualism and Possible Worlds." *Theoria* 42: 139–160.
Plantinga, Alvin. 1983. "On Existentialism." *Philosophical Studies* 44: 1–20.
Putnam, Hilary. 1971. *Philosophy of Logic*. New York: Harper & Row.
Quine, W. V. O. 1951. "Two Dogmas of Empiricism." *Philosophical Review* 60: 20–43.
Quine, W. V. O. 1960. *Word and Object*. Cambridge, MA: MIT Press.
Rosen, Gideon. 2010. "Metaphysical Dependence: Grounding and Reduction." In *Modality: Metaphysics, Logic, and Epistemology*, edited by Bob Hale and Aviv Hoffmann, 109–136. Oxford: Oxford University Press.
Salmon, Nathan. 1998. "Nonexistence." *Noûs* 32: 277–319.
Schaffer, Jonathan. 2009. "On What Grounds What." In Chalmers et al. (2009), 347–383.
Sider, Theodore. 2009. "Ontological Realism." In Chalmers et al. (2009), 384–423.
Sider, Theodore. 2011. *Writing the Book of the World*. Oxford: Clarendon Press.
Sider, Theodore. 2013a. "Reply to Kit Fine." *Philosophy and Phenomenological Research* 88: 738–746.
Sider, Theodore. 2013b. "Reply to Jonathan Schaffer." *Analysis* 73: 760–770.
Stalnaker, Robert. 1968. "A Theory of Conditionals". In *Studies in Logical Theory: American Philosophical Quarterly Monograph Series, No. 2*, edited by James Cornman, 98–112. Oxford: Blackwell.
Stalnaker, Robert. 1986. "Possible Worlds and Situations." *Journal of Philosophical Logic* 15 (1): 109–123.
Tomberlin, James, ed. 1994. *Philosophical Perspectives 8: Logic and Language*. Atascadero, CA: Ridgeview.
Weatherson, Brian. 2003. "What Good Are Counterexamples?" *Philosophical Studies* 115: 1–31.
Williamson, Timothy. 1998. "Bare Possibilia." *Erkenntnis* 48: 257–273.
Williamson, Timothy. 2000. "Existence and Contingency." *Proceedings of the Aristotelian Society* 100: 321–343.
Williamson, Timothy. 2002. "Necessary Existents." In *Logic, Thought and Language*, edited by A. O'Hear, 233–251. Cambridge: Cambridge University Press.
Williamson, Timothy. 2007. *The Philosophy of Philosophy*. Malden, MA: Blackwell.
Williamson, Timothy. 2013. *Modal Logic as Metaphysics*. Oxford: Oxford University Press.

Reply to Sider

Timothy Williamson

Wykeham Professor of Logic, University of Oxford, Oxford, UK

In 'On Williamson and simplicity in modal logic' (Sider 2016), Ted Sider makes a case against my application of an abductive methodology to modal logic, and in particular against my use of simplicity and strength as joint criteria for preferring some theories over others in that field.

1. Fundamental and non-fundamental abductions

My disagreement with Sider is best understood in the context of our agreement on much else concerning methodology. We agree that a broadly abductive methodology is appropriate for theory choice in many fields, including physics, chemistry, archaeology, criminology, fundamental metaphysics and fundamental logic. We also share a full-bloodedly realist conception of those fields as primarily engaged in the pursuit of truth: the abductive methodology is independent of any pragmatist or otherwise anti-realist view of what is at stake in theory choice. Sider is willing to apply a modestly realist conception to quantified modal logic too: either necessitism is true and contingentism false or *vice versa*, and we may hope to find out which. He even seems ready to grant that *if* we apply the abductive criteria of simplicity and strength to theory choice in modal logic, they will favour necessitism over contingentism.

However, in Section 4 of his paper, Sider argues that we should *not* apply the abductive criteria of simplicity and strength to theory choice in modal logic. For present purposes, □ is to be read as expressing metaphysical necessity, and he has argued elsewhere that metaphysical necessity does not cut reality at a fundamental joint, but rather makes a messy, arbitrary, disjunctive distinction (Sider 2011, 266–291). In the present paper, he argues that metaphysically modal inquiry lacks the sort of successful track record needed to back up the claim that it asks fundamental questions. For Sider, modal logic, metaphysically interpreted, does not investigate the sort of area where we can expect exceptionless strong and simple laws.

Sider concedes that the legitimate application of an abductive methodology is not restricted to fundamental science, since it may legitimately be applied in many non-fundamental sciences too, such as chemistry, archaeology and criminology, and even to everyday practical problems. In all those areas, we need to use some form of inference to the best explanation, in order to dismiss infinitely many gruesomely complicated alternative explanations that ingenuity could devise but good sense would not take seriously. However, Sider argues, the applicability of abduction in those areas depends on their relation to fundamental physics, because violations of simple non-fundamental principles would require violations of simple fundamental principles too. By contrast, he assumes, modality lacks that connection with fundamental physics. Thus, on Sider's view, modal logic is methodologically worse off than criminology.

One may question Sider's justification of abduction for non-fundamental science in terms of abduction for fundamental physics. He suggests that 'a sudden, huge, isolated exception to otherwise simple physical generalizations [...] would also constitute a sudden, huge, and isolated exception to otherwise simple chemical generalizations'. Perhaps so. But his next step is to claim: 'More generally, *any* sudden, huge, and isolated exception to otherwise simple chemical generalizations would require some such physical complexity or other; thus we assume a priori that these [chemical] exceptions are unlikely' (because the physical exceptions are unlikely). That generalization goes from chemical anomalies to physical anomalies, whereas the premise it is supposed to generalize went from physical anomalies to chemical anomalies; it is unclear just what connection between them Sider intends. In any case, how do such considerations help a detective rank an elaborate conspiracy theory involving international Freemasons below more mundane hypotheses as explanations of what appears to be a local gang killing? The conspiracy theory entails no violation of accepted physical laws.

The gap between fundamental physics and many legitimate applications of abduction to non-fundamental matters is much wider than Sider suggests. In particular, given that there are exceptionless strong and simple generalizations in fundamental terms, why should we also expect there to be exceptionless strong and simple generalizations in non-fundamental terms? The non-fundamental nature of the latter terms might unsuit them for capturing such generalizations. As Sider notes, it is not enough for the non-fundamental terms to be based somehow or other on fundamental terms, since if 'green' and 'blue' can be reduced to fundamental physical vocabulary, so too can 'grue' and 'bleen', yet the latter are far less suited to abduction than the former. Sider suggests that the disjunctive nature of 'grue' and 'bleen' is the problem. However, if 'gang killing' is reducible to the language of fundamental physics, it will presumably be as something like an infinite disjunction of infinite conjunctions whose disjuncts are not unified by any distinctive similarity at the fundamental physical level. Analogous issues arise even concerning the relation of macroscopic physics to microscopic physics. A proper account of abduction for non-fundamental branches of science may need to pay more attention to their quite significant degree of methodological autonomy.

2. Modal physics

Sider concedes the legitimacy of an abductive methodology in both fundamental and non-fundamental physics. He takes for granted that doing so implies nothing good about the application of such a methodology to modal inquiry, perhaps because he takes the language of physics to be purely non-modal. Since so much of physics is expressed in mathematical notation, without modal operators, one could easily get that impression. But mathematics becomes physics only when the variables are assigned physical meanings. Sometimes those physical meanings are themselves implicitly or explicitly modal.

For instance, a *phase space* consists of the set of all *possible* states of a physical system at a time, mathematically structured in some way (geometrically, topologically, or the like), typically including a dynamics to describe how the system evolves through a series of states over time. With pardonable exaggeration, a recent history of the idea of a phase space begins: 'Listen to a gathering of scientists in a hallway or a coffee house, and you are certain to hear someone mention phase space' (Nolte 2010, 33). As examples, it cites stock prices in economics, the dust motes in Saturn's rings, and high-energy particles in an accelerator. Phase spaces are used in even the most fundamental physics, such as quantum mechanics. When scientists reason about phase space, they often talk simply of 'states' without making the qualification 'possible' explicit, but it is tacitly understood, because no trajectory traced out by the system according to the dynamics over a complete history goes through every state in the space (typically, each trajectory is of lower dimension than the space itself). The key mathematical structure is defined over the space as a whole, not trajectory by trajectory. Each state may be represented by the values of the physically relevant parameters. For instance, if the system comprises *n* particles, each of which varies on six abstract dimensions, then a given state may be represented by a 6*n*-tuple, and the whole system is 6*n*-dimensional. In reasoning about the system, a scientist uses ordinary non-modal mathematics, but once the mathematics is applied the intended interpretation is modal. For example, let X be the set of states in which the system meets a condition C, and Y the set of states in which it meets a condition D. Then the scientist may write '$X \subseteq Y$', which looks purely non-modal, but the intended reading is 'The system meets D in every possible state in which it meets C', in other words, 'Necessarily if the system meets C then it meets D'.

Reasoning about a phase space is like reasoning about the set of worlds in a standard model of modal logic, and even more like reasoning about the set of ordered pairs of possible worlds and times in a standard model of modal temporal logic, which has temporal as well as modal operators. Since the standard model theory of modal logic and modal temporal logic is done in a metalanguage without modal or temporal operators, it too uses ordinary non-modal mathematics (set theory), but once the mathematics is applied to an intended model over a set of genuine worlds or ordered pairs of genuine worlds and times, the intended interpretation is modal.

States, like worlds or world-time pairs, are understood as jointly exhaustive and mutually exclusive. They are maximally specific in all relevant respects, for although they are typically states of a local physical system, not the whole universe, that system is treated in isolation. Just as we can equate propositions in a model of modal logic with sets of worlds, and propositions in a model of modal temporal logic with sets of world-time pairs, we can equate propositions in a phase space with sets of states. The set X of states corresponds to the proposition that the physical system is in a state belonging to X.

Even though the system does not traverse all the states in the phase space in any given history, they are all treated as possible states of the system absolutely, not relative to the standpoint of one state or another. A state space has no proper analogue of the accessibility relation between worlds in a model of modal logic, by which only worlds in a restricted subset are treated as possible from the standpoint of a given world. Of course, some states and not others are future from the standpoint of a given state, but that is a temporal rather than modal matter. In effect, for a phase space the only necessary proposition is the set of all states, and any nonempty set of states is possible. It follows that the propositional modal logic of phase spaces is S5, on which possibility and necessity are not themselves contingent matters. Thus mathematical reasoning about a phase space is not modally neutral; there is an implicit commitment as to which modal logic is appropriate.

The analogy between states in phase space and world-time pairs is not quite identity, because the dynamics may cycle: in some cases, if all the relevant parameters take given values at one time, they will take all those same values again a fixed length of time later (and so on ad infinitum, if the dynamics is deterministic). Thus the system may be in the same state at different times. Nevertheless, the analogy is strong enough to bring out the modal aspect of phase spaces. It is more than a purely formal analogy, because it is rooted in the physicists' own conception of the states as the possible states of a physical system.

The analogy can be carried into higher-order logic. Just as in modal temporal logic one can interpret quantification into sentence position, given a model, as quantification over propositions (sets of world-time pairs), so in the corresponding modal quasi-temporal logic for phase spaces one can interpret quantification into sentence position, given a phase space, as quantification over propositions (sets of states), and thereby formalize quantification over sets of states in the language. The logic of such quantification will be (higher-order) necessitist rather than contingentist, because which sets of states one is quantifying over does not depend on which state is current. Thus analogues of the Barcan and converse Barcan formulas will hold. Elsewhere, I develop the modal quasi-temporal logic corresponding to phase spaces in dynamical systems theory in much more detail (Williamson 2016). More generally, physicists' free use of mathematical reasoning over phase spaces no more indicates neutrality on modal matters than does metaphysicians' free use of mathematical reasoning over intended possible world models of quantified S5.

For present purposes, there is no need to insist on the point that physicists' reasoning about phase spaces embodies implicitly necessitist assumptions. What matters most is that when they investigate the phase space of a physical system, they are asking and answering implicitly modal questions. They are exploring the dynamical structure of possible states of the system. Physics is not a modality-free zone.

A potential source of confusion is that when scientists investigate the phase space of a physical system, they often do so in the spirit of *model-building*, following the widespread scientific practice of making simplified mathematical models of the target system in order to facilitate understanding of its key features. Thus they may not believe that the dynamics is perfectly accurate; they may not even believe that every state in the model is really physically possible. For instance, some standard models of predator–prey interaction specify the evolution of population numbers for the two species by differential equations, which treat the population numbers as continuous variables, even though the answers to 'how many?' questions should be discrete. A more general kind of simplification is that interactions between the target system and the rest of the universe are ignored. However, this kind of indirect model-building strategy is widespread throughout most sciences. It is applied to both modal and non-modal questions, and is quite consistent with a full-bloodedly realist attitude towards the target physical system. It does not detract from the concern of physics with modal matters.

3. Physical and metaphysical modalities

At this point, Sider might concede a modal dimension to physics, but deny that the modality is of the right kind for my purposes. Indeed, when physicists speak of 'all possible states' of a system, they may well not intend all *metaphysically* possible states. For instance, physicists may build into the dynamics various physical assumptions about the working of the system without assuming them to be metaphysically necessary. They may not even intend them to be physically necessary, in the strong sense of following from the laws of physics; they may instead embody more specific assumptions about, say, the dust motes in Saturn's rings that can be held fixed for their purposes. The issue is not whether they comprehend all metaphysically possible states of the system but whether they are relevantly related to its metaphysically possible states.

If the states in phase space were only intended to be *epistemically* possible, in other words, consistent with what we know about the system, they would not be relevant in the right way for my purposes. For suppose that some scientists conjecturally propose a dynamics for the phase space of a given system (perhaps in the form of some differential equations governing the specified parameters). The dynamics implies, for instance, that *necessarily* if the system is in state s_0 then in one second it will be in state s_1. The scientists cannot simply replace the strict conditional 'necessarily if' by a material one (perhaps governed by 'always'), for that would make the conjecture about s_0 vacuously true if the system actually never is in s_0: but if they happen to know that for some accidental reason the system actually never is in s_0, that does not verify the conjectured dynamical relation between s_0 and s_1. But it is also not to the point for them to conjecture that it is *epistemically* necessary that if the system is in s_0 then in one second it will be in s_1, which is in effect to conjecture that they *know* that if the system is in s_0 then in one second it will be in s_1. They know that they do not know so much (by hypothesis the dynamics is conjectural), and anyway their aim is to investigate the physical system itself, not their knowledge about it. They are, after all, physicists, not epistemologists. It is clear that the relevant modality for their conjectures to be about is not epistemic necessity but some sort of *physical* necessity, perhaps weaker than nomic necessity.

It is widely agreed that what is physically possible is also metaphysically possible. Someone might doubt the entailment on the grounds that logical consistency with the laws of nature suffices for nomic possibility but not for metaphysical possibility. However, such a view of nomic possibility makes it unduly sensitive to syntactic form, for instance by implausibly making it nomically possible for Hesperus to be distinct from Phosphorus. I argue in more detail elsewhere against such a conception (Williamson 2016). Anyway, whether 'Physical system S is in state *s*' is logically consistent with the laws of nature depends on delicate issues about the semantics of the terms 'S' and '*s*', of a kind with which physicists are hardly bothering when they consider the states in the phase space as the possible states of S. It is reasonable to assume that physical possibility in the physicists' sense entails metaphysical possibility. I will make that assumption henceforth, but the arguments below could also be adapted to the weaker assumption that physical possibility of the relevant sort is just *good evidence* for metaphysical possibility.

By Sider's own lights, when we read □ and ◇ as expressing *physical* necessity and *physical* possibility, in a sense appropriate to phase spaces as studied in physics, we can legitimately apply abductive criteria such as simplicity and strength to our choice of theory in quantified modal logic, because such theories are formulated purely in terms drawn from core non-modal logic and from physics, which he treats as unexceptionable for such purposes. Thus we can argue abductively in the manner of *Modal Logic as Metaphysics* for a restricted version of necessitism: it is *physically* necessary that everything is *physically* necessarily something. Call that principle 'physical necessitism', and its negation 'physical contingentism'; call the original version of necessitism in the book 'metaphysical necessitism', and its negation 'metaphysical contingentism'. By what was assumed above, metaphysical necessitism entails physical necessitism, but there is no obvious argument for the converse. Why should not some truths be physically necessary but metaphysically contingent? Ontology might be constant across the inner realm of physical possibilities, yet start varying in the outer realm of physically impossible metaphysical possibilities. But let us consider what physical necessitism already gives us.

The first point to notice is that the usual alleged common sense counter-examples to (metaphysical) necessitism all involve cases that are *physically* possible, not just metaphysically possible—indeed, if they were physically impossible, that would severely weaken their claim to common sense status. For instance, it was physically possible for Wittgenstein to have had a child, not just metaphysically possible. Similarly, it was physically possible, not just metaphysically possible, for any of us never to have been conceived or born. Thus, if the usual common sense cases are counterexamples to metaphysical necessitism, they are equally counterexamples to *physical* necessitism. Conversely, if physical necessitism is true, then most metaphysical contingentists are wrong about what is going on in the cases they regard as the most glaring counterexamples to metaphysical necessitism, and so far their necessitist opponents are right. That would already be a disaster for metaphysical as well as physical contingentism, since it would deprive both views of their strongest and most natural motivation.

A further corollary of physical necessitism is that metaphysical contingentists would still be stuck with much of the ontological inflation to which they object in metaphysical necessitism. For there would still be all the non-concrete entities that in some physical possibility are children of Wittgenstein, and so on. That is so whether or not one accepts Sider's claim that 'there is surely some sort of presumption to minimize ontology'. My own view is that one should minimize ontology only insofar as doing so contributes towards more general abductive values, such as simplicity and strength. For example, in the foundations of set theory, the tendency is towards new axioms that *maximize* rather than minimize set-theoretic ontology, because they make for a simpler, stronger, and more unified conception of the set-theoretic universe. But there is no need to press issues about ontological parsimony here.

In brief, Sider's methodological reservations leave unscathed an abductive case like that in *Modal Logic as Metaphysics* for physical necessitism in place of metaphysical necessitism, and once one accepts physical necessitism, one has already paid most of the price of metaphysical necessitism in pre-theoretically surprising consequences. In those circumstances, the point of holding out against metaphysical necessitism is obscure.

4. Modal logic's track record

There are also reasons independent of physical modality to reject Sider's methodological strictures on modal logic. He depicts it as having a poor track record, one 'mostly within philosophy'. That is clearly false of modal logic as a technical discipline, which is a flourishing branch of mathematical logic now being advanced more by mathematicians and computer scientists than by philosophers. Under dynamic, temporal and epistemic interpretations it has been extensively applied in both computer science and theoretical economics. The phrase 'mostly within philosophy' is apt for modal logic only when the modal operators are read metaphysically. It is hardly surprising that philosophers are those most interested in the metaphysical reading of the modal operators, just as they are those most interested in the unrestricted reading of the quantifiers. Sider accepts the latter reading as fundamental.

Of course, *Modal Logic as Metaphysics* adopts the metaphysical readings of the modal operators. What is the track record of modal logic with those readings? A formula is *metaphysically universal* if and only if its universal generalization is true on the metaphysical reading of the modal operators and the unrestricted reading of the quantifiers (Williamson 2013, 93–95). At the level of ordinary propositional modal logic, it is clear beyond reasonable doubt that all theorems of the modal system T are metaphysically universal formulas and that all metaphysically universal formulas are theorems of the stronger modal system S5. For T imposes only the compelling requirements that what obtains *can* obtain and that necessity is closed under logical consequence (in a language without tricky modal operators such as 'actually'), while anything stronger than S5 puts a specific finite upper bound to the number of mutually exclusive possibilities (Williamson 2013, 111). I will not argue here against sceptics who doubt that there are non-actual possibilities, just as I will not argue against sceptics who doubt that there is a physical world. When philosophers discuss metaphysical necessity and possibility, much of the time they use only formulas of modal depth 1, that is, formulas in which modal operators occur, but not in the scope of further modal operators. But every theorem of S5 of modal depth 1 is already a theorem of T, so all the candidate logics agree on such formulas. There is no serious dispute as to which formulas of modal depth 1 are metaphysically universal. Although some formulas of greater modal depth are more contentious, there is considerable convergence amongst modal metaphysicians on S5 as the correct propositional modal logic for metaphysical modality—for good reason, as I have argued elsewhere (Williamson 2016).

For quantified modal logic, contingentists and necessitists disagree even over formulas of modal depth 1 (such as $\forall x \Box \exists y\, x=y$). But those theoretical disagreements should not obscure the extensive agreement on the basics. For instance, necessitists and contingentists may well agree that any first-order modal formula true in all variable domain Kripke models is metaphysically universal (since constant domain models are a special case of variable domain models). Although counterpart theorists may reject a few of those theorems, their revisionism does not go far, on pain of discrediting counterpart theory. Of course, agreement is not truth, but there is no serious reason for suspecting that modal logicians have got the modal logic of metaphysical modality radically wrong. Even on its metaphysical reading, modal logic has a far more successful track record than do most branches of metaphysics.

5. Metaphysical modality as the maximal objective modality

I have not discussed Sider's disjunctive conception of metaphysical necessity, as a ragbag of disparate kinds of truth. If such a conception were correct, metaphysical modality might well be an unfit subject for systematic theorizing, abductive or otherwise. My own understanding of it is quite different. In ordinary thought we can get at it through counterfactuals, as what would be the case no matter what was the case (Williamson 2007). At a more theoretical level, we can approach the same metaphysical modality as the limiting case of a kind of modality we may call *objective*, by contrast with other categories such as *epistemic* modality. Physical possibility is an example of an objective modality; so are various grades of practical possibility. At least to a first approximation, objective modalities correspond to what linguists classify as *circumstantial* readings of natural language modal auxiliaries such as 'can' and 'must'. The category is similar to that of objective interpretations of probability, by contrast with epistemic, subjective and purely logical interpretations. Metaphysical possibility is simply the maximal type of objective possibility, in the sense that something is metaphysically possible if and only if it has at least one objective type of possibility. Physical possibility as investigated by physics is an objective type of possibility and so automatically entails metaphysical possibility, in vindication of a claim made earlier.

The objective modalities are closed under various operations. For instance, if $\Diamond_1$ and $\Diamond_2$ are types of objective possibility, $\Diamond_1\Diamond_2$ is also a type of objective possibility. In particular, therefore, if $\Diamond$ is metaphysical possibility, which is an objective type of possibility, then $\Diamond\Diamond$ is also a type of objective possibility, and so entails metaphysical possibility because the latter is the maximal type of objective possibility. That vindicates the S4 axiom $\Diamond\Diamond p \rightarrow \Diamond p$, perhaps the most controversial theorem of S5, which goes some way towards vindicating the convergence noted above on S5 as the propositional logic of metaphysical modality.

I defend the conception of metaphysical modality as the maximal objective modality in more detail elsewhere (Williamson 2016). It is arguably closer to what many metaphysicians have understood by the term than is Sider's disjunctive conception.

If metaphysical modality is just the limiting case of a category of modality many types of which are explored by the natural and social sciences, the presumption should be that we may legitimately apply the same sort of abductive methodology in theorizing about metaphysical modality that is applied in theorizing all over the natural and social sciences, just as *Modal Logic as Metaphysics* did. In particular, that gives us two converging lines of argument for *metaphysical* necessitism. We can give a direct abductive argument for it, along the lines of the book. But we can also give a more indirect argument, by first arguing abductively for various types of *physical* necessitism, and then arguing by a further abductive step for metaphysical necessitism as the strong and simple principle that unifies all the necessitist principles for more specific types of objective modality.

References

Nolte, David. 2010. "The Tangled Tale of Phase Space." *Physics Today* 63 (4): 33–38.
Sider, Theodore. 2011. *Writing the Book of the World*. Oxford: Clarendon Press.
Sider, Theodore. 2016. "On Williamson and Simplicity in Modal Logic." *Canadian Journal of Philosophy* 46 (4–5): 684–699.
Williamson, Timothy. 2007. *The Philosophy of Philosophy*. Oxford: Blackwell.
Williamson, Timothy. 2013. *Modal Logic as Metaphysics*. Oxford: Oxford University Press.
Williamson, Timothy. 2016. "Modal Science." *Canadian Journal of Philosophy* 46 (4–5): 453–492.

Models and reality*

Robert Stalnaker

Linguistics and Philosophy, Massachusetts Institute of Technology, Cambridge, MA, USA

ABSTRACT

Kripke models, interpreted realistically, have difficulty making sense of the thesis that there might have existed things that do not in fact exist, since a Kripke model in which this thesis is true requires a model structure in which there are possible worlds with domains that contain things that do not exist. This paper argues that we can use Kripke models as representational devices that allow us to give a realistic interpretation of a modal language. The method of doing this is sketched, with the help of an analogy with a Galilean relativist theory of spatial properties and relations.

> The philosopher of 'possible worlds' must take care that his technical apparatus not push him to ask questions whose meaningfulness is not supported by our original intuitions of possibility that gave the apparatus its point.
>
> Saul Kripke[1]

1. Introduction

'Actualism' is an old label for a thesis in modal metaphysics. The general thesis of actualism is that to be real and to be actual are the same thing. All that exists is what actually exists. This was not intended as a substantive thesis, setting limits on one's ontological commitments, since it says nothing about what actually exists, and the thesis would seem trivial if it were not for David Lewis's modal realism, which rejected it. On Lewis's picture,[2] reality consists of a plurality of universes that are spatio-temporally disconnected from each other. 'Actual' is

*Apologies to Hilary Putnam for appropriating the title of one of his famous papers (Putnam 1980). My topic is not unrelated to his, but I am not going to talk about his paper. I chose the title because it says what my paper is about.

an indexical term applying only to a proper part of reality, the part that is spatio-temporally connected with us (or when used by speakers of our language in other parts of reality, to refer to the parts spatiotemporally connected with them).

'Contingentism' is a more recent label for a thesis in modal metaphysics, a term coined by Timothy Williamson to contrast with 'necessitism', which labels a doctrine he defends. The necessitist's thesis is simple: everything that exists exists necessarily, and nothing could exist except what actually does exist. Contingentism is the contrary thesis, the denials of the two parts of necessitism: There are things that exist, but might not have, and there might have been things other than those there are. Necessitism is not modal realism, a doctrine that Williamson rejects, and it has no need for a distinction between what really exists and what actually exists. But the thesis does allow for the existence of things that might be people or physical objects, but in fact are not: things whose only properties are modal properties, such as the property of being possibly (but not actually) a table, a cabbage, an aardvark, or a person. And while the thesis holds that actual persons, tables, cabbages, and aardvarks are necessary beings, it allows that these things might have existed only in this form of pure potentiality.

Lewis acknowledged that his modal realism conflicts with unreflective common sense, and he took unreflective common sense seriously. (He thought the 'incredulous stare' that his doctrine elicited to be the strongest argument against it.) His defense of the doctrine was highly theoretical: he judged that despite the counterintuitive character of the doctrine, it offered the best theory for unifying a rich system of concepts involving modality, and for explaining the roles of those concepts in our cognitive and practical lives. The conflict with unreflective common sense is perhaps not quite so strong in the case of Williamson's necessitism as it is with Lewis's modal realism, but I think it should be acknowledged that it seems prima facie reasonable to think that the ordinary things we find in the world, including ourselves, are things that might not have existed at all, and that there might have existed things that in fact have no existence at all. It is, according to an intuitively natural view, contingent not only how existing things are arranged; it is also contingent what things there are to be arranged. So I think one needs a strong theoretical reason to reject these contingency intuitions, but it also must be acknowledged that the tensions between the contingency intuitions and a natural theoretical account of modality lie close to the surface. Orthodox Kripke semantic models for modal logic are hard to reconcile with the thesis that there might have been things that do not actually exist, but that semantic framework nevertheless seems to provide both intuitive insight into the relations between a diverse range of modal concepts, and powerful formal tools for getting clear about compositional semantics for the languages we use to describe modal phenomena. If the contingency intuitions are to be successfully defended, we need to find a way to interpret the orthodox models in a way that reconciles them with the contingency intuitions, and that explains their conceptual and technical success. I think the contingentist theses can be defended, and reconciled with orthodox Kripke semantics (extended to higher order modal languages, as Williamson's theory does). My book, *Mere Possibilities*, is an attempt at such a defense, but that defense was not as explicit as it should have been about the status that I take Kripke models to have, and about the relation between these models and the reality that one is using them to model. This paper is an attempt to spell out in a little more detail what I take this relation to be.

Here is my plan: In section 2 I will describe the orthodox semantics for modal languages, and say what the tension is between the contingency intuitions and this way of modeling modal statements. In section 3 I will make some general remarks about models and model structures, the idea of an *intended* model, and the usual way of understanding the contrast between realistic and instrumental uses of models. I will then suggest the possibility of a more complex relation between models and reality, an interpretation of models that treats them neither as merely instrumental, nor as intended models in the simple sense. The claim will be that we can use models to give a fully realistic interpretation of a modal language without giving a realistic interpretation of the models we are using. In section 4 I will try to clarify the general strategy with an analogy with the use of geometric models in a theory of physical space. I will suggest that the way the Galilean relationist about space uses Newtonian geometric models is analogous to the strategy I will propose for reconciling modal semantics with the contingency intuitions. In section 5, I will sketch the strategy as applied to Kripke models, and argue that it helps, not only to reconcile the contingency intuitions with the model theory, but also to motivate an attractive general metaphysical view of the nature of modal phenomena. In section 6, I will respond to some objections, and then close, in section 7, with some general methodological remarks.

2. The contingentist intuitions and Kripke models

The contingentist intuitions consist of the following two claims

(1) Some things exist only contingently.
(2) There might have been things that do not in fact exist.

It is (2) that is the more problematic, but there are good reasons to think that the two claims must go together. In any case, (2) seems as intuitively compelling as (1). Just as it seems plausible that Hillary and Bill Clinton might have had no children, in which case Chelsea Clinton would not have existed, so it seems equally plausible that they might have had a son rather than, or in addition to, a daughter. But it is at least prima facie reasonable to think that there is no actual thing that could have been Bill and Hillary Clinton's son.

The contingency intuitions proliferate: If Hillary Clinton is a contingently existing object, then it seems reasonable to think that the properties of being identical to Hillary, or being the daughter of Hillary are also contingent. And if there are singular propositions about Hillary (for example, that she is running for President in 2016) then one might think that these propositions are *object-dependent*, and so also things that exist only contingently. This extension to higher order contingency is not compulsory: Alvin Plantinga accepts the contingency intuitions about individuals, but rejects the extension to properties and propositions. But the generalization of contingentism is natural, once the first step is taken, and the tension between the orthodox semantics and the contingency intuitions is there even if one's contingentism stops at the individual level.

So what is the tension? The idea motivating possible-worlds semantics is that a proposition will be possibly true only if it could be realized in a specific way, which is to say only if there could be a complete possible situation in which that proposition is true. So the formal semantics for a modal language is given in terms of a structure that includes a set of possible worlds (usually with one of them representing the actual world of the model), thought of as maximal ways things might be. A statement will be possibly true if and only if it is true in at least one of these maximal possibilities.

A semantic framework of this kind provides a natural generalization of the semantics for an extensional language. Whatever structure the extensional semantics provides for the interpretation of the extensional language, the modal version of that logic will provide a structure of that kind for each possible world. So, since a model for first-order extensional quantification theory postulates a domain of individuals, the semantic strategy suggests that a model for first-order modal quantification theory should postulate a domain of individuals for each of the possible worlds.

The contingency intuitions imply (in the context of models of this kind) that the domains will vary from world to world. In particular, (2) implies that there will be counterfactual worlds with domains that contain things that are not in the domain of the actual world. But on the intended interpretation, the domain of the actual world is the domain of everything there is. For those who reject Lewis's modal realism, there is no room for anything outside of this domain. Our theorizing about modality is taking place in the actual world, and the materials available for theorizing – the things we appeal to in interpreting the modal language – must all be actual things. So what are these mere possibilia that inhabit the domains of the possible worlds that verify examples of our second contingency intuition?

It should be noted that the contingency intuitions, stated in ordinary modal language, do not, on the face of it, imply that there are merely possible things – things that might, but don't, exist. What must exist, according to claim (2) is only the possibility of things that do not in fact exist, and we have no reason to deny that this is an actual possibility. The move from the possibility of things of a certain kind to the existence of things that might be of that kind comes from the model-theoretic representation. But that representation has both a compelling intuitive motivation, and considerable success in clarifying the structure of modal discourse, so we should be reluctant to give it up, or even to treat it as a merely instrumental device. I think we can reconcile the orthodox model theory with a realistic account of the language that the models are used to interpret, but doing so requires a more complex understanding of the representational role of models. In the next section I will sketch the standard simple account of the relation between models and reality, as I understand it, and the complications that I think need to be added to this simple account.

3. Models, intended models, and instrumental uses of models

A model (in the model-theoretic sense) for a language has two parts: first, a structure of some kind that is specified independently of any language; second, a valuation function from expressions of the language to objects that are defined in terms of elements of the structure. In first-order *extensional* quantification theory, the structure is just a domain of individuals. The language contains n-ary predicates for each n, and sometimes names as well. The function assigns individuals from the domain to the names, and subsets of the set of all n-tuples of members of the domain to the n-ary predicates. In the case of a model for first-order modal quantification theory, the model structure consists of a non-empty set (the possible worlds), with one of them designated as the actual world of the model, sometimes a binary relation on the set,[3] and a domain of individuals for each possible world. The valuation function assigns to each primitive descriptive expression an *intension*, which is a function from possible worlds to extensions, where extensions are as they are in extensional quantification theory. So, for example, the intension of a binary relational predicate is a function from possible worlds to sets of ordered pairs of members of the domain of that world. In general, the definition of a model structure is abstract: the set of 'possible worlds', for example, might be any arbitrary non-empty set – say the numbers 1, 2 and 3. Models of this kind might be used for various purposes, for example to show the logical consistency of some set of sentences of the language, but for realistic interpretations of a language, one talks of *intended* models. On the usual way of understanding intended models, the component of the model that is the structure is supposed to be the actual subject matter of the language – what the language is being used to talk about. It is assumed that a use of a model is either intended, or merely instrumental, which in the latter case means that the model is used to show something about the logical capacities of a language to represent something, but not anything about what the language is actually used to represent.

Some may talk of *the* intended model of a language, or *the* intended structure. One may, for example, take the intended domain of extensional quantification theory to be the domain of absolutely everything that exists. Jon Barwise and John Perry take this view of what is required by a realistic interpretation of possible worlds semantics: 'If the model theoretic structures of possible worlds semantics . . . are supposed to be a model of something, say superreality . . . then there ought to be *one* that is an intended or standard model, *the* one that really corresponds to superreality.'[4] (They might better have said, 'the one that *is* superrreality. The correspondence is between the expressions of the language and its subject matter – a structure that constitutes modal reality itself.) But realism about model structures does not require completeness or comprehensiveness, and does not even require that there be a coherent notion of metaphysical completeness or comprehensiveness. So one who rejects the intelligibility of the idea of a domain of absolutely everything can still give a realistic interpretation of extensional quantification theory – an interpretation that specifies or presupposes a domain that is what the quantified sentences of the language are taken to be about. A model, to be an intended model in the standard sense, could be a model with a domain that is potentially extendable, but it must be a domain that includes *only* things that exist. (Does that imply that if gods do not in fact exist, the theist who intends a realistic interpretation of her language cannot include gods in her domain of discourse? Yes it does imply that, though of course the theist can coherently claim that there are gods. It is just that if there in fact are no gods then these claims will be false, on the theist's intended interpretation.)

But the situation of the theorist who accepts both the contingency intuitions and the Kripkean semantics is worse than that of a theist in a godless world. This theorist acknowledges the non-existence of some of the things that his model theory seems to be quantifying over (the merely possible things that inhabit the domains of some counterfactual possible worlds), and this seems to foreclose the possibility of a realistic interpretation of the model theory. I think this is right: the contingentist cannot take the model structures that are appropriate for modeling metaphysical modality to *be* the reality, or the part of reality, that our modal languages are talking about. But the relationship between models and the reality that is modeled may in some cases be more complex.

The central claim of this paper is that one may intend a realistic interpretation of a *language*, but use nonrealistically interpreted models as aids in stating the compositional semantics for the language. That is, one may use a model as a representational device, interposed between the language being interpreted and the reality that is its intended subject matter. So there will be a correspondence between the language and the model structure, and a further correspondence between certain features of the model structure and the reality being modeled. The interpretation of the modal object language will be fully realistic in the following sense: Each closed expression of the language (predicates, names, and sentences) will denote something that is real (at least according to the contingentist's theory of what there actually is.) This will be true, not only of the first-order modal language, but also of the higher order extensions of the language of the kind that Timothy Williamson discussed in his book, *Modal Logic as Metaphysics*. But, with the more complex use of models, there will be two steps to be taken on the road from an expression of the language to the entity that it denotes: from expression to entity in the model structure to which it corresponds, and from entity in the model structure to object, property, relation, proposition that it denotes. I will try to explain how this works in the next two sections, starting in section 4 with an analogy that provides a precedent for the strategy I want to promote, and then in section 5 by sketching the application of this strategy to the use of Kripke models to interpret modal languages in a way that is compatible with the contingency intuitions.

4. Geometric models and relativism about space

Consider the contrast between a Newtonian conception of absolute space and a Galilean conception of relative space. The relativist claims that there is no fact of the matter about absolute motion, no difference between an inertial frame moving at constant velocity through space and the frame at rest. The Newtonian acknowledges that according to the physical theory they both accept, there is no difference in the way bodies behave in an inertial frame at rest and the way they behave in one moving at constant velocity. The issue between the absolutist and the relativist is whether there is a *metaphysical* difference. According to the Galilean, physical geometry is a system of spatial relations between bodies, but the relations are not grounded in intrinsic spatial properties. There are no *locations* in space: the spatial points in a Euclidean geometric model of space are just devices for representing the relations. There is, for example, a real relation that holds between two bodies that are 10 m apart, but the fact that this relation holds is not grounded in the fact that one of the bodies is located at a certain spatial point that is 10 m from the point where the other is located. The Galilean can model his relativism by using the same models that the Newtonian absolutist uses, but he supplements the models with a family of permutation functions each of which maps space-time points onto space-time points. The set of mappings will be defined so that the distance relations between spatial points are preserved. Suppose we use quadruples of real numbers to represent space-time points, and that there is a permutation function in the family that maps each point $\langle t, x, y, z \rangle$ to a point $\langle t, x^*, y^*, z^* \rangle$. (Since time is absolute in the physics shared by the Newtonian and the Galilean, the time coordinate will always map to itself.) This will be an admissible permutation function only if the distance from the point $\langle x, y, z \rangle$ to point $\langle u, v, w \rangle$ in Euclidean space is the same as the distance from $\langle x^*, y^*, z^* \rangle$ to $\langle u^*, v^*, w^* \rangle$. But the points themselves may be different. If, in the model, the center of mass of the sun is located, at a certain time t_0 at point $\langle x, y, z \rangle$, then in the permutation of the model, it will be located (at time t_0) at point $\langle x^*, y^*, z^* \rangle$. The sun will be at rest from t_0 to t_1 in the basic model if it is located at the same spatial point throughout this interval (that is, if it is located at $\langle x, y, z \rangle$ for all t from t_0 to t_1). But it might be in motion (at constant velocity) in an admissible permutation of the original model, since it might be that as the time shifts from t_0 to t_1, the permuted spatial location at that time, $\langle x^*, y^*, z^* \rangle$, moves uniformly in some direction in a straight line. The Galilean's claim is that the original model and the permuted model represent the same factual situation. The facts about the motion of bodies – the changing spatial relations between them – are the facts that are invariant under the admissible permutation relations.[5]

Notice that the Galilean is not aiming to give a general criterion for distinguishing features of the geometric model that are artifacts from features of the model that will be shared by *any* model representing the same reality. I am not sure what it would mean to do that. The aim is the more modest one of representing the distinction between certain spatial properties and relations (represented in the models by sets of locations, or sets of ordered n-tuples of locations) that are real from objects of this kind that are artifacts of the model. So consider the fact that in our basic model, we used quadruples of real numbers as our space-time points. The model presupposes an arbitrary coordinate system. We might connect our model with physical phenomena by stipulating that the point $\langle 0, 0, 0, 0 \rangle$ shall be the center of mass of the sun at the moment of Isaac Newton's birth. The fact that this particular spatial point is the origin of our coordinate system, of course, has no representational significance. But once we have made this stipulation in our basic geometric model, the fact that the center of mass of the sun at the time of Isaac Newton's birth is the origin, $\langle 0, 0, 0, 0 \rangle$, will be invariant under the admissible permutation functions. The general idea behind the permutation strategy is to use models (or model structures) as representational devices, and it is the model plus the family of permutation functions that is the representational device. *Any* representational device will have features that distinguish it from other representational devices for representing the same thing. If the model structure were the reality itself, as it is on the simple picture, then (trivially) nothing about it would be an artifact of the model. The Newtonian absolutist might say that in the intended model, the points are represented, not by quadruples of real numbers, but by the locations themselves. The Galilean can't say this, since he claims there are no such things as locations themselves, but he can still use mathematical models (with a particular conventional coordinate system) to say in a precise way which of the properties and relations of a particular kind (as represented in the model) are real, and which are only virtual properties and relations.

One can generalize about the kind of situation in which the permutation strategy used by the Galilean is appropriate. Leibniz stated and defended the principle that *all relations between substances are reducible to (supervenient on, grounded in) the intrinsic properties of the relata*, but this is a substantive metaphysical thesis. Galilean relationism[6] about space is a paradigm example of a theory that rejects this Leibizian thesis,[7] since it holds that there are spatial relations that are not grounded in the intrinsic locations of the related bodies. There are also other structures that seem to have this feature. Preference relations, for example, seem to be essentially comparative. One might try to ground them in some measure of the absolute degree to which one likes or wants something, but this would not be very plausible. Measures of the extent of preference (as in utility theory) are usually taken to be irreducibly relational, and one uses the permutation strategy to make this way of understanding utility values explicit. All the positive linear transformations of a utility function are taken to be equivalent. A second example: I suggested once that a functional theory of phenomenal qualia might be an example of an ungrounded relational structure, since the capacity to discriminate between experiences is the central capacity on which such a theory is based.[8] The inverted spectrum example is a prima facie problem for a functional theory of qualia, since it suggests the possibility of a perfectly symmetrical system of the relations between types of qualitative experience. The permutation strategy helps to explain how symmetrical types of qualitative experience that are in a sense functionally equivalent, can still be different. There is a permutation that maps red to green and green to red, preserving discriminatory capacities, but it is an invariant feature of the structure that red is discriminable from green. (Whether such an account is a plausible account of qualia is another question, but at least it illustrates the abstract idea.)

5. Modal reality, and model structures

As I said in the last paragraph of section 3, the contingentist's strategy is to use a model structure, not understood as the structure of an *intended* model in the usual sense, but instead as a representational device interposed between the language and the reality that is being modeled. So let me start with some general remarks about the subject matter that we will be intending our modal language to be about. The type theory developed by Daniel Gallin[9] and discussed by Williamson in his book,[10] is a useful framework for describing the subject matter of a modal language, for both necessitists and contingentists. The type theory is a typology of entities that can be specified independently of any language that the type theory might help to interpret. There are two basic types, first, *e* for 'entity' (the domain of individuals)[11] and second a type that is mysteriously labeled $\langle\rangle$ (a pair of empty brackets) for propositions.[12] The other types are defined by a single recursive rule: if $t_1, \ldots t_n$ are types (basic or derived), then $\langle t_1, \ldots t_n \rangle$ is a type. The interpretation of the derived types is this: $\langle t_1, \ldots t_n \rangle$ is the domain of n-ary relations between things of types $t_1, \ldots t_n$, respectively. So for example, $\langle e, e \rangle$ is the domain of binary relations between individuals, $\langle e, \langle\rangle \rangle$ is the domain of binary relations between an individual and a proposition, $\langle \langle e, e \rangle \rangle$ is the domain of monadic properties of binary relations.

The basic type e has no intrinsic structure – that is, no structure that is imposed by the type theory itself. But the abstract interpretation of the type theory does impose a structure on the other types, and on the relations between the entities of the different types. The domain of propositions, for example, brings with it some basic properties and relations that are essential to the idea of a proposition, for example the relation of entailment that holds between a class of propositions and a proposition. The general abstract theory of propositions will impose certain closure conditions on the class of propositions, for example that every proposition has a contradictory (where the relation of being a contradictory is definable in terms of entailment). The derived types will also have related essential structural properties, required by the most general assumptions of what properties and relations are, and there will be a rich structure of interrelations between the members of the different types. An individual (member of type e) and a monadic property of individuals (member of type ⟨*e*⟩) will determine a (presumably unique) proposition (member of ⟨⟩) – the proposition that is true if the individual instantiates the property.

The *categories* of the type theory are built recursively from the basic categories, and the theory will impose an elaborate structure on these categories, specifying certain closure conditions, and relations between elements of the different categories. But we need not assume that the membership of the domain that is specified by a derived type is determined by the membership of the domains of the types from which it is derived. That is, it could be that the structural conditions imposed on the types and the relationships between them do not ensure that once we have specified the domain of individuals, and the domain of propositions, we have thereby determined the domains of all properties and relations, even on the most abundant construal of properties and relations. If our metaphysics makes certain necessitist assumptions, then a reduction of the domains of properties and relations to the domains of individuals and propositions may be possible. Consider the class of all functions from maximal consistent propositions to subsets of the domain of individuals (things of type e). The class of maximal consistent propositions is definable in terms of the domain of propositions, given its essential structure, so the necessitist might identify the domain of monadic properties of individuals (things of type ⟨*e*⟩ with this set of functions, and so reduce the domain of monadic properties to a domain definable in terms of the basic types. If this works, one could reduce all the higher types in a similar way. But the contingentist cannot do this, since for the contingentist there may exist a property F that is possibly exemplified by something that does not actually exist (and so is not in the domain of the type e). For the contingentist, the relational structure exhibited by the type theory will be ungrounded in a way that it would be grounded if the necessitist's metaphysical assumptions were true. This kind of ungrounded relational structure is, in general, the kind of phenomenon that motivates the permutation strategy illustrated by the Galilean theory of space.

Recall that the tension between the contingentist intuitions and Kripke models arose because the possibility that there is a thing of a certain kind is modeled by a thing (a member of a domain of a nonactual world) that is possibly of that kind. The necessitist takes this commitment at face value, but the contingentist notes that (aside from what the models seem to require) the inference does not seem to be mandatory. That is, for example, it does not seem incoherent on its face to assume that there really is the possibility that Hillary Clinton have had a son, even though there is nothing that might have been Hillary Clinton's son. The aim is find a way to use orthodox models to clarify modal phenomena without justifying this kind of inference.

So how can we use Kripke models to interpret a language for talking about a modal reality that has this kind of ungrounded relational structure? Just as the Galilean uses exactly the same geometric models as the Newtonian, so the contingentist uses the same Kripke model structures as the necessitist – model structures that would be necessitist if they were interpreted realistically. The materials in a model structure (a set of ***possible worlds***, with a domain of ***individuals*** for each ***world***) suffice to determine ***propositions*** (represented by subsets of the given set of ***possible worlds***) and ***properties*** and ***relations*** (represented by functions from ***possible worlds*** to subsets of the domain of the ***world***, or more generally to subsets of the set of n-tuples of members of the domain of the ***world***.) Since the plan is to use the model structure itself merely as a representational device, there is room for confusion in the use of terms for the elements of the model structure. So when the terms ('property', 'relation, 'possible world' 'proposition', etc.) are used to refer to elements of the model structure, I write them in bold italics, while when they are used as terms for components of the reality we mean to talk about (members of the categories in the type structure) I will use them unaccented. ***Possible worlds*** and ***possible individuals*** are just arbitrary objects – perhaps they are all real numbers. Some, but not all, of these things, and some but not all of the functions definable in terms of the primitives of the model structure – the ***propositions***, ***properties*** and ***relations***, will represent real entities – real propositions, properties, and relations. Which ones? That question is answered by classes of permutation functions mapping the ***possible worlds*** of the model structure onto the ***possible worlds***, and mapping the members of the domains of those ***worlds*** onto the domains of corresponding ***worlds***. The representational device we are interposing between the language and the alleged reality will consist of a Kripkean model structure plus a class of permutation functions of this kind associated with each ***possible world*** of the structure. A ***proposition p*** (a set of ***possible worlds***) will represent a real proposition (relative to ***possible world w***) if and only if it is invariant under all of *w*'s permutation functions, which is to say if and only if each of the permutation functions takes the set *p* to itself. The ***proposition p*** will be merely virtual if it is not invariant. Of course the set ***p*** itself, whether invariant in this sense or not, is a real set – perhaps, as I suggested, it is a set of real numbers. By calling some of these sets virtual, I am saying that they do not represent genuine propositions, where genuine propositions are the members of the domain of propositions in the type theory that we are taking to be the reality that our model language represents. We interpret the model structure with a correspondence relation between the invariant elements of the domains of ***individuals***, ***properties***, ***relations*** and ***propositions*** and elements of the appropriate kind in the corresponding categories in our type theory.

One of the ***possible worlds*** in our model structure is identified as the ***actual world*** of the model, and this component of the model structure will correspond to the maximal true proposition – the way the world actually is.[13] The classes of permutation functions are defined relative to each ***possible world***, but it is the ***actual world***'s permutation functions that are relevant to determining the actual commitments of our theory. (The other classes of permutation functions will be relevant to the interpretation of iterated modal claims.)

We have said nothing yet about a modal language, or about any correspondence between a modal language and the model structure. Our focus has been on the characterization of the subject matter that our language will be talking about, on a model structure (which is specifiable independently of any interpretation of a modal language), and on a correspondence between the model structure and the elements of the different types of our type structure. The language will be interpreted in the standard way by a correspondence between the primitive expressions of the language and semantic values that are definable in terms of the basic components of the model structure. It will be a constraint on a realistic interpretation of the modal language that each name and each primitive sentence letter or predicate be interpreted (in the model structure) by an ***individual***, ***proposition***, ***property***, or ***relation*** that is invariant, relative to the actual world's class of permutation functions. So, since the invariant elements of the model structure correspond to elements of the appropriate kind in the domains of the type structure, all of these primitive expressions will correspond indirectly to elements of things of the appropriate kind in the intended subject matter of the theory. And the structural conditions on the relational structure exhibited by the type-theory will ensure that each of the closed complex expressions of the language will correspond to real properties, relations, and propositions if the primitive ones do.[14]

My aim in this paper is to clarify in a general way the role of Kripke models in giving a realistic semantics for modal concepts that is compatible with a contingentist metaphysics. Discussion of the formal details are for another occasion, but the appropriate constraints on the permutation functions have been worked out, and result in interesting structures.[15] The relation between models and reality is complex on the contingentist picture, but I think the tensions in our intuitive conception of modal phenomena are real: both the contingency intuitions and the Kripke structures have features that need to be reconciled. The overall comparative evaluation of necessitism and contingentism will have many dimensions, but I believe there is a coherent contingentist picture that is attractive and worth developing. There may be some deep reasons for resisting the picture, but there are also some objections that are based on misunderstandings. In the next section I will respond to two of these.

6. Objections and replies

Nicholas Jones, in a recent paper,[16] argues that the invariance strategy for reconciling the contingency intuitions with Kripke models 'is flawed: [Stalnaker's] characterization of the representational/non-representational distinction does not cohere with his metaphysical view.' His argument for this conclusion presupposes that the aim of this strategy is to give a general characterization of all the features of a model that are artifacts of the model – all the features of a model that distinguish it from any model that is representationally equivalent to it. I doubt that this is a coherent aim, but in any case, as I have tried to make clear in sections 4 and 5 above, it is not the aim of the invariance strategy. Jones argues that the strategy fails in the representation of the Galilean relationist theory of space as well as in the attempt to reconcile the use of Kripke models with contingentism. Let me look first at that simpler geometric case:

'Stalnaker's relationist denies that spatial locations exist: "there are really no such things as spatial locations – there are just spatial relations between things." But it is an invariant feature of *all* the spatial models in any permutation class that there are spatial locations. Because Stalnaker's relationist rejects the existence of spatial locations, she faces a choice: (a) reject every spatial model as inaccurate; or (b) reject the Invariance-based analysis of non-representational artefacts.' I choose (b), given Jones's interpretation of the invariance-based analysis of nonrepresentational artifacts, but we can still use the invariance strategy to distinguish the spatial properties and relations between bodies that are real (according the theory) from those that are virtual. That is all the invariance strategy is trying to do.

Jones's argument for the failure of the invariance strategy in the modal case has the same structure, with possible individuals playing a role analogous to locations in the spatial case. Before looking at the argument, let me make a terminological clarification. The representational device that we are interposing between the language and the modal reality described by the type theory is a single Kripke model structure, supplemented with a family of permutation functions. Both the structure and the permutation functions are specified independently of any language that they might be used to interpret. So models, in the model-theoretic sense, that include, as one component, a valuation function mapping expressions to entities definable in terms of the elements of the model structure are not involved at this point. So we are not talking about invariance across a class of models, but invariance of permutation classes of ***properties***, ***relations***, ***propositions***, etc. that are features of the one model structure. But the claim that some sentence of a modal language is satisfiable will have implications for the existence of model structures meeting certain conditions.

Jones's argument concerns the interpretation of a sentence of the modal language that he calls '∃con': $\Diamond\exists x @\forall y\, x{\neq}y$, (where '@' is the actuality operator). The contingentist claims that ∃con is consistent, and this will be true only if there is a model structure in which there is a ***possible world*** with a domain that contains an ***individual*** (call it '***n***': that is, let '***n***' be a metalinguistic name for that component of the model structure) that is not in the domain of the ***actual world***. But the constraints on admissible permutation functions will ensure that every permutation takes ***n*** to a possibly different ***possible individual*** that is also not in the domain of the actual world. So (Jones argues) it is an invariant feature of the model structure, supplemented with the permutation structure, that, as he puts it, ∃con has a 'witness': a merely possible individual in virtue of which it is true. The witness will be different for different permutations, but there will always be one. This fact about invariance is true, but irrelevant to the question whether the invariance strategy commits the theorist to the existence, in modal reality, of merely possible individuals. A ***possible individual*** of the model structure corresponds to a real individual (a member of the domain of individuals in modal reality) only if the relevant permutation functions are invariant (all take the ***individual*** to itself.) The ***individual n*** does not pass the test, nor do any of the different ***possible individuals*** that are the images of ***n*** for some admissible permutation function. So, Jones's argument does not show that the invariance strategy commits the contingentist to the existence of merely possible individuals.

Williamson, in his book, gives a somewhat different argument for a closely related claim: that the contingentist cannot coherently reject the Barcan formula, (BF) $\Diamond\exists x\phi \rightarrow \exists x\Diamond\phi$. Williamson also seems to assume that the ambition of the invariance strategy is to give a general criterion of representational significance, but I think the main problem of his argument is that it conflates the ***individuals*** that are the components of a model structure with the individuals that, on the intended interpretation, are the range of the quantifiers of the language, and it conflates the classes of permutation functions on a model structure with different models.[17]

Here is the argument, as I understand it: Suppose we have a model M that invalidates BF. This requires a model structure in which there is a ***possible world w*** that has in its domain an ***individual o*** that is not in the ***domain*** of the ***actual world***. But independently of the model, "***o*** is an actual individual, just like anything else. . . . ***o*** is an actual individual that represents a non-actual individual (but not any particular non-actual individual).'[18] (This last remark needs interpretation. It is right that *o* does not represent any particular non-actual individual, since there are no non-actual individuals. What it represents is the possibility that there be an individual that does not in fact exist.)

Now (the argument continues), since ***o*** is an actual individual, there must be another model, M* in which ***o*** is in the domain of the actual world. This is true, but M* will not be relevant to the class of permutation functions that map the ***worlds*** of the structure onto the ***worlds***, and the ***possible individuals*** onto the ***possible individuals***, since all of the permutations of ***o*** will take it to an ***individual*** in the domain of a non-actual world. But Williamson seems to be assuming that since ***o*** is an actual individual, it must represent an actual individual (itself) in some admissible permutation on the model structure, and he is right that this would not be compatible with the use of ***o*** to represent the possibility of an individual that does not in fact exist. It is also right that if ***o*** is in the domain of individuals in our type structure that represents the intended subject matter of our language, then there will have to be something in the ***domain*** of the ***actual world*** of our model structure that corresponds to ***o***. But there is no reason why it has to be ***o*** itself.

7. Concluding methodological remarks

The necessitist's theory is simpler than the contingentist's, and some may take this to be a reason to be a necessitist. Simplicity is a theoretical virtue, so long as the simpler theory accounts as well for the data, but it is less clear in metaphysics what the data are than it is in science. The contingency intuitions are not unassailable data. We do need to account for the distinctions that they reflect, but the necessitist has a story to tell about the special status of the things that the vulgar are inclined to say are merely contingent things, and about the facts that ground the possibilities of things that we may be inclined to say do not in fact exist. Why jump through the complicated hoops that the contingentist metaphysics forces us to jump through? One possible answer is that the questions that contingentism raises are good questions, and that being forced to answer them brings out features of reality, and of our conceptual resources for talking and thinking about reality. Whether this is right, or whether alternatively the questions are forced on us only by a false metaphysical picture is a delicate question to be answered only by developing the alternative conceptions of metaphysical modality.

The invariance strategy is an appropriate general strategy for representing relational structures that are in a sense ungrounded in the intrinsic properties of the things that are related. The Galilean relativist grants that the Newtonian's models are simpler, but insists that reality does not provide the distinctions that would ground those models. We can nevertheless use those same models, she claims, to represent physical reality, and to factor out the distinctions to which nothing in reality answers. I think the general strategy works, and that it helps us to understand any structure of relations that do not satisfy the Leibnizian metaphysical grounding principle. It is, however, a further metaphysical question whether there are ungrounded relational structures of this kind, and if there are, where they are to be found.

The necessitist picture is an extensionalist picture, taking at face value the extensional, set-theoretic structures that modal model theory uses to clarify the compositional semantics of modal languages. Properties and relations are explained in terms of the individuals that instantiate them, or that might instantiate them, and propositions are explained in terms of sets of complete ways that the domain of all the things there are might be arranged. On the contingentist picture, possibilities, properties and relations are not reducible to the ways the things there are might be arranged. The hierarchies of propositions, properties and relations cannot be built with the materials found at the ground level: the individuals and possible worlds. But the contingentist can still the use the same extensional set-theoretic tools that the necessitist uses, with the help of the permutation strategy. The necessitist picture may in the end be defensible on metaphysical grounds, but I think the contingentist picture provides a coherent and attractive vision of modal reality that is worth taking seriously.[19]

Notes

1. Kripke 1980.
2. See Lewis 1986.
3. I will ignore the binary accessibility relation that is a constituent of a frame in a general Kripke model structure for modal languages, restricting attention to the S5 case where the accessibility relation is universal; necessity (in a given world *w*) is truth in all possible worlds, and not just in all that stand in some accessibility relation to *w*.
4. Barwise and Perry 1985. 120.
5. The kind of permutation strategy I am sketching is familiar in mathematics, and in measurement theory, but it is important to distinguish the use of this strategy by the Galilean from its use to bring out the conventionality of units of measurement. Newtonians and Galileans will agree, for example, that the choice of a coordinate system for physical space and the unit of measurement for representing distance relations are conventional. Any permutation of the numbers used to represent the distance between two bodies that preserves the ratios between distances will be an equivalent representation. What is permuted, in this case, is the relation between a number and the relational property that holds between two bodies iff they are a certain fixed length apart. The relation, *x being 2 meters from y* is the same relation as *x being 4 half-meters from y*, but the Newtonian and the Galilean agree that this is a real relation. Consider a possible world like ours, except that all the distance relations between bodies were doubled. This is very different from considering a world in which we meant by 'one meter' a distance relation that was what we actually mean by 'one half meter' (cf. Grünbaum 1964). The Newtonian claims that there are location properties, even though it is conventional what quadruples of real numbers we use to represent them. The Galilean, in contrast, denies that there are such properties. This disagreement is reflected in the permutation functions that are permitted by the Galilean's theory, which cannot be interpreted as permuting just the numerical representation of spatio-temporal

locations. The disagreement is about whether there is a real difference between rest and uniform motion at constant velocity.

6. It is an irony that Leibniz is both the one who propounded this grounding thesis, and also a prominent defender of a relational theory of space. His views are reconcilable, at least on the surface, since on his metaphysical theory, substances do not stand in spatial relations. His relationism about space is more radical (at least on one way of reading him) than that of the Galilean: space itself is merely ideal. The view seems to be that spatial relations between phenomena are ultimately reducible to the intrinsic properties of substances.
7. I think one should think of physical space itself as a structure of properties and relations of the things in space. On this way of thinking, for the absolutist, spatial locations, and regions defined by sets of specific locations, are intrinsic properties of the things at the location, or in the region. Spatial relations are determined by these intrinsic properties: once you have identified the specific locations of the relata, you have determined how they are spatially related to each other. But for the Galilean, the Leibnizian principle fails.
8. Stalnaker 2000.
9. Gallin 1975.
10. See Williamson 2013, 221ff.
11. As I understand him, Williamson interprets this domain as the domain of absolutely everything, including all of the members of the domains that are the other types. But as I suggested in section 3, we need not take a realistic specification of the subject matter of some language – even a language for doing metaphysics – to be comprehensive. So we could understand the type e as just a class of basic individuals, and take the other types to be classes of entities that are disjoint from the things of type e.
12. The motivation for the label, and for the idea that this type is really a derived type, is the idea that just as n-ary relations can be thought of as properties of *n*-tuples, so propositions can be thought of as 0-ary relations, properties of 0-tuples.
13. So our model structure is what Williamson calls an 'inhabited model structure'. There is a 'fixed point' constraint on the permutation functions, ensuring that in any of ***world w***'s permutation functions will take ***w*** to itself, so the ***actual world*** of the model structure , with its domain, will be, by definition, invariant.
14. I say that if the primitive predicates sentence letters and names, all denote real entities of the appropriate kind, then so will the complex ***closed*** predicates and sentences. As Bruno Jacinto has shown (in as yet unpublished work), this is not true for ***open*** expressions, which have as their semantic values ***propositional functions***, ***propositional function-functions***, functions from ***individuals*** to ***properties***, etc. Virtual propositional functions seem to play an ineliminable role in the compositional process by which complex predicates and quantificational constructions are interpreted, even though all of the complex predicates and closed quantificational constructions get an invariant interpretation. I should perhaps not have found this as surprising as I did, since the whole detour through virtual models (that in a sense, represent reality as more fine-grained than it is), together with permutation functions (that serve to bring us back to the right grain in our account of what is real) would be unnecessary if we could do the compositional semantics directly in terms of the real entities that our language is talking about.
15. I spell out some of the details in two appendices to my book, Stalnaker 2012, but the appendices are overly compressed, and contain some errors, which have been pointed out and corrected in work by Peter Fritz (Fritz, forthcoming).

16. Jones 2016.
17. I take some responsibility for the latter conflation, since as Williamson notes in a footnote (Williamson 2013, 189, note 48), I confusingly use the label 'model' for what is in fact a model structure. He assumes I must really have meant to talk about a model (the structure plus a valuation function), but what I meant was just a model structure.
18. Williamson 2013, 191.
19. Thanks to Peter Fritz, Bruno Jacinto, Nicholas Jones and Tim Williamson for very helpful discussion and correspondence about these issues.

References

Barwise, J., and J. Perry. 1985. "Shifting Situations and Shaken Attitudes." *Linguistics and Philosophy* 8: 105–161.
Fritz, P. (forthcoming). "Propositional contingentism," *Review of Symbolic Logic*.
Gallin, D. 1975. *Intensional and Higher-Order Modal Logic: With Applications to Montague Semantics*. Amsterdam: North-Holland.
Grünbaum, A. 1964. "Is a Universal Nocturnal Expansion Falsifiable or Physically Vacuous?" *Philosophical Studies* 15: 71–79.
Jones, N. 2016. "The Representational Limits of Possible Worlds Semantics." *Philosophical Studies* 173: 479–503.
Kripke, S. 1980. *Naming and Necessity*. Oxford: Blackwell.
Lewis, D. 1986. *On the Plurality of Worlds*. Oxford: Blackwell.
Putnam, H. 1980. "Models and Reality." *The Journal of Symbolic Logic* 45: 464–482.
Stalnaker, R. 2000. "Comparing Qualia Across Persons." *Philosophical Topics* 26: 385–405.
Stalnaker, R. 2012. *Mere Possibilities: Metaphysical Foundations of Modal Semantics*. Princeton: Princeton University Press.
Williamson, T. 2013. *Modal Logic as Metaphysics*. Oxford: Oxford University Press.

Reply to Stalnaker

Timothy Williamson

Wykeham Professor of Logic, University of Oxford, Oxford, UK

In 'Models and reality' (Stalnaker 2016), Robert Stalnaker responds to the tensions discerned in *Modal Logic as Metaphysics* between contingentism in modal metaphysics and the use of Kripke models in the semantics of modal logic. Stalnaker holds that 'the tensions in our intuitive conception of modal phenomena are real: both the contingency intuitions and the Kripke structures have features that need to be reconciled' (Section 5), and that 'The contingency intuitions are not unassailable data' (Section 7). His strategy for achieving reconciliation involves accepting 'the contingency intuitions' while arguing that 'we can use models to give a fully realistic interpretation of a modal language without giving a realistic interpretation of the models we are using' (Section 1). The paper follows up the strategy of Stalnaker's earlier work (2010, 2012). Of his defence of contingentism and its compatibility with orthodox Kripke semantics, he writes (in Section 1):

> that defense was not as explicit as it should have been about the status that I take Kripke models to have, and about the relation between these models and the reality that one is using them to model. This paper is an attempt to spell out in a little more detail what I take this relation to be.

It is easy enough to make out the broad outlines of Stalnaker's big picture, but much harder to bring it sharply into focus. Recent work by Peter Fritz and Jeremy Goodman has revealed some of the complexities and difficulties beneath the surface of Stalnaker's account (Fritz 2015; Fritz and Goodman 2016). Stalnaker's present paper does not fully resolve them. Since aspects of his account remain unclear, my comments will be tentative.

1. Stalnaker's realism and instrumentalism in model-theoretic semantics

Consider this passage from Stalnaker's earlier book (2012, 41):

> the merely possible individuals, and the points in logical space used in Kripke models as I am interpreting them, are like the spatial points in a relativist's model

> of spatial structure. The intended subject matter of our modal theory consists of the actual individuals, the (actual) properties and relation[s] that they might exemplify, and the (actual) higher-order properties and relations that might be exemplified by properties, relations, and propositions, as well as by individuals. About all these things, our theory can be resolutely realistic.

The reader may get the impression here that Stalnaker is contrasting the actual individuals in a Kripke model (the members of the domain of its actual world) with the merely possible individuals in the model (the non-members of that domain that are members of the domain of other worlds in the model): the merely possible individuals play an instrumental role, while the actual individuals themselves belong to the intended subject matter of the modal theory. In my discussion of Stalnaker's criterion of representational significance, I quoted a similar passage from his slightly earlier article (Stalnaker 2010, 24; Williamson 2013, 189). However, in his response to my critical discussion of his earlier work, Stalnaker now insists that there is 'no reason' why actual individuals in the Kripke model have to represent themselves (Section 6). Thus, even the identities of the members of the domain of the actual world of the model become representationally insignificant; what matters is only which individuals they 'correspond to'. Is that what Stalnaker means by being 'resolutely realistic' about such aspects of the model?

Rather than worry about whether Stalnaker has modified his view, I will concentrate on his account in the present paper. He formulates his overall view of the model theory's role thus (Section 3):

> The central claim of this paper is that one may intend a realistic interpretation of a *language*, but use nonrealistically interpreted models as aids in stating the compositional semantics for the language. That is, one may use a model as a representational device, interposed between the language being interpreted and the reality that is its intended subject matter. So there will be a correspondence between the language and the model theory, and a further correspondence between certain features of the model structure and the reality being modeled.

To specify *which* features of the favoured model structure are to correspond to features of the modal reality being modelled, Stalnaker associates each world in the model with a group of permutations of the worlds and individuals, which induce permutations of all the set-theoretic entities based on the model too. Then a set-theoretic feature of the model should correspond to something real according to a world *w* in the model if and only if the feature is invariant under all the permutations associated with *w*. Since being real according to the actual world should coincide with being real (*tout court*), a set-theoretic feature of the model should correspond to something real if and only if the feature is invariant under all the permutations associated with the actual world of the model. Depending on the nature of the invariant set-theoretic feature, it will correspond to a real object, property, relation, proposition or whatever.

To see how this might work, we can use a toy model. For simplicity, we may first consider a model of propositional modal logic, with worlds but no domains of individuals. For the set W of worlds, we take a set of just six ordered pairs: {⟨0, 0⟩, ⟨0, 1⟩, ⟨1, 0⟩, ⟨1, 1⟩, ⟨2, 0⟩, ⟨2, 1⟩}. The designated actual world of the model is ⟨0, 0⟩. As Stalnaker makes clear, there is no requirement for the 'worlds' of a model to be worlds in any serious informal sense. Let the permutation group associated with the world ⟨*i*, *j*⟩ comprise just those permutations of W that map ⟨*i*, *j*⟩ to itself (as Stalnaker's account requires) and leave the first component of each pair fixed. One can think of ⟨*i*, 0⟩ and ⟨*i*, 1⟩ as qualitatively identical twin worlds, or mirror images of each other. Thus, the permutation group associated with the actual world ⟨0, 0⟩ contains only four permutations: the identity permutation, the permutation that just swaps ⟨1, 0⟩ with ⟨1, 1⟩, the permutation that just swaps ⟨2, 0⟩ with ⟨2, 1⟩, and the permutation that makes both swaps. Since all those permutations leave each of ⟨0, 0⟩ and ⟨0, 1⟩ fixed, both those worlds in the model should correspond to real worlds: in the model, ⟨0, 1⟩, though actually real, does not actually obtain. Each other world in the model is shifted by some of the permutations, so it should not correspond to a real world.

If sets of worlds in the model correspond to anything real, they correspond to propositions. For instance, each of the sets {⟨1, 0⟩, ⟨1, 1⟩} and {⟨2, 0⟩, ⟨2, 1⟩} is invariant under all the actual world permutations, so it should correspond to a real proposition. By contrast, each of the sets {⟨1, 0⟩, ⟨2, 0⟩} and {⟨1, 0⟩, ⟨2, 1⟩} is shifted by some of the permutations, and so should not correspond to a real proposition. But *which* real proposition is the set {⟨1, 0⟩, ⟨1, 1⟩} supposed to correspond to? Obviously, nothing has yet been done to determine that. We cannot say that it is the proposition true in the worlds ⟨1, 0⟩ and ⟨1, 1⟩ and false in all other worlds, because neither ⟨1, 0⟩ nor ⟨1, 1⟩ is supposed to correspond to a real world. The set {⟨1, 0⟩, ⟨1, 1⟩} is supposed to correspond to something real even though none of its members does. This is presumably the sort of phenomenon Stalnaker has in mind when he writes that 'we need not assume that the membership of the domain that is specified by a derived type is determined by the membership of the types from which it is derived' and that 'For the contingentist, the relational structure exhibited by the type theory will be ungrounded in a way that it would be grounded if the necessitist's metaphysical assumptions were true' (Section 5). But then how *is* the correspondence to real propositions to be set up? Of course, in this particular case we could just *stipulate* a proposition for {⟨1, 0⟩, ⟨1, 1⟩} to correspond to, preferably one that could be true in just two ways. But we cannot solve the problem for the whole infinite type-theoretic hierarchy by ad hoc stipulation alone. A more systematic approach is needed.

Similar issues arise once we move to models of quantified modal logic and give each world a domain of individuals. Consider another toy model, this one with just two worlds, *w* and *w**. By stipulation, the domain of individuals for *w* is {0, 1, 2, 3}; the domain of individuals for *w** is {4, 5, 6, 7}. Each world is associated with the permutations of individuals that fix each individual in its domain and map 'even' individuals to 'even' ones and 'odd' individuals to 'odd' ones. Of course, this is just a convenient way of specifying the permutations; the *real* individuals are not supposed to be numbers, and the numbers representing them have no intrinsic significance. Let *w* be the actual world of the model. Thus, the group of permutations of individuals associated with the actual world contains only four permutations: the identity permutation, the permutation that just swaps 4 with 6, the permutation that just swaps 5 with 7, and the permutation that makes both swaps. Since only individuals in the domain of the actual world, are invariant under all the associated permutations, only they should correspond to real individuals. Let F be the monadic property intension whose extension at *w* is {0, 2} and whose extension at *w** is {4, 6}. F is invariant under all four permutations, so it should correspond to a real property. But *which* real property should it correspond to? Even once we have associated the invariant individuals 0, 1, 2 and 3 in the model with real individuals, and the invariant worlds *w* and *w** with real worlds, we cannot say that F should correspond to the property possessed by just the real individuals corresponding to 0 and 2 at the real world corresponding to *w* and by just the real individuals corresponding to 4 and 6 at the real world corresponding to *w**, for neither 4 nor 6 is supposed to correspond to a real individual. The supposed correspondence between F and a real property is not determined by correspondences lower down the type-theoretic hierarchy.

An *ad hoc* stipulative approach rapidly becomes infeasible even for very low levels of the type-theoretic as the models increase towards more realistic levels of complexity, in both the number of worlds and the associated domains of individuals. But the problem goes deeper than that. For Stalnaker wants Kripke models to play a key role in mediating between the modal language and the realistic interpretation we eventually give it. He says that his 'claim will be that we can use models to give a fully realistic interpretation of a modal language without giving a realistic interpretation of the models we are using' (Section 1) and that 'one may intend a realistic interpretation of a *language*, but use nonrealistically interpreted models as aids in stating the compositional semantics for the language' (Section 3). Moreover, he regards such a key role as well-motivated: he says that the model-theoretic representation 'has both a compelling intuitive motivation, and considerable success in clarifying the structure of modal discourse, so we should be reluctant to give it up, or even to treat it as a merely instrumental device' (Section 2). Thus, the model-theoretic intermediaries should be properly integrated into the final 'fully realistic interpretation' of the modal language.

The problem is less urgent for the interpretation of the non-logical primitive constants of the language, because they are normally interpreted by one-off stipulations. It is much more serious for the interpretation of the quantifiers and variables of the various types. It is that which is at stake when Stalnaker writes 'We interpret the model structure with a correspondence relation between the invariant elements of the domains of ***individuals***, ***properties***, ***relations*** and ***propositions*** and elements of the appropriate kind in the corresponding categories in our type theory' (Section 5; the boldface italic expressions apply to appropriate set-theoretic ersatz items in the model, while the 'elements of the appropriate kind in the corresponding categories in our type theory' are the real individuals, properties, relations and propositions). We are told very little about *which* correspondence it is. Of course, we have some prior informal understanding of what real individuals, properties, relations and propositions are, and so of what it means to generalize over them, independently of the model theory. But if we fall back on that prior informal understanding to interpret the modal language, we are in effect *bypassing* the formal model theory, contrary to Stalnaker's stated intention of using the model theory to mediate between the modal language and its realistically interpreted compositional semantics.

In his earlier book, Stalnaker has an appendix (B) discussing an approach to possible worlds semantics in terms of propositional functions, whose upshot is supposed to be that

> We can talk with a clear conscience, in the metalanguage, about a domain of possible individuals because we have shown how to reconcile that talk with more austere ontological commitments and how to do the compositional semantics in a way that assigns as values only properties, relations, and functions that actually exist, according to the metaphysics that is presupposed. (2012, 147–148)

One might hope that this alternative semantics would somehow bear on the problem. But on examination it does not help. In particular, it leaves the role of the worlds unreduced, and the problem already arises for worlds, as we saw above. Nor does it help explain how the intension F in the second toy model gets to correspond to a real property. Since Stalnaker does not talk about propositional functions in the present paper, I will say no more about them here.

The problems just discussed do not prevent Stalnaker from using the model theory as an algebraic device with which to specify a formal consequence relation for the modal language. But doing so falls far short of providing a compositional semantics for the modal language, since it does not provide its sentences with truth conditions, even relative to specifications of the real objects, properties, relations and propositions associated with its non-logical constants. For instance, if someone tells you which sentences are logical consequences of which sets of sentences, and that the 1-place predicate *S* corresponds to the property of being spherical, you cannot determine from that alone that $\Box\exists x\, Sx$ is true if and only if it is necessary that something is spherical; inequivalent readings of $\Box$ and different domains for the quantifier may yield the same consequence relation. Of course, you may happen to know the truth condition of the formula independently, from your informal understanding of the modal language, but that is not what Stalnaker needs, since it involves no mediating role for the model theory. Moreover, if the model theory is only an algebraic device for specifying a formal consequence relation for the modal language, no other link having been specified to the intended interpretation of the language, why should we be interested in that consequence relation? We have not yet even been told why we should expect it to preserve truth from premises to conclusion, let alone why we should expect its logical truths (the consequences of the empty set of premises) to be all and only the metaphysically universal formulas.

As far as I can see, Stalnaker has not yet elucidated an integral role for Kripke models to play in a realistically interpreted compositional semantics for a modal language.[1]

2. Comprehension principles for Stalnakerian modal logic

One of the important points of modal metaphysics on which Stalnaker and I agree, and on which we disagree with some other contributors to this volume, is that higher-order contingentism is a natural — though not automatic — generalization of first-order contingentism. As he says, 'If Hillary Clinton is a contingently existing object, then it seems reasonable to think that the properties of being identical to Hillary, or being the daughter of Hillary are also contingent' (Section 2). Stalnaker endorses such higher-order contingentism.

As *Modal Logic as Metaphysics* notes, higher-order contingentism involves restrictions on comprehension principles for higher-order modal logic. For instance, in the quoted passage, Stalnaker allows that there is the property of being the daughter of Hillary Clinton, which we may express in the higher-order modal language, thus (reading the 1-place predicate *D* as 'is the daughter of Hillary Clinton'):

(1) $\exists X \Box \forall x(Xx \leftrightarrow Dx)$

But Stalnaker denies that there is *necessarily* such a property; in other words, he denies the necessitation of (1):

(N1) $\Box \exists X \Box \forall x(Xx \leftrightarrow Dx)$

In effect, Stalnaker rejects the natural modal comprehension principle Comp_M for quantification into monadic predicate position, which provides necessitated instances such as (N1) (Williamson 2013, 262–263).

Restricting Comp_M has a cost: it obstructs normal second-order reasoning in modal contexts. For instance, suppose that in reasoning about counterfactual circumstances we assume the second-order generalization $\forall X\phi(X)$. We may wish to infer the instance $\phi(D)$; but we have no general right to do that, given the failure of (N1); (1) is insufficient, since we have it only for the actual circumstances. Of course, first-order contingentism faces a similar problem. For instance, suppose that in reasoning about counterfactual circumstances we assume the first-order generalization $\forall x\, \phi(x)$. We may wish to infer the instance

$\phi(d)$; but we have no general right to do that, given the failure of $\Box\exists x\, d=x$ under first-order contingentism; $\exists x\, d=x$ is insufficient, since we have it only for the actual circumstances. What my book emphasizes is that the failures of comprehension undermine the free use of the kind of reasoning that provides the most powerful reason for introducing higher-order logic in the first place (Williamson 2013, 282–288). I now elaborate the point by showing how much higher-order modal logic we in effect assume when applying non-modal mathematics in scientific reasoning about phase spaces of possible states of a physical system (Williamson 2016b, 2016c).

In the most natural formalization of reasoning about phase spaces, specific states are treated like worlds, and sets of such states like propositions, so generalizing about sets of states corresponds to quantifying into sentence position. Stalnaker is a contingentist about propositions too, so similar issues arise. For instance, he allows that there is the proposition that Hillary Clinton is running for President in 2016, which we may express using quantification into sentence position thus (reading *P* as 'Hillary Clinton is running for President in 2016'):

(2) $\exists X\Box(X \leftrightarrow P)$

But Stalnaker denies that there is *necessarily* such a proposition; in other words, he denies the necessitation of (2):

(N2) $\Box\exists X\Box(X \leftrightarrow P)$

In effect, Stalnaker rejects the natural modal comprehension principle Comp_P for quantification into sentence position, which provides necessitated instances such as (N1) (Williamson 2013, 290). But the obvious semantics for interpreting a modal language with quantification into sentence position over phase spaces verifies principles generalizing (2) and (N2).

It is not even clear whether Stalnaker's principles entitle him to instances of the comprehension principles *without* the initial occurrence of the necessity operator. For instance, he will presumably allow that there is the property of *not* being the daughter of Hillary Clinton (permutation-invariance is preserved under negation). We might formalize that thus:

(3) $\exists X\Box\forall x(Xx \leftrightarrow \neg Dx)$

For Stalnaker, such a property exists only contingently: in some possible circumstances, there is no Hillary Clinton and no property of not being her daughter; there is just no distinction between being her daughter and not being her daughter. By classical modal reasoning that Stalnaker accepts, in those circumstances many things would have not been her daughter. By (3), they would in effect have had the property of not being her daughter — even though there would have been no such property for them to have had. In other words, a predication can be true with respect to circumstances in which the property designated by the predicate does not exist. But this involves Stalnaker in treating predications as asymmetrical between subject and predicate with respect to existential commitment, for his preferred first-order modal logic includes the principle that a predication *cannot* be true with respect to circumstances in which the object designated by the subject term does not exist (Stalnaker 2003, 148; see Williamson 2013, 183–188 for discussion). In brief, predications strictly imply the existence of the object but do not strictly imply the existence of the property. By contrast, an attractive feature of Stalnaker's approach in the present paper is the symmetry of treatment between objects, properties, relations and propositions.

In this reply, I have raised several challenges to Stalnaker's account of the semantics, logic and metaphysics of modal discourse. I hope that they will encourage him to develop a more explicit theory, so that the strengths and weaknesses of his version of contingentism can be properly assessed.

Note

1. For a closely related debate on how the model theory of modal logic connects to modal metaphysics with reference to Williamson (2013) see deRosset (2016), Kment (2016), and Williamson (2016a).

References

Fritz, Peter. 2015. "Propositional Contingentism." *The Review of Symbolic Logic* 9: 123–142.

Fritz, Peter, and Jeremy Goodman. 2016. "Higher-order Contingentism Part 1: Closure and Generation." *Journal of Philosophical Logic*.

Kment, Boris. 2016. "Model Theory and Contingent Existence." *Analysis* 76: 172–190.

deRosset, Louis. 2016. "Modal Logic and Contingentism: A Comment on Timothy Williamson's Modal Logic as Metaphysics." *Analysis* 76: 155–172.

Stalnaker, Robert. 2003. *Ways a World Might Be: Metaphysical and Anti-Metaphysical Essays*. Oxford: Clarendon Press.

Stalnaker, Robert. 2010. "Merely possible propositions." In *Modality: Metaphysics, Logic, and Epistemology*, edited by Bob Hale and Aviv Hoffman. Oxford: Oxford University Press.

Stalnaker, Robert. 2012. *Mere Possibilties: Metaphysical Foundations of Modal Semantics*. Princeton: Princeton University Press.

Stalnaker, Robert. 2016. "Models and Reality." *Canadian Journal of Philosophy* 46 (4–5): 710–727.

Williamson, Timothy. 2013. *Modal Logic as Metaphysics*. Oxford: Oxford University Press.

Williamson, Timothy. 2016a. "Replies to King, deRosset and Kment." *Analysis* 76: 201–222.

Williamson, Timothy. 2016b. "Modal Science." *Canadian Journal of Philosophy* 46 (4–5): 453–492.

Williamson, Timothy. 2016c. "Reply to Sider." *Canadian Journal of Philosophy* 46 (4–5): 700–709.

An A-theory without tense operators

Meghan Sullivan

Philosophy Department, University of Notre Dame, Notre Dame, IN, USA

ABSTRACT
A-theorists think there is a fundamental difference between the present and other times. This concern shows up in what kinds of properties they take to be instantiated, what objects they think exist and how they formalize their views. Nearly every contemporary A-theorist assumes that her metaphysics requires a tense logic – a logic with operators like $\mathcal{P}$ ('it was the case that ...') and $\mathcal{F}$ ('it will be the case that ...'). In this paper, I show that there is at least one viable A-theory that does not require a logic with tense operators. And I will argue that three common indispensability arguments for tense operators are unsound.

1. Introduction

A-theorists think there is a metaphysical difference between the present and other times. This difference influences how many A-theorists understand the nature of change, what objects they think exist and what kinds of properties they take to be instantiated. It also influences how they formalize their views. Nearly every contemporary A-theorist assumes that her metaphysics requires a tense logic with operators like $\mathcal{P}$ ('it was the case that ...') and $\mathcal{F}$ ('it will be the case that ...'). In semantics, a tense operator takes a well-formed formula to a 'time' (or 'times') in a specified model, where the formula is then evaluated. In metaphysics, A-theorists often take indispensable operators to reflect a primitive 'tensed' structure of reality. Still it remains obscure what precisely this structure is. Moreover, Priorian tense operators have logical and metaphysical complications which have long been an albatross for A-theorists. Do we have any alternatives?

In this essay, I aim to question the operator dogma. I will show that there is at least one well-defined ideology for the A-theory that does not require tense operators, and I will raise objections to three common arguments for the indispensability of tense operators. Here is how the essay will proceed. Section 2 outlines the typical commitments of an A-theory of time and contrasts it with the two most common formulations of the B-theory. Section 3 considers three arguments that A-theories require tense operators and discusses some of the costs associated with irreducible tense operators. Section 4 proposes a theory of 'was' and 'will' as predicate modifiers and outlines ways we could adapt predicate logic to include the modifiers. Section 5 uses this apparatus to introduce an operator-free A-theory and argues that the primary indispensability argument is unsound. It also addresses some metaphysics-based objections to the proposal. Section 6 discusses ways the operator-free theory might be expanded to accommodate more robust A-theoretic assumptions, in particular the assumption that propositions change truth-value and the assumption that objects come into and out of existence. As we'll see, the operator-free A-theory fits best with a permanentist ontology. In the conclusion, I'll raise questions about whether necessitists about modal ontology should likewise dispense with modal operators.

2. Three accounts of property change

As I write this, Barack Obama is President. He won't always be President. The Obama administration is temporary. For some, this induces anxiety. For others, it comes as a great relief. But only the most downtrodden Republican, the most pie-in-the-sky Democrat, or the most inflexible nominalist would deny that being President is some kind of temporary property of Obama. More generally, almost all of us believe in:

> Property Change: There is at least one object, *x*, and at least one property, *C*, such that *x* is *C* only temporarily.

Philosophers nevertheless disagree about how this principle is to be expressed most precisely and what the principle entails about objects, properties and the structure of time. One major fault line in this debate is between the *A-theorists of time* and *B-theorists of time*. In a nutshell, the A-theorists assume that there is some objective, metaphysically privileged distinction between the present and other times. Accordingly, some kinds of change in the world can only be expressed by distinguishing ways an object is from ways it was or will be. Obama is President, but *was* not always president and *will* not always be President. B-theorists, in contrast, think of space and time as components of a multi-dimensional manifold, with objects 'spread out' in the manifold much the way we ordinarily think objects are spread out in space. Just as there is no objective, privileged 'here' in space, there is no objective, privileged present, past and future in the B-theory manifold. There are merely regions of space and time and objects that are located earlier or later with respect to others. So for B-theorists, marking any kind of change involves nothing more than marking how an object is with respect to different parts of the manifold. For example, the B-theorist may describe Obama as someone who is President at our part of the manifold, but isn't President at some other parts.

This is an admittedly rough and pictorial way of characterizing the A-theory/B-theory distinction. We can make the debate even more precise by considering how the different parties explicitly formulate the property change principle. A-theorists think that there are some temporary properties that a single object can have or lack without any relativization to a space–time manifold. We will call these *A-properties*. For a toy example, suppose we treat *is President* as an A-property. Then Obama is temporarily President because (i) Obama has the property *is President*, (ii) he is an enduring object (not an object spread out in a manifold) and (iii) he – the very same, numerically identical object – has lacked that property or will lack it. More generally, A-theorists think the following principle is true of our world and time:

> A-Property Change: There is at least one enduring object, x, and at least one non-time-relational property, C, such that x is C only temporarily.

Let us say that if A-property change is, was or will be true of some world, then it is an *A-theoretic world*. There are different ways the A-property change principle might be satisfied. Many A-theorists think most or all objects change by enduring and gaining or losing non-time-relational properties. Others think A-property change is limited to special properties like *is present* which a particular time can have or lack. For these A-theorists, A-property change occurs for regions of space–time rather than for particular objects. What unites A-theorists is the view that at least some kinds of change in our world cannot be explained as mere variation in a spatiotemporal manifold.[1]

The B-theorists, in contrast, explain *all* change as variation in a space–time manifold. There are differences of thought within the B-theory about how objects and properties relate to the manifold. One camp insists that all temporary properties are relations between an object and a region of space–time. Such relational B-theorists understand property change as:

> Relational Change: An object o changes with respect to a property, C, if and only if there are at least two times, t_1 and t_2, C is a time-relational property, o is C at t_1 and o is not C at t_2.

In other words, for relational B-theorists, objects change their properties by varying the relations they have to different times. If Obama is temporarily President, this means he is *President-at–2012* but is not *President-at-some-other-time.*[2]

Other B-theorists explain change using the doctrine of temporal parts. Suppose we took the multi-dimensional manifold and 'sliced' it down one of its time indices. According to temporal parts theorists, if an object has any parts at that slice, then it has a temporal part at that slice. The temporal part is the maximal region of that slice that the object fills. We can use this notion of a temporal part to give the informal temporal part principle of property change:

Temporal Part Change: An object *o* changes with respect to a property, *C*, if and only if *o* has at least two temporal parts, p_1 and p_2, such that p_1 is *C* and p_2 is not.

On this account, if Obama is temporarily President, this is because he has a temporal part located at 2014 which is President, and he has at least one other temporal part that is not President. According to some adherents, an upshot of the temporal parts account over the relational account is that it does not require the changing property *C* to be a time-relation. At least some temporary properties can be intrinsic properties.[3]

Each of these three principles – A-property change, relational change and temporal parts change – carry particular commitments about the structure of changing properties and the objects that instantiate them. For example, if the relational B-theorists are right, many properties have hidden structure. Nothing just has a property *C*; something is only *C-at-a-time*. If the temporal parts theorists are right, property change entails that objects have multiple temporal parts. If the A-theorists are right, objects don't need to have temporal parts or relational properties to undergo property change – some change is not to be described relative to a space–time manifold. Some philosophers doubt that these differences between the theories run very deep. But for the purposes of this essay, I will assume that the endurantism/temporal parts and time-relational/non-time-relational debates are substantive.[4]

So much for sketching the metaphysical terrain of this debate. Many A-theorists and B-theorists think that they also have distinctive *logical* commitments. (Indeed, retreating to a less ambiguous formal language is one way to reply to charges of non-substantivity.) Say we wanted to express the different principles of property change in a dialect of logic-ese. And say we wanted the logical representation of our theory to provide a somewhat accurate guide to our ontological and ideological commitments. Both kinds of B-theorist would only need the resources of predicate logic to express their principles. Relational B-theorists can express their view as:

Relational Change 1: An object *o* changes with respect to a property, *C* iff $\exists t_1 \exists t_2 (C(o, t_1) \wedge \neg C(o, t_2))$.

And temporal parts theorists will make use of a precise, primitive parthood relation. Here is one way to formalize the temporal parts view. Let *TPART(x,y)* abbreviate the regionally maximal parthood relation: *x* is a regionally maximal part of *y* iff for some arbitrary time-like slice of the space–time manifold, if *y* has parts located at the slice, then every part of *y* at that slice is also a part (proper or improper) of *x*. The view can then be expressed in predicate logic as:

Temporal Part Change 1: An object *o* changes with respect to a property, *C* iff $\exists p_1 \exists p_2 (TPART(p_1, o) \wedge TPART(p_2, o) \wedge C(p_1) \wedge \neg C(p_2))$.

In English: Some object has temporal parts and one part instantiates *C* while another does not. As with the relational theory, the logical backdrop for the temporal parts theory is purely extensional. To know whether there is change, we just need to know what objects are temporal parts of other objects and which objects instantiate *C*.

In contrast to the B-theorists, A-theorists assume that they need a formal language with tense operators to capture their view. Most commonly, they use a logic based on the four Priorian operators:

$\mathcal{P}$: 'it was the case that';

$\mathcal{F}$: 'it will be the case that';

$\mathcal{H}$: 'it has always been the case that'; and

$\mathcal{G}$: 'it is always going to be the case that'.

These operators behave like the $\Diamond$ and $\Box$ of modal logic; they shift formulas in their scope to be evaluated at other 'times' just as modal operators shift formulas to other 'worlds', where times and worlds are points in the model theory. For example, $\mathcal{P}\alpha$ is true if and only if for some past time, f was true then. In a quantified tense logic, the A-property change principle is expressed as:

A Property Change 1: For some property *C*, $\exists x(C(x) \wedge (\mathcal{P}\neg C(x) \vee \mathcal{F}\neg C(x)))$.

In English, there is some object and some property *C* such that the object has the property and either it was the case that the object lacks the property or it will be the case that the object lacks the property.[5]

From the earliest days of presentism, this added commitment to tense operators has been widely assumed to be part and parcel of adopting A-theoretic metaphysics. Prior – the godfather of the contemporary A-theories – declares in *Past, Present and Future*: 'The formation rules of the calculus of tenses are not only a prelude to deduction but a stop to metaphysical superstition'.[6] And elsewhere: 'Tense logic is for me, if I may use the phrase, *metaphysically fundamental,* and not just an artificially torn-off fragment of the first-order theory of the earlier-later relation'.[7] But more recently, the link between A-theories and Priorian tense logic has come into question. Zimmerman argues persuasively that a commitment to irreducible tense operators in one's metaphysics is not sufficient for having an A-theory of time, since some B-theorists may need tense operators to express their views about propositional content.[8] Could we argue the other direction? Are tense operators even *necessary* for expressing an A-theory? More precisely, can we describe the A-property change principle in a rigorous logical system without using tense operators? To answer this, we must consider arguments for the indispensability of tense operators. There are at least three such arguments suggested in the literature, and as we will see, each depends on some problematic background assumptions. Consider each in turn.

3. Three indispensability arguments

3.1. The argument from expressive adequacy

The first indispensability argument for tense logic requires the fewest assumptions but will serve as the touchstone for most of this essay. In Section 2, I assumed that what distinguishes A-theorists is their commitment to the A-property change principle. Most A-theorists will want to express their distinctive metaphysical claim in a formal language. As we saw, if the A-theorist adopts some version of Priorian tense logic, this is easy enough to do. And it does not seem possible to express that an object has a temporary, non-time-relational property with just a version of predicate logic. For suppose an A-theorist tried to work without tense operators. She could either use time-indexed predicates to formulate her view or not. If she doesn't use time-indexed predicates, it seems she'll be forced to say that an object has contradictory properties. She'll have to formulate A-property change as:

A Property Change 2: For some property $C, \exists x(C(x) \wedge \neg C(x))$.

This principle is never true. She could avoid the problem by relativizing C to particular times and expressing the principle as:

A Property Change 3: For some property $C, \exists x \exists t_1 \exists t_2 (C(x, t_1) \wedge \neg C(x, t_2))$.

But then her view collapses into the relational B-theory. Here, then, is the *argument from expressive adequacy.* A-theorists can either choose some logic with tense operators as the background for their metaphysics or they can choose some version of predicate logic. If they choose tense operators, they can express their distinctive metaphysical commitment in a consistent and interesting way. If they choose predicate logic, they cannot. If an ideology is required for giving a consistent and interesting version of a theory, then that ideology is indispensable. Therefore, tense operators are indispensable to the A-theory. The problematic assumption, as we will see, concerns the expressive limitations of an operator-free logic.[9]

3.2. The argument from propositional temporalism

A second, related argument motivates the link between A-theories and tense logic by way of propositions. Many A-theorists are serious about propositions in the following sense. First, they believe propositions are the bearers of truth and falsity. Second, they believe true propositions accurately represent the actual world, and false propositions do not. As we've defined it, an A-property is a property which is not held relative to a space–time manifold. Suppose *is President* is one of Obama's temporary A-properties. It seems the accurate way to represent this state of affairs is to use a proposition like [*Pres*(*o*)]. (I will use brackets when I intend to be explicit about the logical form of a proposition.) Call this the *Obama proposition.* As I write this sentence, the Obama proposition is true. And given that *is President* is a temporary property, the Obama proposition is only temporarily true. The Obama proposition will not always accurately reflect the world. How do we express that the proposition is only temporarily true? It seems we can either build a time-index into the proposition or append a tense operator. If we take the former option, we'd think sentences like 'It is temporary that Obama is President' really express the same proposition as 'Obama is President at 2014, and there is some time such that he is not President then'. (In logic-ese: $[Pres(o, 2014) \wedge \exists t \neg Pres(o, t)]$.) But then the Obama proposition does not accurately represent the actual world because, according to the A-theorists, the property *is President* is *not* time-relational. So it must be that we express the temporariness of the Obama proposition using a tense operator, i.e. $[Pres(o) \wedge \mathcal{F} \neg Pres(o)]$. Likewise for any other temporary A-property. Therefore, the A-property change principle and seriousness about propositions entail a commitment to tense logic. Call this the *argument from propositional temporalism*.[10] The problematic assumption, as we'll see, is that only tense operators or time-indexing can capture temporary truth.

3.3. The argument from ontological change

The last argument for the indispensability of tense operators also has the most background assumptions. Many metaphysicians nowadays are neo-Quineans.[11] Neo-Quineans endorse a package of views about the existence:

(1) There is a single sense of 'exists' of interest to metaphysics.
(2) Existence is not a monadic property of objects.
(3) Existence is properly expressed by the existential quantifier ($\exists x$) and the identity relation (=).

Neo-Quineanism is a popular view and with good reason. It precisely captures widely held convictions about the univocity of existence, and it provides adherents with a defense against charges that their ontological debates are non-substantive. When deflationists complain that A-theorists and B-theorists are merely talking past one another, neo-Quineans can insist that if the debates are appropriately regimented in a quantified logic, they are carrying on a substantive debate in a common, unambiguous language.[12]

Neo-Quineanism coupled with a further view about change seems to entail the indispensability of tense operators. The most common A-theories are presentism and the growing block theory. Presentists think that only the present is real. On one way of precisifying the view, presentists think that every non-abstract object is located at, at most, one region of space–time, and over time objects and/or regions of space–time come to exist and cease to exist. There are no merely past or future objects.[13] Growing blockers think that present *and* past objects are real. They think that no concrete objects are located at future times, and as time passes, objects and/or regions of space–time come to exist.[14]

Both kinds of traditional A-theory supplement the A-property change principle with an additional principle:

> Temporary Existence: Some object exists only temporarily.[15]

Growing blockers think some objects haven't always existed. Presentists think some objects haven't always existed and some objects won't always exist.[16] Notice, for Neo-Quineans, this temporary existence principle must be an *extra* assumption in addition to the A-property change principle, because neo-Quineans do not think existence is a property that objects gain and lose.

Like every other sensible person, presentists and growing blockers think there are truths expressed by claims like 'There was a dinosaur' or 'There will be a 64th President'. But neither thinks that the future President exists yet, and presentists further deny there are any dinosaurs. Call any merely past or merely future object a *temporal alien*. To handle what seem to be true existential claims about temporal aliens, neo-Quinean A-theorists turn to the resources of tense logic. It is widely thought that tense operators like $\mathcal{F}$ and $\mathcal{P}$ have a prophylactic effect on quantifiers in their scope, akin to the effect that possibility and fiction operators have on quantifiers. Suppose Ruth and Jake are planning to have a child in a few years. Presentists and growing blockers will maintain that a de dicto claim of the form $\mathcal{F}\exists x Child(x,r,j)$ is true. But they will deny this entails any de re claim like $\exists x Child(x,r,j)$ or $\exists x \mathcal{F} Child(x,r,j)$, because they deny that any merely future object exists. Their child will come to exist, but nothing *is* their future child. The wide-scope prophylactic tense operators block unwanted ontological commitments, and there is no other explicit mechanism in the neo-Quinean framework for doing this. So given neo-Quineanism, temporary existence and truths about temporal aliens, traditional A-theorists conclude that tense logic is indispensable. We see such reasoning explicitly in Zimmerman: 'The presentist must, I think, be a serious tenser. At the very least, tenseless statements that require ontological commitment to past and future things must be treated as equivalent to tensed truths that do not'.[17] Call this the *argument from ontological change*. The problematic assumption, as we'll see, is that prophylactic tense operators can help traditional A-theorists capture truths about temporal aliens.

3.4. But tense operators have baggage ...

If the three foregoing arguments succeed, tense operators are a necessary component of the ideology for A-theories of time. This is not a universally welcome result, for at least three broad reasons. First, some metaphysicians think that their logical primitives ought to reflect their views about the structure of reality. The A-theories are often characterized as theories which hold that tense operators reflect something metaphysically fundamental. But – we might wonder – what could it be for an intensional operator to reflect some aspect of reality? Wouldn't it be better if, like the B-theories, every logical device required by the A-theory mapped to more familiar categories like objects and properties? So the first major problem with Priorian tense logic is that indispensable tense operators seem objectionably mysterious when compared with the B-theoretic ideology, which merely requires the more familiar object/property distinction.[18]

Second, Priorian tense logic has well-known difficulties capturing the appropriate truth conditions for tensed sentences in natural language. In particular, Priorian tense logic has difficulty accounting for true claims involving cross-time relations and inferences that involve temporal anaphora. For these reasons (among others), the overwhelming trend in natural language semantics is to model tense as quantification over times, just the way B-theorists do. Shouldn't it give the B-theorists some advantage if their ideology maps more closely to the framework that best predicts the validity of natural language inferences about time?[19]

The third problem is, in my opinion, the most severe. According to the usual dogma about tense operators, they are supposed to block unwanted ontological commitments. Operators allow A-theorists to theorize about merely future or past objects without entailing that those objects exist. But Priorian tense logic inherits all of the problems of the modal logic it is based on. In particular, if you add the standard logic of quantifiers to even the weakest versions of tense logic, then you can prove that the temporary existence principle is false. (This is the well-known problem of the Barcan formulas, which can be derived in tense logic just as in modal logic). And to date, no alternative version of tense logic has been defended that upholds the neo-Quinean scruples. So while tense operators are supposed to be a refuge for certain A-theorists – the presentists and growing blockers – in practice, tense logic has never fulfilled this promise. And in the absence of an appropriate quantified tense logic, a major premise of the argument from ontological change is unjustified.[20]

These problems with tense operators should lead us to question whether A-theorists need even bother with them. For too long, the Priorian dogma has gone unquestioned. Here is the proposed engineering project: can we build a theory of time – a metaphysical theory and corresponding ideology (i.e. formal language)–that upholds the distinctive commitments of an A-theory of time without making use of tense operators? In the remainder of this paper, I will show that we can. First I will devise a version of predicate logic with a different tool for expressing tense. Then I will use this logic to give an interpretation of the A-property change principle, showing that the argument from expressive adequacy is unsound. In the final part of the paper, I will describe how the theory could be extended to handle propositional temporalism and ontological change, ultimately concluding that the theory fits best in a permanentist setting.

4. Step 1: Tense as a predicate modifier

4.1. The basic idea

The first step in devising the new A-theory requires us to expand the language of predicate logic. This section will provide a broad overview of the expanded logic and some independent motivations for it. It is a straightforward matter to develop an explicit version of the system using type-theoretic semantics or lambda abstraction.[21] But here I will use a simpler notation more suited to understanding the key philosophical ideas.

We start with standard predicate logic. Standard predicate logic contains a set of terms (constants and first-order variables) $\{a_1, \ldots, a_n\}$ and $\{x_1, \ldots, x_n\}$, a set of predicates each with positive arity n, $\{C^n \ldots\}$, truth-functional operators $(\neg, \rightarrow, \wedge, \vee, \leftrightarrow)$ quantifiers $(\forall, \exists)$ and parentheses and brackets. The grammar for standard predicate logic is familiar: the simplest, atomic well-formed formula results from a predicate with n arity being satisfied by n terms; $\neg$ or any quantifier placed to the left of a wff yields a wff; and any two wffs connected with a $\rightarrow$, $\wedge$, $\vee$, or $\leftrightarrow$ yields a wff.

We are going to add a new component to this language, a *predicate modifier*. Predicate modifiers have been studied in some detail by philosophers and linguists as a formal tool for capturing reasoning with adverbs, prepositions and relative adjectives.[22] We will symbolize predicate modifiers using capital letters with non-negative superscripts and subscripts $\{M^n_m \ldots\}$. Semantically, predicate modifiers behave as functions from predicates to predicates. In my notation, the superscript, n, indicates the arity of the predicates that the modifier can modify. Call this the modifier's *degree*. The subscript, m, indicates the arity of the complex predicate that the modifier returns. Some predicate modifiers are what I will call *simple*: they return a complex predicate of the same arity as the one that they modify. For example, suppose the English modifier 'slowly' is simple. It takes intransitive verbs like 'runs' and results in complex one-place predicates like 'slowly runs'. 'Runs' and 'slowly runs' each need just one variable or constant in order to produce a well-formed formula. But other predicate modifiers are *expanders*: they return a predicate with a higher arity than the predicate that they modify. Suppose the English preposition 'in' is an expander. The modifier 'in' takes a one-place predicate like 'runs' and returns a relation of the form 'runs in (...)' where the new argument place is meant to be filled with a location. Some predicate modifiers might even be *reducers*, reducing the arity of predicates they modify. We have no need for reducers here, so we will ignore this option.[23]

We'll need a more systematic account of the formation rules for modifiers, which is easily provided. First, for any simple predicate modifier M^n_n and any wff α such that α does not contain any quantifiers or logical constants, $M^n_n\alpha$ is a wff. Second, for any expander M^n_{n+i} and any wff α such that α does not contain any quantifiers or logical constants, $M^n_{n+i}[\alpha](x_1, \ldots, x_i)$ is a wff. Here the notation $[\alpha](x_1, \ldots, x_i)$ abbreviates α with i argument places added to the right and satisfied with *i* variables or constants. In this notation, we use the brackets to distinguish the new argument places from the argument places that were already in the unmodified predicate. Finally, all of the usual grammatical rules for quantifiers and logical constants hold. It is easiest to grasp how these grammatical rules work with examples. Let Y^1 be the predicate 'yells'. In our toy version of predicate logic with modifiers, we can express 'John yells furiously' as $F_1^1Y^1(j)$. In this case, 'furiously' behaves as a simple modifier. We can express 'John yells at Tony' by treating the preposition 'at' as an expander: $A_2^1[Y^1(j)](t)$. We can express 'John yells at something' using a quantifier as $\exists x A_2^1[Y^1(j)](x)$. We can express 'John yells at something furiously' as $\exists x A_2^1[F_1^1Y^1(j)](x)$. And we can express 'John does not yell furiously at anything' as $\neg\exists x A_2^1[F_1^1Y^1(j)](x)$.

4.2. Motivations for predicate modifier logic

Why go through all of the trouble to include predicate modifiers in our logic? One historically important motivation comes in explaining the logic of adverbs, relative adjectives and prepositional phrases. Other approaches to formalizing these features of language in predicate logic are awkward and have unacceptable entailments. Reichenbach and Davidson developed the first approaches to adverbial phrases in predicate logic. They did not treat modifiers as functions from first-order predicates to first-order predicates, but rather treated them as second-order predicates or first-order predicates of events.[24] Begin by considering Reichenbach, who treats modifier phrases as second-order predicates. In Reichenbach's system, 'John yells' expresses the same proposition as 'There is a property John has and it is a yelling property'. Let μ be the second-order predicate that denotes *is a yelling property*. In predicate logic-ese, Reichenbach formulates 'John yells' as: $\exists C(C(j) \wedge \mu(C))$. Adverbs and prepositions are then treated as additional second-order predications. For example, let σ be a second-order predicate for *is a furiously property*; in Reichenbach's system, 'John yells furiously' is $\exists C(C(j) \wedge \mu(C) \wedge \sigma(C))$. Or let σ be a second-order relation for *is a property directed at (...)*. 'John yells at Tony' is $\exists C(C(j) \wedge \mu(C) \wedge \sigma(C, t))$. Inspired by Reichenbach, Davidson takes a similar approach to adverbs and prepositions. But instead of using second-order properties, Davidson treats modifiers as predicates of events. He formulates sentences like 'John yells furiously' as $\exists x(Y(j, x) \wedge F(x))$ where the quantifier ranges over events. In English: There is an event of John yelling and the event has the property of furiousness.

There are at least three difficulties for these early proposals. First, the approaches require quantification over either properties or events. Some metaphysicians may find this controversial insofar as they take the quantifiers to reveal ontological commitments. Second, the instantiation relations seem wrong – in the 'John yells furiously' case it seems odd to think of a property or event as instantiating furiousness rather than John. Third and most important, neither system can straightforwardly handle non-standard predicate modifiers.

Consider a sentence like 'John is apparently drinking'. We can formalize the sentence in Reichenbach's system just as before: $\exists C(C(j) \wedge \mu(C) \wedge \sigma(C))$. Here we treat the modifier 'apparently' and verb 'drinking' as distinct second-order properties, denoted by μ and σ, respectively. This is equivalent to the claim that John has a property that is a drinking property and is an apparent property. And this entails $\exists C(C(d) \wedge \mu(C))$. In English: John has a drinking property. But if one is apparently drinking, this in no way entails that one *is* drinking. Something's gone wrong with the Reichenbachian translation. Similarly, a naive Davidsonian treatment translates 'John is apparently drinking' as $\exists x(D(j, x) \wedge A(x))$; there is an event of John drinking and it is an apparent event. And this entails $\exists x(D(j, x))$; there is an event of John drinking. Once again, the entailment is wrong.

The predicate modifier approach does not have these problems. There is no need for explicit quantification over properties or events, rather there is only an implicit commitment to whatever properties modified and unmodified predicates pick out. There is no mis-ascription: in the case of 'John yells furiously', the modifier approach represents John as having the complex property *is yelling furiously*. And there is no problem reasoning with non-standard modifiers like 'apparently', since the modifiers can map to a set of entities completely distinct from the set of entities picked out by the unmodified predicate. The set of objects who are drinking can be completely distinct from the set of objects who are apparently drinking.

It is an interesting question whether modifiers give a successful treatment of adverbs and prepositions in natural language, or whether some updated version of the Davidsonian system should be adopted. But I am less interested in the role modifiers play in philosophy of language and more interested in the role they might play in theorizing about time and change. Can we use modifiers to draw the past–present–future distinction that is central to the A-theory?

4.3. *Options for a tense modifier logic*

Here is the proposal. Let P^n_m be the predicate modifier 'was' and F^n_m be the predicate modifier 'will'. These modifiers act on a stock of tenseless or present-tensed predicates like 'is running' or 'is President' to yield past or future-tensed predicates like 'was running' or 'will be President'. Just as before, we have options for typing these modifiers:

> Option 1: 'was' and 'will' are simple modifiers – they do not change the arity of a predicate that they modify.

Suppose John was running. If P^n_m is a simple modifier, we would express this as $P^1_1R^1(j)$. Or:

> Option 2: 'was' and 'will' expand the arity of the predicate they modify by one argument place–a time slot.

If P^n_m is a one-place expander, we would express 'John was running' as $\exists t P^1_2[R^1(j)](t)$. In this case, we assume that the sentence 'John was running' elides a preposition phrase. It expresses the same proposition as 'There is some time such that John was running at that time'.

Each option guides how we reason about time in interesting ways. I prefer Option 2, for reasons I will elucidate. But Option 1 is also viable, with some provisos.[25]

Option 1 is awkward in cases when we want to describe a series of changes that are entirely past or entirely future. For example, suppose *is alive* is a temporary monadic property. Understated historical fact: Caesar was alive and then he wasn't alive. How would an Option 1-er formalize this? On first pass, she might try: $P^1_1 A^1(c) \wedge \neg P^1_1 A^1(c)$. But then she encounters a major problem: the formula ascribes inconsistent properties to Caesar.

Can an Option 1-er avoid this by embedding predicate modifiers? Note that the formation rules as stated at the beginning of this section allow this. So she might try to express Caesar's life and death as: $P^1_1 P^1_1 A^1(c) \wedge \neg P^1_1 A^1(c)$. This brings new problems for Option 1. First problem: doubly tense modified predicates like $P^1_1 P^1_1 A^1(x)$ don't translate well back to English – 'Caesar was was alive' seems badly formed. So Option 1 requires a more dramatic break from English grammar. Second problem: we will want to distinguish sentences like 'John was not drinking' from 'John never drank'. To do this, the Option 1-er could introduce additional primitive predicate modifiers: N^1_1 for 'not' and A^1_1 for always. Then she could formalize these sentences as $P^1_1 N^1_1 D^1(j)$ and $P^1_1 A^1_1 N^1_1 D^1(j)$, respectively. But if she treats 'always' and 'not' as primitive modifiers, then she will also need to introduce new postulates to explain the obvious entailment relations between sentences like 'John never drank' and 'John was not drinking', since sentences with mixed embedded modifiers do not automatically entail sentences with fewer modifiers. This is one of the lessons of the apparent drinking cases.

Option 2 has none of these problems, though the notation is somewhat more cumbersome. An Option 2-er would express the sentence 'Caesar was alive and then wasn't' as follows: $\exists t_1 P^1_2[A^1(c)](t_1) \wedge \exists t_2 \neg P^1_2[A^1(c)](t_2) \wedge (t_1 < t_2)$ In English: There is a time when Caesar was alive, a time when he wasn't and the first time is earlier than the second. The sentence is logically consistent and has a natural English translation. And with Option 2, we need never multiply embed the same tense modifiers, since the sequence of events is determined by the earlier/later-than clause rather than the order of the modifiers. With Option 2, we can also easily distinguish 'John was not drinking' from 'John never drank'. We formalize 'John never drank' as: $\forall t \neg P^1_2[D^1(j)](t)$. We formalize 'John was not drinking' as $\exists t \neg P^1_2[D^1(j)](t)$. And the former entails the latter assuming our modifier logic keeps the usual axioms and rules for quantifiers.

Still there is a philosophical worry with Option 2 – what should we make of these times that are being quantified over? Some presentist A-theorists might balk at the approach, because it requires quantification over past and future times, and they do not believe the past or future exist. And some growing blockers will worry about the future times introduced by the 'will' modifier. This worry is most serious if you are an A-theorist who thinks *any* quantification over times must be illegitimate. But many A-theorists are happy to quantify over times so long as they are treated as abstract states, or something other than regions of a B-theoretic manifold. They should have no problem, in principle, with the suggested system.[26] We'll come back to what presentists and growing blockers should make of the proposal in Section 6.

Let's take stock. I've offered a proposal for thinking of tenses as special kinds of predicate modifiers that expand the arity of a predicate by one argument place – a time slot. And I have given a sketch of how a logic of tense modification would work. The logic will simply be a special case of a more general version of predicate logic with modifiers. Note that there are a wide range of options for formally implementing a tense modifier logic, and the basic proposal is flexible depending on your other commitments. We won't further pursue the issue of whether a logic of tense modifiers is adequate for capturing natural language tenses. Instead, I want to return to the primary philosophical ambition for this project. In the next section, I will argue that predicate logic expanded with tense modifiers is expressive enough to capture the most basic commitments of an A-theory of time. So if the modifier approach is successful, we have a viable alternative to Priorian tense logic.[27]

5. Step 2: A-theory metaphysics without operators

5.1. Describing change with tense modifiers

The argument from expressive adequacy in Section 3 contends that there is no way to consistently describe A-property change without resort to tense operators. With tense modifiers in hand we can show that this argument is unsound, by stating a version of the A-property change principle that is consistent and does not make use of tense operators. Initially, our operator-free A-theory will consist of three principles:

(1) A-Property Change: There is at least one enduring object (possibly a region of spacetime), *x* and at least one non-time-relational property, *C*, such that *x* is *C* only temporarily.

(2) Core Properties and Predicates: There is a set of core predicates that denote properties. These predicates are either tenseless or present-tensed. (It does not matter for our purposes.) At least some of these predicates do not have time slots, and they denote properties that are not held relative to a particular time. And at least some of the properties denoted by these predicates are temporary properties. An example of a core property might be *is charged*, denoted by $C^1(\ldots)$.

(3) Past and Future Analogue Properties and Predicates: 'Was' and 'will' are predicate modifiers, as described in Section 4. The expressions formed by joining modifiers to core predicates denote past and future analogue properties of objects.

An example of a past analogue property might be *was charged at a time*, denoted by $\exists t P_2^1[C^1(\ldots)](t)$.

The operator-free A-theorist then formalizes the A-property change principle as:

A-Property Change 4: For some property C^n, $\exists x(C^n(x) \wedge \exists t(\neg P_{n+1}^n[C^n(x)](t) \vee \neg F_{n+1}^n[C^n(x)](t)))$

According to this version of the A-theory, there is A-property change because something has a core property, endures and lacks either the past or future analogue of that property. The minimal A-theoretic commitments are both respected: (1) the change only involves a single enduring object – the object assigned to x – and (2) the temporary property, C^n, need not be time-relational. Only C^n's past or future time analogues are time-relational. So the tense modifier logic can express a consistent A-theory without collapsing into a version of the relational B-theory.

5.2. *Objections to the operator-free A-theory*

Unlike the relational B-theory, the operator-free A-theory holds that (1) some temporary properties are non-relational and (2) properties ought to be differentiated based on whether they are core, past or future properties. A question arises at this juncture: don't the arguments that support treating past and future analogue properties as time relations also support treating the core properties as time relations? This would eliminate at least one of the two differences between the proposed A-theory and the relational B-theory. One might argue as follows. Before we have given it much thought, we might think a sentence like 'Peter is ready' conveys that Peter has the monadic property *is ready*. But when we probe further, this expression appears incomplete. Peter might be ready *for work* but not ready *for a softball game*. How we fill in the missing argument place will determine the truth-value of the sentence. And once we realize that the predicate 'is ready' often suppresses a prepositional phrase, our natural reaction is to suppose that there simply is no monadic property *is ready*. So why don't we draw the same conclusion about core properties once we realize that their tensed analogues require a further argument place? According to the *collapse objection*, the same considerations that lead us to treat tenses as expanders ought to motivate us to treat all temporary properties as time relations.

There are two reasons we should reject the collapse objection. First and most important, while there is no deep philosophical reason to postulate a monadic readiness property, A-theorists typically think there *is* good metaphysical reason to take some temporary properties as non-time-relational. A-theorists have philosophical arguments for why core properties are indispensable. One important set of arguments concerns the central role that temporalist propositions play in how A-theorists since Prior conceive of temporariness and the flow of time.[28] According to one prominent understanding of A-theory, the most accurate description of reality is subject to change – indeed, this is part of what it is to take the flow of time seriously. But if all changing properties are fixed relations between an object and a time, then the most accurate description of reality never changes. Second, there are arguments against the assumption that predicates like 'is ready' are always context-sensitive in the way described. The most important work in philosophy of language on this topic has come from Lepore and Cappelen.[29]

One might also worry that the operator-free A-theory does not adequately capture property change because it does not express change with respect to a *single* property, C^n; rather, it only expresses that an object varies with respect to two distinct properties – the core property and the past or future analogue. Call this the *too many properties objection*. The objection closely parallels one we might raise against the temporal parts B-theory. Recall the temporal parts formulation of property change:

Temporal Part Change 1: An object *o* changes with respect to a property, *C* iff

$$\exists p_1 \exists p_2 (TPART(p_1, o) \wedge TPART(p_2, o) \wedge C(p_1) \wedge \neg C(p_2)).$$

According to the view, the object which changes does not directly instantiate the changing properties; rather, its distinct temporal parts stand-in as the instantiators. One might object that this account of change fails to capture our intuition that a single object changes by gaining and losing properties over time. Call this the *too many objects objection* to temporal parts theories.[30]

We shouldn't take the too many properties or too many objects objections seriously. Every theory of change surveyed so far has a surprising entailment. If supporters of the operator-free A-theory are right, having a temporary property entails lacking at least one of the temporal analogue properties. If the temporal parts camp is right, having a temporary property entails having multiple temporal parts. If the relational B-theories are right, change involves having and lacking relations to regions of space–time. If the Priorian A-theorists are right, having a temporary property entails some fact that can only be expressed with tense operators. How should we weigh these different ideological commitments? On this point, I agree with Sider: 'What is certain is that things persist somehow, that things change somehow, and that things have properties at time somehow. It is not a part of reasonable common belief just how this occurs'.[31]

Related to the too many properties objection is a worry that past and future analogue properties are incoherent. At first pass, the temporal analogue properties seem to be properties that are either ungrounded or – at best – grounded in past or future facts about an object. But this account of analogue properties violates a typical A-theoretic assumption that the present is most fundamental. If you think the present is most fundamental, then presumably present facts ought to ground all other facts about an object. Call this the *grounding objection* to analogue properties.

I think operator-free A-theorists should reject the assumption that the present is most fundamental. The characteristic feature of A-theories is that they postulate a fundamental *distinction* between the present and other times: if some A-theorists think the present is *more* or *most* fundamental, then this requires further explanation and argument. Of course, there is a sense in which past and future analogue properties are related to core properties; they are related in whatever sense properties denoted by modified predicates are related to properties denoted by unmodified predicates. Surely there is some sense in which *is drinking* and *is apparently drinking* are importantly similar properties. Perhaps we only grasp what it is to have the latter property in virtue of understanding what it is to have the former. But there is no sense in which something's being apparently drinking is grounded in that object drinking *simpliciter*. Likewise, core properties and their past and future analogues are importantly similar, but the former do not ground the latter. And without some strong grounding assumption, there is no reason to think that operator-free A-theorist is cheating if she postulates a primitive similarity between core properties and their past and future analogues.

6. Accommodating temporalism and ontological change

I've argued that A-theorists do not need tense operators to express the distinctive kind of property change they believe in. I did this by proposing an alternative, operator-free A-theory. But most self-described A-theorists think they are committed to much more than a theory of property change. As we saw in Section 3, many think tense operators are indispensable because they are propositional temporalists or believe in ontological change. Can an operator-free A-theory describe these other kinds of change? Here the verdict is mixed. Predicate logic with modifiers has the resources to support propositional temporalism, but only if we are willing to introduce substantive truth predicates. And predicate logic with modifiers can express a kind of temporary existence claim, but one that requires a departure from neo-Quineanism. I'll close by sketching these extensions of the operator-free A-theory, raising objections to the remaining indispensability arguments.

According to the argument from propositional temporalism, tense operators are indispensable for capturing the temporary truth of propositions about objects and their A-properties. Propositional temporalists think that a proposition α's temporary truth is expressed by the truth of the following schema: $\alpha \wedge (\neg\mathcal{P}\alpha \vee \neg\mathcal{F}\alpha)$. For example, the proposition that Obama is President is temporarily true. We express this as $[Pres(o) \wedge (\neg\mathcal{P}Pres(o) \vee \neg\mathcal{F}Pres(o))]$. Do we have a schema for expressing temporary truth that eliminates the tense operators? We do if we add the following assumption to the operator-free A-theory's metaphysics:

(4) Truth Property: At least one core truth predicate, T^1, denotes a property of propositions.

With the truth property principle, we can express the temporariness of certain propositions using tensed truth predicates:

Propositional Temporalism 2: For some proposition $\alpha, T^1(\alpha) \wedge \exists t(\neg P^1_2[T^1(\alpha)](t) \vee \neg F^1_2[T^1(\alpha)](t)))$

On this construal, propositional temporalism is the view that there is some proposition which is true, and which either was not true at some time or will not be true at some time. This version of the proposal requires not only a core truth predicate (and corresponding property) but also past and future truth predicates (and corresponding properties). Some A-theorists may welcome this alternative for expressing temporalism, and with the truth predicates we can recover many of the inference patterns typically formalized by tense operators.[32] But others may balk at the truth property principle. Presumably one could take propositions seriously, endorse the A-property change principle, but nevertheless contend that truth-predicates are extraneous or problematic. The usual tense logics, in contrast, can be developed with or without an explicit truth predicate. While temporalist deflationism seems to be Prior's mature view, I am not aware of any other A-theorists tempted by truth-predicate deflationism.[33] Barring another way to capture propositional temporalism, the operator-free A-theory will not be of much use to temporalist deflationists. Happily, this doesn't exclude many A-theorists after Prior.

What about A-theorists who believe in ontological change? Expressing the temporary existence principle in the operator-free A-theory is more problematic. The system lacks any tense operators, so *a fortiori* it lacks prophylactic tense operators. But we can express temporary existence claims if we assume:

(5) Existence Property: A first-order existence predicate, $E!^1$, denotes a property.

With this assumption, we can express a version temporary existence using tense modifiers:

Temporary Existence 2: $\exists x(E!^1(\alpha) \wedge \exists t(\neg P^1_2[E!^1(\alpha)](t) \vee \neg F^1_2[E!^1(\alpha)](t)))$.

On this construal, ontological change occurs if there is some object which exists, endures and for some time lacks the property of existing at that time.[34] We'd express truths about temporal aliens as truths about objects which have *only* past or future analogue existence properties. For example, 'Caesar existed' is $P^1_2E!^1(c)$. From this we cannot infer that Caesar exists any more than we can infer 'John is drinking' from 'John is apparently drinking'. So there is no worry about facts about temporal aliens entailing that those objects still exist *simpliciter*.

The trouble is, as noted earlier, many A-theorists (and B-theorists) are neo-Quineans, and neo-Quineans reject the existence property assumption. So expressing temporary existence in the operator-free A-theory requires a departure from the neo-Quinean program. How bad is this? Not as bad as you might think. Many A-theorists are forced away from neo-Quineanism anyways. Because of the Barcan formulas, there is no acceptable version of Priorian tense logic that uses the standard univocal existence quantification theory *and* upholds the temporary existence assumption. So presentists and growing blockers already have good reason to look for new ways to express their temporary existence claim. The operator-free A-theory provides such an option.

7. Conclusions: narrow and more ambitious

We see that fundamental tense operators – at least the operators given to us by Prior – are not an automatic commitment of an A-theory of time and change. We might wonder further whether the theory of tense modifiers is just another animal in Russell's 'logical zoo' or whether it more accurately represents the commitments of our best theory of time and change. And as we have seen, answering this question requires some heavy philosophy. Are there past and future analogue properties? Do objects endure or perdure? Are all temporary properties relations between an object and a region of space–time? I submit that the pressing work for A-theorists is finding principled answers to these questions, rather than refining Priorian tense logic or looking for metaphysical structure that corresponds to tense operators.

More ambitiously, A-theorists who doubt either neo-Quineanism or temporary existence should be especially interested in developing this new system. Predicate logic with tense modifiers perspicuously and parsimoniously reflects the basic ontological and ideological commitments of an A-theory, without forcing adherents to believe in anything more than new sets of properties. And treating tense modifiers as expanders further gives the A-theorist the semantic benefits of quantification over times, without collapsing her view into a B-theory.

Still more ambitiously, developments in the logic of time often go hand in hand with developments in the logic of possibility and necessity. If we determine that predicate logic with modifiers provides a suitable background for the metaphysics of time, it may also provide an alternative to modal logic. On this approach, we would treat 'possibly' and 'necessarily/essentially' as predicate modifiers. A system like this seems most congenial to necessitists, like Williamson.[35] These are issues worth pursuing.[36]

Notes

1. Examples of various kinds of A-theories include Broad (1923), Prior (1967), Adams (1989), Zimmerman (1998), Crisp (2003), Markosian (2004), Forrest (2006), and Merricks (2007).
2. Thomson (1983) and van Inwagen (1990) give early formulations of the relational B-theory, though van Inwagen does not treat times as physical regions of a manifold. For a more contemporary, relativistic version see Gibson and Pooley (2006).
3. Examples of temporal part theorists include Russell (1915), Quine (1950), Lewis (1986), Price (1996), and Sider (2001). Some, like Sider, are stage theorists – instead of using a primitive temporal parthood relation, they use a primitive counterpart relation. These differences in formulation do not matter for our purposes.

4. For the opposing, deflationist view, see (Hirsch , 2009). For a response to various deflationist views about change, see Chapter 11 of Sider (2011).
5. Throughout what follows, I will understand an object *x*'s lacking a property *C* just as it's being false that *x* is *C*. *x* lacks *C* iff $\neg C(x)$.
6. Prior (1967, 19).
7. Prior and Fine (1977, 32). Emphasis is his.
8. Zimmerman (2005). In particular, Zimmerman identifies Lewis and Mellor as 'serious tenser B-theorists'. See Lewis (1983) and Mellor (1998).
9. Zimmerman offers something like this argument in response to Lewis's charges that only a four-dimensionalist theory can consistently accommodate changing, non-time-relational properties. See Zimmerman (1998). In that paper, he is responding on behalf of presentist A-theories to the problem of temporary intrinsics developed in Lewis (1986, 202–204). For Lewis's reply (which specifically targets presentism) see Lewis (2002, 2).
10. For more on the connection between A-theories and propositional temporalism, see Sullivan (2014).
11. See van Inwagen (1998) and the introduction to Sider (2001).
12. At least they agree on the language of quantification, and so are having a genuine debate about claims like 'There are dinosaurs' or 'There are temporal parts'. The usual A-theorists will have more fundamental ideology than B-theorists, since they postulate primitive tense operators and the B-theorists do not. See Sider (2011) for considerations in favor of the substantivity of this debate over tense operators.
13. Some examples of presentist views are Prior (1998), Zimmerman (1998), Crisp (2003), Markosian (2004), and Merricks (2007).
14. Some examples of growing block views include Broad (1923), Adams (1989), Tooley (1997), and Forrest (2006). Note that many growing blockers are perdurantists about ordinary objects – they may think all ordinary objects only have B-properties. But they must hold that the growing edge of the block – the distinguished, present space–time region – changes in accord with A-property change, since it goes from *being present* to *being past*.
15. I treat A-property change as the distinctive commitment of the A-theories of time. But could there be a meaningful dispute between A and B-theorists even if there was no property change in the world? Suppose there is a world where bare particulars come into and out of existence, but no other change occurs. I'm inclined to say that neither the A-theory nor the B-theory is true in such a world, since nothing has the property of being located in space or time. Compare: we typically do not classify worlds with only abstract objects as either A or B-theory worlds. If the particulars do have locations, then one or the other of the property change principles will hold. Still, if you think worlds where only temporary existence holds count as A-theoretic worlds, this should not affect the arguments of this paper, since the indispensability arguments are arguments about characterizing actual change.
16. Compare this to moving spotlight theorists, who think the only form of A-property change is times gaining or losing the properties of being past, present or future. Or eternalist A-theorists who believe objects change with respect to properties of being spatiotemporally located or concrete, but do not gain or lose existence. Both of these versions of the A-theory deny ontological change, but they deny it in different ways. For more on eternalist A-theories, see Sullivan (2012a).
17. Zimmerman (1998, 211).
18. For discussion of this, see Chapter 11 of Sider (2011).

19. For a survey of some of the difficulties using tense operators to capture intuitive truth conditions, see Kamp (1968), Partee (1973), Enc (1986), Lewis (2004), Szabo (2007) and King (2007).
20. See Sullivan (2012b) and Chapter 2 of Burgess (2009) for discussion of the Barcan formulas in tense logic. See Williamson (2013) for a general defense of Barcan formulas in modal logic and tense logic. These problems vexed Prior throughout his career, and ultimately resulted in his awkward System Q tense logic; see Prior (1967).
21. For instance, one could adapt the logic of predicate modifiers outlined in Chapter 4 of Gamut (1991). The main change will be offering a functional application clause in the formal semantics for what I call 'expanders' below – functions that map well-formed formulae of arity n to formulae of arity $n+1$.
22. Much of the initial philosophical work on these kinds of predicate modifiers was done by Parsons and Clark in the 1970s. See Clark (1970), Parsons (1970), Clark (1975), and Clark (1974). I'll roughly follow their notation here with some of my own adjustments. Parsons later developed his theory into a book: Parsons (1990). For a rigorous treatment of the logic of predicate modifiers using the theory of types, see Chapter 4 of Gamut (1991). I follow Gamut's system in the formal appendix.
23. Which isn't to say reducers aren't philosophically interesting. For example, reducers would help with a program for eliminating constants, if we reinterpret constants as predicate reducers of the form C^{n}_{n-1}. On this approach, we could define an atomic sentence as any complex predicate with arity 0. The proposal has a certain affinity with Quine's program for eliminating variables with 'de-relativizing' operators in Quine (1960).
24. Reichenbach (1947) and Davidson (1980).
25. The main benefit of Option 1 may come in formalizing reasoning about changing properties of times themselves. Suppose we wanted to express '2008 was an election year'. With a simple past modifier, we might express this as $P^1_1 E^1(2008)$. With expanders, we would have to relativize to a time. For example, we might express it as $P^1_2[E^1(2008)](2008)$. But this construction looks strange – it expresses that 2008 was an election year at 2008. Some past and future properties of times themselves seem non-relational, and so might be better captured with simple tenses along the lines of Option 1.
26. For examples of A-theorists who offer theories for quantifying over times or past/future states, see Prior and Fine (1977), Chisholm (1978), Davidson (2003), Crisp (2007), Zimmerman (2011) and Meyer (2012). Parsons (1990) and Szabo (2007) give semantic arguments for quantifying over merely past states. It is also somewhat common to think of times as abstracta when giving a semantics for tense logic, akin to the way we might treat worlds as abstracta in modal logic.
27. I am not the first philosopher to focus on the role that tensed copulae might play in a theory of time. Johnston, for example, develops a theory where we understand tensed copulae as denoting an instantiation relation with three argument places. Sentences like 'Obama was President' are to be understood as expressing the same proposition as 'Obama has-at-2008 the (non-relational) property *is President*'. Haslanger (1989) develops a similar theory. For a critical discussion of these proposals, see Lewis (2002). Lewis objects that their approach to tensed copulae over-emphasizes the instantiation relation, inviting Bradley's regress.
28. See, for example, Prior (1996), Williamson (2002), Sullivan (2014) and Chapter 11 of Sider (2011).

29. See Cappelen and Lepore (2005).
30. Haslanger calls it the 'proper subject' objection and expresses some sympathy for it. See Haslanger (2003, 331–334).
31. Sider (2001, 215).
32. We could, for instance, treat the axioms of our favored 'old' tense logic as meaning postulates. For example, suppose you wanted to recover the old System K axiom $\alpha \rightarrow \mathcal{H}F\alpha$ in the meaning postulates of the new, operator-free theory. A temporalist of the kind discussed above could do this by quantifying over propositions: $\forall\alpha(T(\alpha) \rightarrow \forall t_1(P(T)(\exists t_2(F(T(\alpha))(t_2)))(t_1)$. In English, for any proposition α, if α is true then the proposition that there will be a time when α will be true has always been true. The main difference is this isn't an axiom of a proof theory anymore.
33. Prior (1971).
34. We might also choose to modify the identity relation, if we assume that the identity predicate can be used to express the sense of existence of interest to metaphysics.
35. Williamson (2013). Though note, Williamson is a primitivist about modal operators.
36. Note that Wiggins develops a theory of 'necessarily' as a predicate modifier and uses it to defend his essentialism from Quinean objections to quantifying into modal contexts. See Wiggins (1976). And in an appendix to the Wiggins paper, Peacocke develops a formal semantics for the necessity modifier. See Peacocke (1976). But Wiggins' system is supposed to be added to a logic with modal operators; it is not intended as a replacement.

Acknowledgements

Thanks to Andy Egan, Ted Sider, Jeff Speaks, Dean Zimmerman, Tim Williamson and audiences at Rutgers University, University of Toronto, University of Virginia, University of Iowa, University of Arizona, the Metaphysical Virtues Conference at WMU, the Carolina Metaphysics Workshop, the Birmingham Time Workshop, and the Barcelona As and Bs of Time Workshop.

References

Adams, R. M. 1989. "Time and Thisness." In *Themes from Kaplan*, edited by H. W. Joseph Almog, and John Perry, 23-42. Oxford: Oxford University Press.
Broad, C. 1923. *Scientific Thought*. London: Routledge and Keegan Paul Ltd.
Burgess, J. 2009. *Philosophical Logic*. Princeton: Princeton University Press.

Cappelen, H., and E. Lepore. 2005. *Insensitive Semantics: A Defense of Semantic Minimalism and Speech Act Pluralism*. Oxford: Blackwell.
Chisholm, R. 1978. "Objects and Persons: Revisions and Replies." *Grazer Philosophishe Studien* 7 (8): 317–361.
Clark, R. 1970. "Concerning the Logic of Predicate Modifiers." *Nous* 4 (4): 331–335.
Clark, R. 1974. "Adverbial Modifiers." In *Ontological Commitment*, edited by R. H. Severens, 22–35. Athens: University of Georgia Press.
Clark, R. 1975. "Facts, Fact-correlates, and Fact-surrogates." In *Fact, Value, and Perception: Essays in Honor of Charles A. Baylis*, edited by P. Welsh, 3–18. Durham NC: Duke University Press.
Crisp, T. 2003. "Presentism." In *The Oxford Handbook of Metaphysics*, edited by M. J. Loux and D. W. Zimmerman, 211–246. Oxford: Oxford University Press.
Crisp, T. 2007. "Presentism and the Grounding Objection." *Nous* 41 (7): 90–109.
Davidson, D. 1980. "The Logical Form of Action Sentences". In *Essays on Actions and Events*, 105–121. Oxford: Oxford University Press.
Davidson, M. 2003. "Presentism and the Non-present." *Philosophical Studies* 113: 77–92.
Enc, M.1986. "Referential Analysis of Temporal Expressions." *Linguistics and Philosophy* 9 (4): 405–426.
Forrest, P. 2006. "General Facts, Physical Necessity, and the Metaphysics of Time." *Oxford Studies in Metaphysics* 2: 137–154.
Gamut, L. 1991. *Language, Logic and Meaning: Volume II Intensional Logic and Logical Grammar*, Chicago: University of Chicago Press.
Gibson, I., and O. Pooley. 2006. "Relativistic Persistence." *Philosophical Perspectives* 20: 157–198.
Haslanger, S. 1989. "Endurance and Temporary Intrinsics." *Analysis* 49 (3): 119–125.
Haslanger, S. 2003. "Persistence through Time." In *The Oxford Handbook of Metaphysics*, edited by M. J. Loux and D. W. Zimmerman, 315–354. Oxford: Oxford University Press.
Hirsch, E. 2009. "Ontology and Alternative Languages." In *Metametaphysics: New Essays on the Foundations of Ontology*, edited by D. J. Chalmers, D. Manley and R. Wasserman, 231–259. Oxford: Oxford University Press.
van Inwagen, P. 1990. "Four-dimensional Objects." *Nous* 24 (2): 245–255.
van Inwagen, P. 1998. *Meta-ontology. Erkenntnis* 48: 233–250.
Kamp, H. 1968. "Tense Logic and the Theory of Linear Orders". Ph. D. thesis, UCLA.
King, J. C. 2007. *The Nature and Structure of Content*. Oxford: Oxford University Press.
Lewis, D. 1983. "Attitudes De Dicto and De Se." In *Philosophical Papers*, Vol 1, 133–155. Oxford: Oxford University Press.
Lewis, D. 1986. *On the Plurality of Worlds*. Oxford: Blackwell.
Lewis, D. 2002. "Tensing the Copula." *Mind* 111: 1–14.
Lewis, D. 2004. "Tensed Quantifiers." In *Oxford Studies in Metaphysics*, edited by D. W. Zimmerman, 3–14. Oxford: Oxford University Press.
Markosian, N. 2004. "A Defense of Presentism." *Oxford Studies in Metaphysics* 1: 83–106.
Mellor, D. 1998. *Real Time II*. London: Routledge.
Merricks, T. 2007. *Truth and Ontology*. Oxford: Oxford University Press.
Meyer, U. 2012. "Times as Abstractions." In *The Future of Philosophy of Time*, edited by A. Bardon. Oxford: Routledge, pp. 41-55.
Parsons, T. 1970. "Some Problems Concerning the Logic of Grammatical Modifiers." *Synthese* 3 (4): 320–334.
Parsons, T. 1990. *Events in the Semantics of English*. Cambridge, Mass: MIT Press.
Partee, B. 1973. "Some Structural Analogies between Tenses and Pronouns in English." *Journal of Philosophy* 70: 601–607.

Peacocke, C. 1976. "An Appendix to David Wiggins' 'Note'." In *Truth and Meaning: Essays in Semantics*, edited by G. Evans and J. McDowell, 313–324. Oxford: Clarendon Press.
Price, H. 1996. *Time's Arrow and Archimedes' Point: New Directions for the Physics of Time*. Oxford: Oxford University Press.
Prior, A. 1967. *Past, Present, and Future*. Oxford: Clarendon Press.
Prior, A. 1971. *Objects of Thought*. Oxford: Oxford University Press.
Prior, A. 1996. "Some free Thinking about Time." In *Logic and Reality: Essays on the Legacy of Arthur Prior*, edited by B. Copeland, 47–51. Oxford: Clarendon Press.
Prior, A. 1998. "The Notion of the Present." In *Metaphysics: The Big Questions*, edited by P. van Inwagen and D. W. Zimmerman, 80–82. Oxford: Blackwell.
Prior, A., and K. Fine. 1977. *Worlds, Times and Selves*. Amherst, MA: University of Massachusetts Press.
Quine, W. 1950. "Identity, Ostension, and Hypostasis." *The Journal of Philosophy* 47 (22): 621–633.
Quine, W. 1960. "Variables Explained Away." *Proceedings of the American Philosophical Society* 104 (3): 343–347.
Reichenbach, H. 1947. *Elements of Symbolic Logic*. London: Macmillan.
Russell, B. 1915. "On the Experience of Time." *The Monist* 25: 212–233.
Sider, T. 2001. *Four-Dimensionalism*. Oxford: Oxford University Press.
Sider, T. 2011. *Writing the Book of the World*. Oxford: Oxford University Press.
Sullivan, M. 2012a. "The Minimal A-theory." *Philosophical Studies* 158: 149–174.
Sullivan, M. 2012b. "Problems for Temporary Existence in Tense Logic." *Philosophy Compass* 7: 43–57.
Sullivan, M. 2014. "Change We Can Believe In (and Assert)." *Nous* 48 (3): 474–495.
Szabo, Z. G. 2007. "Counting across Times." *Philosophical Perspectives* 20: 399–426.
Thomson, J. J. 1983. "Parthood and Identity across Time." *The Journal of Philosophy* 80 (4): 201–220.
Tooley, M. 1997. *Time, Tense and Causation*. Oxford: Clarendon Press.
Wiggins, D. 1976. "The De Re 'Must': A Note on the Logical Form of Essentialist Claims." In *Truth and Meaning: Essays in Semantics*, edited by G. Evans and J. McDowell, 285–312. Oxford: Clarendon Press.
Williamson, T. 2002. "Necessary Existents". In *Logic, Thought, and Language*, edited by A. O'Hear, 233–251. Cambridge: Cambridge University Press.
Williamson, T. 2013. *Modal Logic as Metaphysics*. Oxford: Oxford University Press.
Zimmerman, D. W. 1998. "Temporary Intrinsics and Presentism." In *Metaphysics: The Big Questions*, edited by P. van Inwagen and D. W. Zimmerman, 206–218. Oxford: Blackwell.
Zimmerman, D. W. 2005. "The A-theory of Time, the B-theory of Time, and Taking Tense Seriously." *Dialectica* 59: 401–457.
Zimmerman, D. W. 2011. "Presentism and the Space-time Manifold." In *The Oxford Handbook of Philosophy of Time*, edited by C. Callendar, 163–245. Oxford: Oxford University Press.

Reply to Sullivan

Timothy Williamson

Wykeham Professor of Logic, University of Oxford, Oxford, UK

In 'An A-theory without tense operators' (Sullivan 2016), Meghan Sullivan vigorously contests the received view that an A-theory of time is adequately expressible only in a language with sentential tense operators. She develops and defends an interesting alternative A-theory in a language with tense predicate modifiers instead. Her argument intersects *Modal Logic as Metaphysics* at several points. The proper formulation of A-theoretic doctrines such as presentism is sensitive to the background quantified temporal logic, and in particular to the dispute between permanentism and temporaryism, the temporal analogue in the book of the modal dispute between necessitism and contingentism: the permanentist asserts, and the temporaryist denies, that always everything is always something. Moreover, I sketch a conception of radical change, analogous to radical contingency, adequately expressible only in A-theoretic language (Williamson 2013, 403–422).

I agree with Sullivan that an A-theory is adequately expressible in a language with temporal predicate modifiers. However, I will question whether the move to such a language makes as much difference as Sullivan suggests, and whether it has the advantages she claims for it.

1. Operators and modifiers

The first point to notice is that sentence operators are in effect a special case of predicate modifiers. In section 4 of her paper, Sullivan symbolizes predicate modifiers 'with non-negative superscripts and subscripts'; M^n_m takes an *n*-place predicate as input and returns an *m*-place predicate as output. So, in particular, a predicate modifier M^0_0 takes a 0-place predicate as input and returns a 0-place predicate as output. But a 0-place predicate is tantamount to a sentence: for an *n*-place predicate yields a sentence when completed with *n* singular terms, so in particular a 0-place predicate yields a sentence when completed with 0 singular terms, a null operation. Thus in effect the modifier M^0_0 takes a sentence as input and returns a sentence as output. In other words, M^0_0 is a sentence operator.

The upshot is that Sullivan's preferred form of language with temporal predicate modifiers already permits the category of temporal operators.

Of course, Sullivan could add the artificial stipulation that the predicate modifier M^n_m is permitted only if m and n are positive, not just non-negative. But how much difference would that make? Suppose for the time being that the language contains a λ-operator that enables us to form a one-place predicate $\lambda v(A)$ given an individual variable v and a formula A. Sullivan omits such an operator from her language for simplicity, but seems to have no objection to it in principle: 'It is a straightforward matter to develop an explicit version of the system using type-theoretic semantics or lambda-abstraction'. Indeed, languages without something like a λ-operator are drastically restricted in the use they can make of their predicate modifiers, since the latter cannot be applied to complex predicates. A λ-operator would solve problems of formalization Sullivan discusses in section 4.

Let us then temporarily admit an atomic sentence operator O into the language, in order to show that its effect can be simulated by a predicate modifier O^1_1. For we can define the 1-place predicate $O^1_1(P)$ as equivalent to $\lambda v(O(Pv))$ for all 1-place predicates P. Here I depart from Sullivan's practice of writing the argument of a predicate modifier as if it were the formula consisting of the predicate and its arguments, which makes the predicate modifier look like a sentence operator, contrary to her intention; the present notation is more perspicuous. Let A be any formula, and v any individual variable. Thus the predicate $O^1_1(\lambda v(A))$ is equivalent to $\lambda v(O(\lambda v(A)v))$, which is in turn equivalent to $\lambda v(O(A))$, because $\lambda v(A)v$ is equivalent to A by the standard λ-conversion principle. Consequently, the formula $O^1_1(\lambda v(A))v$ is equivalent to the formula $\lambda v(O(A))v$, which is in turn equivalent to $O(A)$ by the λ-conversion principle. In other words, the effect on a formula of applying λv, followed by the predicate modifier O^1_1, followed by predication of v, is equivalent to that of the original sentence operator O. If we now expel O from the language, because it is a sentence operator (and non-truth-functional: Sullivan's language still has the usual truth-functional sentence operators), and even drop the λ-operator, we can still retain O^1_1 as an atomic predicate modifier with the same semantic effect as before. In brief, a sentence operator can always be dressed up as a predicate modifier.

The point can be illustrated with an example that Sullivan gives for another purpose, the sentence 'John is apparently drinking'. She treats 'apparently' as a predicate modifier, but we can think of it as derived from the sentence operator 'it appears that' (O). Then we can understand her sentence as 'John is such that it appears that he is drinking', with the overall form $\lambda v(O(Dv))j$.

Of course, the application of λv to A is vacuous when v does not occur free in A, but such vacuous applications are harmless and normally permitted; any standard semantics for the λ-operator handles them smoothly. Even if they were banned, we could easily circumvent the difficulty by using a trivial equivalent of A in which v does occur free, such as $v = v \rightarrow A$.

A philosopher who took the sentence operator O to be *meaningless* would protest that the original definition of O_1^1 in terms of O conferred no meaning on O_1^1. But Sullivan nowhere suggests that temporal operators are meaningless, and that extreme view is hardly plausible.

A more subtle objection is that λ-conversion may fail in a contingentist or temporaryist setting when a modal or temporal operator intervenes between an occurrence of the λ-operator and an occurrence of a variable that it binds (Stalnaker 2003; intentional operators such as 'John believes that' may also make difficulties for λ-conversion, but are not the main concern of Sullivan's paper). For instance, where P is a past tense operator, it may be held that $\lambda v(P(Dv))j$ ('John is such that he was drinking') entails $\exists v\, j=v$ ('John is something'), because predication entails being, but that $P(Dj)$ ('In the past, John was drinking') has no such entailment (see Williamson 2013, 148–158 and 172–188 for discussion). Fortunately, we can finesse that issue by assigning a necessary and permanent being such as the null set as the value of the variable v, in which case $O_1^1(\lambda v(A))v$ is anyway equivalent to $O(A)$.

Thus predicate modifiers are hardly an *alternative* to sentence operators, for they subsume the latter. It is therefore no surprise that predicate modifiers give A-theorists the expressive power they need. Conversely, predicate modifiers by themselves give no more expressive power than variable-binding temporal operators do. For consider the predicate modifier M_1^1. Let Ov be a sentence operator binding the variable v such that $Ov(A)$ is equivalent to $M_1^1(\lambda v(A))v$. Then, for any 1-place predicate F, since $\lambda v(Fv)$ is equivalent to F by a trivial instance of λ-conversion, $M_1^1(F)v$ is equivalent to $M_1^1(\lambda v(Fv))v$, which is in turn equivalent to $Ov(Fv)$. The argument can easily be generalized to predicate modifiers M_m^n for any natural numbers m and n.

The foregoing arguments do not involve assimilating sentence operators to predicate modifiers applied to a truth predicate for propositions, a device Sullivan contemplates on behalf of A-theorists who wish to characterize changes in ontology without treating them as changes in a given thing. No truth predicate was used in the constructions above; nor was any quantifier over propositions. If one wants to generalize about change, one can go second-order and quantify into predicate or sentence position without assuming that there are such *things* as properties or propositions (Williamson 2013, 257–261).

The foregoing arguments also do not involve treating existence as a monadic property of objects. Thus they do not violate the view that 'Existence is not a monadic property of objects'. That is the second in a package of three views Sullivan describes as 'neo-Quinean', but since Quine rejected the postulation of properties, he would not have regarded the formulation as apt: given that there are no properties, it is trivial that existence is not a property. However, with the λ-operator, we can of course formulate the monadic predicate of objects $\lambda v(\exists x\, v=x)$, which deserves to be called an 'existence predicate' given the third view in the 'neo-Quinean' package, that 'Existence is properly expressed by the existential quantifier ($\exists x$) and the identity relation (=)'.

So far, nothing much seems to be at issue in the A-theorist's choice between sentence operators and predicate modifiers. We must now examine Sullivan's reasons for thinking otherwise.

2. Objections to operators

Sullivan gives three reasons why A-theorists might be dissatisfied with using temporal sentence operators to articulate their account of time and change. I will consider each reason in turn.

Sullivan's first reason is that A-theorists may hold that 'their logical primitives ought to reflect their views about the structure of reality', but it is not clear what it could be 'for an intensional operator [such as a temporal operator] to reflect some aspect of reality'. Thus, by using an ideology of temporal operators, A-theorists risk losing out to B-theorists, whose ideology 'merely requires the more familiar object/property distinction'.

If there is a problem here, the shift from sentence operators to predicate modifiers does not solve it. For it is no clearer how an intensional predicate modifier can reflect some aspect of reality than it is how an intensional sentence operator can do so. Of course, if predicates stand for properties or relations, then since the result of applying a predicate modifier to a predicate is itself a complex predicate, it stands for a property or relation. But that does not meet the original challenge, which concerned 'logical primitives', not complex expressions. The logical primitive here is the predicate modifier itself, not the result of applying it to a predicate. The predicate modifier does not stand for a first-order property or relation. Perhaps it stands for a higher order property or relation. But any such account that works for predicate modifiers M^n_m will generalize to sentence operators. For it will explain how the relevant aspect of reality relates the input n-place property or relation uniquely to the output m-place property or relation, for various natural numbers m and n. A corresponding account will then explain in particular how the aspect of reality relevant to a sentence operator relates an input 0-place property or relation to an output 0-place property or relation. What is a 0-place property or relation? It is a *state of affairs*, in a sense to be refined by analogy with whatever theory of properties and relations is in play. If aspects of reality can map properties or relations to properties or relations, then an aspect of reality can map states of affairs to states of affairs. This is just the metaphysical analogue of points made about semantics in the previous section.

The contrast with the B-theoretic ideology is in any case overblown. For the B-theorist's 'logical primitives' include not only singular terms and predicates, standing for objects and properties or relations (or indeed states of affairs, in the case of primitive sentence constants), but also primitive logical devices such as conjunction, negation and the universal quantifier. It is not immediately obvious which aspects of reality those logical devices reflect. But suppose that, from a sufficiently abstract perspective, we can understand extensional sentence operators like conjunction and negation as reflecting aspects of reality. Then, from an equally abstract perspective, why can't we understand intensional sentence operators such as the temporal ones as also reflecting aspects of reality?

Thus Sullivan's first reason for the A-theorist to switch from sentence operators to predicate modifiers is not compelling.

Sullivan's second reason for the switch is that 'Priorian tense logic has difficulty accounting for true claims involving cross-time relations and inferences that involve temporal anaphora'. There are indeed such difficulties, but how does the shift to predicate modifiers help resolve them? As she explains in section 4 of her paper, Sullivan's preferred method is to add quantification over times to the language. I agree with her that, on an appropriate understanding of what times are, such an addition is consistent with the spirit of an A-theory. But then what is to stop A-theorists who stick with temporal sentence operators from adding quantification over times to their language too?

Of course, adding quantification over times will be of little help if the time variables do not engage with the atomic sentences in any non-trivial way. But one can solve that problem by also adding to the language temporal sentence operators of the form 'at *t*', where '*t*' is a time variable open to quantification, as in hybrid logic (Williamson 2013, 417–422). Such an A-theorist understands 'John is sitting at *t*' as 'at *t*: John is sitting'. The point of the operator 'at *t*' is *not* to supply a value for an implicit 'when?' argument place in 'John is sitting', just as the point of the operator 'according to Mary' in 'According to Mary, John is sitting' is not to supply a value for an implicit 'according to whom?' argument place in 'John is sitting'. For an A-theorist such as Prior, the sentence 'John is sitting' expresses the *same* complete proposition when uttered at different times (holding other things fixed, such as the reference of 'John'). The hybrid sentence operator 'at *t*' maps one complete proposition to another, just as the sentence operator 'according to Mary' does. By contrast, a B-theorist may postulate an implicit time variable in 'John is sitting', whose value 'at *t*' can fix. Once one has such hybrid temporal operators, their temporal variable can be bound and quantified on, to solve problems of cross-time relations and temporal anaphora. What matters is the availability of quantification over times and devices with an effect like that of 'at *t*', not the shift from sentence operators to predicate modifiers. Thus Sullivan's second reason for the A-theorist to make the shift is also not compelling.

Sullivan's third reason for the switch, and the problem which she regards as 'the most severe' of the three, is the derivability of Barcan formulas and their converses in quantified modal and temporal logics, which makes such logics unsuitable for contingentists. Here she rather overstates her case. Some axiomatized quantified modal logics lack the Barcan and converse Barcan theorems and are sound and complete with respect to a possible worlds model theory with variable domains (see Hughes and Cresswell 1996, 289–311 for an introduction). For the temporal case, Sullivan adverts to derivations of the Barcan and converse Barcan formulas in a natural axiomatic system of quantified temporal logic by John Burgess, which make such logics unsuitable for temporaryists (2009, 33–34). But of course Burgess does not claim, absurdly, that those formulas are derivable in *all* axiomatic systems of quantified temporal logic. He sketches a Kripke-style model theory for quantified temporal logic with variable domains and notes that 'a perspicuous axiomatic proof procedure delivering as theorems just the closed formulas valid for this model theory is lacking' (2009, 37). The overall situation in quantified modal and temporal is that excluding the Barcan and converse Barcan formulas tends to make for more messy complications than including them, at the level of proof theory, though not at the level of model theory. I argue that contingentists are forced to take a more instrumentalist line towards the Kripke-style model theory than are necessitists (Williamson 2013, 134–139), but that is not the issue Sullivan is raising.

Since the problem Sullivan is raising for quantified temporal logic with sentence operators concerns proof systems, one might expect her to present a proof system for quantified temporal logic with predicate modifiers, to show how the switch enables one to avoid the Barcan and converse Barcan theorems perspicuously, without messy complications. However, her paper does not attempt to sketch or cite any proof system. It is thus unclear what bearing the switch from sentence operators to predicate modifiers is supposed to have on the proof-theoretic problem.

After various reflections, Sullivan concludes that 'the operator-free A-theory fits best with a permanentist ontology' (as she puts it in her section 1), and hints at an analogous operator-free form of necessitism. But since the proof-theoretic problem only arose in the first place for those unwilling to accept necessitism and permanentism, even under pressure, this concession seems to undermine her third reason for switching from sentence operators to predicate modifiers. For once one has accepted those metaphysical views, the proof-theoretic problem should not worry one, so one can comfortably return to the operator formulations after all.

In consequence, Sullivan's three reasons for A-theorists to switch from sentence operators to predicate modifiers are not compelling, individually or together. That supports the conclusion of section 1 above, that the switch does not make very much difference. We can stick with the operator formulations.

References

Burgess, John P. 2009. *Philosophical Logic*. Princeton, NJ: Princeton University Press.
Hughes, George, and Max Cresswell. 1996. *A New Introduction to Modal Logic*. London: Routledge.

Stalnaker, Robert. 2003. "The Interaction of Modality with Quantification and Identity." In his *Ways a World Might Be: Metaphysical and Anti-metaphysical Essays*. Oxford: Clarendon Press.

Sullivan, Meghan. 2016. "An A-Theory without Tense Operators." *Canadian Journal of Philosophy* 46 (4–5): 736–759.

Williamson, Timothy. 2013. *Modal Logic as Metaphysics*. Oxford: Oxford University Press.

Williamsonian modal epistemology, possibility-based

Barbara Vetter

Institut für Philosophie, Friedrich-Alexander-Universität Erlangen, Erlangen, Germany

ABSTRACT
Williamsonian modal epistemology (WME) is characterized by two commitments: realism about modality, and anti-exceptionalism about our modal knowledge. Williamson's own counterfactual-based modal epistemology is the best known implementation of WME, but not the only option that is available. I sketch and defend an alternative implementation which takes our knowledge of metaphysical modality to arise, not from knowledge of counterfactuals, but from our knowledge of ordinary possibility statements of the form 'x can F'. I defend this view against a criticism indicated in Williamson's own work, and argue that it is better connected to the semantics of modal language.

1. Williamsonian modal epistemology

This paper is an exercise in broadly Williamsonian modal epistemology (WME): while I will disagree with Williamson on some important points, I share his general outlook, which I take to be characterized by two basic commitments. The first is a commitment to *realism*: what we know about, when we know about metaphysical modality, is an entirely objective or mind-independent phenomenon that has nothing whatsoever to do with the human mind (except, of course, where the possibilities or necessities concern the human mind). The second commitment is *anti-exceptionalism*[1]: in Williamson's own words,

> a plausible non-skeptical epistemology of metaphysical modality should subsume our capacity to discriminate metaphysical possibilities from metaphysical impossibilities under more general cognitive capacities used in ordinary life. (Williamson 2007, 136)

According to anti-exceptionalism, our knowledge of metaphysical modality is *continuous* with our everyday knowledge about the world.

The two commitments are not entirely independent: as Williamson notes, an exceptionalist modal epistemology, with its 'apparent cognitive isolation of metaphysically modal thought' (Williamson 2007, 136) has a difficult time explaining how such isolated thought is to hook on to reality. An anti-exceptionalist has a much better chance of explaining the relation of our metaphysically modal thought to an independent metaphysically modal reality in terms of the relation between our ordinary thought to ordinary (modal and non-modal) reality.

WME, as I understand it, is intended to answer two distinct but related questions.

The first question concerns our grasp of the *concepts* of metaphysical modality: how do we even get to think about *metaphysical* modality (as opposed to some more mundane forms of modality, say practical or epistemic modality)? The second question concerns our knowledge of the *extension* of those concepts: how do we know what is metaphysically possible or impossible? It would appear that the two questions, and Williamson's answers to them, are closely related: knowledge of the extension of the metaphysical modalities, the subject of the second question, requires that we have – in whatever form, under whatever name – some concept of these metaphysical modalities; and if the concept of metaphysical modality is 'the exclusive preserve of philosophers, then so is knowledge of metaphysical modality' (Williamson 2007, 135). In what follows, I will take care to address both questions: the question where we get our concept of metaphysical modality, and the question how we know about its extension.

The *locus classicus* for WME is chapter 5 of Williamson (2007). But arguably there is some reliance on WME in the methodology of Williamson's own more recent work on modality, culminating in Williamson (2013) (see especially the 'methodological afterword'). However, in what follows I will focus on Williamson (2007) and some more recent papers that pick up threads from it.

Williamson's own account of modal knowledge is, of course, more specific than the broad outlines that I have sketched so far. It is his *implementation* of these broad outlines that I will take issue with in what follows. Section 2 presents his implementation, and some preliminary reasons for exploring the kind of alternative that I prefer. Section 3 presents my preferred implementation of WME, which I call a 'possibility-based approach'. Sections 4–6 argue for my account by defending it against a challenge of Williamson's, and then turning the challenge around and arguing that it is in fact a problem for Williamson's implementation, not for mine. My argument will rely to a large extent on the findings of contemporary semantics, following Williamson's dictum that '[p]hilosophers who refuse to bother about semantics . . . resemble scientists who refuse to bother about the theory of their instruments' (Williamson 2007, 284f).

2. Williamson's modal epistemology

Williamson's own account, as is well known, is twofold. In a first step, Williamson identifies the *entry point* into the relevant kind of modal thought and knowledge within our ordinary thought and knowledge about the world. In a second step, he explains how we extend modal thought and knowledge beyond that entry point to its *limiting case*: metaphysical modality.

First, the *entry point* into modal thought and knowledge, for Williamson, are counterfactual conditionals such as 'If the bush had not been there, the rock would have ended in the lake' (Williamson 2007, 142). Our knowledge of such counterfactual truths, Williamson argues, does not require any outlandish cognitive capacities. In evaluating a counterfactual, we typically use 'offline' the very capacities that we would use 'online' in perception and expectation-forming. Moreover, it is not surprising that we have the capacity to evaluate counterfactuals, since having such capacities brings with it clear evolutionary advantages.

Second, Williamson argues that given our knowledge of counterfactual conditionals, we are in a position to know about metaphysical necessity as a kind of *limiting case* because necessities are logically equivalent to certain counterfactual constructions:

(V□) $\Box A \equiv (\neg A \,\Box\!\!\rightarrow \bot)$
(V*□) $\Box A \equiv (\neg A \,\Box\!\!\rightarrow A)$
(Q□) $\Box A \equiv \forall p(p \,\Box\!\!\rightarrow A)$

(And likewise for possibility, applying the standard definition in terms of necessity.)

In what follows, I want to develop and defend a different kind of WME. Like Williamson's own approach, mine will be realist and anti-exceptionalist. It will differ from Williamson's primarily in the 'entry point' that it stipulates for our modal thought and knowledge: not counterfactual conditionals, but restricted possibility statements of the form 'x can F' and their equivalents in other languages. A different entry point will, of course, also require a different story about how we reach the limiting case of metaphysical modality (see Section 3). As will become clear later, my proposed entry point into modal thought and knowledge differs from Williamson's in two ways: it is a form of possibility, and it is an ascription of modal properties. To have labels for the two approaches, I will refer to Williamson's own modal epistemology as 'counterfactual-based' and mine as 'possibility-based'.[2]

Why even look for an alternative? For one thing, there is clearly room within WME for different approaches, and it should be of interest to everyone who is sympathetic to the general approach what the available implementations are. Moreover, there have been a fair amount of concerns raised about Williamson's own implementation. If those concerns do not apply to my proposed implementation, that should make it a worthwhile option to explore.

One kind of worry concerns whether the equivalences offered by Williamson are a realistic description of how we do (as opposed to how we might in principle) gain modal knowledge (see Vaidya 2015; Jenkins 2008).[3] Perhaps most relevant to our present purposes, Sonia Roca Royes argues that Williamson's equivalences are too demanding to provide a plausible account for our very easy knowledge of a great many *de re* possibilities, such as the possibility of my desk breaking or the possibility of my speaking French (Roca Royes 2011; Roca Royes forthcoming). Roca Royes herself suggests that a possibility-based modal epistemology does much better justice to the ease with which we come by some such modal knowledge. I hope it will become clear in Section 3 that this is true for my own preferred account.

A second kind of worry that has been raised about using Williamson's equivalences to explain modal knowledge is that the equivalences require the vacuous truth of counterpossibles, that is, counterfactuals with a metaphysical impossibility as their antecedent.[4] But of course this commitment to vacuous truth is highly controversial (see e.g. Nolan 1997). I myself tend to be convinced by Williamson's argument for the vacuity thesis (Williamson 2007, 171–175; Williamson forthcoming-a). But even so, a worry about the use of the equivalences in accounting for our modal knowledge remains. This worry is not about the *truth* of the vacuity claim, but about its non-obviousness. Recognition (implicit or explicit) of this non-obvious fact, the vacuous truth of counterpossibles, should not count as a precondition for knowledge of metaphysical modality – after all, the latter is presumably much more widespread than the former, and it certainly has been throughout the history of philosophy.

My own approach, it will be seen, proceeds without such a commitment. It is, I take it, less demanding, and more in line with the way in which philosophers have historically thought about metaphysical modality, than Williamson's own implementation. Nevertheless, it is firmly Williamsonian in the sense outlined in Section 1. It is time now to have a closer look at it.

3. A possibility-based approach

In recent work, Williamson (ms) proposes that we understand 'metaphysical modality as the maximal objective modality' (Williamson ms, 5), where objective modalities include practical and physical modality, and in general everything that is classified as 'circumstantial' modality in linguistics. A proposition, on this proposal, is 'metaphysically possible if and only if it has at least one sort of objective possibility' (Williamson ms, 2).

The proposal in Williamson (ms) is about the nature, not the epistemology, of metaphysical modality. But my claim is that it is very close to a promising version of WME. This version, which I will sketch in the present section, will deviate from Williamson's model in Williamson (ms) in two respects. First, Williamson thinks of metaphysical possibility as the 'union of *all* types of objective possibility' (Williamson ms, 3). I, on the other hand, will introduce metaphysical possibility as a gradual extension from a single kind of possibility (or, at any rate, a single kind of expression), the restricted possibility expressed by ordinary uses of the modal auxiliary 'can'. Second, Williamson tends to use the past-tense/subjunctive form 'could have' as a linguistic starting point, while I use the indicative 'can'. According to Williamson, '[a]lthough 'could' is the past tense of 'can', English permits us to recruit the past tense to express something more purely modal' (Williamson ms, 3). Whether or not this is true,[5] I think that it is not a suitable entry point into modal thought and knowledge – presumably we need to already have some understanding of metaphysical modality to recruit the past tense in this way. Moreover, it is 'can' (in the indicative form) that is *the* paradigmatically circumstantial (i.e. objective) modal (see Collins 2009; Vetter 2013) – a fact to which I will return below.

Let me, then, sketch my own version of a possibility-based WME. Like Williamson's own counterfactual-based proposal, my possibility-based proposal requires two steps.

First, I need to identify an *entry point* into modal thought and knowledge from our ordinary thought and knowledge about the world. I have already indicated that the entry point consists in ordinary statements of the form 'x can F', which I will from here on call 'can statements'.

Second, I need to show how we extend our understanding of this entry point so as to reach the *limiting case* of metaphysical modality. My suggestion will be that this works in much the same way as it does in other cases, such as ontology: we gradually extend the contexts of utterance, we abstract from times, and we then apply any other methods of systematic metaphysical theorizing that we have at our disposal.

The main claim that I will be defending in this paper is the claim that the entry point to thought and knowledge about metaphysical modality consists in can statements. My defence of this claim comes in Sections 4–6. In this section, I will briefly address the more immediate questions about this entry point: how do we know the truth of can statements? And how do we manage to extend from them to the limiting case of metaphysical modality?

How do we know the truth (or falsity) of can statements? As with counterfactuals (see Williamson 2007, 152), there is presumably no unique way of properly evaluating can statements. But it is abundantly clear that we do know, of a great many can statements, whether they are true or false, and that we must have such knowledge prior to philosophical reflection. It is crucial for practical deliberation that we know what we can do (if I cannot jump over the river, there is no point in deliberating about whether to attempt such an action), how things can be manipulated (if the stone is so heavy that it cannot be lifted, then again there is no point in deliberating about whether or not I should pick it up to use it as a tool) and how other people and animals can react (out in the wild, it may be important to know that a tiger can run faster than I can). It would not be surprising, therefore, if at least some of our ways of knowing about what objects can do were hard-wired.

Margot Strohminger (forthcoming) argues that we can know by perception, in at least some cases, what things can do and what we can do with them: I perceive, and thereby come to know, that I can reach the mug, that the mug can shatter, and that the tomato is edible. According to a roughly Gibsonian theory of perception, such modal properties of objects are even directly represented in perception (see Nanay 2012, 2011, 2013).[6]

Other ways of knowing about what can do what are certainly available, and need not consist in explicit theoretical reasoning. Imagining may well play a role here, as it does with counterfactuals (see Williamson 2007; Williamson forthcoming-b[7]). Sonia Roca Royes (forthcoming) argues that much of our knowledge of simple *de re* possibility claims such as 'my desk can break' is obtained by induction on actualized possibilities; such induction, of course, is often performed implicitly without much theoretical thought being given to the matter at all.

It may seem that my entry point into thought about metaphysical modality already is a form of knowing about metaphysical modality. It is very nearly; but not quite. Each can statement entails a metaphysical possibility;[8] by knowing that entailment, our knowing about the more mundane objective possibilities expressed by can statements puts us in a position to know about metaphysical possibilities as well. But in order to know about the entailment, we need to have some grasp (implicit or explicit, under this name or any other) of metaphysical possibility. And of course, since the possibilities in question are restricted, our knowledge about the falsity of a can statement does not yield knowledge of metaphysical impossibilities and necessities. What we need is some sort of grasp on metaphysical, as opposed to the more restricted forms of, possibility. Having some such grasp should enable us to turn our knowledge of restricted possibilities into knowledge of metaphysical possibilities, and it should also ensure that we do not turn knowledge of restricted impossibilities into false beliefs about metaphysical impossibilities.

This takes us to the second task of this section, an account of how we reach the *limiting case* that is metaphysical modality.

My proposed response is this. We begin with can statements such as the ones that I have mentioned above: I can reach the mug, I can speak French, the desk can break, the tomato can be eaten. Can statements are our paradigmatic forms of expression for objective, or in linguistic terms: circumstantial modality. (See Kratzer (1991), and the corpus-linguistic surveys Coates (1983) and Collins (2009); according to Collins, 'can' is used circumstantially in 81% of its uses, which is far more than any other modal auxiliary.) In order to get from ordinary can statements to metaphysical modality we do what we generally do when we go from ordinary thought to thought in metaphysics: we generalize. In particular, we generalize away from times and from contextual restrictions.

Here is an analogous case. When we make or consider existence claims (in the present tense) in ordinary contexts, we talk about what there is *in some relevant domain* and *at the time of the utterance*. When we think about existence claims in metaphysics (also usually phrased in the present tense), we want to speak timelessly and unrestrictedly. If our theories about existence and about existents did not cover dinosaurs because they no longer exist,[9] or if they did not cover distant stars or elementary particles because those are not generally relevant to us, then that would constitute a problem for those theories; metaphysics is about all there is, at any time, whatever our interests.[10]

Likewise, when we make or consider possibility claims, in particular can statements (in the present tense) in ordinary contexts, we talk about *relevant* possibilities obtaining *at the time of the utterance*. When we think about possibility claims in metaphysics (also usually phrased in the present tense), we want to speak timelessly and unrestrictedly. A metaphysics of modality is about all possibilities whatsoever, at any time, whatever our interests. Let's see how we might implement this thought.

First, times. When we think about metaphysical possibility, we don't just ask ourselves whether this can be or that can be. We think (metalinguistically) about whether a sentence of the form '*x* can F' is, was or will ever be true; or we think (in object language) about whether it is, was, or ever will be the case that *x* can F. This idea is nicely captured by Dorothy Edgington:

> The [. . .] pre-philosophical notion [behind the philosophical notion of metaphysical modality] is that certain things can happen, certain things can't; people and other objects can do certain things and can't do others. What can and can't happen, in this sense, is a matter of empirical discovery. This car can do a hundred miles per hour (though it never will), this other car can't – as they are presently constituted. Later, when the first has deteriorated and the second hotted up, the position may be reversed. Diseases which were once incurable no longer are. [. . .] Call something absolutely metaphysically impossible if it is metaphysically impossible at all times. Its negation is absolutely metaphysically necessary. What is not absolutely metaphysically impossible is absolutely metaphysically possible. (Edgington 2004, 6)

Second, contexts. Can statements, like modal statements in general, are notoriously context-sensitive (Kratzer 1977; Kratzer 1981). Take a well-known example from Lewis 1976: compared to an ape, I can speak Finnish; I have the requisite larynx, brain and nervous systems. An ape can't speak a human language such as Finnish; but I can. However, don't expect me to speak Finnish. I can't speak Finnish; I've never learned it. Many more examples of this kind can be constructed. Fortunately, it is the very context-sensitivity of can statements that allows us to overcome their restricted nature and ultimately reach an unrestricted notion of metaphysical possibility.

Lewis's example points the way. My friend Mari can speak Finnish, but I can't; or so we would say in one context. Compared to an ape, of course, I *can* speak Finnish – I have the requisite physiological and neural make-up. And what about the ape? He doesn't have the right larynx or neural wiring to speak a human language such as Finnish. But with surgical alteration, he could in principle be changed in those respects (I think) and become able to speak. He contrasts with, say, a rock: no amount of surgery could make a rock able to speak. Compared to a stone, the ape *can* speak – it just takes more changes for him to realize that possibility than it does for me.[11]

Can statements made in ordinary language are context-sensitive; with a little work on the context, we can generally extend the realm of true possibility statements. It seems, though, that this ends somewhere. I cannot think of a context that would make me assent to 'Mount Everest can speak Finnish' (and if you can, then try 'the number two can speak Finnish'). The concept of metaphysical possibility simply marks the result of the extension, whatever exactly it may be; the concept of metaphysical necessity marks the limits of such an extension, whatever they may be.

We do not, of course, always actually go through some such process of gradual extension. We use terms such as 'absolutely' (as in Edgington's passage above) to mark the end-point of such extensions directly; sometimes the work is done simply by the fact that we are using modal terms while doing metaphysics. (Compare again the case of existence: we could, but we do not usually, extend the context for our quantificational statements gradually; rather, we tend to use adverbs such as '*simpliciter*' to mark the end-point of such an extension.) The important point for present purposes is that there *are* natural limits to contextual extension, though it is not always easy, and almost never uncontroversial, to state where they are located.

Note that all this is not intended as a definition or an analysis of metaphysical possibility or our concept of it, but simply as an account of how we come to think about (and consequently to know about) metaphysical possibility. I am not saying that *p*'s being metaphysically possible consists in the truth of 'it can be that *p*' in some (actual? or possible?) context. The proposal is merely that this kind of relaxation of actual contextual restrictions is what leads us to thinking about metaphysical possibility, and ultimately also about metaphysical necessity. (And as the examples illustrate, we need not have a notion of linguistic context in order to perform this kind of relaxation of contextual restrictions.) An analogous claim, I have suggested, is plausible for the metaphysical notion of existence.

We need careful attention to individual cases and contexts to see just how far such an extension can go. But that too is unsurprising: figuring out the extension of the metaphysically possible is hard; every metaphysician can attest to that. Note, however, that even absent a demarcation of the limits of metaphysical possibility, we know a great deal about metaphysical possibility on this approach. Can statements entail statements of metaphysical possibility, and given an understanding of metaphysical possibility along the lines I have sketched, this entailment is now something that we are in a position to know.

It is important here to keep in mind the distinction between the two questions that I have distinguished in Section 1: how we get to think about metaphysical modality, and how we manage to extend our knowledge about it. I am suggesting that the two extensions I have outlined – relaxing context, abstracting from times – are enough to get us thinking about metaphysical modality proper (as opposed to some more restricted form of modality): metaphysical possibility is, as Williamson has said, the 'maximal' or most general form of objective possibility, that is, of the kind of possibility that we express in ordinary talk with can statements. I emphatically do not say that the two extensions I have outlined are enough to get us all the knowledge we need about metaphysical modality. It is just a start. (Again, the comparison with existence claims holds: relaxing context and abstracting from times is not enough to know what the right ontology is.) But I take it that what we need to make further progress is not all that different from what we do when theorizing in metaphysics about other phenomena: we need to carefully reflect on features of the everyday phenomenon with which we started (here: our understanding of can statements) and apply systematic metaphysical theorizing to them. For instance, reflection on certain temporal asymmetries in can statements, explicit or implicit, may give rise to the idea that an object could not have originated otherwise than it did. We can use all the resources of philosophical (and scientific) theorizing: considerations of simplicity and elegance, parsimony, the internal coherence or instability of a theory, formal representations and models, and so forth.

What is distinctive about knowledge of metaphysical modality is not so much how we get on once we have started the inquiry; it is rather how we get into an inquiry of *this* phenomenon, it is the entry point from which we generalize. It is this claim about the entry point – can statements, rather than counterfactual conditionals – that I will defend in the remainder of this paper.

Williamson discusses, albeit very briefly, a possibility-based alternative to his own modal epistemology (Williamson 2007, 177f.) which differs from my proposal in this section by taking 'could have' statements, rather than can statements, as its starting point. He takes the two approaches as 'not mutually exclusive . . . but rather as different cases in which the cognitive mechanisms of the one already provide for the other' (Williamson 2007, 178). My discussion in this paper should go some way towards showing that there are indeed substantial differences between the two approaches. Williamson then goes on to raise two worries about the envisaged possibility-based approach.

First, it is less clear how the possibility-based approach, starting as it does from more ordinary cases of circumstantial possibility, ever reaches the 'limiting case' of metaphysical possibility. Williamson's own epistemology, of course, does so smoothly: by incorporating contradictions (or a quantifier), it takes us immediately to the limiting case. My favoured approach, proceeding by gradual extension, gives no guarantee of ever reaching the limiting case. I take this to be part of what is otherwise an attractive feature of the approach: it provides a degree of continuity between ordinary and metaphysical modal thought that seems truer to the spirit of anti-exceptionalism than Williamson's leap from ordinary counterfactuals to the rather unordinary counterfactuals in his equivalences. Still, I agree that there is a worry here; to point out just one case, I have not even said anything thus far on how we get to *de dicto* possibility statements. Important as the worry is, I will only address it very briefly in my concluding remarks in Section 7. I want to focus, instead, on Williamson's second challenge to a possibility-based approach.

Here is the second challenge. A possibility-based approach, Williamson says, needs to provide 'some account of what demarcates the relevant forms of possibility from irrelevant ones, such as epistemic possibility' (Williamson 2007, 178). The question arises twice over: first with our entry point, and second with the extension to the limiting case.

First: starting, as I have suggested we do, from can statements, how do we know that some of those can statements do not express epistemic possibility? It is well known that epistemic possibility, unlike circumstantial possibility, does not entail metaphysical possibility: it is epistemically possible that Goldbach's conjecture is true and epistemically possible that it is false, but one of these epistemic possibilities corresponds to a metaphysical impossibility. So if some of the can statements that we start from expressed epistemic possibility, we would have no guarantee that by generalizing from can statements we would not end up including metaphysical impossibilities.

Second: even if it were granted that we start only with properly objective can statements, how can we be sure that in generalizing from them we do not generalize too far, thus again including some epistemic possibilities that are metaphysically impossible? I suggested above that there is a natural limit of our contextual extension; yet how can we be sure that the limit is firmly located between metaphysical and epistemic possibility?

Let me refer to this double challenge as the *demarcation challenge*. I will argue in the remainder of this paper, first, that my own possibility-based approach easily meets the demarcation challenge (Section 4); and second, that it is in fact the counterfactual-based account that is faced with a serious demarcation challenge regarding its own entry point, counterfactual conditionals (Sections 5 and 6). The considerations on which my argument is based will also bring into sharper focus the second aspect of my entry point that has so far been somewhat in the background – the fact, that is, that we use 'can' to ascribe modal *properties* to objects.

4. The demarcation challenge

What, then, demarcates 'the relevant forms of possibility from irrelevant ones, such as epistemic possibility'? My answer to this question, in the present section, will appeal to some recent work in linguistics which indicates that we very clearly distinguish circumstantial modality (the relevant form for our purposes) from irrelevant, in particular epistemic, modality at the level of logical form. We have a deeply engrained, if mostly implicit, knowledge of that distinction; and respecting the relevant features of the distinction, as (we will see) my sketch in the previous section did, we run little to no risk of confusing the circumstantial possibility expressed by can statements with epistemic possibility, either at the level of the entry point or in properly extending from there to the limiting case of metaphysical modality.

In the now classic approach to modal expressions (henceforth abbreviated to 'modals'), shaped especially by Angelika Kratzer (see Kratzer 2012), all modals are alike at the level of logical form: they are sentence operators that take a whole sentence to form a new sentence. The resulting sentence is true or false depending on whether the embedded sentence is true or false in all or some of the selected worlds. Epistemic modals differ from circumstantial ones merely in their selection of worlds.

More recent work in linguistics, though still very much in the Kratzerian spirit, presents a more nuanced picture. It is now widely recognized that 'different semantic categories of modals are located in different positions in syntactic structure', and in particular that 'epistemic modals reside higher in the tree than non-epistemic ones' (Portner 2009, 143), at any rate higher than circumstantial modals.[12] Specifically, it can be seen across languages that epistemic modals take scope over the tense, aspect, and sometimes over quantifiers of the embedded sentence; while circumstantial modals scope below any tense, aspect, and quantifiers, which is to say, their complement does not itself contain any tense, aspect, or quantifiers.

There is a growing linguistic literature on these contrasts and how to interpret them (see, for instance, Brennan 1993; Cinque 1999; Viebahn and Vetter forthcoming; Hacquard 2006; Hacquard 2009; Hacquard 2010, and for an overview Portner 2009). In the present context, I will only be able to illustrate them briefly before continuing to discuss their implications for our present purposes. My illustrative examples will all be English, but the phenomenon is well attested across many languages of different families (Cinque 1999).

Circumstantial modals scope below, and epistemic modals above, tense and aspect. A first, though still somewhat superficial observation in this direction is that we can embed verbs with explicit tense and aspect under an epistemic, but not a circumstantial modal, as testified by the following sentences ('may' is a standard expression of epistemic, and 'can' of circumstantial possibility):

(1) (a) He may have travelled to Paris last summer. (epistemic)

(b) *He can have travelled to Paris last summer. (circumstantial)

(2) (a) He may be travelling to Paris right now. (epistemic)

(b) *He can be travelling to Paris right now. (circumstantial)

We can also see how the same form of a modal, when read either epistemically or circumstantially, elicits different readings:

(3) (a) Mary had to be home at the time of the crime. (Epistemic: it is now necessary that in the past Mary was home at the time of the crime.)

(b) Mary had to sneeze. (Circumstantial: in the past, it was necessary for Mary to sneeze.)

We get both readings also with the past-tense/subjunctive 'could':

(4) (a) Mary could have committed the murder; we certainly have no evidence to rule it out. (Epistemic: it is now possible that in the past Mary committed the murder.)

(b) Mary could have committed the murder; she had a weapon at her disposal and plenty of opportunity – but I'm sure she didn't do it. (Circumstantial: in the past, Mary had the means and the opportunity to commit the murder.)

The availability of epistemic readings as witnessed by (4-a) is one reason why I used the indicative 'can', rather than 'could (have)', in my positive proposal in Section 3.

There is some debate concerning whether or not epistemic modals can be in the scope of another tense operator: a well-known counterexample is

(5) A: Why did you look in the drawer?

B: My keys might have been in there. (=It was possible that my keys were in there)

(The example is from Hacquard (2006, 121), who adapts the example from von Fintel and Gillies 2008. For discussion, see Hacquard (2006, 120–123) and von Fintel (2012, 476–477)). In fact, we might even get a tense-over-modal reading of our earlier examples: imagine, for instance, (3-a) uttered by a detective who is asked why she ceased to investigate Mary as a possible suspect and uses (3-a) to explain her past state of evidence. But what is uncontroversial in the linguistic literature is that circumstantial modals always scope below, and never above, any temporal operator, while epistemic modals always scope above a temporal operator (whether or not they may also scope below another). Hacquard goes so far as to say that '[o]ne way to unambiguously distinguish an epistemic from a root [including a circumstantial] interpretation is to look at the modal's interaction with tense' (Hacquard 2009, 289).

As for quantifiers, Brennan (1993, 93) notes that the following two sentences differ in the availability of a *de dicto* reading:

(6) Every radio may get Chicago stations and no radio may get Chicago stations.
(7) Every radio can get Chicago stations and no radio can get Chicago stations.

(6), featuring the epistemic modal 'may', exhibits both a *de re* reading that (on the obvious background assumption that there are radios) sounds contradictory and a *de dicto* reading that is consistent; while (7), featuring the circumstantial modal 'can', cannot be heard as consistent (on the same background assumption), showing that it has only the contradictory *de re* reading.

On the basis of this and other evidence, Brennan generalizes to the claim that circumstantial modals allow only for *de re* reading, taking scope below any quantifiers (unless they are part of the verb phrase), while epistemic modals can scope either above or below any quantifier. (Both epistemic and circumstantial modals can scope above quantifiers that are part of the verb phrase, e.g. as part of the object of a transitive verb: cp. 'She can keep every ball in the air simultaneously'.[13]) Some even argue that quantifiers can never take scope over epistemic modals (see, for instance, von Fintel and Iatridou (2003); Swanson (2010) disputes the view but notes that is it a 'common view', Swanson (2010, 529)). For our purposes, again, it is enough that circumstantial modals always scope below, and never above, quantifiers that are not part of the verb phrase, while epistemic modals may uncontroversially scope above such a quantifier (whether or not they can also scope below one).

How is this claim to be accommodated in a semantic theory? Generally speaking, the idea is this: circumstantial modals ascribe *modal properties* such as abilities, dispositions, and capacities to objects, typically to the subject of the sentence; such properties may be gained and lost, hence their ascription is tensed. Typically, therefore, they function by relating a sentence's subject to an (untensed) predicate. Epistemic modals express the epistemic standing (typically, at the time of the utterance) of an entire proposition, hence they apply to a sentence that expresses a full-blown proposition complete with tense, aspect and possibly quantifiers.

Exactly how this general idea is implemented, is somewhat controversial. Brennan (1993) argues that while epistemic modals are sentence operators just as Kratzer's semantics has it, circumstantial modals are predicate modifiers, taking a verb phrase to form another verb phrase expressing a modal property, which is then attributed to the sentence's subject. Potential counterexamples to this claim include such apparently circumstantial sentences as 'there can be rain in California', which lack a semantic subject. Brennan (1993) and Portner (2009) respond by introducing an additional category of 'quantificational modals'. Others modify the original proposal, while upholding the general idea. Hacquard, in particular, holds that circumstantial modals apply to a tense- and aspectless 'proto-sentence', which expresses an event, and are 'anchored to' some object participating in the event – typically, but not always, the object referred to by the sentence's grammatical subject – to which a modal property is ascribed. Epistemic modals, on the other hand, apply to a full-blown sentence and are anchored to the 'local knowledge bearer' – typically, but not always, the speaker of the sentence. (Hacquard 2010, 92)

For our purposes, these differences do not matter. What matters is the general diagnosis: circumstantial modals such as 'can' and epistemic modals differ significantly in their syntactic behaviour. Circumstantial modals are restricted to a certain place in the sentence structure: below tense, aspect, and any quantifiers (outside the verb phrase). Epistemic modals appear freer in their syntactic position, but they must have tense and aspect in their scope, and they certainly can scope above any quantifier. Moreover, these differences are best explained by the semantic fact that we implicitly recognize a significant semantic distinction: circumstantial modals ascribe modal properties to the objects that the sentence is about; epistemic modals express, in some way or another, the epistemic standing of a proposition relative to a subject that the sentence need not in any sense be about.

We are now in a position to answer the demarcation challenge, both as it applies to the entry point itself and as it applies to the process of generalizing from it.

How do we manage to start with a non-epistemic modal concept as our *entry point*? The answer is, first, that we start with an understanding of expressions of circumstantial modality, the modal 'can' and its cognates in other languages, and that we have a firm if usually implicit grasp of the distinction between such circumstantial modality and epistemic modality, as evidenced by our mastery of the subtle syntactic differences between them. Secondly, we can now see that this syntactic implementation of circumstantial and epistemic modality is no accident. We use circumstantial modals such as 'can' to ascribe modal properties to objects; apart from some special cases,[14] we do not generally use epistemic modals to ascribe properties to objects (or only, perhaps, to a special kind of object: propositions). This is not an arbitrary distinction: we tend to think about objective, mind-independent reality as a matter of objects having properties. It is no wonder that we use that same way of thinking for objective, as opposed to epistemic, modality.

Further, given this entry point, how do we avoid *generalizing* too far and ending up with mere epistemic possibilities after all? The answer should now be clear. The meanings of circumstantial modals are sufficiently different from each other that we should expect there to be a natural boundary between them when we generalize beyond the restrictions of ordinary circumstantial modals. In fact, they are so different from each other that there is little reason to expect a natural passing from one to the other; so the question of a boundary between them may not even arise.

To be sure, we have modals that express both circumstantial and epistemic modality[15]); but as we have seen in this section, there are important disanalogies with respect to both syntax and semantics, suggesting that such modals in fact exhibit a kind of ambiguity, or more precisely, polysemy.[16]

This is emphatically not to say that philosophers never do, in fact, mistake mere epistemic possibilities for metaphysical possibilities; I think they often do, in particular when they simply examine intuitions about sentence-operators such as 'it is possible that . . .', which are epistemic in all ordinary contexts (see Kratzer 1981; DeRose 1991). It is to say, rather, that there is a good way of not making that mistake: by linking metaphysically modal thought, however implicitly and however indirectly, to thought about the modal properties of objects.

So far, I have argued that the possibility-based approach has the resources to meet Williamson's demarcation challenge. Williamson's phrasing of the challenge suggests that he believes no analogous problem to apply to his own, counterfactual-based account. In the next two sections, I will argue that this is not so. The counterfactual-based account begins with our understanding of linguistic constructions like the English 'if it were the case . . . then it would be . . .'. I will argue for two claims. First, counterfactual conditionals have epistemic as well as circumstantial readings. And second, it is much less clear that we can implicitly demarcate these readings for counterfactuals, making counterfactuals a less suitable entry point into metaphysical, non-epistemic, modal thought and knowledge than can statements. This is my own demarcation challenge to Williamson's modal epistemology, more specifically, to its entry point. Section 5 will defend the first claim, Section 6 the second.

5. Epistemic counterfactuals

Williamson (2007) does not explicitly state but clearly assumes that the counterfactuals to which he appeals have circumstantial and not epistemic readings (see also Williamson ms, 10). In a more recent paper, he explicitly states that '[n]ormally, counterfactuals are on the objective side of this contrast [between objective/circumstantial and epistemic]' (Williamson forthcoming-a, 14); he does not say anything about any abnormal readings.

I begin by arguing that there are counterfactuals, i.e. sentences of the form 'if it were/had been that . . . then it would be/would have been that . . .', which have epistemic readings. A nice array of examples can be found in Edgington (2008) (numbers and paragraphs are inserted by me):

(i) There is a treasure hunt. The organizer tells me 'I'll give you a hint: it's either in the attic or the garden.' Trusting the speaker, I think 'If it's not in the attic it's in the garden.' We are competing in pairs: I go to the attic and tip off my partner to search the garden. I discover the treasure. 'Why did you tell me to go to the garden?' she asks. 'Because if it hadn't been in the attic it would have been in the garden: that's (what I inferred from) what I was told.' That doesn't sound wrong in the context.

(ii) Or consider: 'Why did you hold Smith for questioning?' 'Because we knew the crime was committed by either Jones or Smith – if it hadn't been Jones, it would have been Smith.'

(iii) There's also a nice example of van Fraassen's (. . .): the conjuror holds up a penny and claims he got it from the boy's pocket. 'That didn't come from my pocket', says the boy. 'All the coins in my pocket are silver. If that had come from my pocket, it would have been a silver coin.'

(Edgington 2008, 16f.)[17]

In all three cases, it seems clear that the connection between the antecedent and the consequent of the conditional is not 'worldly' but rather relative to our state of information: the utterance in (i) is true even if the organizer never had the intention, or even the option, of hiding the treasure in the garden; it is true simply by virtue of the speaker's having had the evidence that she had. The utterance in (ii), likewise, is true even if there was never objectively any significant objective chance of Smith committing the murder; rather, it is related to the evidence that the speaker had in the past. Finally, the boy in (iii) makes no claim to the effect that the coin would have turned into silver had it been in his pocket. Rather, it is relative to his knowledge that all the coins in his pocket are silver that this coin would have to (epistemic *have to*!) be silver if it had come from his pocket.

Edgington (2011) and Khoo (2015) have both pointed out that we can produce epistemic readings along the lines of (ii) for just about every counterfactual, even the paradigmatic example 'if Oswald hadn't shot Kennedy, someone else would have': just imagine a detective explaining why she was investigating other suspects after it has become clear that Oswald was indeed the shooter (see Edgington 2011, 239; Khoo 2015, 25f.). This should not come as much of a surprise: we noted a corresponding reading for non-conditional modals in Section 4.

I would like to add a hypothesis of my own to extend the class of epistemically-read counterfactuals. Williamson has argued in a number of places (Williamson 2007, 171–175; Williamson forthcoming-a) that counterpossibles are vacuously true. Imagine that, while walking in the wilderness, we enter a clearing and something runs past us from an unexpected angle. In fact, it is a harmless gazelle; but we both know that there are dangerous tigers about. Shocked, you say to me:

(8) If that had been a tiger, we would be dead now.

This seems a reasonable thing to say under the circumstances, and a useful statement reminding us to be more careful in the future (next time it might be a tiger instead of a gazelle). On the contrary,

(9) If that had been a tiger, it would have had tea with us.

Seems, on the face of it, clearly false. Of course, given the necessity of kindhood (or at least the necessity of not belonging to a very different biological kind), both are counterpossibles, and hence both are true, according to Williamson. The difference between them is one of assertability, not of truth (Williamson 2007, 173; Williamson forthcoming-a, 9). Arguably, however, if we had a way of making (8) true and (9) false without getting into trouble, that would be the preferable reading. Williamson, of course, has forceful arguments to show that we *do* run into trouble unless we accept the truth of both (8) and (9). But, as I have pointed out elsewhere (Vetter forthcoming) and as Williamson readily acknowledges (see Williamson 2007, 175; Williamson forthcoming-a, 14), his arguments turn on the idea that the counterfactuals in question are objective, that is, circumstantial. If (8) and (9) are read epistemically, then the metaphysical impossibility of their antecedents has little bearing on their truth-value.

Compare: on a circumstantial (objective) reading of 'could', the following are both false in the scenario described for (8) and (9):

(10) This could have been a tiger.
(11) This could have been a cloud.

On an epistemic reading, this is not so: (10) is true, while (11) is, presumably, false. Epistemic readings allow us to make distinctions among metaphysical impossibilities that are important, and to allow those distinctions to be just what they appear to be: differences in truth-value, not in assertability. I suggest, therefore, that we read (8) and (9), as well as other (metaphysical) counterpossibles, as epistemic.[18] (We will meet the two tiger conditionals again later.)

It seems, then, that there are a reasonably wide range of epistemically read counterfactuals. So, in gaining knowledge about metaphysical modality from our knowledge of counterfactuals, we must make sure that the counterfactuals we start with are understood circumstantially, not epistemically.

Khoo (2015) appears to offer an easy way out arguing that the circumstantial reading is the default reading of a counterfactual. According to Khoo, epistemic counterfactuals all share with circumstantial counterfactuals (but not with indicative conditionals) a syntactic feature that we have seen to be typical of circumstantial modals: they are embedded under a (past) tense. The epistemic counterfactual in Edgington's (ii) is properly read as saying that *in the past* it was epistemically necessary, under the supposition that it wasn't Jones, that it would have been Smith. Khoo takes this to explain why circumstantial readings are the default reading for this kind of construction: we rarely 'care about past possibilities which are no longer open possibilities (having learnt something new which rules them out)' (Khoo 2015, 27).

However, I take Edgington's example (iii) to constitute a counterexample to this general claim. In (iii), what we have is an epistemic counterfactual that is relativized to the boy's current evidence (in particular, his knowledge about the coins in his pocket all being silver, and having been so when the conjuror did his trick). It *is* necessary (at the time of utterance: present tense), given his present evidence, that if in the past the coin had come from the boy's pocket, it would then in the past have been silver. The interaction between the epistemic counterfactual and the past tense appears to be more complex than Khoo's hypothesis allows; and therefore his explanation for, and his claim of, a default circumstantial reading does not apply in all cases. (Moreover, as we will see shortly, there may well be reason to care about counterfactuals even when they concern past possibilities that are no longer open: like circumstantial counterfactuals, they help us learn from experience.)

Of course, to close off the easy way out is not yet to show that there is no other way out. However, I will now argue that it is not at all clear that we are as good at distinguishing the relevant forms of counterfactuals from the irrelevant ones, in particular the epistemic ones, as we are with the simpler modal constructions, such as expressions of possibility.

6. Demarcating counterfactuals

How are we to demarcate circumstantial from epistemic counterfactuals, so as to make sure that we do not infect our metaphysically modal thought with epistemic modality? This section will offer no positive answer to the question, but rather point out that an answer is not as easy to come by as in the case of can statements: we are, I suggest, not all that good at drawing the required distinctions, at least not without some theoretical background. This makes counterfactuals less suited as an *entry point* into metaphysically modal thought, although it may be that with more advanced theorizing about these kinds of modality, we come to understand and apply the distinction competently and successfully. But that understanding has to start elsewhere; my suggestion has been that it starts with can statements.

I offer three kinds of consideration to support my claim. First, the syntactic considerations that I used for the case of can statements do not carry over to counterfactuals. Second, there is evidence that speakers with the relevant linguistic and metalinguistic competence do indeed disagree dramatically about the readings of standard counterfactuals. And third, given the practical and evolutionary role that counterfactuals play, there is no reason to expect that we should be very good at distinguishing between their epistemic and circumstantial readings, for they serve much the same purposes.

6.1. First consideration: syntax

To begin with, note that the considerations that I have adduced for the case of can statements do not straightforwardly apply to counterfactuals. As we saw in Section 4, there is good syntactic and semantic evidence that we use 'can' to ascribe modal properties to objects, and that we are implicitly very aware of the distinction between the (circumstantial) ascription of modal properties on the one hand and epistemic modality on the other. After all, we consistently apply it in the syntactic construction of modals. Counterfactuals, whether epistemic or circumstantial, do not appear to ascribe modal properties to individual objects; they appear rather to link two states of affairs to each other.[19] This is mirrored in their syntactic structure.

The exact connection between the syntactic features of the different modal flavours, as described in Section 4, on the one hand and counterfactuals on the other hand has not received (or has, perhaps, resisted) the kind of attention and theoretical work that would provide us with a systematic understanding of them. But it is relatively easy to see that counterfactuals, even if they are plausibly read circumstantially, do not exhibit the syntactic marks of circumstantial modals. First, the counterfactual 'would' – unlike any of the standard circumstantial modals – can embed both aspect (as in (12)) and quantifiers (as in (13)). Consider:

(12) (a) If you were in Italy, you would be having dinner right now.
(b) If you had gone to Italy, you could have been eating pasta all week.
(13) (a) If that rock had fallen down, somebody would have been hurt.

The relation between counterfactuals and tense is somewhat more complicated (and contested). Hacquard (2006, 160), for instance, argues that counterfactuality is a syntactic element in its own rights, situated below epistemic modals and tense, but above aspect and circumstantial modals. A further complication arises from what is sometimes called 'fake tense' (Iatridou 2000): an apparent past tense which, as in (12-a) ('you were'), does not serve to indicate a past time at all but rather appears to have a modal function itself. I cannot even begin to disentangle these various issues here. But there are at least some cases where the circumstantial, counterfactual 'would' must take scope over a past tense: backtracking counterfactuals. (Here I use 'backtracking' in a more narrow sense than is usual in the literature, for counterfactuals whose consequent is about a time prior to the time of the antecedent; those are the only cases that I am interested in at present.) It would appear that there are true instances of backtracking counterfactuals with circumstantial readings, such as

(14) If she had a twin sister, her mother would have had at least two children.
(15) If she had measles now, she would have got it from someone else.
(16) If Jim had asked Jack for help, they wouldn't have quarreled yesterday.

((14) is from Arregui (2005), (16) is adapted from Lewis (1979)). Each of these sentences, in at least some contexts, are circumstantial: they are about the necessary conditions for someone to have a twin sister, about the causal history of a child's disease, and about Jim's character; and each of these sentences is or at least may well be true in such a context. But in each of these sentences, the past tense of the consequent does not take scope over the counterfactual 'would': it is *now* the case that a certain present state counterfactually requires a certain past state.

If this is right, then the counterfactual 'would', unlike standard modals, can in principle scope above tense, as well as over aspect and quantifiers (outside the verb phrase). The syntactic-cum-semantic considerations that I have used to counter the demarcation challenge to my own possibility-based approach does not, then, seem to be available for a counterfactual-based account.

6.2. Second consideration: linguistic intuitions

Perhaps Williamson will respond that we nevertheless have a good intuitive grasp of the distinction between epistemic and circumstantial readings in the case of counterfactuals; perhaps that is all we need. I respond that there is little reason to think that we do have such a grasp. An indication of this can be found in classifications made by professional linguists, certainly reflective and informed speakers of their native language *par excellence*.

While the semantic literature on the various flavours of modals tends to neglect the counterfactual 'would', there have been two comprehensive corpus-linguistic studies on the English modals including the uses of 'would' in counterfactual conditionals (Coates 1983; Collins 2009). Both are characterized by their authors' close and careful attention to the semantic features, in particular the modal flavours, of the enormous number of real-life examples that they examine. Both largely classify the 'would' of the counterfactual conditional as epistemic. Coates (1983, 213), in fact, classifies *all* occurrences of 'would' where it occurs as a 'general hypothetical marker' (rather than a past-tense or hypothetical form of 'will') as epistemic. She does so, unlike with other modals, not so much on the grounds of her linguistic intuitions about particular cases, but because the counterfactual 'would' aligns syntactically with epistemic, not circumstantial or deontic modals, with respect to several syntactic markers that she found to be 100% correlated with epistemic modality in her sample.[20] I myself would not accept her sweeping classification of all hypothetical 'would' as epistemic. But it is interesting that Coates herself, a trained linguist with a keen ear for semantic classifications (much of her study relies on her classifying occurrences of various modals according to semantic intuition) has no qualms in so classifying it. In a more recent study based on larger, computer-generated text corpora, Collins (2009) found that the hypothetical 'would' was used epistemically in 64.3 % of cases, and circumstantially in only 22.9% (Collins 2009, 140), though he does not discuss his reasons for classifying the occurrences as he does.

Now, as philosophers, we need not be silenced by these kinds of syntactic observations. We need not agree with either Coates or Collins; we may argue, on a case-by-case basis, that the most natural reading of 'would' in the counterfactuals in question is circumstantial. (I, for one, think that it often is.) But the fact that qualified linguists, who certainly understand what they are talking about, disagree with us does show that there is some uncertainty about whether or not a given counterfactual is to be read epistemically or circumstantially. With 'can', on the other hand, there are no advocates of widespread epistemic readings; 'can' is firmly and uncontroversially circumstantial. I take this to be a clear advantage of my proposed approach vis-á-vis the demarcation challenge.

6.3. *Third consideration: evaluating counterfactuals*

I now turn to a third kind of consideration: there is no reason, given our practice of evaluating counterfactuals and the evolutionary explanations for those practices, why we *should* be particularly good at keeping them apart. Our evaluations proceed in much the same way, and serve much the same purposes, whether the counterfactuals in question are circumstantial or epistemic.

How do we evaluate counterfactuals with circumstantial readings, according to Williamson? There is no uniform answer to this question: 'In general, our capacity to evaluate counterfactuals recruits *all* our cognitive capacities to evaluate sentences. . . . There is no uniform epistemology of counterfactuals'. (Williamson 2007, 152) Nevertheless

> [w]e can still schematize a typical overall process of evaluating a counterfactual conditional thus: one supposes the antecedent and develops the supposition, adding further judgements within the supposition by reasoning, offline predictive mechanisms, and other offline judgements. . . . To a first approximation: one asserts the counterfactual conditional if and only if the development eventually leads one to add the consequent. (Williamson 2007, 152f.)

It is natural to think that, if this picture is right for (many) circumstantial counterfactuals, it is right also for (many) epistemic counterfactuals. This may be easiest to see when we eliminate the role of past tense in epistemic counterfactuals. For instance, imagine a group of police officers considering possible suspects for a murder. The weapon has not been found, but the flat of suspect A has been thoroughly searched. If in this context one of the officers says,

(17) If A were the murderer, she would be hiding the weapon in a very safe place,

then this is plausibly read epistemically.[21] How would the police officers reason about (17)? Plausibly, by supposing A to be the murderer and developing the supposition further, using the evidence that is currently available to them (e.g. that the weapon has not been found).[22] But this is just what one would do in the circumstantial case. Admittedly, the kinds of evidence that are considered in the two cases seem to differ, and I am not able to give, nor do I know of anyone giving, a systematic account of exactly how they differ.

Nevertheless, my point stands: what we do in the two cases is a very similar kind of process.

There is one relatively clear disanalogy between how we (properly) evaluate circumstantial and epistemic counterfactuals, which is clear when there are indexicals involved. Consider again the counterfactual (8): 'if this had been a tiger, we would be dead now'. In evaluating this counterfactual circumstantially, we would have to develop the impossible supposition that this very thing, which is a gazelle, was a tiger. In evaluating it epistemically, we can focus on the way in which 'this' is presented to us, rather than on what the word 'this' refers to. We might say, then, that we are developing a different supposition. (In two-dimensional terms, we are dealing with the A-intension, not the C-intension, of the antecedent.) We might also say that we are holding fixed different pieces of knowledge, or belief, in evaluating the counterfactual circumstantially or epistemically. When evaluating the circumstantial reading, we hold fixed what Williamson calls 'constitutive matters' (Williamson 2007, 170) such as the identity and kind of the object involved. When evaluating the epistemic reading, we hold fixed the more superficial matters of how things appear to us; their constitution need not be of concern to us in this case. This is a substantial difference; but it is not one, I submit, to which we pay much attention in reasoning with counterfactuals. For most practical intents and purposes, we can switch between these two kinds of considerations just as easily as we switch between holding fixed different sets of facts. This is because both circumstantial and epistemic counterfactuals serve more or less the same purposes in everyday life – or so I am now going to suggest.

Williamson stresses that there is a valid evolutionary explanation available for our reliability in evaluating (objective) counterfactual possibilities. In general, he says,

> [o]ur overall capacity for somewhat reliable thought about counterfactual possibilities is hardly surprising, for we cannot know in advance exactly which possibilities are or will be actual. We need to make contingency plans. In practice, the only way for us to be cognitively equipped to deal with the actual is by being cognitively equipped to deal with a wide variety of contingencies, most of them counterfactual. (Williamson 2007, 137)

Note that this, too, applies to the epistemically possible if it applies to the circumstantially possible. In practice, we may add, it makes little difference whether a certain contingency is objectively possible given how things are, or whether it is epistemically possible, not ruled out by what we know. We prepare for contingencies that we know have some chance of coming about; and we prepare for contingencies that we know we can't well rule out. Theoretically, there is a world of a difference between the two; practically, that difference vanishes because we cannot easily tell the difference between facts and known facts – the known facts are, after all, the only facts that we know of.[23]

What goes for circumstantial and epistemic modality in general, goes for circumstantial and epistemic counterfactuals. Williamson points out that counterfactuals can provide abductive and deductive evidence for their antecedents: the counterfactual 'if Jones had taken arsenic, he would have shown just exactly those symptoms which he does in fact show' can be used to infer (abductively) that Jones has taken arsenic, while 'If Jones had taken arsenic, he would not have shown just exactly those symptoms which he does in fact show' can be used to infer (by modus tollens) that Jones hast not taken arsenic (Williamson 2007, 137). Likewise, Edgington's epistemic counterfactual 'If that (coin) had come from my pocket, it would have been silver' is used by the boy in Edgington's example to infer, by modus tollens, the falsity of its antecedent. In a different scenario, it might well be used the other way around, to infer abductively from the coin's being silver to its having come from the speaker's pocket.

Counterfactuals are crucial for learning from mistakes, as Williamson (2007, 140) notes (for a more extensive exposition of the point, see Krödel (2012) and Krödel (forthcoming)). To use Thomas Krödel's example,

> suppose that I am attacked by a tiger while asleep during an expedition and only narrowly escape death. I reason that I would not have been attacked by a tiger if I had lit a campfire, and resolve to light a campfire on similar occasions in the future. Doing so, I avoid tiger attacks for the remainder of the expedition and return home safely. (Krödel 2012, 4)

Now consider again our earlier example of what I have argued to be an epistemic counterfactual: 'if that had been a tiger, we would be dead now'. This, too, enables us to learn from mistakes: I will resolve never again to tread so carelessly in the wilderness, because from what I know, that might end very badly.

I am not saying that epistemic counterfactuals share all the useful features of circumstantial ones; many circumstantial counterfactuals do not share all those features either. (Epistemic counterfactuals presumably do not link to causation in the way that many circumstantial ones do, to mention an obvious point of departure; but many circumstantial counterfactuals are not linked to causation either, see Williamson 2007, 140f). But given that the two kinds of counterfactuals share important functions for practical and practice-oriented reasoning – delineating options for action, providing a basis for inductive inference, learning from mistakes – there is no reason to believe in any evolutionary pressure for a sharp distinction between the relevant cognitive capacities.

This completes my third, and longest, consideration. Given both the available practices for evaluating counterfactuals, and the evolutionary pressures under which those practices may be supposed to have evolved, there is no reason to expect that we are particularly good at telling them apart. Their syntactic and semantic structure does not help either, as we have seen in my first consideration; and, as my second consideration showed, there is empirical evidence from corpus linguistics that even reflective, competent native speakers do not firmly have the right intuitions (whether it is Coates or us – one side has to be wrong).

Do similar considerations not apply to my own approach? Like circumstantial counterfactuals, can statements presumably share with their epistemic counterparts a great deal of their practical and evolutionary significance. But this fact does not have the same implications in the case of can statements that it does in the case of counterfactuals. For we have, in the case of non-conditional modal statements, the syntactic-cum-semantic considerations outlined in Section 4. And moreover, the methods for assessing a can statement may differ drastically from those used for assessing a statement of epistemic possibility – after all, one is a method for evaluating the ascription of modal properties to objects, while the other is a method for evaluating the attribution of an epistemic standing to a proposition.

7. Conclusion

WME is characterized by realism about metaphysical modality, along with anti-exceptionalism about our ways of knowing about it. In this paper I have suggested a form of WME that differs from Williamson's own version in that it takes can statements, not counterfactual conditionals, as the entry point into modal thought. We come to understand and know about metaphysical modality, on the view that I have suggested, by gradually extending from can statements and understanding the phenomenon about which we are reasoning to be the limiting case of that extension. I have argued that, *contra* a suggestion of Williamson's, my own approach does better justice to the deep division between metaphysical, that is objective, and epistemic modality. It does so precisely because can statements are, *qua* attributions of modal properties, syntactically and semantically set apart from statements of epistemic modality in a way in which circumstantial counterfactuals are not.

How much weight we are willing to attribute to this depends on how much weight we are willing to attribute to the demarcation challenge in general – that is, it depends on how concerned we are (or think we should be) about confusing metaphysical with epistemic modality. Given the history of philosophy, I think there is reason to take the demarcation challenge very seriously. (Recall that WME, in both versions, is intended as a descriptive thesis, but descriptive only of what happens in the best case: when we do acquire knowledge.)

My concern, as I have stressed throughout, has been with arguing that can statements provide a better entry point into metaphysically modal thought and knowledge than counterfactuals. I have left open a great many questions concerning the positive nature of my proposal. In concluding, I want to briefly mention (if not properly address) two limitations of my proposed account that may seem troubling.

An important factor in my argument was that can statements attribute modal properties to objects at times. This may seem to be limiting twice over.

First, we may feel that even abstracting from times in the way that I did (whatever can be the case at *some* time, is metaphysically possible at all times) does not yield properly timeless modality. What about the truths of mathematics, whose necessity seems prior to any time at which they are necessary? And more pressingly, what about possibilities that were not possible at any time – for instance, that the world had always contained fewer objects or more dimensions?

In Section 3, I said that we reach a notion of metaphysical possibility by abstracting from circumstantial possibility both concerning times and concerning contexts. I did not claim that we abstract *only* from times and contexts, in the way I outlined there. Perhaps we can abstract further still and thereby yield a less limited conception of metaphysical possibility, one not subject to temporal asymmetry in the way just sketched. Perhaps appeal to the apparently timeless modal truths of mathematics helps us get on this route. But given our observations about the syntax of circumstantial and epistemic modals, I want to sound a cautionary note here. We should not leap too hastily from the syntactic construction provided by circumstantial modals to a full-blown sentence operator along the lines 'it is possible that . . .', on pain of losing the features that guide our linguistic understanding of the distinction between the relevant (circumstantial) and irrelevant (epistemic) kinds of modality. Perhaps such a move is ultimately needed, for general metaphysical considerations of some sort; but it is one that tends to switch our linguistic intuitions over to the epistemic side. It may be that some of our intuitions about metaphysical possibility are indeed generated by relying too much on our intuitions concerning the truth of sentences which, in ordinary language, are understood as epistemic. What is needed, then, is not reliance on those particular linguistic intuitions, but simply metaphysical theorizing. What I have said here leaves wide open the question how far beyond our entry point such theorizing can take us.

Second, we may feel that the focus on modal properties of objects yields an undue limitation to *de re* possibilities. How, then, do we get a grasp on metaphysical modality *de dicto*?

The most conservative response would be to simply adopt the Barcan formulas (in both directions): *de dicto* possibilities are just *de re* possibilities in a different guise. The consequences of such a stance are well known; it is intuitive for a great deal of *de dicto* possibilities (such as the possibility that there be a woman President of the U.S.), but is committed to either rejecting others (such as the possibility of there being an identical twin of mine) or accepting *mere possibilia* and thereby a significant increase in ontology. Williamson, of course, takes the latter route (see, for instance, Williamson 1999, 2002, 2013). But his argument is metaphysical; so the question remains whether his metaphysical commitment is properly incorporated into an epistemology of modality.

Again, we may take a less conservative stance here and propose further extending our notion of modality from the circumstantial to the metaphysical.[24] But here as before, we must be careful not to leap into a generalization that may turn the expressions we use into epistemic modals.

The account I have given was, of course, one of our modal *knowledge*, not of modality itself. So even if those limitations persist, they may be epistemic rather than metaphysical. But WME, by its nature, need not be overly conciliatory. Given realism about modality, we may just get things very wrong. Given anti-exceptionalism, we may get things wrong especially when we try to obtain modal knowledge by some exceptionalist route, such as conceiving (in one of the special senses sometimes given to the term by rationalists about modal knowledge). For any putative counterexample, we may reasonably ask what the evidence for the particular possibility (or necessity) is. Appeal to conceivability will not do. [25]

Notes

1. Anti-exceptionalism about philosophical knowledge is, of course, one of the main tenets of Williamson 2007; Williamson uses the term 'exceptionalism' to describe his opponents, see e.g. Williamson(2007, 3).
2. I adapt these labels from Hale (2003). Note, however, that Hale only distinguishes between possibility-based and necessity-based modal epistemologies. Williamson's modal epistemology would be necessity-based by Hale's standards (if it is asymmetric at all) because the natural, and probably the epistemically primary, equivalences in Williamson's picture link necessity to counterfactuals. Moreover, I would like to stress that the possibility-based epistemology that Hale (2003) discusses is quite different from the one that I will sketch in Section 3, precisely because my own version is broadly Williamsonian.
3. In addition, as Krödel (2012) points out, our credence in a necessity claim will normally be higher than that in the corresponding counterfactual on the right-hand side of an instance of (V□) or (V*□), so our knowledge of the former can hardly be explained in terms of knowledge of the latter. I agree; but as Krödel also points out, this objection does not apply to (Q□).
4. Here is why. If counterpossibles are sometimes false, and if A is a necessary truth, then $(\neg A \;\Box\!\!\rightarrow \bot)$ and $(\neg A \;\Box\!\!\rightarrow A)$ should be prime examples for false counterpossibles: if there is anything that the impossibility $\neg A$ does *not* counterfactually imply, it's its own contradiction (A) or indeed any contradiction ($\bot$). So both (V□) and (V*□) require that all counterpossibles be true. And so does (Q□); otherwise replacing p with $\neg A$ on its right-hand side would yield a counterexample to the universal quantification despite the necessary truth of A.
5. Some linguists claim that we can use tense to express a modal meaning; cf. Iatridou (2000) on 'fake tense'; see also Section 6.
6. Strohminger speaks about 'abilities' where I speak of can statements; Nanay is concerned with 'action properties', a notion that is close to Gibsonian affordances. Both switch between explicit can statements and dispositional idioms

such as 'edible', 'reachable', 'breakable', etc. I take such idioms to be equivalent, probably even synonymous, with corresponding can statements (see Vetter 2014), so I am happy to include them as well. See Nanay (2012, 431f) for some differentiations between his own view, which I have here labelled 'Gibsonian', and Gibson's affordance theory of perceptual content.

7. Williamson (forthcoming-b) can be read as claiming that imagination is a route to knowledge about can statements – but there he appears to think of can statements along the lines of a version of the conditional analysis of ability ascriptions, see Williamson (forthcoming-b, 6, 9). I treat can statements as a kind of restricted possibility statement.
8. *Pace* Spencer forthcoming, who argues against that entailment. My response to his arguments would take us too far afield for present purposes.
9. Of course, we may conclude that dinosaurs do not exist if we are presentists or temporaryists. But the relevant claim then would be that it is *always* that case that what exists is what presently exists (in some sense of that frustrating term), so we must still go beyond the temporal restriction of ordinary existence claims.
10. I am assuming here that we *can* quantify unrestrictedly (see Williamson 2003). If not, the claims in the text could be reformulated accordingly.
11. Am I using counterfactuals in extending beyond ordinary can statements here? If so, do I not require the same resources as Williamson's epistemology, in addition to those of my own? – I am not entirely convinced that the envisaged extension must use counterfactuals. But note that even if it does, it uses them only to get from one objective possibility to another objective possibility. Counterpossibles, or the Williamsonian equivalences, do not feature in the picture at all; and it is the modality expressed by 'can', not that expressed by the counterfactual conditional, which settles the nature of the modality at issue.
12. Deontic modals are harder to classify, but I will focus on the distinction between circumstantial and epistemic modals, since I agree with Williamson that there is less danger of confusing our target type of modality, circumstantial modality, with deontic modality; see Williamson ms, 2.
13. Thanks to Tim Williamson for prompting this clarification and several others.
14. Such as: 'These cities share one feature: Napoleon might have visited them'.
15. 'Can' has hardly any epistemic uses in its positive, present-tensed form. But its negation may be used epistemically, as in 'This cannot be true!'; and of course, 'could' has an established epistemic use. Other languages, such as German, have equivalent expressions which, while predominantly circumstantial, are more easily recruited for epistemic uses.
16. Polysemy is, in short, non-accidental ambiguity – the kind of ambiguity exhibited by 'since' (temporal or causal relation) but not by 'bank'. Emanuel Viebahn and I provide an extended argument for the polysemy of modals in Viebahn and Vetter forthcoming.
17. See also Edgington 2007; Edgington 2011.
18. A more detailed argument can be found in Vetter (forthcoming).
19. I have elsewhere argued that a dispositionalist account for the truth-conditions of counterfactuals can be defended; see Vetter (forthcoming). This is compatible with saying that we do not syntactically or semantically construe counterfactuals in this way.
20. The markers may be somewhat surprising: Coates found that only epistemic, and no circumstantial, modals had (i) inanimate sentence subjects, (ii) existential sentence subjects, (iii) stative verbs, (iv) progressive aspect, and (v) quasi-modals in their scope. As a corpus linguist, Coates makes no claim that

these markers hold for all possible expressions of epistemic and circumstantial modality respectively, and philosophers may find it straightforward to produce counterexamples to some of them (e.g. 'the machine can crush oranges' as against (i)).

21. Add 'have to' after 'would' to make the reading even clearer; I claim that the addition makes no difference to the meaning of (17). Contrast a context where it is A's character that is being discussed: A is extremely cautious and thorough. To illustrate, someone utters (17). That context, I submit, would elicit a circumstantial reading.
22. There are very tricky issues about contextualism vs. relativism here, which I am trying to avoid by imagining the context of utterance and the context of evaluation to coincide. I think that this is fair in the present context because we are not so much interested in agreement and disagreement, but rather in one speaker/thinker coming to a conclusion about whether or not she takes a certain counterfactual to be true.
23. I am not sure that Williamson would disagree with any of these remarks. The role that he sketches for the imagination in Williamson (forthcoming-b), for instance, while tailored to knowledge of objective possibilities, seems to work just as well for epistemic possibilities. Williamson (forthcoming-b, 10) uses an indicative conditional – which is standardly thought to be epistemic – in tandem with a counterfactual one – which is, of course, standardly thought to be circumstantial – as two analogous applications of knowing by imagining, noting only that there are 'subtle cognitive differences in the cognitive process'.
24. I have done so, with only a slightly less conservative point of view but without commitment to the Barcan formulas, in Vetter (2015, ch.7.5).
25. For very helpful comments and discussion, I would like to thank Simona Aimar, Max Bohnet, Mathias Böhm, Luca Castaldo, Steffen Koch, Thomas Krödel, Christian Nimtz, Sonia Roca Royes, Margot Strohminger, Kilu von Prince, and Jonas Werner as well as the audience of a workshop on modality and causation at UCL on 19 December 2015 and my research colloquium at Humboldt-University, Berlin. Special thanks to Mathias Böhm for a very helpful overview of the linguistic literature on conditionals, and of course to Tim Williamson for his extremely helpful comments on an earlier version of this paper.

References

Arregui, Ana. 2005. *Backtracking Counterfactuals and Iterated Modalities*. Handout.

Brennan, Virginia. 1993. "Root and Epistemic Modal Auxiliary Verbs." Ph.D. thesis. University of Massachusetts.

Cinque, G. 1999. *Adverbs and Functional Heads: A Crosslinguistic Perspective. Oxford Studies in Comparative Syntax*. Oxford: Oxford University Press.

Coates, Jennifer. 1983. *The Semantics of the Modal Auxiliaries*. London: Croom Helm.
Collins, Peter. 2009. *Modals and Quasi-Modals in English*. Amsterdam/New York: Rodopi.
DeRose, Keith. 1991. "Epistemic Possibilities." *Philosophical Review* 100: 581–605.
Edgington, Dorothy. 2004. "Two Kinds of Possibility." *Proceedings of the Aristotelian Society, Supplementary* 78: 1–22.
Edgington, Dorothy. 2008. "Counterfactuals." *Proceedings of the Aristotelian Society* 108: 1–21.
Edgington, Dorothy. 2007. "On Conditionals." In *Handbook of Philosophical Logic*, edited by Gabbay, D. M. & Guenthner, F., Vol. 14, 127–221. Dordrecht: Springer.
Edgington, Dorothy. 2011. "Causation First: Why Causation is Prior to Counterfactuals." In *Understanding Counterfactuals, Understanding Causation: Issues in Philosophy and Psychology, 230–241*, edited by C. Hoerl, T. McCormack and S. R. Beck. Oxford University Press.
von Fintel, Kai. 2012. "Subjunctive Conditionals." In *The Routledge Companion to the Philosophy of Language*, edited by Fara, Delia Graff & Russel, Gillian, 466–477. New York: Routledge.
von Fintel, Kai, and Anthony S. Gillies. 2008. "CIA Leaks." *Philosophical Review* 117: 77–98.
von Fintel, Kai, and Sabine Iatridou. 2003. "Epistemic Containment." *Lingistic Inquiry* 34: 173–198.
Hacquard, Valentine. 2006. "Aspects of Modality." Ph.D. thesis, Massachusetts Institute of Technology.
Hacquard, Valentine. 2009. "On the Interaction of Aspect and Modal Auxiliaries." *Linguistics and Philosophy* 32: 279–315.
Hacquard, Valentine. 2010. "On the Event Relativity of Modal Auxiliaries." *Natural Language Semantics* 18: 79–114.
Hale, Bob. 2003. "Knowledge of Possibility and Knowledge of Necessity." *Proceedings of the Aristotelian Society*, 103, 1/20.
Iatridou, Sabine. 2000. "The Grammatical Ingredients of Counteractuality." *Linguistic Inquiry* 31: 231–270.
Jenkins, C. S. 2008. "Modal Knowledge, Counterfactual Knowledge and the Role of Experience." *Philosophical Quarterly* 58: 693–701.
Khoo, Justin. 2015. "On Indicative and Subjunctive Conditionals." *Philosopher's Imprint* 15(32): 1–40.
Kratzer, Angelika. 1977. "What 'Must' and 'Can' Must and Can Mean." *Linguistics and Philosophy* 1: 337–355.
Kratzer, Angelika. 1981. "The Notional Category of Modality." In *Words, Worlds, and Contexts*, edited by Eikmeyer, H. J. & Rieser, H., New Approaches in Word Semantics. Berlin/New York: de Gruyter.
Kratzer, Angelika. 1991. "Modality." In *Semantik: Ein internationales Handbuch der zeitgenössischen Forschung, 639–650*, edited by Arnim von Stechow and Dieter Wunderlich. de Gruyter: Berlin/New York.
Kratzer, Angelika. 2012. *Modals and Conditionals: New and Revised Perspectives*, Oxford: Oxford University Press.
Krödel, Thomas. 2012. "Counterfactuals and the Epistemology of Modality." *Philosopher's Imprint* 12(12): 1–14.
Krödel, Thomas. Forthcoming. "Modal Knowledge, Evolution, and Counterfactuals." In *Modal Epistemology after Rationalism*, edited by Fischer, R. W. & Leon, F.
Lewis, David. 1976. "The Paradoxes of Time Travel." *American Philosophical Quarterly* 13: 145–152.
Lewis, David. 1979. "Counterfactual Dependence and Time's Arrow." *Nous* 13: 455–476.

Nanay, Bence. 2011. "Do We See Apples as Edible?" *Pacific Philosophical Quarterly* 92: 305–322.

Nanay, Bence. 2012. "Action-Oriented Perception." *European Journal of Philosophy* 20: 430–446.

Nanay, Bence. 2013. *Between Perception and Action*. Oxford: Oxford University Press.

Nolan, Daniel. 1997. "Impossible Worlds: A Modest Approach." *Notre Dame Journal of Formal Logic* 38: 535–572.

Portner, P. 2009. *Modality*. Oxford: Oxford University Press.

Roca Royes, Sonia. 2011. "Modal Knowledge and Counterfactual Knowledge." *Logique et analyse* 54: 537–552.

Roca Royes, Sonia. Forthcoming. "Similarity and Possibility: An Epistemology of de re Possibility for Concrete Entities." In *Modal Epistemology after Rationalism*, edited by Fischer, B. & Leon, F.

Spencer, Jack. Forthcoming. "Able to Do the Impossible." *Mind*.

Strohminger, Margot. Forthcoming. "Perceptual knowledge of nonactual possibilities." *Philosophical Perspectives*, 29.

Swanson, Eric. 2010. "On Scope Relations between Quantifiers and Epistemic Modals." *Journal of Semantics* 27: 529–540.

Vaidya, Anand. 2015. "The Epistemology of Modality." In *The Stanford Encyclopedia of Philosophy*, edited by Summer, Edward N. Zalta. chapter http://plato.stanford.edu/archives/sum2015/entries/modality-epistemology/.

Vetter, Barbara. 2013. "Can' without Possible Worlds. Semantics for Anti-Humeans." *Philosophers* 13: 1–27.

Vetter, Barbara. 2014. "Dispositions without Conditionals." *Mind* 123: 129–156.

Vetter, Barbara. 2015. *Potentiality: From Dispositions to Modality*. Oxford: Oxford University Press.

Vetter, Barbara. Forthcoming. "Counterpossibles (not only) for Dispositionalists." *Philosophical Studies*.

Viebahn, Emanuel, and Barbara Vetter. Forthcoming. "How Many Meanings for 'May'? The Case for Modal Polysemy." *Philosopher's Imprint*.

Williamson, Timothy. 1999. "Existence and Contingency." *Proceedings of the Aristotelian Society* 73: 181–203.

Williamson, Timothy. 2002. "Necessary Existents." In *Logic, Thought and Language*, edited by A. O'Hear, 233–251. Cambridge: Cambridge University Press.

Williamson, Timothy. 2003. "Everything." *Philosophical Perspectives* 17: 415–465.

Williamson, Timothy. 2007. *The Philosophy of Philosophy*. Oxford: Blackwell.

Williamson, Timothy. 2013. *Modal Logic as Metaphysics*. Oxford: Oxford University Press.

Williamson, Timothy. Forthcoming-a. "Counterpossibles." In *Philosophical Fictionalism*, edited by Armour-Garb, Brad & Kroon, Fred.

Williamson, Timothy. Forthcoming-b. "Knowing by Imagining." In *Knowledge through Imagination*, edited by Kind, Amy, Oxford University Press.

Williamson, Timothy. ms. "Modal Science." To appear in *The Philosophy of Modality: Beyond the Standard Model*, edited by Vaidya, Anand & Prelević, Duko. [Internet document].

Reply to Vetter

Timothy Williamson

Wykeham Professor of Logic, University of Oxford, Oxford, UK

In 'Williamsonian modal epistemology, possibility-based' (Vetter 2016), Barbara Vetter proposes a modification of the approach to the epistemology of metaphysical modality developed and defended in *The Philosophy of Philosophy*. As she emphasizes, her approach is as realist and anti-exceptionalist in spirit as mine: it treats metaphysical modality as just as objective and mind-independent, and roots our knowledge of it just as securely in our ordinary ways of knowing. The main difference between us is that in that book I give the starring role to ordinary ways of knowing about what would have happened if something had been different, whereas Vetter gives it to ordinary ways of knowing what things and people can or cannot do, where 'can' entails a restricted but still objective sort of possibility.[1]

In my more recent work on modality, I have emphasized more systematic and rigorous ways of learning about objective modalities, in particular through logic and natural science (Williamson 2013, 2016b). In this reply, however, I will follow Vetter in focussing on our everyday knowledge of objective modalities.

On folk knowledge of modality, is there much more between Vetter and me than a difference of emphasis? In *The Philosophy of Philosophy*, I note the entailment from restricted forms of possibility such as physical possibility to metaphysical possibility, and comment:

> the connections with restricted possibility and with counterfactual conditionals are not mutually exclusive, for they are not being interpreted as rival semantic analyses, but rather as different cases in which the cognitive mechanisms needed for one already provide for the other. (Williamson 2007, 178)

Indeed, the availability of various alternative routes to knowledge of the same thing is just what is to be expected on a realist, anti-exceptionalist epistemology.[2] However, I also observed in passing about the possibility route that 'we would need some account of what demarcates the relevant forms of possibility from irrelevant ones, such as epistemic possibility' (ibid.). Vetter's paper constitutes a detailed and helpful response to that demarcation challenge. According to Vetter, her possibility-based approach is indeed better placed than my counterfactual-based approach to deal with the challenge. I discuss her arguments for that claim in the next section. In a brief final section I take up some of her remarks on the metaphysical ramifications of the time-dependence of ability ascriptions.

1. Circumstantial and epistemic readings

I applaud much of what Vetter says in her paper. Modal auxiliaries such as 'can' are near-ubiquitous in the ordinary use of natural languages, very often marking our engagement in thought and action with various kinds of practical and physical possibilities, some actual, many counterfactual, and with corresponding sorts of practical and physical impossibilities. If you can lift the stone, then it is possible for you to lift the stone in some objective sense that entails that it is metaphysically possible for you to lift the stone, because metaphysical possibility is the maximal kind of objective possibility; by contrast, if you can't lift the stone, it does *not* follow that it is metaphysically *im*possible for you to lift it (Williamson 2016b). I see no reason to regard the use of such auxiliaries as less basic in any sense — metaphysical, epistemological, psychological, or semantic — than the use of counterfactual conditionals.

As Vetter documents, linguists have accumulated a considerable body of evidence for a syntactically significant distinction between epistemic and circumstantial modals, where the linguists' term 'circumstantial' corresponds closely to my term 'objective' (I prefer the latter term only because it is characteristic of metaphysical modality to be independent of circumstances). In particular, epistemic modals tend to scope higher than circumstantial modals with respect to temporal operators and quantifiers, even if it is tricky to state the difference accurately in precise and general terms: as so often with natural languages, the data are messy. In any case, whether the unpreferred readings of the sentences at issue are impossible or just harder to hear, I accept that there are genuine syntactic differences between epistemic and circumstantial modals. The distinction is at work in natural languages themselves, not merely projected onto them from the outside by philosophers.

It does not follow, of course, that the syntactic differences are the *means* by which speakers or hearers themselves differentiate between epistemic and circumstantial readings. Sometimes, the speaker's choice of modal auxiliary may communicate to the hearer which reading is intended: 'Jane can solve the equation' (circumstantial) or 'Jane may solve the equation' (epistemic). In many other cases, even when the string of words is the same, the conversational context makes it clear which reading is relevant. To adapt one of Vetter's examples, 'Mary could have done it' may state a present epistemic possibility (in answer to the question 'Who are the suspects?'); in another context, it may attribute a past ability to Mary (in rebutting the claim 'Only John had the nerve to stab him in front of so many onlookers'). In such cases, the conversational context (including the speaker's intentions) determines the relevant reading of the modal and its appropriate scope relative to the past tense. The relative scope is not independently given to the hearer as a datum from which to work out whether an epistemic or a circumstantial reading is intended.

That epistemological point bears on the argument of Vetter's paper. In Section 5, she adduces good evidence that counterfactual conditionals sometimes receive epistemic readings (see also Edgington 2008). In *The Philosophy of Philosophy*, I considered only the circumstantial readings of counterfactual conditionals. Their epistemic readings will not serve my purposes there: they cannot be used to define the metaphysical modalities. Therefore, as Vetter points out, I too face the challenge in principle of distinguishing circumstantial readings from epistemic ones, in my case with respect to counterfactual conditionals. In Section 6.1, she argues that the syntactic differences between epistemic and circumstantial readings of 'can' are less clear for counterfactual conditionals: specifically, while circumstantial 'can' always scopes below tense, circumstantial 'would' sometimes scopes above tense, for instance in backtracking counterfactuals about necessary preconditions. One of her examples is this ([15] in her numbering):

(1) If she had measles now, she would have got it from someone else.

The counterfactual conditional in (1) is to be read circumstantially. According to Vetter, (1) posits that her now having measles presently requires the past to contain her getting it from someone else, not that her now having it once required the then present or future to contain her getting it from someone else. But, even if she is right about this syntactic asymmetry between 'can' and counterfactual conditionals, that would not stop speakers and hearers from distinguishing between them by the conversational context. Indeed, in presenting her examples of counterfactual conditionals with epistemic readings, Vetter relies on her readers to distinguish those readings by using her sketches of the relevant conversational contexts.

It is in any case not clear that circumstantial 'can' always scopes below tense. Consider this variant on (1), set at a time when theories of measles were just starting to be developed. A child has measles. The doctor is explaining what, according to a speculative new theory, is physically possible:

(2) She can only have got it from someone else.

The 'can' in (2) is to be read circumstantially. Moreover, (2) excludes as now physically impossible her not having got measles in the past from someone else. It does not exclude as once physically impossible her not getting measles in the then present or future from someone else. Thus the syntactic tests may be fallible for Vetter's 'can', just as they are for my counterfactual conditionals. Section 2 below will consider related examples for other purposes.

In Section 6.3 of Vetter's paper, she discusses another obstacle to recognizing whether a counterfactual conditional is being read epistemically or circumstantially. She argues that, although the readings are distinct in principle, in practice the cognitive processes of assessing the truth-value of a given counterfactual conditional tend to go much the same. Here is her example ([17] in her numbering):

(3) If A were the murderer, she would be hiding the weapon in a very safe place.

(3) has both an epistemic and a circumstantial reading. Vetter plausibly suggests that the investigation the police will go through in evaluating (3) on one reading is very similar to, though not exactly the same as, the investigation they will go through in evaluating it on the other. We may assume that she is right about that. But consider an analogous example for the possibility-based view:

(4) A could have hidden the weapon in a very safe place.

(4) has both an epistemic and a circumstantial reading. But it is also plausible to suggest that the investigation the police will go through in evaluating (4) on one reading is very similar to, though not exactly the same as, the investigation they will go through in evaluating it on the other. Thus, her possibility-based approach seems to face an objection analogous to the one she levels against my counterfactual-based approach: the two readings are hard to tell apart in terms of the associated processes of cognitive evaluation.

However, we should not underestimate the cognitive differences between evaluating the epistemic and circumstantial readings. For instance, in the case of (3), suppose that the police know from taunting messages from the murderer, whoever it is, that he or she is hiding the weapon in a very safe place. They also have strong, though not decisive, evidence that A is careless, lazy, and stupid, not at all the sort of person to be hiding the weapon in a very safe place. Then, by deduction, the police know (3) on its epistemic reading. By contrast, on its circumstantial reading, (3) is very improbable on the police evidence. For, when evaluating (3) on the epistemic reading, the police can exclude worlds in which A is the murderer and is not hiding the weapon in a very safe place; they cannot exclude those worlds when evaluating (3) on the circumstantial reading. Again, in the case of (4), suppose that A has confessed to the murder and revealed that she had found a very safe place in which to hide the weapon but decided at the last minute instead to hand it over to the police and confess. Then (4) is true on its circumstantial reading but false on its epistemic reading.

The upshot is that, with respect to distinguishing the cognitive processes of evaluation for the two readings, my counterfactual-based approach and Vetter's possibility-based approach are in roughly the same position. For both, the distinctions are subtle, but can be made.

In her Section 6.2, Vetter points out that some trained linguists have classified a high proportion of occurrences of counterfactual conditionals in corpora of real-life uses of natural language as epistemic, whereas philosophers of language tend to assume that most such occurrences are circumstantial. Without endorsing the linguists' classifications, she offers the divergence as indicating the difficulty of identifying which reading is intended. One wonders whether there are similar divergences for 'could'. But she takes her stand on 'can', because for it 'there are no advocates of widespread epistemic readings'. However, 'can' does occasionally receive an epistemic reading, as when we hear a knock at the door and one of us asks 'Who can that be?' But I will not attempt to argue for *widespread* epistemic readings of 'can'.

For what is the variability in classification to which Vetter points significant? It may make for difficulties in applying the counterfactual-based theory to examples, although relevant difficulties have not yet been noticeable. In any case, it does not follow that native speakers of natural languages have difficulty in respecting the distinction between epistemic and circumstantial readings in practice. The language module may employ distinctions whose boundaries are hard to trace from the outside, even if one is a native speaker. We may still use those distinctions in coming to grasp the metaphysical modalities.

The foregoing considerations suggest that neither the counterfactual-based approach nor the possibility-based approach has a clearly decisive advantage over the other, as implementations of a realist and anti-exceptionalist epistemology of metaphysical modality. *The Philosophy of Philosophy* concentrated on the former because the use of counterfactuals with a contradictory consequent promises to take one directly to the outer limit of objective modality, whereas there is no corresponding mechanism for 'can'. But 'can' has its own attractions as a more Aristotelian jumping-off point for metaphysical modality, as Vetter's own work illustrates (Vetter 2015). There is no need to see the two approaches as mutually exclusive. Rather, each of them highlights a different aspect of our rich and highly adaptive engagement in ordinary thought and talk with the objective-circumstantial dimension of modality. They reinforce rather than undermine each other. Moreover, they do not exhaust our ordinary ways of engaging epistemically with that dimension. Recent work by Strohminger (2015) on perceptual knowledge of objective nonactual possibilities and by Roca-Royes (201X) on inductive knowledge of such possibilities contains many insights that fit naturally into this overall picture, despite some significant divergences in their theoretical assumptions. After all, an anti-exceptionalist realist expects things to have a back as well as a front, to be approachable from many sides.

2. Possibility in time

Vetter tentatively associates the distinction between epistemic and circumstantial readings with a structural distinction between *de dicto* and *de re* modal claims. The intriguing idea is that the former ascribe an epistemic modal status to a *proposition*, whereas the latter ascribe a more objective modal status to a *property* with respect to an *object*. This might provide a kind of metaphysical explanation for the tendency of circumstantial modals to scope under temporal operators and quantifiers and for epistemic modals to scope over them.

Vetter may be right about the prototypical applications of epistemic and circumstantial modals, but natural languages go well beyond those prototypical cases. For instance, the *de dicto* construction 'it can happen that' with a sentential complement is common in English, typically with a circumstantial reading. A scientific theory may predict that it can happen that there are infinitely many stars. The 'it' is needed for purely syntactic reasons. Conversely, epistemic modals are sometimes used in *de re* constructions, as when a detective asks 'Which people may have been in the house on the night of the murder?' The syntax of natural languages is not a firm enough basis for metaphysical distinctions.

Vetter is especially uneasy with circumstantial possibilities not rooted in a future-directed potentiality at some time, as in (5) and (6):

(5) History could have been always different from what it actually was.

(6) A particle can have been spin up for infinitely long.

The possibility at issue in (5) is not realized in any possible world that branches off from the actual world at some time. The possibility at issue in (6) may just be realized in the actual world but otherwise it too is not realized in any possible world that branches off from the actual world at some time. In both cases, the circumstantial modal takes scope over the temporal operator. In her concluding Section 7, Vetter toys with the idea that the apparent possibilities in (5) and (6) may be illusory, generalizations too far from the potentialities possessed at some time by objects in the actual world.

If Vetter's possibility-based approach really cannot make sense of objective possibilities like those in (5) and (6), that may simply be a limitation of the approach, if treated as mediating all our grasp of objective modality. By contrast, the counterfactual-based approach is well-placed to avoid such a limitation, because it does not depend on a *de re* construction, even on the circumstantial reading. Such possibilities may be physical as well as metaphysical, for the fundamental equations of a well-confirmed physical theory may have solutions on which history is different at every time from what it actually was, or a particle is spin up for an infinite initial segment of history. On a realist and anti-exceptionalist approach to the epistemology of objective modality, we cannot expect our ordinary ways of knowing about it to exhaust what we can know about it. Rather, we should expect a more scientific approach to extend and far exceed our ordinary knowledge of objective modality.

Notes

1. Several authors have misinterpreted the account in Williamson (2007) as moving from the logical equivalences between claims of metaphysical modality and

claims involving counterfactual conditionals to the idea that we first know the latter and then use the logical equivalence to come to know the former. That was never part of my view. Rather, the idea was that we *can* use a similar cognitive process in coming to know either side of the equivalence, and that an economical and plausible hypothesis is that indeed we do so. As emphasized in that book, since each claim of metaphysical modality is logically equivalent to several different counterfactual formulations (which are therefore logically equivalent to each other), and the latter will each be known by slightly different cognitive processes, only approximate similarity can in general be postulated. Indeed, each particular counterfactual formulation may be known in various different ways. This is just what one would expect on the counterfactual approach. The main point is that if one has what it takes to evaluate counterfactual conditionals, one already has what it takes to evaluate claims of metaphysical modality, so there is no need to postulate an additional faculty just to explain our ability to do the latter.

2. Vetter says of Williamson (2016a): 'there he appears to think of can statements along the lines of a version of the conditional analysis of ability ascriptions'. But I agree with Vetter in rejecting that analysis. In the relevant passages, I argue that through a single imaginative exercise one sometimes learns *both* that one can do something and that if one were to try to do it one would succeed; I do not claim that the two items of knowledge are equivalent to each other.

References

Edgington, Dorothy. 2008. "Counterfactuals." *Proceedings of the Aristotelian Society* 108: 1–21.

Roca-Royes, Sonia. Forthcoming. "Similarity and Possibility: An Epistemology of *De Re* Possibility for Concrete Entities." In *Modal Epistemology after Rationalism*, edited by Bob Fischer and Felipe Leon. Springer, Synthese Library.

Strohminger, Margot. 2015. "Perceptual Knowledge of Nonactual Possibilities." *Philosophical Perspectives* 29: 363–375.

Vetter, Barbara. 2015. *Potentiality*. Oxford: Oxford University Press.

Vetter, Barbara. 2016. "Williamsonian Modal Epistemology, Possibility-based." *Canadian Journal of Philosophy* 46 (4–5): 767–796.

Williamson, Timothy. 2007. *The Philosophy of Philosophy*. Oxford: Blackwell.

Williamson, Timothy. 2013. *Modal Logic as Metaphysics*. Oxford: Oxford University Press.

Williamson, Timothy. 2016a. "Knowing by Imagining." In *Knowledge through Imagination*, edited by Amy Kind and Peter Kung, 113–123. Oxford: Oxford University Press.

Williamson, Timothy. 2016b. "Modal Science." *Canadian Journal of Philosophy* 46 (4–5): 453–492.

Epistemicism and modality

Juhani Yli-Vakkuri

CSMN, University of Oslo, Oslo, Norway

ABSTRACT
What kind of semantics should someone who accepts the epistemicist theory of vagueness defended in Timothy Williamson's *Vagueness* (1994) give a definiteness operator? To impose some interesting constraints on acceptable answers to this question, I will assume that the object language also contains a metaphysical necessity operator and a metaphysical actuality operator. I will suggest that the answer is to be found by working within a three-dimensional model theory. I will provide sketches of two ways of extracting an epistemicist semantics from that model theory, one of which I will find to be more plausible than the other.

What kind of semantics should someone who accepts the epistemicist theory of vagueness defended in Timothy Williamson's *Vagueness* give a definiteness operator? To impose some interesting constraints on acceptable answers to this question, I will assume that the object language also contains a metaphysical necessity operator and a metaphysical actuality operator. The question, then, concerns the semantics of the interaction of vagueness, as expressed by 'definitely,' and metaphysical modality, as expressed by 'necessarily' and 'actually,' from an epistemicist point of view. I will argue that the kind of two-dimensional (2D) semantics standardly used for investigating the interaction of 'necessarily' and 'actually' cannot handle the three-way interaction between these and 'definitely,' given epistemicism (or indeed supervaluationism). I will suggest that the answer is to be found by working within the three-dimensional (3D) model theory I have developed with Jon Litland. I will provide sketches of two ways of extracting an epistemicist semantics from that model theory, one of which I will find to be more plausible than the other.

1. According to the epistemicist theory of vagueness (*epistemicism*) defended by Williamson (1994), our ignorance of borderline matters — such as of how many grains of sand it takes to make a heap — has a special source, and that source of ignorance or, perhaps better put, that obstacle to knowledge, is all there is to vagueness. For it to be borderline, whether *S* is simply for there to be a special kind of obstacle to us (and to creatures relevantly like us) knowing whether *S*. Williamson's account of this obstacle to knowledge famously makes use of the thesis that, whenever it is borderline whether *S*, the sentence '*S*' is, to use John Hawthorne's term,[1] *semantically plastic* in that, if the global patterns of use involving some of the words occurring in '*S*' had been ever so slightly different in certain ways, then '*S*' would have had a semantic value different from its actual semantic value. (I will henceforth refer to this thesis as *Semantic Plasticity*.) The rough idea of the explanation is that we are insensitive in our judgments to the ways in which certain slight differences in global patterns of language use make a difference to the semantic values of sentences, and this insensitivity explains why we are not in a position to know whether *S* whenever it is borderline whether *S*. The semantic values that the borderline sentence '*S*' could easily have had had the global patterns of use of some of the words occurring in '*S*' been ever so slightly different in certain ways include both true and false semantic values. This, in turn, gives rise to close possibilities of error — possibilities that are 'close' in a sense that matters to knowledge, and that are therefore incompatible with knowing whether *S*.

Epistemicism appears to be the simplest and most conservative theory of vagueness on offer. Its main advertised virtues are that it enables us to keep classical logic and a disquotational conception of truth,[2] and that, '[a]t least in its simplest form, epistemicism can take over and reinterpret the formal apparatus of supervaluationism,'[3] thereby using the familiar formal apparatus of possible-worlds model theory to investigate the logic and semantics of the operators we use for theorizing about vagueness: in the first instance, the definiteness operator, in terms of which the notion of a borderline case can be defined. (It is borderline whether *S* iff it is neither definite that *S* nor definite that it is not the case that *S*.)[4]

From the beginning, epistemicism faced the incredulous stare. During the first decade or so since the publication of *Vagueness*, there seemed to be a broad consensus that the choice epistemicism presented us with was one between some kind of intuitive plausibility (which it lacked) and theoretical virtues (which it did not).[5] Recently, however, epistemicism's theoretical bona fides have been called into question. Since 2006, certain philosophers — in particular, John Hawthorne, Ofra Magidor, Stephen Kearns, and Michael Caie[6] — have begun to engage with the details of Williamson's proposal. In doing so they have discovered what appear to be quite serious problems with the proposed account of ignorance of borderline matters in terms of Semantic Plasticity, with epistemicist interpretations of supervaluationism's formal (model-theoretic) apparatus, and with the compatibility of these with a commitment to disquotational truth. So far, these challenges have gone unanswered. This paper represents a start — but only a start — in attempting to answer them.

I should warn the reader immediately that no review of the recent literature will be provided, although some points of contact with the papers cited above will be noted. I will largely be concerned with challenges to the Williamsonian epistemicist package that are different from, although closely related to, those discussed in the recent critical literature. The main problem I will be dealing with is that of giving an epistemicist semantics for a language that contains a definiteness operator alongside a metaphysical necessity operator and a metaphysical actuality operator. The difficulties one faces in attempting to carry out this task parallel in various ways the difficulties Williamson's recent critics have noticed. This is not particularly surprising, because the critics, with the exception of Caie, set up their problems using the metalinguistic notion of a sentence (or a sentence in a context, or an utterance of a sentence) being borderline, as opposed to an object language borderline operator (which is definable in terms of a definiteness operator, as indicated above). These are related to each other in obvious ways: in particular, in that the sentence '*S*' is borderline iff it is borderline whether *S*. Different issues arise when the borderline (or definiteness) operator occurs within the scope of a modal operator, but many of the problems discussed in this paper do not essentially involve such embeddings.

2. Let us return to Semantic Plasticity. My description in Section 1 of Williamson's explanation of ignorance of borderline matters in terms of Semantic Plasticity was deliberately vague (in the ordinary sense). This is because much in the recent literature and in this paper turns on just what the correct account of ignorance of borderline matters in terms of Semantic Plasticity is, if indeed there is one. For example, I left it open whether the kind of sentential semantic value mentioned in Semantic Plasticity is a proposition or something else. But we can begin with the assumption that propositions are the relevant kind of semantic value, since they clearly play that role in Williamson's own expositions of his explanation of ignorance of borderline matters.[7] If there is to be an explanation in terms of Semantic Plasticity of why we cannot know whether *S* when it is borderline whether *S*, that explanation must posit close worlds in which '*S*' expresses a false proposition, as well as close worlds in which '*S*' expresses a true proposition. Only close possibilities of error are incompatible with knowledge; close non-actual possibilities in which one judges something that is merely actually false are not, and neither are actual possibilities in which one judges something that is merely counterfactually false. On the other hand, the relevant possibilities of error must match actuality in certain respects: the existence of a close world in which, e.g. 'Tim is thin' expresses a false proposition other than the proposition it actually expresses and in which Tim's precise physical measurements are different than they actually are may be incompatible with one's knowing that Tim is thin, but it is not an example of the kind of obstacle to knowledge presented by vagueness. Much of the recent critical literature can be read as suggesting that these two demands on an explanation of ignorance of borderline matters in terms of Semantic Plasticity are difficult to reconcile with each other and with the other commitments of Williamson's epistemicism: in particular with disquotational truth and an epistemicist reinterpretation of supervaluationism's model-theoretic apparatus. In fact, Caie appears to think that the two commitments force a dilemma on Williamson's kind of epistemicism, both forks of which lead to clearly incorrect conclusions about what is borderline.

One way of throwing the difficulties I have just gestured at into particularly sharp relief is to ask what kind of a semantics someone committed Williamson's kind of epistemicism should give a definiteness operator. I will pursue this line of inquiry, and I will take Caie's recent discussion as my starting point, although his discussion, unlike mine, is not conducted in an explicitly model-theoretic framework. My discussion will also differ from Caie's (among other respects) in that I will ask not only what kind of semantics the epistemicist should give a definiteness operator — which I'll designate by '$\triangle$' — but a definiteness operator in combination with a metaphysical necessity ($\Box$) and actuality (*A*) operator. It will turn out that the inclusion of $\Box$ and *A* in the object language makes it possible to produce reasonably snappy counterexamples to certain semantic proposals that would otherwise require more elaborate counterexamples.

3. Let us begin by drawing a distinction between a *model theory* and a *semantics*.

A *semantics* for a language, as I will use the term in this paper, is an appropriately informative specification, in a metalanguage, of the function that assigns to each sentence of the language, in each context of utterance, the proposition the sentence expresses in that context. I will assume that propositions are (or are represented by — a distinction of which I will not be making heavy weather) functions from metaphysically possible worlds to truth values. In particular, I will assume that a proposition *p* is the function that assigns truth to a world *w* if *p* is true in *w* and otherwise assigns falsehood to *w*. (I make this assumption in order to simplify the discussion: nothing in this paper turns on whether propositions are more fine-grained than this.) When dealing with a suitably simple language — as in this paper — we can assume that contexts of utterance are also nothing more than metaphysically possible worlds. Metaphysically possible worlds (or simply 'worlds'), then, play two roles in a semantics: the role of contexts and the role of the entities with respect to which propositions are evaluated for truth. I will follow one standard nomenclature (that of Kaplan 1977) in calling worlds *circumstances* when they play the second role. A semantics for a language, then, is an appropriately informative specification of a function from the sentences of the language to functions from contexts to functions from circumstances to truth values. Equivalently, we can think of a semantics for a language as associating each of its sentences ϕ with a two-dimensional semantic value that is its *total semantic profile*: a set P of ordered pairs of worlds such that $\langle w, v\rangle \in P$ iff the proposition ϕ expresses in w is true in v.

Three further remarks about the present conception of a semantics are in order.

First, we will have to think of at least some semantic features of the expressions of the object language as being fixed independently of context. At least we will have to assume that the conventional meanings — in Kaplanian terms, *characters* — of the logical constants, which in the present setting are $\triangle$, $\Box$, A, and the truth-functional connectives, do not vary with context.[8] Thus, for example, we assume that, for any context w, the proposition $\neg\phi$ expresses in w will be true in a circumstance iff the proposition ϕ expresses in w is false in that circumstance, regardless of how (and whether) the expression $\neg$ is used in w. I'll leave it an open question what else is fixed. In various other settings it would be natural to follow Kaplan in assuming that the characters of all object language expressions are fixed independently of context. If we did so, we should think of the total semantic profile of a sentence as simply being (or representing) its character. However, certain peculiar features of the task of giving a semantics for a language containing $\triangle$ might be thought to rule out this assumption, so I won't make it. I will return to this theme in Section 4.

Second, it will have to be decided what the proposition expressed by ϕ in w is when ϕ is not used in w. We will have to resolve to use the term 'expresses' in a technical sense according to which the proposition an atomic sentence ϕ expresses in w, in the technical sense, when ϕ is used in w, is the proposition ϕ expresses in w in the ordinary sense, but the proposition an atomic sentence ϕ expresses in w, in the technical sense, when ϕ is not used in w, is something else — for example, it might be the proposition that is false in every world. (For non-atomic sentences the proposition is determined by the recursion clauses of the semantics.)[9] It makes no difference how this question is decided, so I won't decide it.

(A related idealization: I assume that, when ϕ *is* used in w, ϕ expresses a unique proposition in w. Thus, I idealize away, *inter alia*, ambiguity, semantically defective uses of sentences, and dimensions of indexicality other than sensitivity to the world of the context.)

Third, a semantics may, and in almost any interesting case will, do more than give an appropriately informative description of the total semantic profile of each sentence of the object language. It is a natural and widely endorsed methodological precept (which is, however, rarely observed in practice[10]) that a semantics should specify an assignment of semantic values to sentences in a way that is, in the standard sense of the word, *compositional*, and that these semantic values either are or determine the sentences' total semantic profiles. Often the compositional semantic values can only determine, rather than be, the total semantic profiles. And this is how things go, as we will see, with a language containing $\triangle$, $\Box$, and A, when $\triangle$ is given an epistemicist interpretation:

the compositional semantic values of its sentences must be three-dimensional rather than two-dimensional. (This is also true of supervaluationist-friendly semantics for $\triangle$, $\Box$, and *A*: see Litland and Yli-Vakkuri, 2016.) But in such cases we can assume that each context supplies a unique value for each parameter, and we can still take the total semantic profile of a sentence to be a set of ordered pairs of contexts and circumstances, since the contexts in such pairs will determine values for all of the parameters that it takes to assign a sentence a unique truth value at each circumstance.

A *model theory*, on the other hand, is a definition of a relation designated as *truth in a model*, or, in the cases that will interest us, *truth at a point of evaluation in a model*, supplemented by two further definitions: a *model* is defined as a certain kind of set, and a relation $\vDash$ of *logical consequence* is defined as (e.g.) truth-preservation at every (proper) point of evaluation of every model. These definitions supplement some background set theory — ZFC, say — so that anything proved by means of them will be a theorem of that set theory. Among the interesting claims that might be proved or disproved in this way is that the set of sentences designated as *valid* (i.e. the logical consequences of the empty set) is recursively axiomatizable ('completeness'). Such results may be interesting on their own, even if no bridge principles are supplied to connect the model theory to a semantics in the above sense — or to semantics in any natural sense.

The development of a model theory, however, is typically guided by a semantic picture, in a broader sense of 'semantic.' The broadest contours of such a picture are often the following. A model is (or represents) a way of interpreting, in some sense, all of the non-logical expressions of the object language. Furthermore, when models provide points of evaluation, but do not provide a privileged 'actual' point,[11] the 'proper' points of evaluation are naturally taken to be (or represent) contexts. Logical consequence, then, is truth-preservation in all contexts under all ways of interpreting the object language's non-logical expressions.

If we assume this broad-outlines picture, as I will do, we can go on to ask questions about the model that is (or represents) the actual or correct interpretation of the non-logical expressions of the object language — the *intended model*, as it is usually called. If the models come with representatives of contexts as well as circumstances, then a suitable specification of the intended model will be a semantics.[12]

The standard approach to the model theory of languages with $\triangle$ is that associated with supervaluationism.[13] In the simplest case, where we are dealing with a propositional language whose only non-truth-functional logical operator is $\triangle$, a supervaluationist model is a creature familiar from texts on modal logic: a triple $\mathfrak{U} = \langle W, R, [\![.]\!] \rangle$, where W is a non-empty set (of 'valuations'), R a reflexive relation on W (perhaps satisfying some further conditions), and $[\![.]\!]$ a function from the atomic sentences to subsets of W. An atomic sentence ϕ is defined as true in $\mathfrak{U}$ at $w \in W$ iff $w \in [\![\phi]\!]$; we have the usual truth definition clauses for the truth-functional connectives; and $\triangle\phi$ is defined as true in $\mathfrak{U}$ at $w \in W$ iff, for all v such that wRv, ϕ is true in $\mathfrak{U}$ at v. Logical consequence can be defined in several ways, which will not be of interest here. The question I want to ask, rather, is this: how, if at all, could we extract an *epistemicist* semantics for a language with $\triangle$ from the supervaluationist's model theory?

Some remarks by Williamson suggest optimism on this score:

> At least in its simplest form, epistemicism can take over and reinterpret the formal apparatus of supervaluationism, with a Tarskian conception of truth. To say that a valuation V^* is admissible by [R-related to] a valuation V is now to say that V^* is indiscriminable from V in the sense indicated. (Williamson 2003, 710)

On one natural way of extrapolating an epistemicist semantics from these remarks, $\triangle$, as used in a given world w, is, in effect, a device for generalizing over assignments of propositions-*cum*-semantic-values to sentences that obtain in worlds that are *close* to w in such a way that global patterns of the use of language that obtain in them differ from those obtaining in w at most in such minor ways that we are insensitive to any differences these differences in use may make to semantic value — insensitive, furthermore, in a way that makes the semantic value assignment that obtains in any world close to w indiscriminable (to creatures like us) from the semantic value assignment that obtains in w. Caie (2012) uses the term 'semantic indiscriminability' for this kind of closeness, and I will adopt his term: I will assume, for now, that two worlds are close to each other in the sense relevant to an epistemicist interpretation of $\triangle$ iff they are semantically indiscriminable from each other in the rough-and-ready sense just indicated.

(Note that the supervaluationist's models, as described above, do not contain any representatives of propositions: they simply assign truth values to atomic sentences relative to valuations. It is, however, natural to think of the valuations as representing ways of assigning propositions to atomic sentences, and to think of the truth value of an atomic sentence ϕ relative to a valuation v as the truth value determined by the proposition ϕ expresses relative to v.)

Caie, however, is not optimistic that anything like the above sketch can be developed into a satisfactory epistemicist semantics for $\triangle$. In fact, Caie goes as far as to suggest that the difficulties he encounters in attempting to extrapolate an epistemicist semantics for $\triangle$ from the quoted remarks by Williamson support the conclusion that 'one cannot provide an adequate account of what it is for a case to be borderline by appealing to facts about our inability to discriminate our actual situation from nearby counterfactual situations in which our language use differs in subtle ways.'[14]

Let us see, then, why Caie thinks this.

Caie observes that there appear to be two inequivalent accounts of what it takes for $\triangle\phi$ to be true in a world (considered as a context, I assume) that are consistent with Williamson's account of ignorance of borderline matters. According to the account Caie calls 'otherworldly,' $\triangle\phi$ is true in a world w (considered as a context) iff, for every world v close to w, the proposition-*cum*-semantic value

ϕ has or expresses in v (considered as a context) is true in v (considered as a circumstance). According to the other account, which Caie calls 'actualistic,' $\triangle\phi$ is true in a world w (considered as a context) iff, for every world v close to w, the proposition-*cum*-semantic value ϕ has or expresses in v (considered as a context) is true in w (considered as a circumstance). Caie claims that both answers 'face crippling defects.'[15]

One thing to note about Caie's otherworldly and actualistic accounts is that neither provides enough information for a semantics for a language with $\triangle$, even against a background theory that specifies the total semantic profile of every sentence containing no occurrences of $\triangle$. Each account only purports to say what it takes for $\triangle\phi$ to be true in a single world w — considered as a context, I have assumed — which is to say, what it takes for the proposition expressed by $\triangle\phi$ in a given world to be true *in that world*. This is not enough to specify the proposition expressed by $\triangle\phi$ in any context. Now, of course, if Caie is right that absurd consequences follow from the what the actualistic and otherworldly accounts say about what it takes for the proposition expressed by $\triangle\phi$ in a world to be true in that same world, then there is no need to consider separately the semantic proposals consistent with them which specify what it takes for the proposition expressed by $\triangle\phi$ in one world to be true in an arbitrary world. However, I do not find Caie's arguments against the otherworldly and actualistic accounts entirely decisive,[16] and I think we can learn something by regimenting Caie's two options using standard model-theoretic resources. In fact, as promised, I will consider not only what an epistemicist would have to say about the semantics of $\triangle$, but also of its interaction with $\Box$ and A, if the semantic options are limited to those consistent with the otherworldly and actualistic options.

4. Let L_{VM} be a propositional language with an infinite stock of atomic sentences, the usual truth-functional connectives, with $\neg$ and $\wedge$ taken as primitive, $\Box$ ('necessarily'), A ('actually'), and $\triangle$ ('definitely'), and no other logical constants, with the usual formation rules and metalinguistic abbreviations. Thus, in particular, '$\Diamond\phi$' abbreviates '$\neg\Box\neg\phi$' and '$\nabla\phi$' abbreviates '$\neg\triangle\phi \wedge \neg\triangle\neg\phi$.' Let L_M be the language that results from dropping $\triangle$ from L_{VM}. L_{VM} is the language whose semantics this paper will, for the most part, be concerned with. Languages with greater expressive resources will also be briefly considered.

The standard model-theoretic treatment of L_M is two-dimensional: it interprets sentences as sets of two-dimensional points of evaluation drawn from the Cartesian product of a non-empty set W and itself. In what follows I will depart from that standard treatment by including in each model an 'accessibility' relation R which is *not* used for interpreting $\Box$; rather, R will later be used, just as in supervaluationist model theory, for interpreting $\triangle$ in L_{VM}, so R will be idle as far as the treatment of L_M is concerned.

So let us say that a *2D model* is a triple $\mathfrak{U} = \langle W, R, [\![.]\!] \rangle$, where W is a non-empty set, R a reflexive and symmetric relation on W, and $[\![.]\!]$ a function from the atomic sentences of L_M (which are also the atomic sentences of L_{VM}) to subsets of $W \times W$. The truth value $|\phi|^{\mathfrak{U}}_{w,v} \in \{0, 1\}$ of an L_M-sentence ϕ at a point of evaluation $\langle w, v\rangle \in W \times W$ in $\mathfrak{U}$ is defined as follows, where α is any atomic sentence and ϕ and ψ are any sentences.

(i)	$\lvert\alpha\rvert^{\mathfrak{U}}_{w,v} = 1$	iff	$\langle w, v\rangle \in [\![\alpha]\!]$
(ii)	$\lvert\neg\phi\rvert^{\mathfrak{U}}_{w,v} = 1$	iff	$\lvert\phi\rvert^{\mathfrak{U}}_{w,v} = 0$
(iii)	$\lvert\phi\wedge\psi\rvert^{\mathfrak{U}}_{w,v} = 1$	iff	both $\lvert\phi\rvert^{\mathfrak{U}}_{w,v} = 1$ and $\lvert\psi\rvert^{\mathfrak{U}}_{w,v} = 1$
(iv)	$\lvert\Box\phi\rvert^{\mathfrak{U}}_{w,v} = 1$	iff	for all $u \in W$, $\lvert\phi\rvert^{\mathfrak{U}}_{w,u} = 1$
(v)	$\lvert A\phi\rvert^{\mathfrak{U}}_{w,v} = 1$	iff	$\lvert\phi\rvert^{\mathfrak{U}}_{w,w} = 1$

Let's say that a point of evaluation is *proper* iff it is of the form $\langle w, w\rangle$, and that ϕ is *true in w in* $\mathfrak{U}$ ($|\phi|^{\mathfrak{U}}_{w} = 1$) iff ϕ is true at the proper point of w, i.e. at $\langle w, w\rangle$, in $\mathfrak{U}$. Logical consequence is defined as truth-preservation at every proper point of evaluation of every model: that is to say, for any set of sentences Γ and any sentence ϕ, $\Gamma \vDash \phi$ iff, for all models $\mathfrak{U} = \langle W, R, [\![.]\!]\rangle$ and all $w \in W$, $|\phi|^{\mathfrak{U}}_{w} = 1$ if, for all $\gamma \in \Gamma$, $|\gamma|^{\mathfrak{U}}_{w} = 1$. And a sentence ϕ is said to be valid ($\vDash \phi$), as usual, iff it is a consequence of the empty set.

(Below I will tend to drop the model superscript from the expressions '$|\phi|^{\mathfrak{U}}_{w,v}$' and '$|\phi|^{\mathfrak{U}}_{w}$,' writing simply '$|\phi|_{w,v}$' or '$|\phi|_{w}$,' when it is clear from the context what generalization over models is meant or which model is being referred to.)

The above definitions give us the logic of 'real-world validity.'[17] The distinctive feature of this logic is that the principle of Necessitation (if $\vDash \phi$ then $\vDash \Box\phi$) fails. For example, $\phi \equiv A\phi$ is true at every proper point of evaluation of every model, so is valid, but whenever $\neg\Box\phi \wedge \neg\Box\neg\phi$ is true at a proper point in a model, $\Box(\phi \equiv A\phi)$ is false at that point in that model, so is not valid.

The extraction of a semantics for L_M from the 2D model theory, according to the standard view, which we largely owe to Kaplan (1977), proceeds in a straightforward way. The idea is that, in the intended model, W is the set of all metaphysically possible worlds, and the worlds in a point of evaluation $\langle w, v\rangle$ play the two roles described in Section 3: w is a context, and v is a circumstance. $|\phi|_{w,v} = 1$ iff ϕ is true at $\langle w, v\rangle$ in the sense of Section 3: i.e. iff the proposition ϕ expresses in the context w is true in the circumstance v.[18] Thus, the proposition expressed by ϕ in the context w is the function $\lambda v|\phi|_{w,v}$. Since ϕ is true *simpliciter* iff the proposition ϕ actually expresses is actually true, ϕ is true *simpliciter* iff $|\phi|_{@} = 1$, i.e. iff $|\phi|_{@,@} = 1$, where '@' is used, as usual, as a name for the actual world.

It's worth noting — to return to a theme from Section 3 — that the particular version of this semantics that is favored by Kaplan will not do for epistemicist (or, indeed, supervaluationist) applications. According to the orthodox Kaplanian version, the function $\lambda w\lambda v|\phi|_{w,v}$ is the character of ϕ. This would have the consequence that no sentence with a constant character that determines, in every context, the same non-contingent proposition is borderline. For suppose that ϕ is such a sentence. Then ϕ has the same truth value at every point of evaluation, and consequently $\triangle\phi \vee \triangle\neg\phi$ is true at every point. But this is clearly wrong. A sentence with a constant character that expresses, in every context, the same non-contingent proposition could easily be borderline. For example, there are vague identity sentences with proper names flanking the identity predicate. Suppose, for example, that it is borderline whether *S* and that we introduce the name '*N*' with the stipulation that '*N*' refers to Kit Fine if *S* and to Tim Williamson otherwise;[19] then it will be borderline whether N = Kit, and it will be borderline whether N = Tim. However, given the standard assumptions that names are rigid designators in the sense of Kripke (1980) and are not indexicals, the sentences 'N = Kit' and 'N = Tim' will have constant characters each of which determines, in each context, the same non-contingent proposition.[20]

Now let us turn to L_{VM}. To extend the above model theory to deal with L_{VM}, we will have to add a clause for $\triangle$ to the definition of truth at a point in a model. If we restrict our options to clauses that are compatible with one or another of Caie's otherworldly and actualistic accounts of the truth of $\triangle\phi$ at a proper point, we have four options. The otherworldly account is consistent with each of the following.

$$(O_1)\quad |\triangle\phi|^{\mathfrak{M}}_{w,v} = 1 \text{ iff, for all } u \text{ such that } vRu,\ |\phi|^{\mathfrak{M}}_{u,u} = 1.$$

$$(O_2)\quad |\triangle\phi|^{\mathfrak{M}}_{w,v} = 1 \text{ iff, for all } u \text{ such that } wRu,\ |\phi|^{\mathfrak{M}}_{u,u} = 1.$$

And the actualistic account is consistent with each of the following.

$$(A_1)\quad |\triangle\phi|^{\mathfrak{M}}_{w,v} = 1 \text{ iff, for all } u \text{ such that } vRu,\ |\phi|^{\mathfrak{M}}_{u,v} = 1.$$

$$(A_2)\quad |\triangle\phi|^{\mathfrak{M}}_{w,v} = 1 \text{ iff, for all } u \text{ such that } wRu,\ |\phi|^{\mathfrak{M}}_{u,v} = 1.$$

Each of these proposals is unacceptable *qua* semantics. *Qua* model theory, each proposal, it will turn out, results in an unacceptable logic: each either validates some sentences that should not be validated or fails to validate some that should be. And in each case the logical flaw is closely related to a semantic flaw.

To begin, one serious logical shortcoming of (O_1) is that it fails to validate $\Box(\triangle\phi \supset \phi)$ — the principle that, necessarily, whatever is definitely so is so. By (O_1), $\Box(\triangle\phi \supset \phi)$ is false at a world w in a model just in case, for some world v of the model, $|\phi|_{w,v} = 0$, while, for all worlds u such that vRu, $|\phi|_{u,u} = 1$, and clearly there are models with worlds satisfying this condition.

Now a failure to validate something that should be validated does not immediately show that the model theory cannot deliver a correct semantics; for that it would also have to say some false things about the intended model. (Cf. if we let the interpretation of $\neg$ vary from model to model we would fail to validate every truth-functional tautology,[21] but this would not necessarily portend any trouble with the semantics, since it would be consistent with $\neg$ being interpreted as negation in the intended model.) However, it is easy enough to see that the actual world of the intended model must witness the invalidity of $\Box(\triangle\phi \supset \phi)$ if (O_1) is correct, and this is why (O_1) is not acceptable for semantic purposes. For example, let ϕ be the sentence 'Hesperus $\neq$ Phosphorus.' In the actual world of the intended model — that is, actually — the proposition ϕ expresses is false in every world. However, in some other world, ϕ expresses a proposition that is true in every world. (Any world in which '$\neq$' means what it actually means and 'Hesperus' and 'Phosphorus' refer to distinct planets in spite of their reference being fixed in the manner it actually is is such a world.) It follows that $\Box(\triangle\phi \supset \phi)$ is false. But $\Box(\triangle\phi \supset \phi)$ says that it is necessary that if it is definite that Hesperus $\neq$ Phosphorus then Hesperus $\neq$ Phosphorus, which is clearly true, so $\Box(\triangle\phi \supset \phi)$ is true, contrary to (O_1).

(O_2) also does not validate $\Box(\triangle\phi \supset \phi)$, and this portends trouble for its semantic applications. However, (O_2) has an even more serious problem that: it validates $\Box\triangle\phi \vee \Box\neg\triangle\phi$. This is immediately disqualifying, because validity entails truth. It is often a contingent matter what is definitely so. For example, I could have been definitely a pastry chef, and I could have been — because I am — definitely not a pastry chef.

Like (O_1), (A_1) fails to validate $\Box(\triangle\phi \supset \phi)$, and, as in the case of (O_1), this logical failure is accompanied by a semantic failure. By (A_1), $\Box(\triangle\phi \supset \phi))$ is false at any world *w* in any model with a world *v* such that $|\phi|_{w,v} = 0$, for all *u* such that *vRu*, $|\phi|_{u,v} = 1$, and clearly there are models satisfying this condition. Now consider the intended model. As in the discussion of (O_1), we can let ϕ be 'Hesperus $\neq$ Phosphorus.' Then, in the actual world of the intended model — i.e. actually — ϕ expresses a proposition that is false in every world. Thus, if there is also a world *w* such that, in every world close to *w*, ϕ expresses a proposition that is false in *w*, $\Box(\triangle\phi \supset \phi)$ is false. And, clearly that there is such a world. Consider the fact that, if 'Hesperus' and 'Phosphorus' had referred to different objects, ϕ would have expressed the same proposition — the necessary one — in every close world, and that proposition, being the necessary one, would have been true. Thus (since 'Hesperus' and 'Phosphorus' could have referred to different objects), there is a world *w* such that, in every world close to *w*, ϕ expresses a proposition that is true in *w*. Again, we reach the absurd conclusion that $\Box(\triangle\phi \supset \phi)$ — which says that it is necessary that if it is definite that Hesperus $\neq$ Phosphorus then Hesperus $\neq$ Phosphorus — is false.

Finally, while (A_2) does validate $\Box(\triangle\phi \supset \phi)$, it suffers from another logical flaw, which it shares with (A_1): it violates the principle I call *Definitization*, namely:

If $\vDash \phi$, then $\vDash \triangle\phi$.

Definitization says, in effect, that each valid sentence is (validly) definite. The models that are counterexamples to Definitization under (A_2) also turn out to include the intended model, and this dooms (A_2) as a semantic proposal. $\phi \equiv A\phi$ is true at every proper point of every model, but, by (A_2), in any model,

$$
\begin{aligned}
|\triangle(\phi \equiv A\phi)|_w = 0 \quad & \text{iff} \quad \text{for some } v \text{ such that } wRv,\ |\phi \equiv A\phi|_{v,w} = 0 \\
& \text{iff} \quad \text{for some } v \text{ such that } wRv,\ |\phi|_{v,w} \neq |A\phi|_{v,w} \\
& \text{iff} \quad \text{for some } v \text{ such that } wRv,\ |\phi|_{v,w} \neq |\phi|_{v,v}
\end{aligned}
$$

For any sentence ϕ, then, $\triangledown(\phi \equiv A\phi)$ is true iff ϕ expresses, in some world *w* *R*-related to the actual world, a proposition that has different truth values in *w* and in the actual world. And, it is reasonably clear that there are sentences that satisfy this condition, at least according to Williamson (1994): the worlds close to the actual world differ from the actual world with respect to various semantic states of affairs, and, provided that the semantic supervenes on the non-semantic, also with respect to various non-semantic states of affairs. If ϕ expresses some such state of affairs (proposition) both actually and in some close world, then $\triangledown(\phi \equiv A\phi)$ is true. But this is absurd: we are in a position to know $\phi \equiv A\phi$ to be true, no matter what sentence ϕ might be, so, because borderline matters are unknowable, $\phi \equiv A\phi$ is not borderline.

In my view the above considerations are decisive against both of the otherworldy and both of the actualistic semantics for $\triangle$. There are, however, other ways of arguing against each option — ways that do not discriminate between (O_1) and (O_2), or between (A_1) and (A_2) — which it may be instructive to survey.

To begin with the otherworldly semantics, these share a flaw that can be leveraged to produce a variety of counterexamples. The shared flaw is the following. On any reasonable view of vagueness, not just Williamson's, it is not sufficient for it to be borderline whether *S* that it could have (easily or otherwise) not been the case that *S*. The actualistic semantics get this right; the otherworldly ones do not.[22]

One way to appreciate this flaw is to note that, according to the otherworldly semantics, a certain kind of metaphysical contingency is sufficient for borderlineness: if ϕ expresses the same proposition in all close worlds, and this proposition is contingent in such a way that it is true in some and false in other close worlds, then $\triangledown\phi$ is true according to the otherworldly semantics. This is clearly wrong from the point of view of just about anyone's theory of vagueness. It is also wrong specifically from the point of view of Williamson's epistemicism, because of the latter's commitment to Semantic Plasticity, according to which a necessary condition for the truth of $\triangledown\phi$ is that ϕ fails to express the same proposition in all close worlds. The otherworldly semantics allow $\triangledown\phi$ to be true even when ϕ is not semantically plastic.

Let us next consider the actualistic semantics.

One problem with these is that they misclassify certain definite disquotational sentences as non-definite (both in the actual world and worlds close to the actual world). That they misclassify certain definite T-sentences as non-definite is particularly damning, given that one of epistemicism's advertised virtues was its ability to respect the disquotational conception of truth.[23]

To present the problems that disquotational sentences pose for the actualistic semantics properly it will be necessary to extend the 2D model theory to deal with a language that results from adding some predicates and singular terms to L_{VM}. To that end, let a 2D model now be a quadruple $\langle W, R, D, [\![.]\!] \rangle$, where W and R are as before, D is a non-empty set representing the domain of individuals, and $[\![.]\!]$ is otherwise as before, except that, for all singular terms $\tau, \tau_1, \ldots, \tau_n$, and all n-place predicates ϕ,

(vi) $[\![\tau]\!]$ is a function from $W \times W$ to D

(vii) $[\![\phi]\!]$ is a function from $W \times W$ to $\mathcal{P}(D^n)$

(viii) $|\phi(\tau_1, \ldots, \tau_n)|^{\mathfrak{U}}_{w,v} = 1$ iff $\langle [\![\tau_1]\!](w, v), \ldots, [\![\tau_n]\!](w, v)\rangle \in [\![\phi]\!](w, v)$

In order to deal with disquotational sentences, we'll assume that, for each expression ξ, there is a singular term $\ulcorner\xi\urcorner$ (a quote name) such that, in the intended model,

(ix) If $@Rw$, then $[\![\ulcorner\xi\urcorner]\!](w, v) = \xi$

Now let us begin by considering *T-sentences*, i.e. sentences of the form $T(\ulcorner\phi\urcorner) \equiv \phi$, where T is the enriched language's truth predicate. For paradox-avoiding reasons we had better not assume that all T-sentences are true, so, since definiteness is factive, we had better not assume that all T-sentences are definite either. But let us say that a T-sentence is *healthy* when the sentence whose quote-name occurs in it is suitable in the sense of note 2. Certainly, we can safely assume that all healthy T-sentences are definite. And we can safely assume also all healthy T-sentences are definite in all close worlds, as used in those worlds. This is because a world is close to the actual world iff it is semantically indiscriminable from it; a world in which a healthy T-sentence, as used in that world, is not definite in that world, is one in which T-sentences are semantically different, and are semantically different in a discriminable way (since we know all healthy T-sentences are definite), from how they are in the actual world.

However, there is a straightforward argument for the conclusion that, if either of the actualistic semantics is correct, then some healthy T-sentence is not definite in some close world. The argument has two premises. The first is that, in the intended model,

(1) If ϕ is suitable, then $\phi \in [\![T]\!](@, w)$ iff $|\phi|_{w,w} = 1$.

(1) is, I take it, an uncontroversial minimal condition on the behavior of a truth predicate: it says that the truth predicate, as interpreted in the actual world, applies to a suitable sentence in a world w just in case that sentence, as used in w, expresses a proposition that is true in w. This follows from the trifling

observations that, first, necessarily, a suitable sentence is true iff the proposition it expresses is true, and, second, that, necessarily, a suitable sentence is true iff it belongs to the extension determined by the actual semantic value of 'true.'

(1), together with either of the actualistic semantics, implies:

(1*) If $@Rw$, $|\phi|_{@,w} \neq |\phi|_{w,w}$, and ϕ is suitable, then $|\triangle(T(\ulcorner\phi\urcorner) \equiv \phi)|_{w,w} = 0$.[24]

The second premise of the argument is:

(2) For some w and some suitable ϕ, $@Rw$ and $|\phi|_{@,w} \neq |\phi|_{w,w}$.

(1*) and (2) imply that there is a healthy T-sentence that is not definite in some close world, which is absurd.

The case for (1) is overwhelming. What about (2)? Note that anyone who accepts either actualistic semantics must accept, on pain of denying that there are any suitable borderline sentences, that there are worlds v, w, and a suitable sentence ϕ, such that vRw, $|\phi|_{v,w} \neq |\phi|_{w,w}$; otherwise, on either actualistic semantics, $\triangledown\phi$ is false at every proper point of evaluation, so in particular is actually false. Given this, it would be very odd if there were no such cases in which $v = @$ — which is to say, if (2) were not true.

There is a similar argument to be made about disquotational *R-sentences* — sentences like

(E) 'Mount Everest' refers to Mount Everest.

To deal with these, we'll assume that the object language contains a reference predicate *Ref*. An R-sentence, then, has the form $Ref(\ulcorner\tau\urcorner, \tau)$. Clearly (at least, given our idealizations), we want to say that all R-sentences are definite. But there is a straightforward argument for the conclusion that this is not so if either of the actualistic semantics is correct.

This argument, too, has two premises. The first is that, in the intended model,

(3) $\langle x, y\rangle \in [\![Ref]\!](@, w)$ iff $[\![x]\!](w, w) = y$

(3) is minimal condition on *Ref* expressing the relation that 'refers' expresses in (E): (3) says, in effect, that, necessarily, a pair $\langle x, y\rangle$ belongs to the extension determined by the relation *Ref* is as *actually* used to express if and only if x refers to y. Since *Ref* is actually used to express the relation of reference, (3) reduces to the trivial observation that, necessarily, a pair $\langle x, y\rangle$ belongs to the extension determined by the relation of reference if and only if x refers to y.

(3), together with either of the actualistic semantics, implies:

(3*) If $[\![\tau]\!](w, w) \neq [\![\tau]\!](@, w)$ and $wR@$ then $|\triangle Ref(\ulcorner\tau\urcorner, \tau)|_{w,w} = 0$.[25]

The second premise of the argument is:

(4) For some singular term τ and some world w, $[\![\tau]\!](w, w) \neq [\![\tau]\!](@, w)$ and $wR@$.

(4) and (3*) entail that some R-sentence is non-definite in some close world, which is absurd.

The argument had two premises: (3) and (4). (3) was a truism. What about (4)? As in the case of the argument concerning T-sentences, it can be argued that the advocate of an actualistic semantics must accept (4) — in this case, on pain of denying the that there are borderline identity statements with proper names, given the natural way of introducing an identity predicate = into the semantics, which is via the clause

$$|\tau_1 = \tau_2|_{w,v} = 1 \text{ iff } |\tau_1|_{w,v} = |\tau_2|_{w,v}.$$

Assuming, again following Kripke (1980), that proper names are modally rigid, which is to say, in the 2D setting, that $|\tau|_{w,v} = |\tau|_{w,u}$, the only way for $\nabla(\tau_1 = \tau_2)$ to be true, on either actualistic semantics, is for there to be worlds w, v, such that $[\![\tau_1]\!](w, w) \neq [\![\tau_2]\!](v, w)$ and wRv. It would be very odd if there were no such case in which $v = @$.

(Further counterexamples involving other kinds of disquotational sentences — e.g. of the form '$\forall x$('F' applies to $x \equiv F(x)$)' — can be devised, but I'll leave them as an exercise for the reader.)

The problems for the actualistic semantics do not end there. I'll mention one further problem that will make an interesting test case for further semantic proposals. Consider the *Meter Sentence* made famous by Kripke (1980, 54f):

> The Standard Meter is one meter long.

The Standard Meter is, of course, the object whose length is used to fix the reference (and content) of 'meter.' Since the length of the Standard Meter — a platinum bar cast using nineteenth-century technology — could very easily have been different from its actual length, the Meter Sentence could very easily have expressed a proposition that is actually false. It is also highly plausible that some of the worlds in which the Meter Sentence expresses a proposition that is actually false, are worlds that are close to the actual world in the sense of 'close' that matters to the epistemicist interpretation of $\triangle$; but if so, it follows by either actualistic semantics that the Meter Sentence is borderline, so, because borderline status is incompatible with knowledge, no one knows that the Standard Meter is one meter long, which is absurd.

(Counterexamples like the above can be multiplied indefinitely. 'Kilogram,' 'second,' and other measure words whose reference is fixed by paradigm objects whose relevant measurements could very easily have been different than they actually are easily lend themselves to similar counterexamples. So, in general do cases in which we use a reference-fixing description that could very easily have had a referent different from its actual referent.)

I think the arguments given so far establish the incorrectness of both of the otherworldly semantics and both of the actualistic semantics, but they do not come close to showing that there is *no* 2D semantics for L_{VM} acceptable to epistemicists. After all, I have only considered four possible truth definition clauses for $\triangle$ within 2D model theory — the ones corresponding to the two proposals Caie (2012) finds the most 'natural' — but this still leaves a large number of alternative truth definition clauses unexamined.[26] Are we in a position to know that all of the alternatives are also unacceptable?

In fact, we are, at least if we impose some further natural conditions on the adequacy of an epistemicist semantics for L_{VM}. We only need two conditions in addition to Definitization. The first condition is that every instance of the schema

K: $\triangle(\phi \supset \psi) \supset (\triangle\phi \supset \triangle\psi)$

should be true in the intended model. The second is that there should be no vagueness in the world, in a sense that will be made precise below. A result by Peter Fritz, which is included in the Appendix to this paper (Fritz 2016), shows that there is a precise sense in which there is no 2D semantics for L_{VM} that satisfies all three conditions.

The idea that there is no vagueness in the world is most naturally expressed using propositional quantification, although we will later see that this is dispensable. Let us, then, begin by considering the language L_{VMQ}, which results from the addition of propositional variables and the universal quantifier $\forall$ to L_{VM}, with the usual formation rules. In L_{VMQ}, we express the claim that there is no vagueness in the world by the sentence

NVW: $\forall p(p \equiv \triangle p)$,

which says, in effect, that every state of affairs (proposition) obtains iff it definitely obtains. Semantic Plasticity mandates NVW: if every borderline formula could easily have had expressed a proposition other than the proposition it actually expresses, a formula that is a variable cannot be borderline because, under any given variable assignment, it could not have expressed any proposition other than the one it actually expresses. The same goes, *mutatis mutandis*, for variables of all types. (For similar reasons, one would expect a supervaluationist to endorse NVW.)[27]

Now of course an epistemicist who endorses NVW must reject the validity of universal instantiation on propositional variables, as that rule would take us from the plausible NVW to each instance of the absurd schema

TRIV: $\phi \equiv \triangle\phi$,

which says, in effect, that there is no vagueness. But there is nothing new here: it is natural for epistemicists (and indeed supervaluationists) to reject the validity of universal instantiation on variables of all types. For example, in a first-order language with $\triangle$, it is natural for an epistemicist (or supervaluationist) to accept $\forall x \forall y(x = y \supset \triangle x = y)$ while rejecting some instances of $a = b \supset \triangle a = b$, because there are vague names.[28]

Let us next consider how we should augment the 2D model theory in order to handle the L_{VMQ}. Intuitively, the propositional variables range over propositions. In the model theory, however, we must interpret them, under a variable assignment, as entities of the same type as a model's interpretation function assigns to the atomic sentences, which is to say, as sets of points of evaluation of the model. Now clearly, the admissible values of the propositional variables cannot be *all* sets of such pairs: some of them represent context-dependent as well as circumstance-dependent variation in truth value, whereas the truth value of a proposition depends only on the circumstance. The admissible values of the propositional variables, then, must satisfy the condition

$$\langle w, v\rangle \in X \qquad \text{iff} \qquad \langle u, v\rangle \in X.$$

Let's say that any set of points of evaluation satisfying the above condition is a *barcode*.[29] In the intended model, the subsets of $W \times W$ that are barcodes, then, represent the propositions. We'll define a *variable assignment* on a model as any function from the propositional variables to the model's barcodes. Now the notion of the truth value $|\phi|^{\mathfrak{U}^g}_{w,v}$ of a formula ϕ at a point of evaluation $\langle w, v\rangle$ in a model $\mathfrak{U}$ is relativized to a variable assignment g. $|\phi|^{\mathfrak{U}^g}_{w,v}$ will be defined exactly like $|\phi|^{\mathfrak{U}}_{w,v}$ except for when ϕ is a propositional variable or a universally quantified formula; for these cases we add (xii) and (xiii).

(xii) $|p|^{\mathfrak{U}^g}_{w,v} = 1$ iff $\langle w, v\rangle \in g(p)$

(xiii) $|\forall p \phi|^{\mathfrak{U}^g}_{w,v} = 1$ iff, for each variable assignment f on $\mathfrak{U}$ differing from g at most in what f assigns to p, $|\phi|^{\mathfrak{U}^f}_{w,v} = 1$

Logical consequence is defined as truth preservation at each proper point of each model under each variable assignment on that model.

Now it turns out that, no matter what truth definition clause we introduce for $\triangle$ in L_{VMQ}, if the clauses already introduced are in place and the model theory validates every instance of *K* and satisfies Definitization, every model in which NVW is true at every proper point will also be a model in which every instance of TRIV is true at every proper point. So, in particular, this will hold of the intended model, and there is no hope of getting an acceptable epistemicist semantics for L_{VMQ} out of a 2D model theory. This is a consequence of Fritz's Proposition 1, which is more general. Proposition 1 states that there is no way of interpreting $\triangle$ by a Scott–Montague neighborhood function that satisfies our desiderata without validating every instance of TRIV. Each way of interpreting $\triangle$ in a relational model-theoretic semantics of the kind I have been working with in this section corresponds to a neighborhood function, but not conversely. There is, then, a good sense in which Fritz's result shows that there is no 2D model theory that can serve as the basis of an acceptable epistemicist semantics for L_{VMQ}.

Fritz's result does not rely essentially on the presence of propositional quantification in the language. While we cannot express the claim that there is no vagueness in the world in L_{VM}, we can express that claim in the 2D semantics of L_{VM} by requiring that, for each barcode *b*, the neighborhood function that interprets $\triangle$, in effect, classifies *b* as definite at a proper point of evaluation iff *b* is true at that point. Let us call this condition *NVW**. If we assume that the interpretation of $\triangle$ obeys NVW*, and we assume Definitization and the validity of each instance of K, it once again follows that each instance of TRIV is true. For let us say, following Fritz, that a *generalized 2D frame* is a pair $\langle W, D\rangle$, with *W* as before and *D* a neighborhood function that interprets $\triangle$, and let us say that the *quantifier-free logic* of a class of generalized 2D frames is the set of all L_{VM}-sentences true at every proper point of every model based on a frame in that class. Fritz's second result (Proposition 2) states that the logic of any class of generalized 2D frames satisfying NVW* that includes all instances of *K* and is closed under the rule $\phi/\triangle\phi$ (corresponding to Definitization) also includes all instances of TRIV. Now if we make the further natural assumptions that there is an intended generalized 2D frame, on which the intended model is based, and whose logic includes each instance of *K* and is closed under the rule $\phi/\triangle\phi$, it follows that if this frame satisfies NVW*, then each instance of TRIV is true at each proper point of the intended model, so is true *simpliciter*.

5. The failure of the 2D approach to the semantics of L_{VM} speaks in favor of a move to a 3D approach. The model theory turns out to be straightforward, and it has already proved to be fruitful in logical investigations: Litland and Yli-Vakkuri (2016) show how their 3D model theory can be used to obtain completeness results for certain propositional logics of vagueness and modality, at least one of which is plausibly the correct logic from an epistemicist point of view. However, what semantics the epistemicist should marry with this model theory is as yet unclear. In this section I will introduce a 3D model theory for L_{VM} that is a simplified version of one of the model theories discussed by Litland and Yli-Vakkuri, and I will ask how we might go about extracting an epistemicist semantics from it.

Let us say that a *3D model* is a triple $\mathfrak{U} = \langle W, R, [\![.]\!] \rangle$, where *W* is a non-empty set, *R* a reflexive and symmetric relation on *W*, and $[\![.]\!]$ a function from the atomic sentences to 3D points of evaluation in $\mathfrak{U}$, which are subsets of $W \times W \times W$.

The truth value $|\phi|^{\mathfrak{U}}_{w,v,u} \in \{0, 1\}$ of a sentence ϕ at $\langle w, v, u\rangle \in W \times W \times W$ in $\mathfrak{U}$ is defined as follows, where α is an atomic sentence and ϕ and ψ are any sentences.

(i)	$\lvert\alpha\rvert^{\mathfrak{U}}_{w,v,u} = 1$	iff	$\langle w, v, u\rangle \in [\![\alpha]\!]$
(ii)	$\lvert\neg\phi\rvert^{\mathfrak{U}}_{w,v,u} = 1$	iff	$\lvert\phi\rvert^{\mathfrak{U}}_{w,v,u} = 0$
(iii)	$\lvert\phi\wedge\psi\rvert^{\mathfrak{U}}_{w,v,u} = 1$	iff	both $\lvert\phi\rvert^{\mathfrak{U}}_{w,v,u} = 1$ and $\lvert\psi\rvert^{\mathfrak{U}}_{w,v,u} = 1$
(iv)	$\lvert\Box\phi\rvert^{\mathfrak{U}}_{w,v,u} = 1$	iff	for all $u' \in W$, $\lvert\phi\rvert^{\mathfrak{U}}_{w,v,u'} = 1$
(v)	$\lvert A\phi\rvert^{\mathfrak{U}}_{w,v,u} = 1$	iff	$\lvert\phi\rvert^{\mathfrak{U}}_{w,v,u'} = 1$
(vi)	$\lvert\triangle\phi\rvert^{\mathfrak{U}}_{w,v,u} = 1$	iff	for all w' such that wRw', $\lvert\phi\rvert^{\mathfrak{U}}_{w',v,u} = 1$

Logical consequence is defined just like before, except in that one of the terms occurring in its definition is re-refined: we now say that a proper point of evaluation is a point of the form $\langle w, w, w\rangle$. We will also say that ϕ is true in w iff ϕ is true at the proper point of w, i.e. at $\langle w, w, w\rangle$.

As the reader can verify, this model theory does not share any of the logical flaws of the otherworldly and actualistic 2D model theories discussed in Section 4. In fact, there is some plausibility to the idea that the relation of logical consequence that it — or some close variant of it[30] — delivers is the correct logic of vagueness and modality from an epistemicist point of view. I won't press the case for that here, however. I do so in my joint work with Litland, which discusses both epistemicist and supervaluationist applications of the 3D model-theoretic framework.[31]

It is also worth noting that the 3D model theory can easily be extended to handle L_{VMQ} from Section 4. The extension is carried out using exactly the same definition of the notion of truth at a point in a model under a variable assignment. Only the definition of a barcode will be different: in the 3D setting a barcode of a model is any set X of points of evaluation of the model such that $\langle w, v, u\rangle \in X$ iff $\langle w', v', u\rangle \in X$. This model theory validates NVW, as desired, but it does not validate every instance of TRIV.

So far we only have a 3D model theory for L_{VM}, but what we want is a way of extracting a semantics for L_{VM} from that model theory. This requires us to make sense of the idea of a sentence being true at a triple of worlds. I will discuss two ways of doing so.

I will call first way of extracting a semantics for L_{VM} from the 3D model theory the *metasemantic interpreration* of that model theory. According to the metasemantic interpretation, ϕ is true at $\langle w, v, u\rangle$ iff the *character* ϕ has in w, when applied to the *context* v yields a proposition that is true in the *circumstance* u. R can be, as before, glossed as a relation of 'semantic indiscriminability,' except in this case the semantic facts with respect to which the worlds it relates are indiscriminable are facts about the characters rather than about the (propositional) contents of sentences.[32] The metasemantic interpretation recognizes two different roles for contexts, which I'll call the role of *metasemantic context* (the role of the first world in a triple) and simply *context* (the role of the second world in a triple). The metasemantic context represents the dependence the characters of L_{VM} sentences on global patterns of language use, and generally on whatever else facts about character supervene on. The context is just a context in Kaplan's (1977) sense: once the character of a sentence is fixed by a metasemantic context, that character is applied to a context to obtain a content. According to the metasemantic interpretation, then, since the character of a sentence is its actual character, the character of ϕ is the function $\lambda w \lambda v|\phi|_{@,w,v}$. A suitable characterization of the function $\lambda\phi\lambda w\lambda v|\phi|_{@,w,v}$ will be a semantics.

If the metasemantic interpretation is correct, then the kind of 'semantic value' mentioned in the statement of Semantic Plasticity must be character rather than (propositional) content. It is a consequence of the metasemantic interpretation that, if $\triangledown\phi$ is true in *w*, then for some *v* close to *w*, the character of ϕ in *v* is different from the character of ϕ in *w*. But it is not a consequence of the metasemantic interpretation that, if $\triangledown\phi$ is true in *w*, then for some *v* close to *w*, the content of ϕ in *v* is different from the content of ϕ in *w*.

Suppose that the metasemantic interpretation is correct. Then it will be tempting to think that there is a straightforward account of why it gets right all the cases that made trouble for the 2D actualistic semantics from Section 4.

Consider first the case of $\triangle(\phi \equiv A\phi)$. Here, the 3D model theory alone solves the problem faced by the actualistic 2D model theories: because the 3D model theory validates each instance of $\phi \equiv A\phi$ and satisfies Definitization, it also validates each instance of $\triangle(\phi \equiv A\phi)$. A valid sentence is true at every proper point of the intended model, so is true *simpliciter*. So, each instance of $\triangle(\phi \equiv A\phi)$ is true.

The metasemantic interpretation, however, also delivers an explanation of *why* the model theory gets the case of $\triangle(\phi \equiv A\phi)$ right, as follows.

Let's say that a character *f* is *diagonally true* iff, for all *w*, $f(w)(w) = 1$. As we have learned from Kaplan, the character of any instance of $\phi \equiv A\phi$ is diagonally true. That the character of each instance of $\phi \equiv A\phi$ is diagonally true is a semantic fact about instances of $\phi \equiv A\phi$, in the sense of 'semantic' relevant to semantic discriminability under the metasemantic interpretation. It is also a known semantic fact. But since it is, in the relevant sense, a known semantic fact that the character of each instance of $\phi \equiv A\phi$ is diagonally true, then each instance of $\phi \equiv A\phi$ has a diagonally true character in every metasemantic context semantically indiscriminable from (*R*-related to) the actual one. It immediately follows that, in every metasemantic context *R*-related to the actual one, each instance of $\phi \equiv A\phi$ has a character that, when applied to the actual world, gives a proposition that is true in the actual world, and this, by the metasemantic interpretation and (vi), is equivalent to saying that each instance of $\triangle(\phi \equiv A\phi)$ is true.

It is tempting to generalize this explanation to all of the problematic cases, and to conjecture that T-sentences, R-sentences, the Meter Sentence, etc., all have characters that are known to be diagonally true (both in the actual metasemantic context and in ones *R*-related to it), so the definitizations of those sentences are true (both in the actual metasemantic context and in ones close to it). Here is why that line of thought is tempting.

Consider, to begin, either kind of disquotational sentence. Even though the propositions expressed by, e.g. '"Tim is thin" is true iff Tim is thin' and '"Tim" refers to Tim' are contingent, each sentence appears to be associated with a kind of semantic guarantee of truth. One natural way to make precise the idea of a sentence enjoying a semantic guarantee of truth is to say that the sentence has a diagonally true character: a diagonally true character guarantees that, in any world, the sentence expresses a proposition that — even though it may be a contingent proposition — is true in that world. Similarly, even though the Meter Sentence expresses a contingent proposition, and one that could very easily have been false, there is something about the convention we use to fix the content of 'meter' that guarantees that the sentence cannot be asserted falsely (as long as that convention is in effect, and as long as the other words in the sentence have their actual conventional meanings). If we take the further step of assuming that the convention in question is the character of 'meter,' then it will be tempting to conclude that the character of the Meter Sentence is also diagonally true. And similarly for other similar sentences involving 'kilogram,' 'second,' etc. If the characters of all of the problematic sentences are diagonally true in all worlds close to the actual one, and we are in a position to know this in all worlds close to the actual one, then they will all be definite, according to the metasemantic interpretation of the 3D model theory, in all worlds close to the actual one.

Superficially, at least, this line of thought seems attractive. Yet it is, I think, quite clearly incorrect. The metasemantic interpretation faces two serious problems, the second of which I find to be an especially decisive consideration against adopting it.

The first problem is this. Under the metasemantic interpretation, the 3D model theory only differs from the otherworldly 2D semantics on whether $\triangle\phi$ is true at the proper point of *w* (i.e. at $\langle w, w\rangle$ according to the 2D model theory and at $\langle w, w, w\rangle$ according to the 3D model theory) when ϕ has a non-constant character (i.e. is an indexical sentence) in some world close to *w*. It follows that disquotational sentences, the Meter Sentence, etc., must be indexical, in the actual world or worlds close to it, in subtle ways we haven't noticed before. Since a sentence is indexical in a world if and only if at least one of its simple constituents is,[33] these sentences must contain indexical words: apparently 'true,' 'refers,' 'meter,' and so on. Yet it is very unclear what the non-constant characters of the semantic predicates might be. In the case of 'meter,' one might think, as suggested above, that the non-constant character is the function that assigns to each world *w* the property of being as long as the Standard Meter is in *w* (or the content that represents that property). But one might also think, more plausibly, that this line of thought confuses semantic with metasemantic matters:

the convention that fixes the content of 'meter' also fixes its character, which is constant. Thus, if the length of the Standard Meter had been different from its actual length, the Meter Sentence would have had both a different character and a different content. And indeed this is the consensus on how so-called 'reference-fixing descriptions' work: they do not introduce indexicals, but rather expressions whose character as well as content depends on what satisfies the description.[34] If the consensus is correct, as I take it to be, then the metasemantic interpretation is incorrect.

(What about the above sketch of an argument for the conclusion that disquotational sentences and the Meter Sentence have diagonally true characters in all worlds close to the actual world? It confuses two properties: that of having a diagonally true character and that of having a character that determines a true proposition. It is plausible that each of the problematic sentences has, in each world *R*-related to the actual one, a character that, when applied to *w*, yields a proposition that is true in *w*. But it does not follow from this — and nothing has been said to support the claim — that each of the problematic sentences has, in each world *R*-related to the actual one, a character that, when applied to an arbitrary world *v*, yields a proposition that is true in *v* — i.e. a diagonally true character.)

The second problem with the metasemantic interpretation was noticed by John Hawthorne. In conversation, Hawthorne observed that we can refer to the actual world by both indexical and non-indexical means, and he used this to make trouble for the metasemantic interpretation. Here, I will present a simplified version of Hawthorne's objection:[35]

The indexical means of referring to referring to the actual world in L_{VM} is, of course, to use *A*. But we could also introduce a non-indexical 'actuality' operator into a language as follows. First, we introduce a non-indexical proper name, say, 'Worldy,' for the actual world into our metalanguage; then we enrich L_{VM} with an operator A^*, such that $A^*\phi$ translates into the metalanguage as 'In Worldy, it is the case that ϕ.' Since we know that A^* has a constant character,[36] A^* must, by familiar reasoning, have a constant character in every world *R*-related to the actual one. In particular, it must be that:

(1) If $@Rw$, then $|A^*\phi|_{w,v,u} = 1$ iff $|\phi|_{w,w,w} = 1$.

It follows that each instance of $A^*\phi \equiv A\phi$ is true, and, having noticed this, we know that each instance of $A^*\phi \equiv A\phi$ is true. But from (1) and the rest of our semantics it follows that an instance of $\triangle(A^*\phi \equiv A\phi)$ is false whenever $|\phi|_{w,w,w} \neq |\phi|_{w,@,@}$, for some *w* such that $@Rw$. That there is a world *w* and a sentence ϕ such that $|\phi|_{w,w,w} \neq |\phi|_{w,@,@}$ is pretty clear. (According to the semantics, there is vagueness only if there are *R*-related worlds *w*, *v* such that $|\phi|_{w,w,w} \neq |\phi|_{w,v,v}$, and it would be odd if there were no such case in which $v = @$.) But then such an instance of $A^*\phi \equiv A\phi$ will be a counterexample to the principle that knowledge entails definiteness.

This latter problem, which I take to be decisive, exposes a weakness that the metasemantic interpretation of the 3D model theory shares with both of the actualistic 2D semantics. According to all of these semantics it is sufficient for the truth of $\nabla\phi$ at a proper point that ϕ is false at an *improper* point suitably related to it. But no one ever uses, or could use, a sentence at an improper point, so, *a fortiori*, no one ever uses, or could use, a sentence to say something false at an improper point. Falsehood at such a point is not a genuine possibility of error, so is not the kind of possibility of error whose closeness precludes knowledge.

The second way of extracting a semantics from the 3D model theory, which I find to be more promising, is the *epistemic interpretation*. According to the epistemic interpretation, the first coordinate of a 3D point of evaluation is not a metaphysically possible world, but an epistemic possibility, in a certain broad sense, about which I will have more to say below. I will use '*e*,' '*e*′,' etc., as variables for these (broadly speaking) epistemic possibilities, and I will call the epistemic possibility that obtains in *w* '*e*(*w*)'; the actual epistemic possibility, then, is designated by '*e*(@).' The rough idea is that ϕ is true at $\langle e, w, v\rangle$ iff the proposition ϕ expresses in *w*, according to *e*, is true in *v*. A proper point of evaluation will be of the form $\langle e(w), w, w\rangle$. (Thus 'epistemic interpretation' turns out to be a bit of a misnomer: it is not an merely an interpretation of the 3D formal apparatus, since it requires our semantics for L_{VM} to draw finer distinctions than that apparatus does, such as the distinction between *e*(*w*) and *w*. I will return to the importance of keeping epistemic possibilities and metaphysically possible worlds separate below.)

The epistemic interpretation makes no use of the Kaplanian distinction between character and content. When an epistemic possibility *e* represents a sentence ϕ as expressing different propositions in worlds *w* and *v*, this could come about in two ways: either ϕ has different constant characters in *w* and in *v* according to *e* or ϕ has the same character in *w* and in *v* according to *e*, but this character is variable and determines different contents in *w* and in *v*. The epistemic interpretation is entirely insensitive to whether a given difference in content is due to a difference in character. For this reason it is not vulnerable to the first problem faced by the metasemantic interpretation: under the epistemic interpretation, that the 3D semantics and the 2D actualistic semantics differ on whether a sentence is definite at a proper point of evaluation tells us nothing about whether the sentence is indexical at any point.

According to the epistemic interpretation, the function $\lambda\phi\lambda w\lambda v|\phi|_{e,w,v}$ takes each sentence to the function that takes each world to the content that sentence expresses in that world, according to the epistemic possibility *e*. Which proposition a sentence expresses in a world is not a contingent matter; therefore (by (vi) and the fact that there is vagueness) some of the epistemic possibilities are metaphysically impossible. *e* will be metaphysically possible only if $\lambda\phi\lambda w\lambda v|\phi|_{e,w,v} = \lambda\phi\lambda w\lambda v|\phi|_{e(@),w,v}$. The only feature of *e* relevant to the envisioned semantics is the way in which content facts supervene some of the facts according to *e*, which is represented by the function $\lambda\phi\lambda w\lambda v|\phi|_{e,w,v}$. The barebones version of the epistemic interpretation offered here is neutral on the question of just which facts constitute the supervenience base in question. One answer, which is in the spirit of Hawthorne (2006), is that they are all of the 'metaphysical groundfloor' facts in some sense — perhaps the microphysical facts. Another possible answer is that they are all of the facts about the use of language, where 'use' is construed broadly enough to guarantee that no two worlds differ with respect to content facts without differing with respect to the use facts. However the details are filled in, the epistemic interpretation is committed to ignorance due to vagueness being explained by a kind of principled ignorance of the way in which semantic (content) facts supervene on certain other facts. In this respect, it represents a departure from the epistemicism of Williamson (1994): it does not include a commitment to Semantic Plasticity (although it is consistent with it) — a matter to which I'll return below.

For this reason, on the epistemic interpretation, *R* can no longer be glossed as 'semantic indiscriminability.' Rather, points of the first dimension that are *R*-related to the actual point *e*(@) represent ways in which, for all we (or creatures relevantly like us) are actually in a position to know, the content facts supervene on certain other facts. The points *R*-related to points *R*-related to *e*(@) represent ways in which, for all we are actually in a position to know, for all we are in a position to know, the content facts supervene on certain other facts, etc. For this reason I called the points of the first dimension epistemic possibilities 'in a broad sense.' Assuming that all of the points are related to *e*(@) by the ancestral of *R*, strictly speaking the only epistemic possibilities are those *R*-related to *e*(@), while the others are either merely epistemically possibly epistemically possible, or merely epistemically possibly epistemically possibly epistemically possible, etc. Nevertheless, for lack of a better term, I will continue to call these points 'epistemic possibilities.'

Now, it is fairly clear that the epistemic interpretation does not have a problem accounting for the definiteness of disquotational sentences, the Meter Sentence, or other similar cases involving reference-fixing convention. For consider the version of the epistemic interpretation according to which the points on the first dimension represent epistemic possibilities for the content facts to supervene on the use facts. It is highly plausible that, while I am largely in the dark about the details of the supervenience function for, say, the Meter Sentence, I am in a position to know enough about that function to rule out its being one that assigns to the actual use facts a proposition that is false in the actual world.[37] And the same goes for disquotational sentences and for the other problematic cases involving reference-fixing conventions, as well as, indeed, for the Hawthorne-inspired example, $A^*\phi \equiv A\phi$.

Now one might worry that the epistemic interpretation nevertheless fails to address Hawthorne's problem: after all, what's to stop someone from adding to L_{VM} an operator $A^\dagger$ for which it is simply stipulated that

(1′) If $e(@)Re$, then $|A^{\dagger}\phi|_{e,v,u} = 1$ iff $|\phi|_{e,e,e} = 1$?

Then, by familiar reasoning, some instances of $A^{\dagger}\phi \equiv A\phi$ will be borderline but known to be true. The answer to this worry is simply that, according to the kind of semantics under consideration, $|\phi|_{e,e,e}$ is undefined, because $|\phi|_{e,w,v}$ is defined only when *w* and *v* are metaphysical, not epistemic, possibilities. (This is where the 3D model theory will require some tweaking if it is to deliver the intended model.)

One might be inspired by Hawthorne's example to have the following further worry. If we introduce an operator $A^{\ddagger}$ for which the condition

(1″) $|A^{\ddagger}\phi|_{e,v,u} = 1$ iff $|\phi|_{e(@),v,u} = 1$

holds, and we know that (1″) holds then, whenever ϕ is borderline, $\phi \equiv A^{\ddagger}\phi$ will also be borderline, yet known to be true.

I am not entirely clear on what the correct answer to this worry is, but, whatever it is, it is not particularly incumbent upon the advocate of the epistemic interpretation to come up with it. An 'epistemic actuality' operator like $A^{\ddagger}$ would wreak havoc on the semantic applications of any standard model theory for epistemic logic.[38] For suppose that there is a knowledge operator *K* such that, as is standard, we treat $K\phi$ as true in an epistemic possibility *e* iff ϕ is true in every epistemic possibility relevantly accessible (by a reflexive relation) from *e*. Then, if there is a unique actual epistemic possibility $e(@)$ such that ϕ is true in $e(@)$ iff ϕ is true, and there is an operator $E^{@}$ such that $E^{@}\phi$ is true in an arbitrary epistemic possibility *e* iff ϕ is true in $e(@)$, then every instance of $K(\phi \equiv E^{@}\phi) \supset (K\phi \vee K\neg\phi)$ is true. But this is absurd: if $E^{@}$ is interpretable in this manner at all, then it will often be much more difficult to know whether ϕ is true than to know that $\phi \equiv E^{@}\phi$ is true. We can come to know the latter simply by coming to know the semantic stipulation by which $E^{@}$ was introduced plus disquotation, but we cannot come to know all truths in this way. Since, according to the epistemic interpretation, the first coordinate of a point of evaluation is an epistemic rather than a metaphysical possibility, the envisaged introduction of $A^{\ddagger}$ by (1″) amounts to the introduction of an 'epistemic actuality' operator, which is presumably impossible for whatever reason the introduction of $E^{@}$, as described in this paragraph, is impossible. And if it is not impossible, then this points to a crisis in the foundations of the standard approach to the model theory of languages with epistemic operators, rather than a problem for the epistemic interpretation of the 3D model theory for L_{VM} in particular.

Finally, I will mention, without attempting to answer, two open questions about the epistemic interpretation. The first concerns Semantic Plasticity. The epistemic interpretation notably differs from the metasemantic interpretation in that the former does not entail Semantic Plasticity, for any natural interpretation of 'semantic value.' It does entail, of course, that when ϕ is borderline, ϕ expresses in the actual world, according to some close epistemic possibility, a proposition other than the one it actually expresses in the actual world. But, as we have already seen, any such epistemic possibility will be metaphysically impossible, because it is a non-contingent matter which proposition is actually expressed by a sentence. Semantic Plasticity requires that a borderline sentence *could have* — in an alethic rather than epistemic modal sense — easily had a semantic value other than its actual semantic value. The epistemic interpretation is consistent with Semantic Plasticity: we could consistently add to it the claim that whenever a sentence ϕ expresses in the actual world, according to a close epistemic possibility, some proposition *p* other than the one it actually expresses in the actual world, there is also a metaphysical possibility that could easily have obtained (in which the use facts are ever so slightly different from the actual use facts) in which ϕ does express *p*. But so far it is unclear why the existence of such a metaphysical possibility would be required for an explanation of ignorance of borderline matters, since the epistemic interpretation already makes available an alternative explanation in terms of some kind of principled ignorance of the supervenience of content facts on certain other facts.

At least one issue in this area seems reasonably clear, which is that an explanation of ignorance of borderline matters in terms of Semantic Plasticity cannot require all of the close possibilities of error that preclude knowledge of borderline matters to be metaphysical possibilities. That is, it cannot be that, whenever ϕ is borderline, there is a close metaphysically possible world *w* such that: in *w* ϕ expresses some proposition *p* other than the proposition ϕ actually expresses, and the truth value *p* has in *w* is different from the actual truth value of the proposition ϕ actually expresses.[39] To see why this cannot be so, note first that the proposed principle entails (*).

(*) Whenever ϕ is true and borderline, there is a close metaphysically possible world in which ϕ expresses a proposition other than the proposition ϕ actually expresses, and in which that proposition is false.

We can use Semantic Plasticity to argue against (*) as follows. Consider the case of a non-semantically plastic, so, by Semantic Plasticity, non-borderline sentence $\#$ that specifies the all of the actual use facts. Let χ be any true borderline sentence, and let $p_@$ be the proposition actually expressed by χ. Then, given the supervenience of content facts on use facts,

(**) It is necessary that, if the proposition actually expressed by $\#$ is true, then χ expresses $p_@$.

Since χ is borderline but $\#$ is not, it follows, on very minimal assumptions about the logic of definiteness,[40] that $\# \supset \chi$ is borderline. $\# \supset \chi$ is also true, because χ is true. By (*), then, there is a close metaphysically possible world in which $\# \supset \chi$ expresses a proposition other than the proposition $\# \supset \chi$ actually expresses, and in which that proposition is false. To see that there is no such world, suppose for a contradiction that there is one — call it 'w^*' — and consider two cases:

First case: the proposition ♯ actually expresses is true in w^*. It follows by (**) that χ expresses $p_{@}$ — i.e. the proposition χ actually expresses — in w^*. Since ♯ is not semantically plastic, ♯ expresses what it actually expresses in w^* as well. Consequently, $\sharp \supset \chi$ expresses what it actually expresses in w^*, contrary to hypothesis.

Second case: the proposition ♯ actually expresses is not true in w^*. Again, since ♯ is not semantically plastic, ♯ expresses what it actually expresses in w^*. It follows that the proposition ♯ expresses in w^* is not true in w^*, and so it also follows that the proposition ♯ expresses in w^* is false in w^*, and that the proposition $\sharp \supset \chi$ expresses in w^* is true in w^*, contrary to hypothesis.[41,42]

The second open question concerns the principled ignorance of the supervenience of content on use that the advocate of the epistemic interpretation must posit.[43] I have no explanation of it to offer, but I will note, following Williamson, that any epistemicist must posit unknowable necessary truths:[44] e.g. if a net worth of $5,000,000 is the cut-off for wealth, then it is a necessary but borderline, and therefore unknowable, truth that to be wealthy is to have a net worth of at least $5,000,000. Of course, Williamson offers an explanation of the unknowability of such truths in terms of Semantic Plasticity, but the above argument casts doubt on the idea that such an explanation will be available in every case of ignorance of borderline matters.

6. I have argued that an acceptable epistemicist semantics for a language containing a definiteness operator $\triangle$ along with operators for metaphysical necessity ($\Box$) and metaphysical actuality (A) cannot be developed using the resources of the kind of 2D model theory that is standardly used in logical and semantic investigations of the interaction of $\Box$ and A alone. This suggests a move to the 3D model-theoretic framework that I developed with Jon Litland. The latter has already proved to be fruitful for logical investigations, but it remains an open question how, if at all, it can serve as the basis for a satisfactory epistemicist semantics for the interaction of $\triangle$, $\Box$, and A. I have sketched two ways in which one might attempt to extract an epistemicist semantics from the 3D model theory. The first of these was the metasemantic interpretation of the model theory, which vindicated a form of Semantic Plasticity, but with Kaplanian characters rather than contents (propositions) in the role of semantic value. The second of these was the epistemic interpretation, according to which the first dimension of semantic evaluation represents, in a certain broad sense, all of the epistemically possible ways for the content facts to supervene on certain other facts. While enjoying some superficial plausibility, the metasemantic interpretation turned out to have implausible consequences concerning indexicals. My own money is on the epistemic interpretation, or something close to it, delivering the correct epistemicist semantics, but if it or something like it does so, it is an interesting open question what role is left for Semantic Plasticity to play in the explanation of our ignorance of borderline matters.

Notes

1. Hawthorne (2006).
2. That is to say, a notion of truth that satisfies each instance of the schema '"S" is true iff S,' where instances are obtained by replacing 'S' with a suitable sentence. A sentence is suitable just in case inserting it into the schema does not result in any trouble, such as, but not limited to, inconsistency (e.g. sentences that purport to ascribe truth to themselves are also not suitable). The relevant notion of suitability is notoriously difficult to make precise.
3. Williamson (2003, 710).
4. An equivalent definition of definiteness in terms of borderline cases is perhaps more intuitive (it is definite that *S* iff *S* and it is not borderline whether *S*), but in this paper I take definiteness as primitive, as this is more natural for logical and semantic investigations, in which 'definitely' is standardly treated like 'necessarily.'
5. For example, Shapiro (2007) has this to say about epistemicism: 'Here I do not muster a sustained argument against that view, and it is not polite to stare' (7). This remark is directed at both Williamson (1994) and Sorensen's (1988) earlier development of an epistemic theory of vagueness.
6. See Hawthorne (2006), Kearns and Magidor (2008), Caie (2012), and Magidor (forthcoming).
7. See Williamson (1994, Section 8) and (2003, 710).
8. The content of *A*, as of any indexical, will vary with context. In this paper, the logical constants are given a syncategorematic treatment — the usual practice — but they could also be assigned characters in obvious ways.
9. A different approach is suggested by a passage in Kaplan's 'Demonstratives': in general we should take the proposition expressed by ϕ in *w* to be the proposition that *would be* expressed by ϕ *if it were* used in *w*. (According to Remark 1 of Section XIX of *Demonstratives*, the 'Content of a sentence in a context is, roughly, the proposition the sentence would express if it were uttered in that context' [Kaplan 1977, 546].) One problem with this suggestion is that, on just about anyone's conception of worlds, it is a non-contingent matter which sentences are used in which worlds (and more generally, what is so in a given world), wherefore a sentence that is not used in a world could not have been used in that world. (Kaplan's contexts, I should note, are not simply worlds: they also include agents, times, and locations. Nevertheless, the same worry applies: given that it is impossible for an utterance to occur in *w* unless it does occur in *w*, it is also impossible for an utterance that doesn't occur in *w* to occur in $\langle w, a, t, l\rangle$ if this means, as it is naturally interpreted, that the utterance occurs in *w*, and is, in *w*, produced by agent *a* at time *t* and location *l*.)
10. One rarely encounters a compositional treatment of variable-binding: see Yli-Vakkuri (2013).
11. As in, e.g. Kaplan (1977).
12. The assumption that there is an intended model is something of an idealization, *inter alia*, for the simple reason that there may be too many entities being theorized about (e.g. points of evaluation) to form a set — a well-known general limitation of set-theoretic semantics, which I will set to one side.
13. See Williamson (1994, Section 5) and (1999).
14. Caie (2012, 365).
15. *Ibid.*
16. See note 23.
17. See Davies and Humberstone (1980).

18. Kaplan's contexts, however, are not simply worlds: see note 9.
19. The example is inspired by Williamson (1994, 253–254). I use it because it is difficult to come up with other kinds of uncontroversial examples of borderline identity statements involving proper names. Many of the standard examples are arguably definite cases where the source of the intuition of vagueness is in the vagueness of, say, 'is located at' or 'is part of.'
20. Thanks to Peter Fritz and Jeremy Goodman for discussion here.
21. Cresswell (1990) takes just this approach. There are no expressions in his quantified modal–temporal language whose interpretation is fixed in all of his models: consequently, 'there will be no wff true in all interpretations, and thus no intensional logic.' Yet he says of this language that 'among its possible interpretations is one which comes closest to reflecting the meanings of particular words in a particular natural language, for convenience English' (7).
22. Caie's (2012) main counterexample to otherworldly semantics exploits this flaw, and some of the arguments in Kearns and Magidor (2008) could be reconstructed as doing so, although the latter are not directly concerned with epistemicist interpretations of $\triangle$.
23. Caie and Hawthorne both argue that Williamson's epistemicism has problems accounting for the definiteness of certain disquotational sentences. Both arguments, however, rely on premises that seem to me less secure than the ones I use.

    Caie (2012, Section 5) does not discuss T-sentences but what I call (below) *R-sentences*: sentences of the form '"*N*" refers to *N*.' The questionable assumption in Caie's argument is that (schematically), for some name '*N*' there is a close world in which '*N*''s referent is different from its actual referent but in which 'refers' expresses the same content as it actually does. (An analogous assumption about sentences and the truth predicate could be used to argue for the non-definiteness of some healthy T-sentences.) But Hawthorne's (2006) 'domestic stability' solution seems to involve rejecting this assumption.

    Hawthorne's (2006, Section 13) argument concerns T-sentences and involves what seems to me a questionable move from the plausible claim (given Semantic Plasticity) that the truth predicate expresses a content different from its actual content in some close worlds to the further claim that there is a close world where the content the truth predicate has in that world determines an extension different from the extension determined in that world by the actual content of the truth predicate (197). There may be a good argument for why an epistemicist committed to Semantic Plasticity would have to accept the latter claim, but Hawthorne does not offer one.
24. Suppose that (a) $|\phi|_{@,w} \neq |\phi|_{w,w}$, (b) $@Rw$, and that (c) ϕ is suitable. Note that

    $|\phi|_{w,w} = 1$ iff $\phi \in [\![T]\!](@, w)$ by (1) and (c)
    iff $[\![\ulcorner\phi\urcorner]\!](@, w) \in [\![T]\!](@, w)$ by (ix)
    iff $|T(\ulcorner\phi\urcorner)|_{@,w} = 1$ by (viii),

    so $|\phi|_{w,w} = |T(\ulcorner\phi\urcorner)|_{@,w}$. Then, by (a), $|\phi|_{@,w} \neq |T(\ulcorner\phi\urcorner)|_{@,w}$, so $|T(\ulcorner\phi\urcorner) \equiv \phi|_{@,w} = 0$. By (b) and either ($A_1$) or ($A_2$), $|\triangle(T(\ulcorner\phi\urcorner) \equiv \phi)|_{w,w} = 0$. Note that this argument requires the assumption that *R* is symmetric. There is, I think, an equally plausible argument for the existence of healthy T-sentences that fail to be definite in some close worlds that does not require the assumption that *R* is symmetric: we can simply assume that, in the case at hand, both *wR@* and *@Rw*. I take it to be fairly clear that there will be such cases if one or another of the actualistic semantics is correct, but I won't supply the philosophical argument for this for lack of space.

25. $|Ref(\ulcorner\tau\urcorner, \tau)|_{@,w} = 1$ iff $\langle [\![\ulcorner\tau\urcorner]\!](@, w), [\![\tau]\!](@, w)\rangle \in [\![Ref]\!](@, w)$ by (viii)
iff $\langle \tau, [\![\tau]\!](@, w)\rangle \in [\![Ref]\!](@, w)$ by (ix)
iff $[\![\tau]\!](w, w) = [\![\tau]\!](@, w)$ by (3)
so
$|Ref(\ulcorner\tau\urcorner, \tau)|_{@,w} = 0$ iff $[\![\tau]\!](w, w) \neq [\![\tau]\!](@, w)$,
so, in particular,
(3*) If $[\![\tau]\!](w, w) \neq [\![\tau]\!](@, w)$ and $wR@$ then $|\triangle Ref(\ulcorner\tau\urcorner, \tau)|_{w,w} = 0$.

26. It is not entirely clear to me that Caie's two options are the most natural ones. In fact, his taxonomy of truth conditions for $\triangle\phi$ at proper points omits the one that seems to me to most closely follow the letter of Williamson's account of ignorance of borderline matters. Consider, for example:

> What distinguishes vagueness as a source of inexactness is that the margin for error principles to which it gives rise advert to small differences in meaning, *not to small differences in the objects under discussion* (Williamson 1994, 231).
>
> Suppose I am on the 'thin' side of the boundary but only just. [...] The sentence 'TW is thin' is true, but could easily have been false *without any change in my physical measurements or those of the relevant comparison class* (Williamson 1994, 231).

The most straightforward way of extrapolating a truth condition for $\triangle\phi$ at a proper point from remarks like these would seem to be this: $\triangle\phi$ is true at $\langle w, w\rangle$ iff, for all v such that wRv and the truth value of the proposition ϕ expresses in w is the same in w and in v, ϕ is true at $\langle v, v\rangle$. This is consistent with two truth definition clauses:

(O_1') $|\triangle\phi|^{\mathfrak{U}}_{w,v} = 1$ iff, for all u such that vRu and $|\phi|^{\mathfrak{U}}_{v,u} = |\phi|^{\mathfrak{U}}_{v,v}$, $|\phi|^{\mathfrak{U}}_{u,u} = 1$.

(O_2') $|\triangle\phi|^{\mathfrak{U}}_{w,v} = 1$ iff, for all u such that wRu and $|\phi|^{\mathfrak{U}}_{w,u} = |\phi|^{\mathfrak{U}}_{w,w}$, $|\phi|^{\mathfrak{U}}_{u,u} = 1$.

(O_1') and (O_2') share the logical shortcomings, respectively, of (O_1) and (O_2), and these are accompanied by similar semantic shortcomings. Unlike (O_1) and (O_2), however, (O_1') and (O_2') both entail that a borderline sentence expresses, in some close world, a proposition different from the one it actually expresses. This is another reason why (O_1') and (O_2') seem to me to be closer matches, within the 2D framework, to the letter of Williamson (1994) than (O_1) or (O_2).

27. See Williamson (2003) for discussion.
28. See Williamson (1994, Section 9.3).
29. Thanks to Peter Fritz for this evocative label.
30. I am inclined to think that we should not require R to be symmetric, but the assumption that R is symmetric helps simplify some arguments in this paper.
31. In addition to note 30, there are some delicate issues in this area, such as the status of the principles $\triangle\Box\phi \supset \Box\triangle\phi$ and $\Box\triangle\phi \supset \triangle\Box\phi$, both of which the model theory validates, which I discuss in my work with Litland. A model theory in which $\triangle$ is interpreted by a relation on $W \times W \times W$ rather than on W offers a more flexible framework for logical and semantic theorizing about L_{MV}. For maximal generality we can also add to the models a second accessibility relation to interpret $\Box$. My work with Litland explores model theories with both features.
32. For this idea to have any hope of working, we must not be too permissive about what to count as a fact 'about character.' For example, it is, in some intuitive sense, a fact about the character of 'Tim is thin' that the character of 'Tim is thin' is its actual character. Furthermore, it is a known fact. But then it cannot be a fact about

character in the intended sense, because if it were, then no world in which 'Tim is thin' has a character different from its actual character will be *R*-related to the actual world; in any such world the known fact that the character of 'Tim is thin' is its actual character fails to obtain.

33. This principle is plausible on it own, but it also becomes a theorem under the metasemantic interpretation of the 3D model theory, if we augment the model theory by assigning metasemantic characters to the logical constants in any of the obvious ways. Then all of the constants except for *A* will have constant characters in every metasemantic context of every model. (We would also have to add functors and an identity predicate to the language to handle the Meter Sentence.)
34. See Remark 11 of Section XIX of *Demonstratives* (Kaplan 1977, 551) for a classic articulation of the consensus view.
35. In Hawthorne's original example, we introduce an indexical singular term, 'Actuality,' and a non-indexical singular term, 'Worldy,' to designate the actual world. The identity 'Worldy = Actuality' is then known but, given natural assumptions about how the 3D model theory would have to deal with singular terms and identity, is not definite according to the metasemantic interpretation.
36. See note 8.
37. It is much less clear that I am in a position to know this much about the supervenience of content facts on 'metaphysical groundfloor' facts, for any natural interpretation of that phrase. For example, it is difficult to see how I could be in a position to know anything at all about how the content facts supervene on the microphysical facts, unless the relevant notion of 'being in a position to know' packs in the idealization I have access to Chalmers' 'cosmoscope' or a similar device (see Chalmers 2012, 114–115).
38. Of course, models for epistemic logic do not usually come with a designated actual epistemic possibility, nor with contexts that determine epistemic possibilities (because they usually do not come with contexts at all), but this is just an artifact of the model theory. A semantics must recognize, for each parameter shifted by an operator, a function that assigns to each context the value of that parameter that is 'realized' or 'present' (etc.) in that context.
39. It would be natural to strengthen this condition by adding the conjunct: 'and *w* matches actuality with respect to the truth value of the proposition ϕ actually expresses,' because possibilities in which one is in error because the facts that actually ϕ concerns (e.g. whether Tim is thin) are different than they actually are irrelevant to vagueness-induced ignorance (see Williamson 1994, 231). However, even the weaker condition does not hold.
40. Specifically, we must assume that the logic of $\triangle$ validates each instance of the *K* axiom schema $\triangle(\phi \supset \psi) \supset (\triangle\phi \supset \triangle\psi)$, and that $\triangle\phi$ is valid whenever ϕ is a truth-functional tautology — call this latter principle *Tautological Definitization*. Because $\triangle\#$ is true and $\triangle\chi$ is not, by K, $\neg\triangle(\# \supset \chi)$ is true. Furthermore, because χ is borderline, $\neg\triangle\neg\chi$ is true, and we can show that $\neg\triangle\neg(\# \supset \chi)$ is true as follows. First we assume for a *reductio*:

    (i) $\triangle\neg(\# \supset \chi)$.

    (ii) is valid by K.

    (ii) $\triangle(\neg(\# \supset \chi) \supset \neg\chi) \supset (\triangle\neg(\# \supset \chi) \supset \triangle\neg\chi)$,.

    and (iii) is valid by Tautological Definitization.

    (iii) $\triangle(\neg(\# \supset \chi) \supset \neg\chi)$.

    Because χ is borderline,

    (iv) $\neg\triangle\neg\chi$.

(i), (ii), and (iii) imply $\triangle\neg\chi$, which contradicts (iv). We get:

$$\neg\triangle\neg(\# \supset \chi) \wedge \neg\triangle(\# \supset \chi),$$

which is definitionally equivalent to $\triangledown(\# \supset \chi)$.

41. This argument does not essentially depend on the assumption that # expresses all of the *use* facts; any kind of fact (e.g. microphysical) on which content facts supervene will do. Nor does the argument require the assumption that # specifies *all* of the facts of the relevant kind; it is enough that # specifies enough of them for it to be the case that, necessarily, if the proposition actually expressed by # is true, then χ expresses $p_{@}$. Thanks to John Hawthorne for discussion here.
42. While I find this argument to be pretty decisive, I should note that at least one philosopher sympathetic to epistemicism rejects two of its assumptions. Magidor (forthcoming) rejects one of the 'very minimal assumptions about the logic of definiteness' mentioned in the main text, namely the *K* axiom for $\triangle$ (see note 40), and Kearns and Magidor (2012) reject the supervenience of content facts on use facts (as well as on microphysical facts, and apparently on any facts other than the content facts themselves), so they would reject (**).
43. Williamson himself appears to posit such ignorance. One relevant passage is this:

    > Since the content of the concept depends on the overall pattern, you have no way of making your use of a concept on a particular occasion sensitive to the overall pattern. Even if you did know all the details of the pattern (which you could not), you would still be ignorant of the manner on which they determined the content of the concept (1994, 231–232).

    See also Williamson (1994, Section 7.4).
44. Williamson (1994, 230).

Acknowledgments

I would like to thank Catharine Diehl, Cian Dorr, Kit Fine, Peter Fritz, Jeremy Goodman, John Hawthorne, Jon Litland, Mark McCullagh, Ofra Magidor, Beau Madison Mount, Jeff Sanford Russell, Gabriel Uzquiano, Sara Kasin Vikesdal, Tim Williamson, and audiences at the University of Oxford and at the 2015 *Williamson on Modality* workshop in Montreal for helpful discussions and for comments on earlier drafts of this paper.

References

Caie, M. 2012. "Vagueness and Semantic Indiscriminability." *Philosophical Studies* 160: 365–377.

Chalmers, D. 2012. *Constructing the World*. Oxford: Oxford University Press.

Cresswell, M. 1990. *Entities and Indices*. Dordrecht: Kluwer.
Davies, M., and L. Humberstone. 1980. "Two Notions of Necessity." *Philosophical Studies* 38: 1–30.
Fritz, P. 2016. "Appendix to Juhani Yli-Vakkuri's 'Epistemicism and Modality'." *Canadian Journal of Philosophy*, this volume.
Hawthorne, J. 2006. "Epistemicism and Semantic Plasticity." In *Metaphysical Essays*, 185–210. Oxford: Oxford University Press.
Kaplan, D. 1977. "Demonstratives." Mimeograph, Department of Philosophy, UCLA. Published in J. Almog et al., eds., *Themes from Kaplan*. Oxford: Oxford University Press, 1989. Page references to the latter.
Kripke, S. 1980. *Naming and Necessity*. Cambridge, MA: Harvard University Press.
Kearns, S., and O. Magidor. 2008. "Epistemicism About Vagueness and Meta-linguistic Safety." *Philosophical Perspectives* 22: 277–304.
Kearns, S., and O. Magidor. 2012. "Semantic Sovereignty." *Philosophy and Phenomenological Research* 85: 322–350.
Litland, J. and J. Yli-Vakkuri 2016. "Vagueness and Modality." *Philosophical Perspectives*.
Magidor, O. Forthcoming. "Epistemicism, Distribution, and the Argument from Vagueness." *Noûs*.
Shapiro, S. 2007. *Vagueness in Context*. Oxford: Oxford University Press.
Sorensen, R. 1988. *Blindspots*. Oxford: Oxford University Press.
Williamson, T. 1994. *Vagueness*. London: Routledge.
Williamson, T. 1999. "On the Structure of Higher-order Vagueness." *Mind* 108: 127–143.
Williamson, T. 2003. "Vagueness in Reality." In *The Oxford Handbook of Metaphysics*, edited by M. Loux and D. Zimmerman, 690–716. Oxford: Oxford University Press.
Yli-Vakkuri, J. 2013. "Propositions and Compositionality." *Philosophical Perspectives* 27: 526–563.

Appendix to Juhani Yli-Vakkuri's 'Epistemicism and Modality'

Peter Fritz

Department of Philosophy, Classics, History of Art and Ideas, University of Oslo, Oslo, Norway

ABSTRACT

A formal result is proved which is used in Juhani Yli-Vakkuri's 'Epistemicism and Modality' to argue that certain two-dimensional possible world models are inadequate for a language with operators for 'necessarily', 'actually', and 'definitely'.

Relative to an assignment function, the two-dimensional models of Section 4 evaluate formulas as true or false at pairs of elements of a set W, and so formulas can be understood as interpreted using sets of pairs of elements of W. Several specific ways of interpreting $\triangle$ based on a binary relation R on W are explored and rejected. The aim of this appendix is to provide the formal basis for generalizations of these arguments. To make these generalizations as strong as possible, $\triangle$ will be allowed to be interpreted in the most general way which respects the following compositionality constraint: under an assignment function, the interpretation of a formula of the form $\triangle\phi$ is determined by the interpretation of ϕ. Thus, each pair $\langle w, v\rangle$ will be associated with a set of sets of pairs – the set of interpretations of formulas to which $\triangle$ is to be applied in $\langle w, v\rangle$. Such functions are commonly studied under the labels of neighborhood and Scott–Montague semantics. So, define a *generalized 2D model* to be a structure $\langle W, D, [\![\cdot]\!]\rangle$ such that W and $[\![\cdot]\!]$ are as in standard 2D models, and $D : W^2 \to \mathcal{P}(\mathcal{P}(W^2))$.

It will be useful to be able to refer easily to the set of pairs at which a given formula is true under a given assignment, so let $[\![\phi]\!]^{\mathfrak{A}}_g = \{\langle w, v\rangle \in W^2 : |\phi|^{\mathfrak{A}g}_{w,v} = 1\}$. With this, extend the interpretation of formulas described in Section 4 by the following clause for $\triangle$:

$|\triangle\phi|^{\mathfrak{A}g}_{w,v} = 1$ iff $[\![\phi]\!]^{\mathfrak{A}}_g \in D(\langle w, v\rangle)$

The first result to be established concerns formulas of the language L_{VMQ} including propositional variables and quantifiers. For a class C of generalized 2D models, let the *logic of* C be the set of L_{VMQ}-formulas true at each proper point under each variable assignment in each model in C. In the following, p will be used to indicate a propositional variable; note that assignment functions map such variables to 'barcodes', whereas atomic sentences which are not variables may be interpreted as arbitrary sets of pairs.

Proposition 1: *If C is a class of generalized 2D models whose logic includes all instances of the schemas*

(1) $\triangle(\phi \supset \psi) \supset (\triangle\phi \supset \triangle\psi)$ *and*
(2) $p \supset \triangle p$,

and is closed under the schematic rule

(3) $\phi / \triangle\phi$,

then the logic of C includes all instances of the schema

(4) $\phi \supset \triangle\phi$.

Proof: Consider any formula ϕ, model $\mathfrak{A} = \langle W, D, [\![\cdot]\!]\rangle$ in C, $w \in W$ and assignment function g such that $|\phi|^{\mathfrak{A}^g}_{w,w} = 1$; we show that $|\triangle\phi|^{\mathfrak{A}^g}_{w,w} = 1$. For any formula ψ, we write $\vdash \psi$ for ψ being in the logic of C.

That $|\triangle\phi|^{\mathfrak{A}^g}_{w,w} = 1$ will be shown as follows: We produce an assignment function f which agrees with g on the free variables in ϕ and a formula ψ such that $|\triangle\psi|^{\mathfrak{A}^f}_{w,w} = 1$ and $|\triangle(\psi \supset \phi)|^{\mathfrak{A}^f}_{w,w} = 1$. It follows by (1) that $|\triangle\phi|^{\mathfrak{A}^f}_{w,w} = 1$, and so with the fact that f agrees with g on the free variables in ϕ that $|\triangle\phi|^{\mathfrak{A}^g}_{w,w} = 1$. The idea behind the construction of ψ and f is to let ψ be a conjunction $p \wedge \chi$, where p is a propositional variable not free in ϕ which f maps to the barcode true at $\langle w, w\rangle$ and no other proper point, and χ is a formula interpreted as the set of proper points. We show that $\triangle$ applies to p and χ at $\langle w, w\rangle$, and conclude that it also applies to their conjunction ψ, which is true only at $\langle w, w\rangle$. Since ϕ is true there, $\psi \supset \phi$ is interpreted as the set of all points, to which $\triangle$ applies, which completes the the proof.

In more detail, let p be a propositional variable not free in ϕ, $b = \{\langle v, w\rangle : v \in W\}$, and f the assignment function which differs from g only in that $f(p) = b$. Since $|p|^{\mathfrak{A}^f}_{w,w} = 1$, it follows with (2) that $|\triangle p|^{\mathfrak{A}^f}_{w,w} = 1$.

To construct χ, let $\delta = \forall p(Ap \supset p)$, and $\chi = A\delta \supset \delta$. By the semantics of A, $\vdash \chi$, and so by (3), $\vdash \triangle\chi$; in particular $|\triangle\chi|^{\mathfrak{A}^f}_{w,w} = 1$. Furthermore, letting $d = \{\langle v, v\rangle : v \in W\}$, $[\![\delta]\!]^{\mathfrak{A}}_f = d$, and so also $[\![\chi]\!]^{\mathfrak{A}}_f = d$.

Since the logic of any class of generalized 2D models includes all substitution instances of tautologies and is closed under modus ponens, it is routine to derive the schema $\vdash (\triangle\theta \wedge \triangle\eta) \supset \triangle(\theta \wedge \eta)$ using (1) and (3). Letting $\psi = p \wedge \chi$, it follows that $|\triangle\psi|^{\mathfrak{A}^f}_{w,w} = 1$.

$[\![\psi]\!]_f^{\mathfrak{A}} = b \cap d = \{\langle w, w\rangle\}$. By assumption $\langle w, w\rangle \in [\![\phi]\!]_g^{\mathfrak{A}}$, which is identical to $[\![\phi]\!]_f^{\mathfrak{A}}$ since f and g agree on the free variables in ϕ. So $[\![\psi \supset \phi]\!]_f^{\mathfrak{A}} = W^2$. By (3), $\vdash \triangle\top$, so $W^2 \in D(\langle w, w\rangle)$. Hence $|\triangle(\psi \supset \phi)|_{w,w}^{\mathfrak{A}^f} = 1$, and so with (1), $|\triangle\phi|_{w,w}^{\mathfrak{A}^f} = 1$. Since f and g agree on the free variables in ϕ, $|\triangle\phi|_{w,w}^{\mathfrak{A}^g} = 1$. □

Although propositional quantifiers are not explicitly mentioned in the statement of Proposition 1, the proof relies on an application of rule (3) to the quantified formula $\chi = A\forall p(Ap \supset p) \supset \forall p(Ap \supset p)$. However, a variant of the result can be established which does not rely on the presence of quantifiers, by considering classes of frames rather than classes of models. Let a *generalized 2D frame* be a structure $\langle W, D\rangle$, with W and D as above; let the *quantifier-free logic of* a class C of generalized 2D frames be the set of formulas in the language L_{VM} (which omits propositional variables and quantifiers) which are in the logic of the class of generalized 2D models whose underlying frames are in C.

Proposition 2: *For every class C of generalized 2D frames $\langle W, D\rangle$ which satisfy*

(2) $\langle w, w\rangle \in b$ only if $b \in D(\langle w, w\rangle)$, for all barcodes b and $w \in W$,*

if the quantifier-free logic of C includes all instances of schema (1) and is closed under schematic rule (3), then it includes all instances of schema (4).

Proof: Like the proof of Proposition 1, but using atomic formulas α and β not occurring in ϕ instead of p and δ, respectively, considering a variant interpretation function mapping α to b and β to d instead of f, and appealing to (2*) instead of (2). □

Further variants of these two results can immediately be obtained by replacing the requirement of the relevant logic being closed under rule (3) with the requirement of containing all instances of the schemas

(3a) $\triangle(A\phi \supset \phi)$,
(3b) $(\triangle\phi \wedge \triangle\psi) \supset \triangle(\phi \wedge \psi)$, and
(3c) $\triangle\top$.

Reply to Yli-Vakkuri

Timothy Williamson

Wykeham Professor of Logic, University of Oxford, Oxford, UK

In 'Epistemicism and modality' (Yli-Vakkuri 2016), Juhani Yli-Vakkuri rigorously treats the problem of developing a model-theoretic semantics for a formal language with the operators 'necessarily' and 'actually', interpreted in terms of metaphysical modality, and an operator 'definitely', interpreted in the spirit of the epistemicist theory of vagueness developed in my book (Williamson 1994). The challenge is especially pertinent because my epistemicist account of definiteness has a modal dimension, whose interaction with other forms of modal variation raises delicate issues for the overall theory.

1. Worldly differences and semantic differences

According to epistemicism, vagueness in a language induces no deviation from classical logic or from bivalent semantics. It is a purely epistemic phenomenon: borderline cases are just cases of unavoidable ignorance for a special sort of reason. The actual meaning of a vague expression differs from some of its potential alternative meanings in ways indiscriminable by native speakers of the language. Following John Hawthorne, Yli-Vakkuri calls this idea 'Semantic Plasticity'. The upshot is a special obstacle to knowing some truths expressed using the vague expression. Very roughly, the definite truths are those knowledge of which is not blocked by an obstacle of that sort. I developed the underlying epistemology in safety-theoretic terms. Knowledge requires safety from error: one knows in a given case only if it is not relevantly close to an error possibility. Actuality is never close in that sense to metaphysical impossibilities, because they present no real danger.

Compare a case where one assents to a true proposition expressed by a vague sentence with a similar case where one assents to a false proposition expressed by the same sentence.[1] The difference in truth value may result from either one of the two kinds of differences between the cases (perhaps from both). The first kind of difference is *worldly*. The vague sentence may express the same proposition in the two cases, while the objects under discussion vary in their properties or relations. For instance, in both cases the sentence 'Jack is bald' expresses the proposition that Jack is bald, but Jack has less hair in one case than in the other. Thus the sentence 'Jack is bald' may express a true proposition in the former case and a false proposition in the latter. The second kind of difference between cases is *semantic*. Even if the objects under discussion have the same properties and relations in the two cases, the sentence 'Jack is bald' may express different propositions because it is used in very slightly different ways. For instance, in one case the sentence 'Jack is bald' still expresses the proposition that Jack is bald, while in the other it expresses the distinct proposition that Jack is bald*, where the standard for being bald* is slightly more demanding than the standard for being bald. Thus, even though the state of Jack's scalp is exactly the same between the two cases, the sentence 'Jack is bald' may again express a true proposition in the former case and a false proposition in the latter. Of course, two cases may also differ in both worldly and semantic respects simultaneously, and in other respects too.

Concerning our ignorance in paradigmatic borderline cases, I proposed that the error possibilities by which vagueness blocks knowledge are the close ones differing from the given case in semantic but not in worldly respects (1994, 231). By contrast, most discussions of safety in epistemology concern worldly but not semantic differences; the proposition at issue is held fixed across the cases under discussion. However, if one wants a non-trivial safety condition for knowledge of necessary truths, as in mathematics, one must look at semantic rather than worldly differences: if a sentence expresses a necessary truth in a given case, then it expresses a truth in all cases differing from the former only in worldly respects (assuming that all cases, or at least all close cases, are metaphysically possible). Thus there are reasons independent of vagueness for using a safety condition that takes account of cases that differ in semantic respects, as well as of those that differ only in worldly ones (Williamson 2000, 181–182). However, the semantic differences relevant to mathematical knowledge will typically be much larger than those relevant to vagueness; the cases will not be semantically indiscriminable to native speakers.

For straightforward examples, we can treat the worldly and semantic dimensions as mutually orthogonal: we can vary a case on either dimension while holding it fixed on the other. For more complex examples, however, the independence assumption breaks down: varying a case on one dimension entails varying it on the other too. For instance, a vague word may be both used and mentioned in the same sentence, as in disquotational biconditionals for truth such as '"Jack is bald" is true if and only if Jack is bald', a type of example much discussed in the recent literature on vagueness and featured in Yli-Vakkuri's paper. A difference between cases in the use of the word 'bald' is semantic in respect of the right-hand side of the biconditional ('Jack is bald') but worldly in respect of the left-hand side ('"Jack is bald" is true') since the latter is about a broadly semantic feature of the former.

It is often unclear how to apply the rough dichotomy between the world and our descriptions of it to examples where we cannot ignore the fact that our descriptions are part of the world. Similarly, it is often unclear how to apply the rough dichotomy between the world and our knowledge of it to examples where we cannot ignore the fact that our knowledge is part of the world, as in the so-called paradox of knowabilty (we cannot know a conjunction of the form: P and it will never be known that P; see Williamson 2000, 270–301). Yet any theory of vagueness must be applicable to such complex examples. Problems of vagueness in the law remind us that the distinction between clear cases and borderline cases arises in practice for arbitrarily complex sentences. One benefit of trying to construct an adequate model theory for a formal language with a 'definitely' operator is that it forces us to think systematically and quite generally about how to apply vagueness-theoretic distinctions to such sentences.

At this point, some philosophers will be inclined to scrap the account of definiteness based on the worldly–semantic distinction as insufficiently general, and start again from scratch. But that is just the sort of methodological impatience that so much impedes progress in philosophy. In particular, the fact that the worldly and semantic dimensions are not fully independent of each other does not entail that an initial assumption of their independence makes a bad starting point. After all, if we are trying to understand a phenomenon in natural science, we may initially make the simplifying assumption that two aspects of it are mutually independent, even though we know that the assumption is not strictly correct. For instance, we may treat propositions about the shape of an object as probabilistically independent of propositions about its colour, so that we can calculate the probability of the conjunction of a shape proposition and a colour proposition as simply the product of the probabilities of the conjuncts. We can then use the results obtained under that simplifying assumption as a starting point for a more general account that allows for the interdependence of the two dimensions. The same goes for the treatment of the worldly and semantic dimensions as mutually independent in the initial explanation of ignorance in borderline cases. Rather than scrapping the first pass explanation, we may do better by refining it. A specific reason for optimism is that vagueness in complex expressions derives somehow from vagueness in their atomic constituents and modes of composition. The worldly–semantic dichotomy works best for atomic expressions, although doubtless not perfectly even there.

A further consideration comes from the nature of the 'definitely' operator. Its treatment as a sentence operator in a formal language with a compositional semantics turns out to involve a significant degree of idealization. Within limits, the theorist has discretion to decide what idealizations of its meaning will conduce to the most illuminating logic of vagueness.

'Definitely' is not just a philosopher's invention. To describe someone as 'neither definitely tall nor definitely not tall' is quite intelligible in ordinary English, even if it sounds a bit pedantic. Although the complement of 'definitely' in English is normally an adjectival phrase rather than a sentence, that goes for modal words such as 'necessarily' and 'possibly' too; it does not prevent them from being most usefully formalized by operators on sentences. Many ordinary uses of 'definitely' are epistemic in a way not specifically related to vagueness. For instance, a detective asking 'Who was definitely in the building last night?' probably means something like 'Who is known to have been in the building last night?'; borderline cases of being in the building last night are probably not at issue. Still, theorists of vagueness can legitimately requisition the word 'definitely' for their more specific purposes. So far, so good.

2. Use and mention

A more pressing concern is that the formalization of 'definitely' as a sentence operator $\triangle$ tailored to suit the theory of vagueness may not fit the structure of a compositional semantics. To start simple, consider a semantic theory that assigns a set of (metaphysically) possible worlds as the semantic value of a sentence. The compositional semantics makes the semantic value of the sentence operator $\triangle$ a function from sets of possible worlds to sets of possible worlds. Let 'small' be a vague predicate of natural numbers whose application to them is nevertheless non-contingent. Thus, we may suppose, 17 is a borderline case for 'small'; it is vague whether 17 is small, but not contingent whether 17 is small. If 17 is small, it is necessary that 17 is small; if 17 is not small, it is necessary that 17 is not small. Consequently, the semantic value of '17 is small' is either the set of all worlds or the set of no worlds. In the former case, '17 is small' has the same semantic value as '$0 = 0$', so by the compositional semantics, '$\triangle$(17 is small)' has the same semantic value as '$\triangle(0 = 0)$', and so is true (definitely 0 is self-identical). In the latter case, the semantic value of '$\neg$(17 is small)' is the set of all worlds and so the same as the semantic value of '$0 = 0$', so by the compositional semantics, '$\triangle\neg$(17 is small)' has the same semantic value as '$\triangle(0 = 0)$', and so is true. Either way, the disjunction '$\triangle$(17 is small) $\vee$ $\triangle\neg$(17 is small)' is true: it is definite whether 17 is small. But that contradicts the hypothesis that it is vague whether 17 is small.

One might simply draw the moral that sets of possible worlds are too coarse-grained to serve as the semantic values of sentences in the semantics for a logic of vagueness and modality. That is hardly surprising. The envisaged semantics is one-dimensional, whereas Yli-Vakkuri argues powerfully that it needs to be three-dimensional rather than two-dimensional, to accommodate both the usual two modal dimensions for the operators 'necessarily' and 'actually' and a further dimension for the operator 'definitely'. But further reflection suggests that the problem goes even deeper. *Any* compositional semantics for $\triangle$ makes the meaning of the sentence $\triangle\phi$ a function of the meaning of the sentence ϕ, in some appropriate sense of 'meaning'. Thus, if ϕ is synonymous with ψ (they have the same meaning in the relevant sense), then $\triangle\phi$ is synonymous with $\triangle\psi$, and moreover $\neg\phi$ is synonymous with $\neg\psi$, so $\triangle\neg\phi$ is synonymous with $\triangle\neg\psi$; thus, ϕ and ψ have the same borderline cases. But is that principle defensible?

To see the difficulty, let us make the standard assumption that the natural kind terms 'furze' and 'gorse' are synonymous. They are also presumably not perfectly precise: for instance, it is slightly vague when furze and gorse first evolved. Nevertheless, for purposes of meaning, the two words are vague in the same way. Let ϕ be the sentence 'All and only furze is furze' and ψ the sentence 'All and only furze is gorse'. On a standard compositional semantics, ϕ and ψ have the same semantic structure and synonymous corresponding constituents, and so are synonymous. Thus, by hypothesis, $\triangle\phi$ and $\triangle\phi$ are synonymous too. Since ϕ is a trivial logical truth, $\triangle\phi$ is true. Hence, $\triangle\psi$ is true too. But is it? We may assume that 'furze' and 'gorse' are understood independently of each other; a speaker may understand either word without understanding the other. No special rule coordinates them; they just happen to have the same meaning. Although the use of 'furze' happens exactly to match the use of 'gorse', there could very easily have been a slight mismatch, where the use of one word was slightly more liberal than the use of the other, so that the extension of the former included slightly earlier samples than the extension of the latter did. The vagueness of 'furze' and 'gorse' makes it unclear to native speakers whether there is indeed such a slight difference in extension. Thus, it seems to be definite that all and only furze is furze, but not definite that all and only furze is gorse. On this view, ϕ and ψ are synonymous, but $\triangle\phi$ and $\triangle\psi$ differ in truth value and so in meaning. If so, the operator $\triangle$ does not have a compositional semantics.

Could a theorist respond by individuating meanings in a more fine-grained way? We may assume that the only relevant difference between 'furze' and 'gorse' is the syntactic difference between the words themselves. A way of individuating meanings on which 'furze' and 'gorse' differ in meaning makes synonymy between distinct words virtually impossible. A supervaluationist might implement such a fine-grained standard by identifying meanings with functions from admissible sharpenings of the language to Kaplanian characters. Suppose that one admissible sharpening assigns the character C to both 'furze' and 'gorse', while another assigns a distinct character C* to both words. Then, if a third admissible sharpening assigns C to 'furze' and C* to 'gorse', while a fourth assigns C* to 'furze' and C to 'gorse', the account makes 'furze' and 'gorse' differ in meaning, even if the overall range of admissible sharpenings associates each of them with exactly the same range of characters. The trivialization of synonymy is a high price to pay for such a view, especially since it threatens to trivialize the requirement of compositionality too.

An alternative way of interpreting the semantics is to retain a conception of meaning that allows 'furze' and 'gorse' to be synonymous, but to treat the $\triangle$

operator as tacitly *quotational*: $\triangle\phi$ is really about the sentence ϕ itself, at least partly (a move Yli-Vakkuri may accept). In that case, the so-called compositional semantics for the language is not what it seems, since it does not determine the meanings of complex expressions just from the meanings of their constituents and the way those constituents are put together. For a genuinely compositional semantics, one might have to make the quotation explicit – and supply a semantics of quotation.

These problems are reminiscent of those facing attempts to give a compositional semantics of variable-binding, on which Yli-Vakkuri has written illuminatingly elsewhere (2013). In the simple case of an extensional, non-indexical language, if one identifies the meaning of a singular term with a function that maps each assignment of values to variables to the denotation of the term on that assignment, the variables *x* and *y* count as different in meaning, simply because they have distinct values on at least one assignment, even though that is just a corollary of their syntactic difference. In one respect, however, the problems raised by $\triangle$ are more acute, since they arise for closed as well as open expressions.[2]

Similar problems arise anyway for epistemic operators, and so pose no threat to the epistemicist view that $\triangle$ *is* an epistemic operator. For instance, even though 'furze' and 'gorse' are synonymous, it may be natural to describe someone as knowing that something is furze without knowing that it is gorse, or as knowing that all and only furze is furze without knowing that all and only furze is gorse. If one is doing the semantics of 'know' in terms of epistemically possible worlds, one may then be driven to assign 'furze' and 'gorse' distinct intensions (here, functions from epistemically possible worlds to extensions), even though they are synonymous.

Of course, one may take a more coarse-grained view of knowledge and simply tough it out, insisting that to know that all and only furze is furze just *is* to know that all and only furze is gorse, even if those who recognize that truth under the guise of the sentence 'All and only furze is furze' cannot recognize it under the guise of the sentence 'All and only furze is gorse'. In principle, someone might take the same line with vagueness and simply tough it out, insisting that for it to be definite that all and only furze is furze just *is* for it to be definite that all and only furze is gorse, even if those who recognize that truth under the guise of the former sentence cannot recognize it under the guise of the latter. However, the tough line is better motivated for 'know' than for 'definite'. When we speak of knowledge and ignorance, we often wish to prescind from the words in which a given piece of knowledge might be expressed. By contrast, when we speak of definiteness and vagueness, the expression in given words is typically central to our interest. To prescind from it would be to miss the very point at issue.

These questions also arise for Yli-Vakkuri's preferred epistemic interpretation of his three-dimensional semantics. To make all the desired distinctions, an intended model will have to assign distinct 3D profiles to sentences that, by a natural compositional standard, are synonymous. That is very far from making the theory useless: it can still serve as a good guide to reasoning about the interaction between vagueness and metaphysical modality. But it will not be a compositional semantics in the sense one might originally have expected.

To the two non-epistemically modal dimensions needed for a standard semantic treatment of 'necessarily' and 'actually', Yli-Vakkuri adds a third modal dimension for the treatment of 'definitely', with promising results.[3] He argues that the third dimension is best understood as epistemically modal. Since he is implementing an epistemicist account of vagueness, that is not altogether surprising. He requires the epistemic dimension to include metaphysically impossible epistemic possibilities as well as metaphysically possible ones. On his interpretation, 'when ϕ is borderline, ϕ expresses in the actual world, according to some close epistemic possibility, a proposition other than the one it actually expresses in the actual world', but 'any such epistemic possibility will be metaphysically impossible, because it is a noncontingent matter which proposition is actually expressed by a sentence'. Such a use of metaphysically impossible worlds in the semantics for 'definitely' poses an apparent threat to epistemicist theories that explain actual borderline cases in terms of semantic plasticity, for surely no metaphysically impossible world is close in the safety-relevant sense to the actual world.

However, we may be able to reconcile Yli-Vakkuri's semantics with the safety-theoretic considerations by elucidating its explanatory ambitions. For just as a theory of the meaning of the word 'light' is not tasked with explaining the underlying nature of light, and a theory of the meaning of the word 'knowledge' is not tasked with explaining the underlying nature of knowledge, so a theory of the meaning of the word 'indefinite' is not tasked with explaining the underlying nature of indefiniteness. The point is easier to miss in the latter case, for while semantics is obviously separate from physics, and only slightly less obviously separate from epistemology, it is not at all obviously separate from the theory of indefiniteness as a linguistic phenomenon. A comprehensive book on semantics *should* say something about vagueness and indefiniteness. Nevertheless, that task must be distinguished from the task of giving the semantics of the word 'indefinite', just as the task of explaining the underlying nature of meaning must be distinguished from the task of giving the semantics of the word 'meaning'. Thus, even if safety-theoretic considerations are central to the underlying nature of indefiniteness, they need not figure in a good semantic account of the word 'indefinite' or the $\triangle$ operator, just as, even if safety-theoretic considerations are central to the underlying nature of knowledge, they need not figure in a good semantic account of the word 'knowledge'. By itself, the appeal to merely epistemically possible worlds in Yli-Vakkuri's semantics for $\triangle$ poses no threat to a safety-theoretic account of the underlying nature of indefiniteness, since the semantics is not expected to describe that underlying level. Of course, if for some sentences in some contexts the semantics predicted truth values incompatible with the safety-theoretic account, that would be a problem, but just appealing to merely epistemically possible worlds forces no such incompatibility. The point can be restated in terms of the distinction between semantics and metasemantics, where a metasemantic theory is supposed to explain why a given semantic theory holds for the language of a given community at a given time. The possible worlds of the safety-theoretic considerations are the concern of the metasemantic theory; Yli-Vakkuri's impossible worlds are the concern of his semantic theory.

Also to be taken into account is the extent to which the △ operator and related English words such as 'indefinite', 'borderline' and 'vague' are all being used as theoretical terms, and rightly so. Although their introduction by examples is good enough for initial working purposes, some refinement is needed once they are used in formulating a positive theory of vagueness. For when we encounter more complex cases than the usual sorites series provide, especially cases where the semantic and worldly dimensions fail to be independent, the initial paradigms give no clear guidance as to how those terms should most fruitfully be applied for theoretical purposes. We see many apparent joints, and are unsure where to cut. For any theory that uses an operator like △ to characterize vagueness-related phenomena, choosing a semantics for △ is a prescriptive as well as a descriptive task. It is a matter of deciding what it is *best* to mean by △. In this respect, the model-theoretic nature of Yli-Vakkuri's semantics strikes a helpful balance between informativeness and flexibility. It is primarily designed to generate a good *logic* for △, especially as it interacts with the modal operators. It validates nice structural principles for △, in particular those corresponding to the modal system T, thereby imposing constructive discipline on our use of △. At the same time, it does not specify in any detail an intended model (or even frame) for the model theory. It thereby leaves considerable scope for fine-tuning the specific application of △ in line with future metasemantic developments.

Like most accounts of vagueness, the epistemicist account in my book focuses on simple borderline cases, and does not provide clear guidance on how to apply an operator like △ to more complex cases, especially those where the semantic and worldly dimensions are not independent. As already noted, that does not make it irrelevant to such cases. Science is full of cases where we know how to apply a theory to simple examples of a phenomenon but not to more complex examples. Typically, progress is then incremental, as models (in the scientific sense, not the semantic one) are developed to permit the theory to be applied to more complex examples, though never all at once, and always under simplifying assumptions. I see no reason why such progress should not also be made in the theory of vagueness, if philosophers are willing to do the hard work of model-building.

3. Semantic plasticity

Where do these general methodological considerations leave semantic plasticity? In Section 5, Yli-Vakkuri argues directly against a version of it that appeals to metaphysical possibilities. He formulates the version thus:

(*) Whenever ϕ is true and borderline, there is a close metaphysically possible world in which ϕ expresses a proposition other than the proposition ϕ actually expresses, and in which that proposition is false.

As a counterexample to (*), he discusses the case where ϕ is the material conditional $\# \rightarrow \chi$, χ is an actually true borderline sentence, and # is a relevantly precise actually non- borderline specification of the actual use facts that determine which proposition χ actually expresses (so # is actually true). Thus $\triangle\#$ and $\neg\triangle\chi$ are true, so $\neg\triangle(\# \rightarrow \chi)$ is also true by the normal logic of $\triangle$. Therefore, since $\# \rightarrow \chi$ is true, it is also borderline. Thus given (*), it follows that there is a close metaphysically possible world *w** in which $\# \rightarrow \chi$ expresses a proposition other than the proposition $\# \rightarrow \chi$ actually expresses, and in which that proposition is false. We may assume that in all close metaphysically possible worlds # and $\rightarrow$ express what they actually express. Thus if # is true in *w**, the actual use facts that determine which proposition χ expresses are the same in *w** as in the actual world, so $\# \rightarrow \chi$ expresses the same proposition as it actually expresses, contrary to hypothesis. On the other hand, if # is not true in *w**, it is false in *w** (given epistemicism, bivalence holds), so $\# \rightarrow \chi$ is true in *w** by the truth table for $\rightarrow$, again contrary to hypothesis. Thus, $\# \rightarrow \chi$ is a counterexample to (*). This argument against (*) does not depend on Yli-Vakkuri's formal semantics, although it does use the standard assumption that the logic of $\triangle$ is like that of $\Box$ in a normal modal logic, which Yli-Vakkuri's semantics validates and I happily grant.

I will not defend (*). However, the falsity of (*) does not mean that the more general phenomenon of semantic plasticity as described in my book plays no role in explaining the borderline status of $\# \rightarrow \chi$, even when semantic plasticity is understood, as here, to be variation in what is expressed in close, so metaphysically possible, worlds (epistemically possible but metaphysically impossible worlds do not count). For the normal logic of $\triangle$ has these theorems, which Yli-Vakkuri's semantics also validates:

$$\triangle\# \rightarrow (\triangle(\# \rightarrow \chi) \leftrightarrow \triangle\chi)$$

$$\triangle\# \rightarrow (\triangle\neg(\# \rightarrow \chi) \leftrightarrow \triangle\neg\chi)$$

$$\triangle\# \rightarrow ((\neg\triangle(\# \rightarrow \chi) \,\&\, \neg\triangle\neg(\# \rightarrow \chi)) \leftrightarrow (\neg\triangle\chi \,\&\, \neg\triangle\neg\chi))$$

In brief, given that # is definite, the question whether $\# \rightarrow \chi$ is borderline reduces to the question whether χ is borderline. We can explain the definiteness of # in terms of its truth and lack of semantic plasticity, both of which Yli-Vakkuri stipulates. Thus, if we can explain the borderline status of χ in terms of semantic plasticity, then we can also explain the borderline status of $\# \rightarrow \chi$ indirectly in terms of semantic plasticity, using our explanations of the definiteness of # and the borderline status of χ in those terms. For instance, if χ is a semantically simple sentence, we may be able to explain its borderline status directly in terms of its semantic plasticity, along the lines I originally proposed, for all that Yli-Vakkuri's arguments show. Of course, χ may itself be complex, and present similar difficulties to those $\# \rightarrow \chi$ did, but then we can apply the same strategy to χ as we did to $\# \rightarrow \chi$, and so on down in a recursive process terminating at the latest in semantically simple constituents.

An encouraging observation is that both the normal logic of $\triangle$ and Yli-Vakkuri's semantics yield the principle that any truth function of non-borderline sentences is itself non-borderline. For instance:

$$(\triangle\phi \vee \triangle\neg\phi) \rightarrow (\triangle\neg\phi \vee \triangle\neg\neg\phi)$$

$$((\triangle\phi \vee \triangle\neg\phi)\ \&\ (\triangle\psi \vee \triangle\neg\psi)) \rightarrow ((\triangle(\phi\ \&\ \psi) \vee \triangle\neg((\phi\ \&\ \psi))$$

The negation of a non-borderline sentence is itself non-borderline, a conjunction of non-borderline conjuncts is itself non-borderline and so on. Contrapositively, if some truth function θ of the simple sentences $\phi_1, \phi_2, \ldots, \phi_n$ is borderline, then at least one of $\phi_1, \phi_2, \ldots, \phi_n$ is borderline too. Thus, if the explanation of borderline status in terms of semantic plasticity works for simple sentences, there is some hope of explaining θ's borderline status indirectly in terms of semantic plasticity through the contributions of its simple constituents. One will not always be able to deduce θ's borderline status from the borderline status of its constituents, for not every truth function of borderline sentences is itself borderline: for instance, $\phi\ \&\ \neg\phi$ is a non-borderline truth function of ϕ, even when ϕ is borderline. Nevertheless, if θ *is* borderline, a reasonable goal is to explain its borderline status in terms of the interrelated ways in which its constituents are borderline, where those ways essentially involve semantic plasticity.

More generally, suppose that *only* semantically plastic sentences are borderline; Yli-Vakkuri's arguments in no way undermine that principle. Let σ be a complex sentence with a compositionally determined meaning. Thus, if all the simple constituents and modes of combination from which σ is composed are semantically *non*-plastic, so is σ itself; by hypothesis, then, σ is non-borderline. Contrapositively, if σ is borderline, then some of its simple constituents or modes of combination are semantically plastic. Thus we can hope to explain σ's borderline status in terms of the semantic plasticity of those simple constituents or modes of combination. Although the present considerations do not guarantee success, they suggest that it is a reasonable strategy.

How does that strategy square with a safety conception of knowledge? For the truth actually expressed by $\# \rightarrow \chi$ is not known, yet in every close metaphysically possible world, the sentence $\# \rightarrow \chi$ expresses a true proposition (by Yli-Vakkuri's argument). Moreover, it is quite compatible with the previous assumptions that the proposition actually expressed by $\# \rightarrow \chi$ is both actually believed and true in every close metaphysically possible world. Invoking epistemically but not metaphysically close worlds looks like abandoning the safety conception. If $\# \rightarrow \chi$ simply refutes that conception, then the original explanation of our ignorance in terms of semantic plasticity fails even for the simplest borderline cases, because it depends on a false epistemology.

At this point, it helps to recall how safety-theoretic considerations would explain the failure of irrational but true mathematical beliefs to constitute knowledge. Someone irrationally plumps for believing a proposition *p* expressed by a precise mathematical sentence π. In fact, *p* is true, and so necessarily true. Thus, in every close world, π expresses *p* (π is not semantically plastic) and *p* is true. However, in some close world *w*, he irrationally plumps in a similar way for believing a different proposition *r* expressed by a different precise mathematical sentence ρ. But *r* is false, and so necessarily false. The sentences π and ρ are not semantically indiscriminable to the subject; he understands both and is aware of their semantic difference. It is just that his plumping does not track their difference in truth value. Since his irrational false belief in the close world *w* is relevantly similar in method to his irrational true belief in the actual world, the latter fails to constitute knowledge.

Now consider Yli-Vakkuri's case. We may assume that the borderline true sentence χ is semantically plastic along the lines of (*), for otherwise his argument against (*) would assume more than it was entitled to by resting on stipulations hostile to (*). We may also assume that the subject actually believes the proposition expressed by the sentence $\# \rightarrow \chi$, for the challenge is to explain in appropriate terms why such a belief does not constitute knowledge. But then we should expect that in some close metaphysically possible world *x*, the subject believes in a similar way the proposition expressed in *x* by the sentence $\#^x \rightarrow \chi$, where $\#^x$ is a relevantly precise non-borderline specification of the use facts in *x* that determine which proposition χ expresses in *x* (so $\#^x$ is true in *x*), while by semantic plasticity, the proposition χ expresses in *x* differs from the one it actually expresses and is false in *x*, even though *x* is just like the actual world in the worldly ways relevant to χ, though not in those relevant to # and $\#^x$. For we were not supposed to assume that the actual world was uniquely special in the subject's believing the proposition expressed by the material conditional with a true relevantly precise non-borderline specification of the use facts that determine which proposition χ expresses as the antecedent and χ as the consequent. But then $\#^x \rightarrow \chi$ expresses a false proposition in *x*, for there $\#^x$ expresses a truth and χ expresses a falsehood. The false belief in the close world *x* in the proposition there expressed by $\#^x \rightarrow \chi$ is relevantly similar in method to the true belief in the actual world in the proposition actually expressed by $\# \rightarrow \chi$, so the latter fails to constitute knowledge. Thus we can explain in safety-theoretic terms the obstacle to knowing the truth expressed by $\# \rightarrow \chi$, and so the borderline status of $\# \rightarrow \chi$. The explanation crucially invokes semantic plasticity in relation to the vague consequent and a different safety-theoretic consideration in relation to the precise antecedent. That is a natural generalization of the original explanation by appeal to semantic plasticity of the obstacle to knowing the truth expressed by a simple vague sentence. The hybrid aspect of the explanans follows the hybrid aspect of the explanandum, where the sentence at issue combines the relevantly vague constituent χ with the relevantly precise constituents # and $\rightarrow$.

The previous paragraph made various simplifying assumptions. We can construct more complicated scenarios. For instance, we can imagine subjects hard-wired to believe propositions expressed by the sentence $\# \rightarrow \chi$ but not those expressed by similar sentences $\#^x \rightarrow \chi$ in any close world. On a suitable development of the story, the true beliefs expressed by $\# \rightarrow \chi$ of such bizarre creatures may bizarrely constitute knowledge.[4] If so, and they also somehow know the truth expressed by the precise but massively detailed specification #, may they not also know the truth expressed by χ inferentially, by modus ponens? But then χ is presumably not borderline for them. That does not mean that it is not borderline for us: our use of χ is different from theirs, even if we do in fact express the same proposition. If # correctly describes their use, then it does not correctly describe ours. If we enter into communication with them, it may be unclear to us whether χ as they use it expresses the same proposition as χ as we use it.

The foregoing line of thought opens up many complex questions that I cannot hope to deal with adequately here. Rich explanatory resources are available to an epistemicist approach to vagueness articulated in terms of safety and semantic plasticity, given a willingness to apply those resources to complex cases in correspondingly complex ways, in the spirit of scientific model building.

Notes

1. For the sake of brevity, when I speak of knowing or believing a proposition expressed by a sentence, I will mean that it is known or believed under the guise of that sentence. Thus, one may know or believe it expressed by one sentence (say, 'Hesperus is Hesperus') but not expressed by another sentence (say, 'Hesperus is Phosphorus').
2. The analogy may suggest a treatment along the lines of Fine (2007). This is no place to discuss such a radical departure from the usual form of compositional semantics; Yli-Vakkuri follows the latter.
3. The technical aspects of the logic were developed in collaboration with Jon Litland (see Litland and Yli-Vakkuri 2016).
4. See Williamson (2016) for an analogous possibility concerning knowledge of mathematical axioms.

References

Fine, Kit. 2007. *Semantic Relationism*. Oxford: Blackwell.

Litland, Jon and Yli-Vakkuri, Juhani. 2016. "Vagueness and Modality." *Philosophical Perspectives*.

Williamson, Timothy. 1994. *Vagueness*. London: Routledge.
Williamson, Timothy. 2000. *Knowledge and Its Limits*. Oxford: Oxford University Press.
Williamson, Timothy. 2016. "Absolute Provability and Safe Knowledge of Axioms." In *Gödel's Disjunction*, edited by Leon Horsten and Philip Welch, 243–253. Oxford: Oxford University Press.
Yli-Vakkuri, Juhani. 2013. "Propositions and Compositionality." *Philosophical Perspectives* 27: 526–563.
Yli-Vakkuri, Juhani. 2016. "Epistemicism and Modality." *Canadian Journal of Philosophy* 46 (4–5): 805–837.

Index

Note: Page numbers followed with "n" refer to endnotes.

actualistic semantics 362–4
ad hoc stipulative approach 278
A-modality 10
assignment function 384
A-theory without tense operators: Barcan formulas 291; change with tense modifiers 296–7; expressive adequacy 288; logical primitives 290–1; ontological change 289–90, 299–301, 302n16; operator-free A-theory 297–9; overview of 283–4; predicate modifier *see* predicate modifier logic; Priorian tense logic 291; property change, accounts of 284–7; propositional temporalism 288–9, 299–300; relational B-theory 297, 301n2; Sullivan, Meghan *see* Sullivan, Meghan

Bacon, Andrew 90–4
Barcan formulas 291, 303n20
Barwise, Jon 262
Boolean algebra 4
Bose–Einstein statistics 34
Bucharest Society of Analytic Philosophy 116n1

Caie, Michael 352, 357–8, 379n23
Cantor's theorem 159
cash-value argument: chunky-style contingentism 168–9; chunky-style necessitism 168–9; higher-order quantification 175–8; mode of argument 171–2; notion of chunkiness 173–5, 185n10, 185n12; philosophical motivation 169–71
causal hypotheses 14–15
CEM *see* conditional excluded middle
chunky-style contingentism (CSC) 168–70, 174, 198
chunky-style necessitism (CSN) 168–70, 198
circumstantial modality 3
circumstantial modals 324–8
Coates, Jennifer 333, 340n20
Collins, Peter 333
Colyvan, Mark 20
compositional semantics 390–2
conditional excluded middle (CEM) 192–3
conservative theory, of vagueness 352
constant domain semantics 27–28
contingentism: in dynamical systems theory 27–32; in probability spaces 32–34
counterfactual-based approach 328, 348
counterfactuals, laws support 13–17; causal hypotheses 14–15; indicative conditional 14; material conditional 15; nomic possibility 16; subjunctive conditional 13–14
counterpart theory 9
CSC *see* chunky-style contingentism
CSN *see* chunky-style necessitism

definitization 361, 366
demarcating counterfactuals: evaluation 334–7; linguistic intuitions 333–4; syntactic considerations 331–2
deontic modals 340n12
dynamical systems theory: mathematical theory 20–22; modal aspect 22–23; necessitism and contingentism in 27–32; 'tensed' modal language for 28–29

Edgington, Dorothy 320, 328–30
epistemic counterfactuals 328–31
epistemicism: actualistic semantics 362–4; advertised virtues 363; generalized 2D frame 368; intended model 359–62; interpretation 373–4; Meter Sentence 365, 371; model theory *see* model theory; overview 351; proposition-*cum*-semantic value 357; *qua* model theory 360; reference predicate 364; and Semantic Plasticity 352–3, 375–6; semantics 354–6; Williamson's kind of 354; Yli-Vakkuri, Juhani *see* Yli-Vakkuri, Juhani
epistemicist theory, of vagueness 351, 378n5
epistemic modalities 2
epistemic modals 324–8
Euclidean geometric model of space 263

Fermi–Dirac statistics 34
Field, Hartry 20
Fine, Kit 9–10, 241; actualism and possibilism 119, 122, 125–6; derivations of 124; modal metaphysics 121, 124; necessitism and contingentism 119–23, 125; standard reduction 126–8; suppositional reduction 128–31
first-order extensional quantification theory 261–3
first-order modal logic: axiomatizations 147–50; higher order logic 153–6; Kreisel's principle 152–3; Kripke model 136–7; logics and consequence relations 141–7; metaphysical universality 135, 136; overview of 132–5; reduction process 137–41; Williamson's argument 150–2
Fritz, Peter 158–60, 176, 193, 194, 366–8, 379n20

Galilean relativist theory 263–4, 266, 272n5
Gallin, Daniel 265
generalized continuum hypothesis (GCH) 142, 144, 145
generalized 2D frame 368
generalized 2D models 384–6
geometric models and relativism 263–5
Goodman, Jeremy 117n18, 158, 223n27, 379n20; capturing distinctions 193–9; first-order contingentism 188–93; higher-order contingentism 188–93; intensionality 199–200
Goodman, Nelson 238

Hawthorne, John 90–4, 352, 372, 374, 375, 379n23
Hawthorn, John 387
higher-order contingentism 280; 'modal tracking' 166–8; parity of motivation 163–6; symmetry considerations 162–3
higher-order free logic: axiomatisation of 83n27; first-order free logic 71, 82n15; Fregean framework 45, 93; grammatical category 85n55, 86n55; limitations 69–71; necessitation and 48–50, 60–2; Russellian picture 43–4; substitutional quantification 83n32; universal instantiation 43, 44; zero-level propositional type 42, 85n51
Hiring Committee 227
Hume, David 8
hyperintensional metaphysics 165

Jacinto, Bruno 273n14
Jones, Nicholas 269–70

Kearns, Stephen 352
Khoo, Justin 330–1
Kratzer, Angelika 2–3, 324
Kreisel's principle 152–3
Kripke model 136, 138, 232; actualism 257–8; contingentism 258; contingentist intuitions and 259–61, 271; first-order extensional quantification theory 261–3; modal reality and model structures 265–9, 274n17; necessitism 258; objections 269–71; orthodox semantics 258–9; permutation functions 273n13; Stalnaker, Robert *see* Stalnaker, Robert
Kripke models 194
Kripke's argument, for rigidity of sets 205–9
Kripke-style model theory 312; finite models 52–4; hyperintension 51; infinite models 55–7; propositional quantifiers 50, 84n42, 84n43, 86n63; rich models and ramification 57–60; variable assignment 51
Krödel, Thomas 336, 339n3

Leibniz's law 82n21, 87n66, 205–6, 208, 209
Lewis, David 9, 240–2, 244n19, 302n9, 303n27, 321; modal realism 257–8, 260
Linnebo, Oystein: plural comprehension principles 228–9; plural extensionality to rigidity 225–8
Litland, Jon 351
logical consequence 359, 369
Löwenheim–Skolem theorem 134, 135, 137, 138, 141
Lyon, Aidan 20

Magidor, Ofra 352
Malament, David 20
Mere Possibilities (Stalnaker) 259
metaphysical modalities, objective modalities *vs.*, 34–35
metaphysical modality 255–6; schematic characterization of 7
metaphysical necessity 6–7
metasemantic interpreration 369–70
Meter Sentence 365, 371
Modal Logic as Metaphysics (MLaM) (Williamson) 41, 92, 122, 123, 132, 158, 241, 254, 280; cash-value argument *see* cash-value argument; contingentism *see* higher-order contingentism; dialectical strategy 178–81; granularity 181–2; non-modal argument 182–4; object-language argument 184n6
modal realism 9
model theory: development of 356; metasemantic interpreration of 369; *qua* semantics 360; relational 368; standard 357; supervaluationism 356–7; three-dimensional 351, 368–9, 377; two-dimensional 358–9, 363, 366, 371
modern set theory 152

necessitism: in dynamical systems theory 27–32; in probability spaces 32–34
neo-Quineanism 289–90, 300
Newtonian geometric models 259
Newton, Isaac 264
nomic modality 10–11
non-epistemic modality 2

Obama, Barack 284–6, 289
objective modalities: category of 2; circumstantial modality 3–4; class of 4; counterpart theory 9; definition 2; essentialism 7–8; folk methods 11–12; Hume's scepticism 8–9; metaphysical modalities *vs.* 34–5; nomic modality 10–11; objective necessity operator 4–7; role in natural science 12; scepticism about 8
objective necessity operators 4–7
objective probabilities: chances 18; microscopic explanation 19–20; non-trivial 19; separation of 3
Occam's razor 235
On the Plurality of Worlds (Lewis) 238, 241
operator-free A-theory: collapse objection 297; grounding objection 298; objects objections 298, 299
orthodox model theory 261

Perry, John 262
Philosophy of Philosophy, The (Williamson) 242, 344, 346, 348
physical necessitism 253
Plantinga, Alvin 260
Plantinga-style contingentist 175
pluralities rigidity and modals: existential assumptions 211–12; extensionality 210, 216–18, 221n16, 222n19, 225–8; Kripke's argument 205–9; mereological analogue 222n26; modal membership profile 210–11; overview of 202–5; partial rigidification 213–14; plural comprehension principles 218–21, 228–9; uniform adjunction 212–13; uniform traversability 214–16
Portner, Paul 2–3
possibility-based approach 317–23, 339n2, 349
predicate modifier logic 303n21, 303n22; description of 292–3; motivations for 293–4; tense modifier 294–6
Prior–Kaplan paradox: closure assumptions 62–4; comprehension and supervenience 64–6; disjunction 85n46; expressing-like relations 75–6; and free quantificational logic 45–8; indefinite extensibility 72; Prior's original argument 77, 81n9; quantification 78, 81n10, 81n11; quantifier

phrase 73, 87n71; ramification and 72–9, 80n2, 86n57; universal instantiation 81n14, 82n21; without universal instantiation 67–9; zero-level proposition 68
Prior's project, of reducing possibility discourse: actualism *vs.* possibilism 97–9; consequences and objections 114–16; criticisms on 96; Fine, Kit *see* Fine, Kit; formulation of 116n2; higher order language 102–5; necessitism *vs.* contingentism 99–100; standard first-order reduction 100–2; suppositional approach *see* suppositional approach; syntactic operations 104
Prior's theorem 45–7, 90–1, 94
probabilities 17
probability spaces, necessitism and contingentism in 32–34
proposition 3, 4

qua model theory 360

reference predicate *(Ref)* 364
relational B-theory 297, 301n2
Roca Royes, Sonia 317, 319

Salmon, Nathan 7
SC *see* suppositional calculus
second-order modal logic: conjunction 234; fundamental and nonfundamental existence 235–7; haecceities 233–4; modal comprehension 233; necessitism 231–2; nonmodal comprehension 233; presumption of 244n16; Sider, Ted *see* Sider, Ted; simplicity and strength arguments 237–43; variable-domain Kripke model 232
'semantic indiscriminability' 374
Semantic Plasticity 352–3, 375–6
semantics 354–6
Sider, Ted: fundamental and non-fundamental abductions 247–9; metaphysical modality 255–6; modal logic's track record 254–5; modal physics 249–51; physical and metaphysical modalities 251–3
Stalnaker, Robert 243; modal comprehension principle 280–2; realism and instrumentalism 269, 275–80
Standard Meter 365, 371
standard model-theoretic resources 358
state spaces 20–21, 27; formal language 25–26; mathematical theory, dynamical systems 20–22; modal aspect, dynamical systems 22–23; valid formulas 24
Strohminger, Margot 319
subjective probabilities 18
Sullivan, Meghan: Barcan formulas 311–12; logical primitives 310–11; operator-free A-theory 312; predicate modifiers 307–10; Priorian tense logic 311; sentence operators 307–10
supervaluationist's model theory 354, 356–7
supervenience 9
suppositional approach: complete first-order description 114; embedded supposition 113; inner and outer quantification 106; logical consequence 107; possible objects 105–6; unembedded supposition 112
suppositional calculus (SC) 107–12, 128–9

Tarski's stratificational approach 85n54
three-dimensional (3D) model theory 351, 368–9, 377
two-cardinal problem, in model theory 137
two-dimensional models 384
two-dimensional (2D) model theory 358–9, 363, 366, 367, 371

uniform traversability (UniTrav) 214–17
universal instantiation (UI) 43–5, 50
Uzquiano, Gabriel 90–4, 212

vagueness: conservative theory 352; epistemicist theory 352, 378n5
variable assignment 367
variable domain semantics 27–28
Vaught, Robert 137–8
Vetter, Barbara: anti-exceptionalism 344; circumstantial and epistemic modals 345–9; possibility-based approach 345, 347

Wiggins, David 304n36
Williamsonian modal epistemology (WME): anti-exceptionalism 339n1; contextualism *vs.* relativism 341n22; counterfactual constructions 316,

340n11; demarcating counterfactuals *see* demarcating counterfactuals; demarcation challenge 323–8; epistemic counterfactuals 328–31; logical equivalence 350n1; metaphysical modality 315; modal thought and knowledge 316–18; possibility-based approach 317–23, 339n2; realism and anti-exceptionalism 314–15
Williamson, Timothy: argument 150–2; infinite sets 134; intended interpretation 133–4; modal logic, simplicity in *see* second-order modal logic; on necessitism *see Modal Logic as Metaphysics*
WME *see* Williamsonian modal epistemology

Yli-Vakkuri, Juhani 387; compositional semantics of 390–2; epistemic interpretation 392; metasemantic theory 394; non-epistemically modal dimensions 393; semantic plasticity 395–8; worldly and semantic differences 387–90

Zimmerman, Dean 287, 302n9

For Product Safety Concerns and Information please contact our EU
representative GPSR@taylorandfrancis.com
Taylor & Francis Verlag GmbH, Kaufingerstraße 24, 80331 München, Germany

www.ingramcontent.com/pod-product-compliance
Ingram Content Group UK Ltd.
Pitfield, Milton Keynes, MK11 3LW, UK
UKHW020023200926
475702UK00014B/2037
9780367229887